TECHNOLOGY
Engineering & Design

Sixth Edition

Sharon A. Brusic, Ed.D.
James F. Fales, Ed.D., CMfgE
Vincent F. Kuetemeyer, Ed.D.

Safety Notice

The reader is expressly advised to consider and use all safety precautions described in *Technology: Engineering & Design* or that might also be indicated by undertaking the activities described herein. In addition, common sense should be exercised to help avoid all potential hazards.

Publisher and Author assume no responsibility for the activities of the reader or for the subject matter experts who prepared this text. Publisher and Authors make no representation or warranties of any kind, including but not limited to, the warranties of fitness for particular purpose or merchantability, nor for any implied warranties related thereto, or otherwise. Publisher and Author will not be liable for damages of any type, including any consequential, special or exemplary damages resulting, in whole or in part, from reader's use or reliance upon the information, instructions, warnings or other matter contained in *Technology: Engineering & Design*.

Brand Disclaimer

Publisher does not necessarily recommend or endorse any particular company or brand name product that may be discussed or pictured in *Technology: Engineering & Design*. Brand name products are used because they are readily available, likely to be known to the reader, and their use may aid in the understanding of the text. Publisher recognizes that other brand name or generic products may be substituted and work as well or better than those featured in *Technology: Engineering & Design*.

The McGraw·Hill Companies

Send all inquiries to:
Glencoe/McGraw-Hill
21600 Oxnard Street
Woodland Hills, CA 91367

ISBN: 978-0-07-876809-5
MHID: 0-07-876809-8
Printed in the United States of America
4 5 6 7 8 9 10 079 12 11 10 09

CONTENTS IN BRIEF

Acknowledgments

About the Authors

Dr. Sharon A. Brusic has been involved in Technology Education for more than 25 years. She currently serves as Associate Professor in the Department of Industry and Technology at Millersville University of Pennsylvania.

Dr. James F. Fales is Loehr Professor Emeritus of Industrial Technology in the Russ College of Engineering and Technology at Ohio University in Athens, Ohio. He is the originator of the CO2-powered race car design and engineering activity commonly used in technology education.

Dr. Vincent F. Kuetemeyer retired after 35 years of secondary and university teaching in Illinois, New York, and Louisiana. Through his company, KE&E, he has continued to consult in construction technology.

Contributing Writers

William A. Atkins
Freelance Writer
Pekin, Illinois

Peggy Hazelwood
Freelance Writer
Denver, Colorado

Robert N. Knight
Freelance Writer
Chicago, Illinois

Steven Miller
Freelance Writer
State College, Pennsylvania

Stuart Soman, Ed.D.
Instructor at Western Suffock BOCES
Dix Hills, New York

Eric Thompson
Technology and Education Consultant
Galesville, Wisconsin

Marlene Weigel
Writing Consultant
Joliet, Illinois

Reviewers

Lane Beard
Agriscience/Industrial Maintenance Instructor
St. James Parish Career & Technology Center
Lutcher, Louisiana

LaMarr Brooks
Project Lead the Way/Engineering Teacher
Pendleton High School
Pendleton, South Carolina

Erica Everett
Science Department Chair
Manchester Essex Regional High School
Manchester by the Sea, Massachusetts

Ed Hamilton
Retired Teacher
Tulsa, Oklahoma

Paul Hayes
Teacher
Woburn Memorial High School
Woburn, Massachusetts

James W. Starling
Electrical Technologies Teacher/Dept. Head
Dunedin High School
Dunedin, Florida

Brooks Wadsworth
Technology Advisor/Instructor
Tarboro, North Carolina

CONTENTS

Unit 2 Communication Engineering & Design....... 76

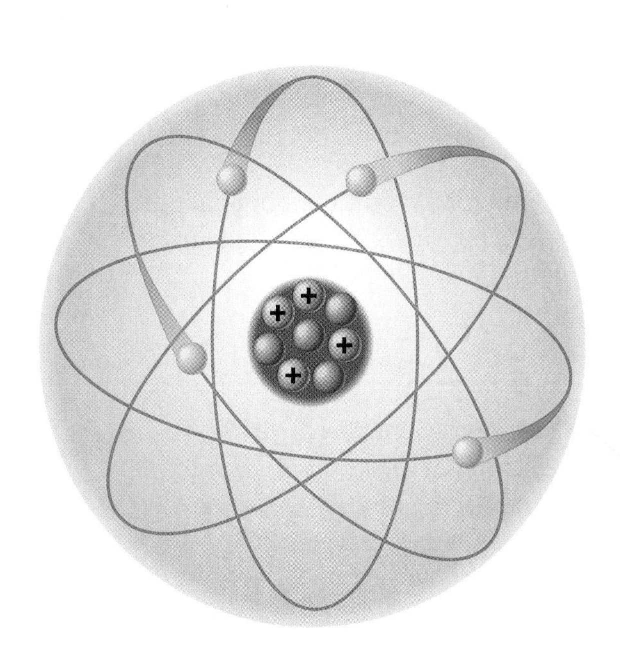

Unit 3 Energy & Power Engineering & Design 162

Unit 5 Construction Engineering & Design.......... 320

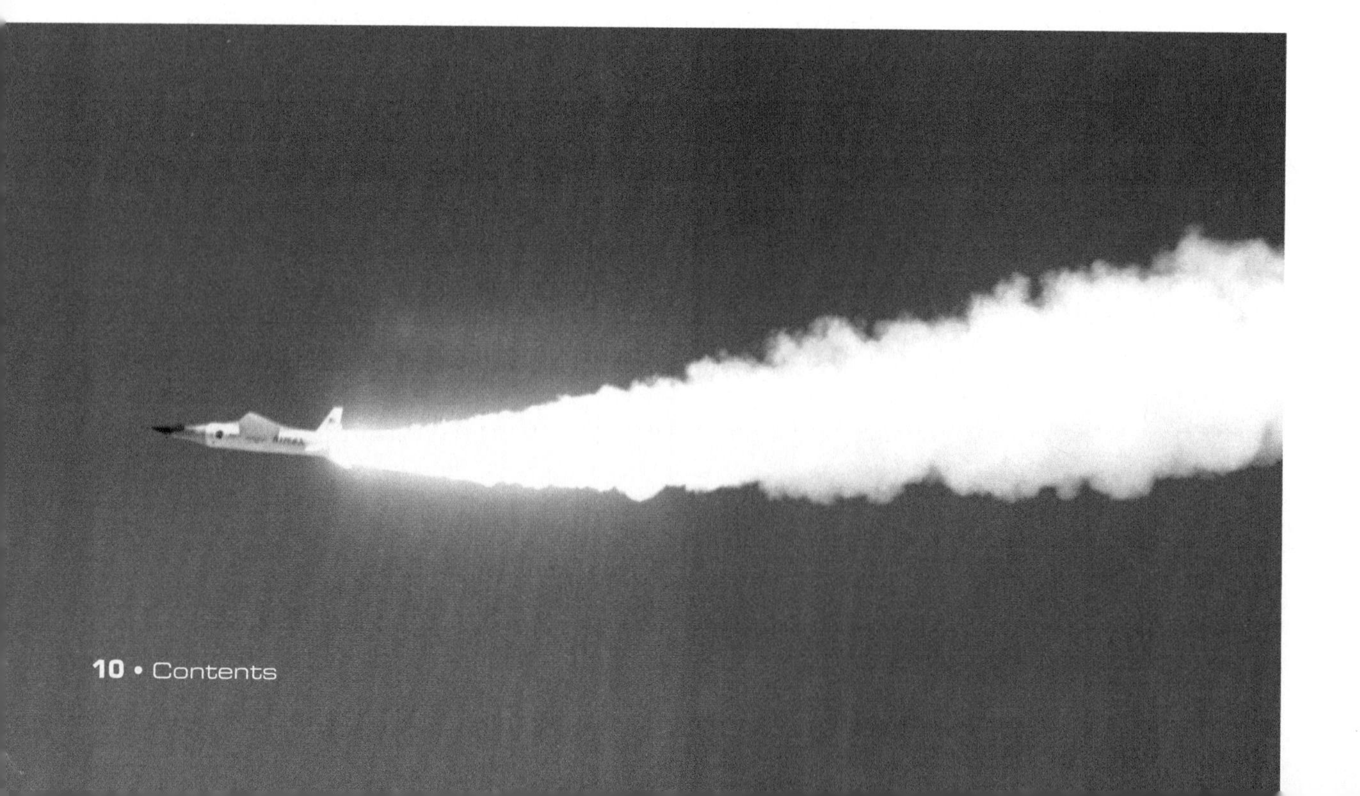

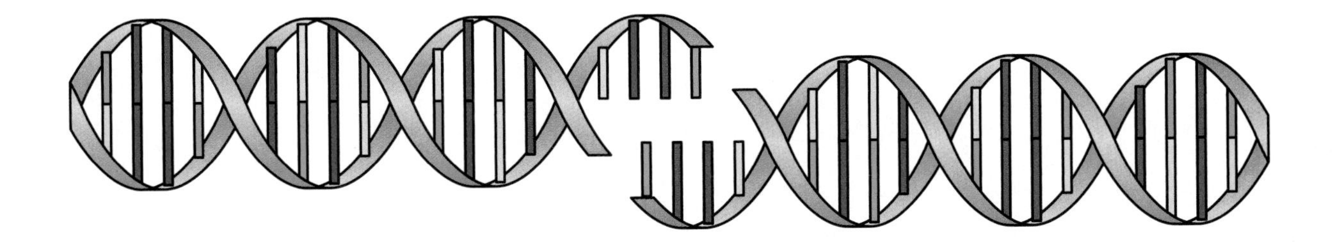

Activities:

Math Application

Science Application

EVOLUTION OF

Young Innovators

CAREERS IN

How to Use This Book

Technology: Engineering & Design will help you understand and apply technology presented in an engineering context. It describes the engineering design process and how it is used to solve technological challenges.

The book is grouped into seven units, each focusing on a different type or field of technology:

Technology Today & Tomorrow covers basic concepts of technology, teamwork, and the engineering design process.

Communication Engineering & Design describes how communication systems work. You'll learn about electronic and printed communication as well as multimedia.

All technology systems need energy to work. In the unit on **Energy & Power**, you will learn about sources of energy and how energy is used to power technological systems.

Almost everything we use in daily life is a manufactured product. In the **Manufacturing** unit, you'll learn how products are designed and made.

The **Construction** unit explains how structures, from tunnels to skyscrapers, are designed and built.

In the **Transportation** unit, you will learn how people and products are moved across a city or around the world.

Bio-Related Engineering & Design describes the relationship of technology and living organisms. You'll learn about medical and agricultural applications of technology.

Each unit opens with **Engineering & Design Frontiers** to focus your attention on the latest developments in technology and engineering.

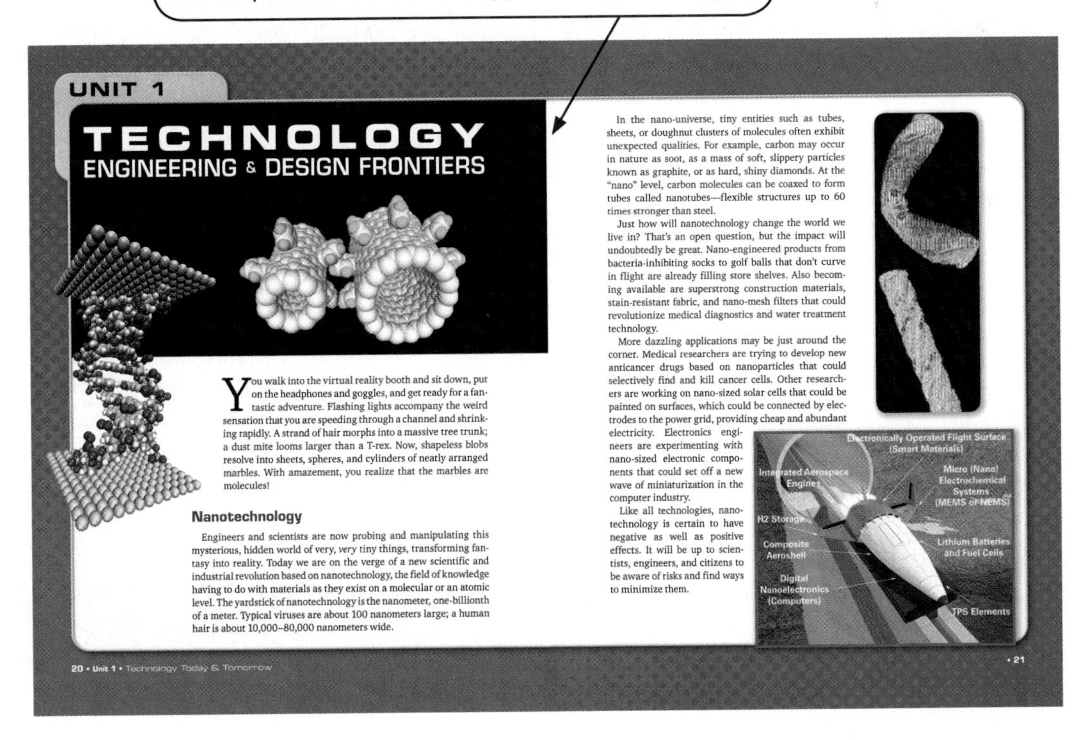
UNIT 1

TECHNOLOGY
ENGINEERING & DESIGN FRONTIERS

You walk into the virtual reality booth and sit down, put on the headphones and goggles, and get ready for a fantastic adventure. Flashing lights accompany the weird sensation that you are speeding through a channel and shrinking rapidly. A strand of hair morphs into a massive tree trunk; a dust mite looms larger than a T-rex. Now, shapeless blobs resolve into sheets, spheres, and cylinders of neatly arranged marbles. With amazement, you realize that the marbles are molecules!

Nanotechnology

Engineers and scientists are now probing and manipulating this mysterious, hidden world of very, *very* tiny things, transforming fantasy into reality. Today we are on the verge of a new scientific and industrial revolution based on nanotechnology, the field of knowledge having to do with materials as they exist on a molecular or an atomic level. The yardstick of nanotechnology is the nanometer, one-billionth of a meter. Typical viruses are about 100 nanometers large; a human hair is about 10,000–80,000 nanometers wide.

In the nano-universe, tiny entities such as tubes, sheets, or doughnut clusters of molecules often exhibit unexpected qualities. For example, carbon may occur in nature as soot, as a mass of soft, slippery particles known as graphite, or as hard, shiny diamonds. At the "nano" level, carbon molecules can be coaxed to form tubes called nanotubes—flexible structures up to 60 times stronger than steel.

Just how will nanotechnology change the world we live in? That's an open question, but the impact will undoubtedly be great. Nano-engineered products from bacteria-inhibiting socks to golf balls that don't curve in flight are already filling store shelves. Also becoming available are superstrong construction materials, stain-resistant fabric, and nano-mesh filters that could revolutionize medical diagnostics and water treatment technology.

More dazzling applications may be just around the corner. Medical researchers are trying to develop new anticancer drugs based on nanoparticles that could selectively find and kill cancer cells. Other researchers are working on nano-sized solar cells that could be painted on surfaces, which could be connected by electrodes to the power grid, providing cheap and abundant electricity. Electronics engineers are experimenting with nano-sized electronic components that could set off a new wave of miniaturization in the computer industry.

Like all technologies, nanotechnology is certain to have negative as well as positive effects. It will be up to scientists, engineers, and citizens to be aware of risks and find ways to minimize them.

20 • Unit 1 • Technology Today & Tomorrow

• 21

Chapter Openers explain what you will learn in each chapter.

Objectives describe what you'll be expected to do and know.

Reading Focus helps you comprehend the ideas presented in each chapter.

Vocabulary lists key terms you'll encounter throughout the chapter.

CAREERS IN

BIO-RELATED TECHNOLOGY

Bio-related technology is one of the fastest-growing career fields. It includes hands-on jobs (farming, for example), high-tech jobs (engineers), jobs helping people who are ill (medical workers), and jobs on the cutting edge of research (scientists).

Medical Lab Technician • Surgeon • Medical Research Scientist Nurse • Emergency Medical Technician • Biomedical Engineer Optometrist • Water Treatment Plant Operator • Ergonomics Engineer Sanitation Engineer • Pharmacist • Pharmaceutical Salesperson Veterinarian • Farm Worker • Milk Truck Driver • Combine Operator Forester • Florist • Butcher • Toxic Waste Technician • Biochemist Waste Management Plant Supervisor • Environmental Engineer Food Plant Machine Operator • Food Safety Inspector • Nutritionist

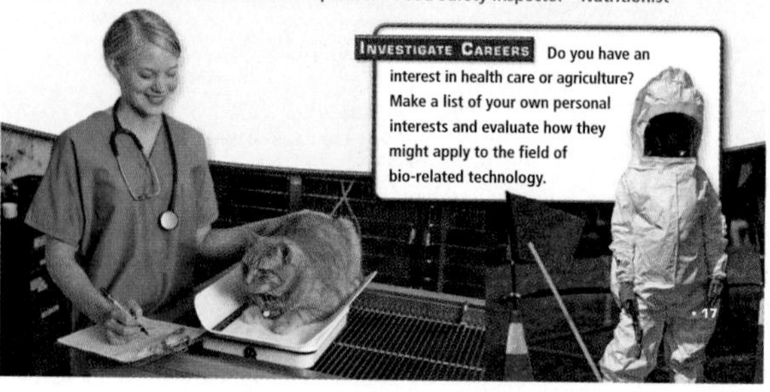

Careers pages provide information about careers in technology and engineering.

Math Application

Reading Graphs A graph is a diagram that usually shows a relationship between quantities. A picture is formed so that a pattern can be seen. Graphs help us understand and solve problems.

The three most common types of graphs are bar, line, and circle graphs. Depending on the information you want to show, one graph might be better to use than others. Often the same information can be shown in all three kinds of graphs.

LINE GRAPH
Annual Sales of the Red Dot Company (2002-06)

Millions of Dollars
900
800
700
600
500
400
375
300
2002 2003 2004 2005 2006

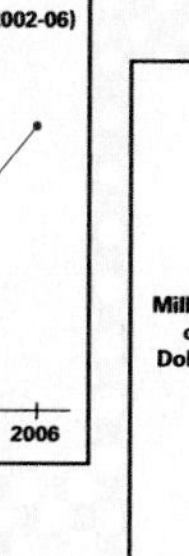

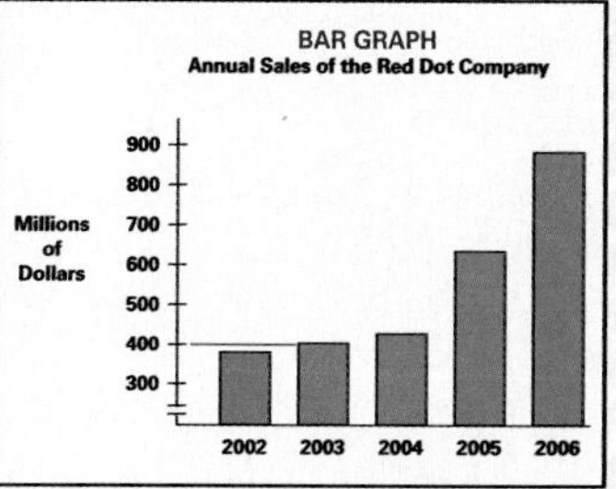

CIRCLE GRAPH
Leading Spring Manufacturers

International Spring Co. 26%
Springo Corp. 14%
R & R Spring Co. 16%
American Spring Co. 24%
Springs Ltd. 20%

FIND OUT Create a bar graph using the information shown on the circle graph. Place the names of the five spring companies at the bottom of the graph and their percentages on the left.

Math Applications and **Science Applications** relate math and science concepts to technology and engineering.

Science Application

Experimenting with Cable Sag Try this experiment and you will be able to feel how the span and sag on a suspension bridge are affected by the distance between the towers.

Attach a weight to the middle of a small rope. Now hold your hands fairly close together, letting the weight make the rope sag. Notice how heavy the weight feels. Now move your hands out wider and notice the amount of thrust it takes and how heavy the weight feels. The weight is the same, but the sag and span have changed. This is why the cables on long spans and small sags must be bigger. On shorter spans with larger sags, the span can be made with thinner cables.

FIND OUT Why do suspension bridges have an extra set of cables at each end?

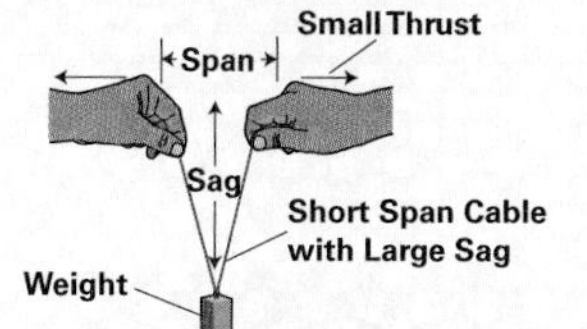

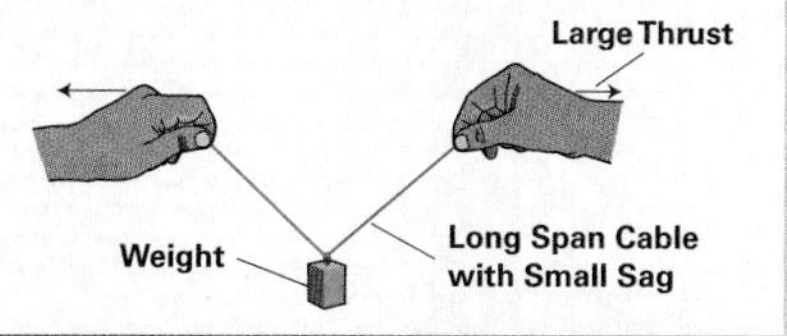

The **Safety First Handbook** addresses general lab safety and the proper handling of tools and materials.

SAFETY FIRST

Before You Begin. Make sure you understand how to use the tools and materials safely. Have your teacher demonstrate their proper use. Follow all safety rules.

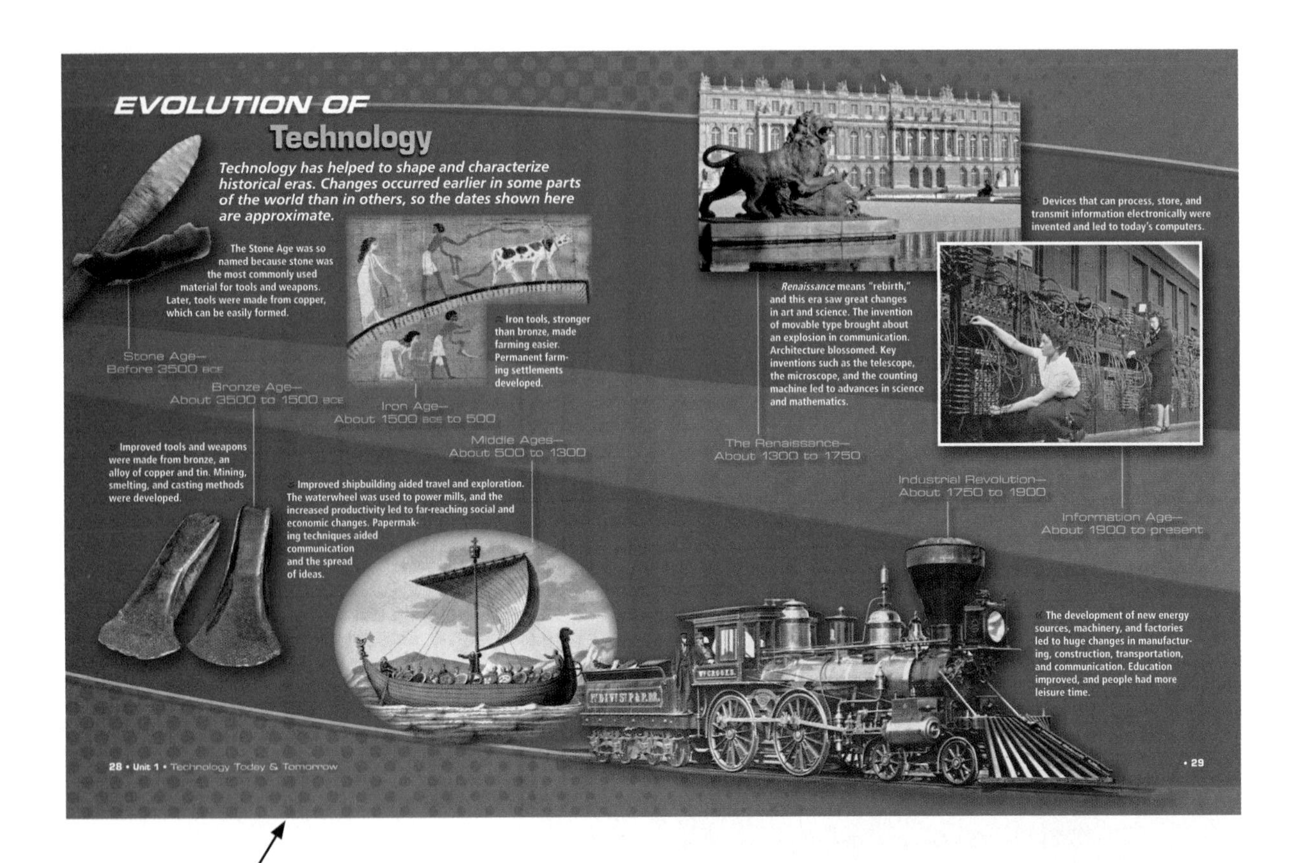

EVOLUTION OF Technology

Technology has helped to shape and characterize historical eras. Changes occurred earlier in some parts of the world than in others, so the dates shown here are approximate.

The Stone Age was so named because stone was the most commonly used material for tools and weapons. Later, tools were made from copper, which can be easily formed.

Stone Age— Before 3500 BCE

Bronze Age— About 3500 to 1500 BCE

Improved tools and weapons were made from bronze, an alloy of copper and tin. Mining, smelting, and casting methods were developed.

Iron tools, stronger than bronze, made farming easier. Permanent farming settlements developed.

Iron Age— About 1500 BCE to 500

Middle Ages— About 500 to 1300

Improved shipbuilding aided travel and exploration. The waterwheel was used to power mills, and the increased productivity led to far-reaching social and economic changes. Papermaking techniques aided communication and the spread of ideas.

Renaissance means "rebirth," and this era saw great changes in art and science. The invention of movable type brought about an explosion in communication. Architecture blossomed. Key inventions such as the telescope, the microscope, and the counting machine led to advances in science and mathematics.

The Renaissance— About 1300 to 1750

Industrial Revolution— About 1750 to 1900

The development of new energy sources, machinery, and factories led to huge changes in manufacturing, construction, transportation, and communication. Education improved, and people had more leisure time.

Devices that can process, store, and transmit information electronically were invented and led to today's computers.

Information Age— About 1900 to present

28 • Unit 1 • Technology Today & Tomorrow

• 29

Evolution of Technology pages describe how technology has changed over time.

Young Innovators

Which Robot Is the BEST?

Your task was to design a robot to service the *Hubble Space Telescope*. With minutes to go before the competition, you run down a final checklist to make sure all systems are working. Then you realize the robotic arm won't move! This is no time to panic. You have spent the last six weeks learning and solving problems, and this is just one more step in the process.

After troubleshooting, you find one of the servo motor connections is loose. You secure the connection and test to make sure the arm works. Now it's time to enter your robot into the BEST Robotics Inc. Middle and High School Robotics Competition!

BEST Robotics Inc. hosts an annual competition where teams from schools all over the country compete to see who can create a robot best suited to a given task. Only one team per school can compete, but the team may be of any size and students are not limited to creating robots. There are any number of jobs for students who want to participate but do not want to build robots. These jobs range from handling publicity to designing team logos and T-shirts; there is something for everyone.

Competing in the BEST Robotics competition is a great way to become technologically literate. Any school can participate and there is no cost to the school or to the students.

24 • Unit 1 • Technology Today & Tomorrow

Young Innovators pages describe competitive events all over the country that allow you to apply your knowledge of technology and engineering.

Chapter Activities provide hands-on experience in applying the engineering design process.

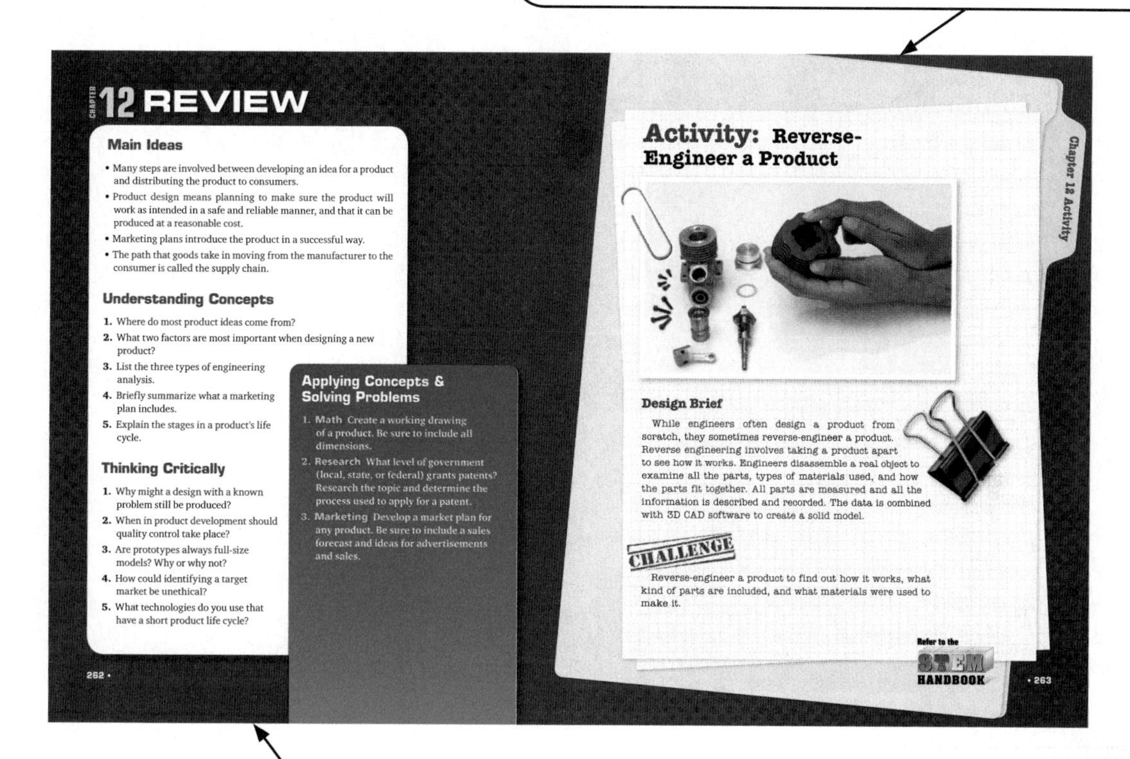
CHAPTER 12 REVIEW

Main Ideas

- Many steps are involved between developing an idea for a product and distributing the product to consumers.
- Product design means planning to make sure the product will work as intended in a safe and reliable manner, and that it can be produced at a reasonable cost.
- Marketing plans introduce the product in a successful way.
- The path that goods take in moving from the manufacturer to the consumer is called the supply chain.

Understanding Concepts

1. Where do most product ideas come from?
2. What two factors are most important when designing a new product?
3. List the three types of engineering analysis.
4. Briefly summarize what a marketing plan includes.
5. Explain the stages in a product's life cycle.

Thinking Critically

1. Why might a design with a known problem still be produced?
2. When in product development should quality control take place?
3. Are prototypes always full-size models? Why or why not?
4. How could identifying a target market be unethical?
5. What technologies do you use that have a short product life cycle?

Applying Concepts & Solving Problems

1. Math Create a working drawing of a product. Be sure to include all dimensions.
2. Research What level of government (local, state, or federal) grants patents? Research the topic and determine the process used to apply for a patent.
3. Marketing Develop a market plan for any product. Be sure to include a sales forecast and ideas for advertisements and sales.

262 •

Chapter 12 Activity

Activity: Reverse-Engineer a Product

Design Brief

While engineers often design a product from scratch, they sometimes reverse-engineer a product. Reverse engineering involves taking a product apart to see how it works. Engineers disassemble a real object to examine all the parts, types of materials used, and how the parts fit together. All parts are measured and all the information is described and recorded. The data is combined with 3D CAD software to create a solid model.

CHALLENGE

Reverse-engineer a product to find out how it works, what kind of parts are included, and what materials were used to make it.

Refer to the STEM HANDBOOK

• 263

Chapter Review lists Main Ideas and provides questions to help you review the main points of each chapter.

A **Science-Technology-Engineering-Math (STEM) Handbook**, at the back of this book, explains the principles related to many applications and activities in the textbook.

HANDBOOK

Science, **T**echnology, **E**ngineering, and **M**ath offer broad career pathways into the high-growth industries competing in the global marketplace. The jobs in these areas are offered by businesses on the frontline of technological discovery. With a solid background in STEM, you'll be following a career path that will offer the opportunity to participate in the development of high-demand and emerging technologies.

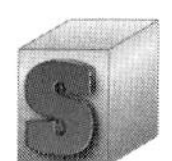

With advances in science driving innovations in technology, materials and procedures researched in the lab are now the backbone of many new processes. Science is used in the design of cell phone circuits, the construction of surfboards, and the processing of food products. While science begins with careful observation, research, and experimentation, it involves more than lab work. Some opportunities will take you into the field. You might also be involved in quality control or technical writing. The range of science career opportunities is extremely large.

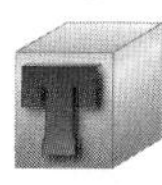

Technology joins science skills with those in engineering and math. It is essential in managing communication and information systems. In manufacturing, technology is used to focus human resources and materials in the production of products ranging from electronic devices to footwear. If you want a career that combines science, math, and engineering, consider a career in technology. In that field, you'll work closely with individuals who have a hands-on knowledge of these subjects. As part of their team, you'll help make the decisions needed to bring a product from design through manufacturing to market.

A career in engineering will involve you in the technological innovations that flow from discoveries in science and math. You'll be incorporating these new discoveries in products that expand our technological reach in many areas, including communications. If you think that engineering is unexciting, name a product or service of interest to you that was introduced within the last year. Then identify the main benefit of that product or service. That benefit probably resulted from an engineering effort that built on developments in science, technology, and math. A career in engineering could place you in a position to participate in the development of such products and services.

Math input is essential to product creation and development—whether the product is real, such as a DVD, or virtual, such as an online video game. If you want to be a valued member of the teams providing the math input for the development of products, you will need solid math skills. You'll need these skills for high-level problem solving and innovation. As you advance along the career pathway, you'll be challenged to apply mathematics to real-world problems. There is a wide variety of math careers from which to choose.

STEM Handbook • 565

UNIT 1

TECHNOLOGY
ENGINEERING & DESIGN FRONTIERS

You walk into the virtual reality booth and sit down, put on the headphones and goggles, and get ready for a fantastic adventure. Flashing lights accompany the weird sensation that you are speeding through a channel and shrinking rapidly. A strand of hair morphs into a massive tree trunk; a dust mite looms larger than a T-rex. Now, shapeless blobs resolve into sheets, spheres, and cylinders of neatly arranged marbles. With amazement, you realize that the marbles are molecules!

Nanotechnology

Engineers and scientists are now probing and manipulating this mysterious, hidden world of very, *very* tiny things, transforming fantasy into reality. Today we are on the verge of a new scientific and industrial revolution based on nanotechnology, the field of knowledge having to do with materials as they exist on a molecular or an atomic level. The yardstick of nanotechnology is the nanometer, one-billionth of a meter. Typical viruses are about 100 nanometers large; a human hair is about 10,000–80,000 nanometers wide.

In the nano-universe, tiny entities such as tubes, sheets, or doughnut clusters of molecules often exhibit unexpected qualities. For example, carbon may occur in nature as soot, as a mass of soft, slippery particles known as graphite, or as hard, shiny diamonds. At the "nano" level, carbon molecules can be coaxed to form tubes called nanotubes—flexible structures up to 60 times stronger than steel.

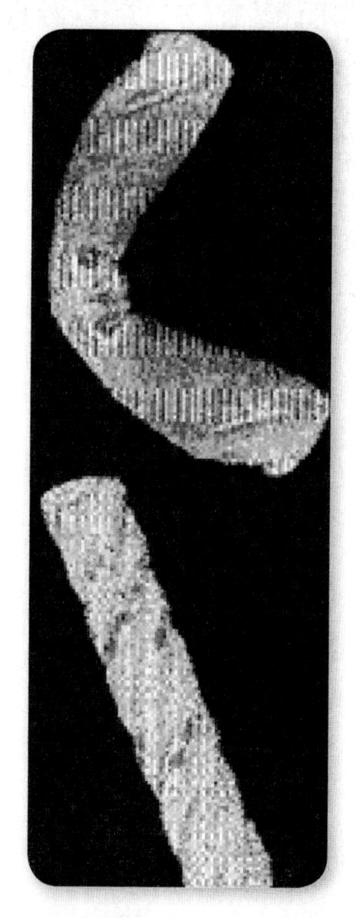

Just how will nanotechnology change the world we live in? That's an open question, but the impact will undoubtedly be great. Nano-engineered products from bacteria-inhibiting socks to golf balls that don't curve in flight are already filling store shelves. Also becoming available are superstrong construction materials, stain-resistant fabric, and nano-mesh filters that could revolutionize medical diagnostics and water treatment technology.

More dazzling applications may be just around the corner. Medical researchers are trying to develop new anticancer drugs based on nanoparticles that could selectively find and kill cancer cells. Other researchers are working on nano-sized solar cells that could be painted on surfaces, which could be connected by electrodes to the power grid, providing cheap and abundant electricity. Electronics engineers are experimenting with nano-sized electronic components that could set off a new wave of miniaturization in the computer industry.

Like all technologies, nanotechnology is certain to have negative as well as positive effects. It will be up to scientists, engineers, and citizens to be aware of risks and find ways to minimize them.

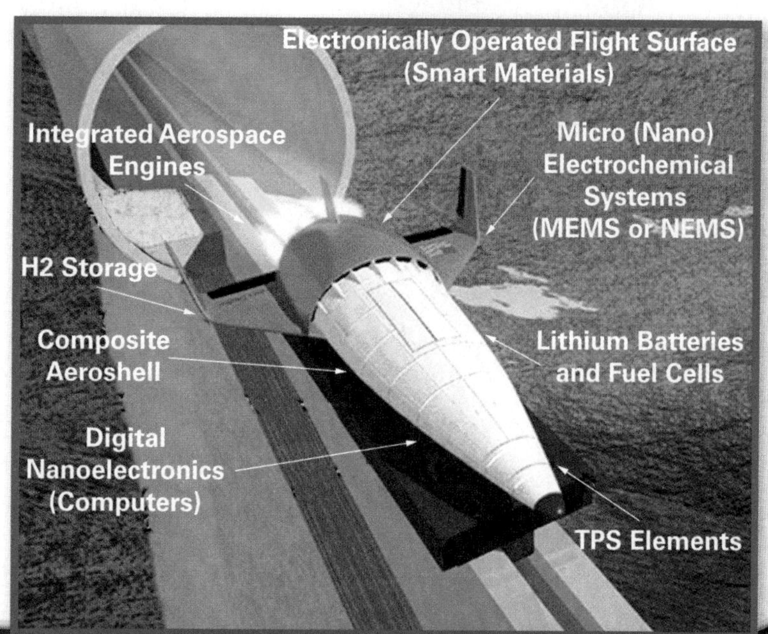

CHAPTER 1

TECHNOLOGY FUNDAMENTALS

Objectives

- Explain why studying technology is beneficial.
- Give examples of ways in which technology evolves.
- Describe technology systems.
- Describe impacts of various technologies.
- Explore ways in which ethics influence technology.

Vocabulary

- technology
- system
- input
- process
- output
- feedback
- impact
- technology assessment
- trend
- ethics

This hot air balloon will ascend rapidly into the cool morning sky because the hot air inside the balloon is less dense than the surrounding cool air. In this chapter's activity, "Design with Math and Science," you will have a chance to demonstrate a principle of math or science by re-creating an invention from the past.

Reading Focus

1. **Read the title of this chapter and describe in writing what you expect to learn from it.**
2. **Write each term in your notebook, leaving space for definitions.**
3. **As you read Chapter 1, write the definition beside each term in your notebook.**
4. **After reading the chapter, write a paragraph describing what you learned.**

What Is Technology?

Technology is all around you. It has always been a part of your daily life. Yet, could you define technology?

Technology consists of processes and knowledge that people use to extend human abilities and to satisfy human needs and wants. In other words, people create technology to solve problems and to make it possible to do new things. People needed a way to keep food cold during hot weather. They invented refrigerators. People wanted to explore the solar system. They invented vehicles that can travel in space. See **Fig. 1-1**.

Technology can be classified into six broad areas: communication, energy and power, manufacturing, construction, transportation, and bio-related technology. In reality, they usually blend together. Each area depends on the others, as you will learn during this course.

Why Study Technology?

When you study technology, you learn about how the world works. Have you ever wondered how an airplane gets off the ground and stays in the air? Studying transportation technology will give you the answer. Have you ever wondered how a flu shot helps you stay healthy? Studying bio-related technology can answer that. Knowing these things can give you confidence. It can also be fun. Watching a NASA probe land on a distant planet is more exciting when you know how it was accomplished.

Fig. 1-1 Shown here is the launch of the Space Shuttle *Discovery,* as the seven-member crew begins a mission to dock with the International Space Station to help with repairs.

Young Innovators

Which Robot Is the BEST?

Your task was to design a robot to service the *Hubble Space Telescope*. With minutes to go before the competition, you run down a final checklist to make sure all systems are working. Then you realize the robotic arm won't move! This is no time to panic. You have spent the last six weeks learning and solving problems, and this is just one more step in the process.

After troubleshooting, you find one of the servo motor connections is loose. You secure the connection and test to make sure the arm works. Now it's time to enter your robot into the BEST Robotics Inc. Middle and High School Robotics Competition!

BEST Robotics Inc. hosts an annual competition where teams from schools all over the country compete to see who can create a robot best suited to a given task. Only one team per school can compete, but the team may be of any size, and students are not limited to creating robots. There are any number of jobs for students who want to participate but do not want to build robots. These jobs range from handling publicity to designing team logos and T-shirts; there is something for everyone.

Competing in the BEST Robotics competition is a great way to become technologically literate. Any school can participate, and there is no cost to the school or to the students.

Studying technology can also help you understand the importance of some of the things you hear about in the news. For example, people disagree about how certain technologies, like genetic engineering, should be used. See **Fig. 1-2**. Studying about them in technology class gives you the facts. It helps you understand what's going on so you can make more informed decisions.

Fig. 1-2 These people are concerned about the safety of new technologies that can alter plant and animal life, affecting the environment and the food we eat.

Math Application

Dimensional Analysis You have probably converted measurements quite often. For example, you know that 12 inches = 1 foot. However, what if there's no formula for a direct conversion? Engineers often use an approach called dimensional analysis to move from one unit to the other until they get the answer they need.

To find the amount of a specific unit using dimensional analysis, multiply your given information (in fraction form) by related common conversions until you get to the unit you're looking for. Common numerators and denominators can cancel each other out when you multiply by the conversion factors.

Let's look at a simple problem using dimensional analysis. How many minutes equal three years? Even if you don't know a direct conversion from minutes to years, you can still use dimensional analysis to calculate the value. Let x represent the number of minutes you want to find.

$$\frac{3 \text{ years}}{x} \times \frac{365 \text{ days}}{1 \text{ year}} \times \frac{24 \text{ hours}}{1 \text{ day}} \times \frac{60 \text{ minutes}}{1 \text{ hour}}$$

Notice how the unit in the numerator of one fraction cancels out the unit in the denominator of the next fraction until you get to the unit you need to find. Multiply the remaining numbers. The answer is x = 1,576,800. There are 1,576,800 minutes in 3 years.

FIND OUT How many feet = 200 centimeters if 1 inch = 2.54 centimeters?

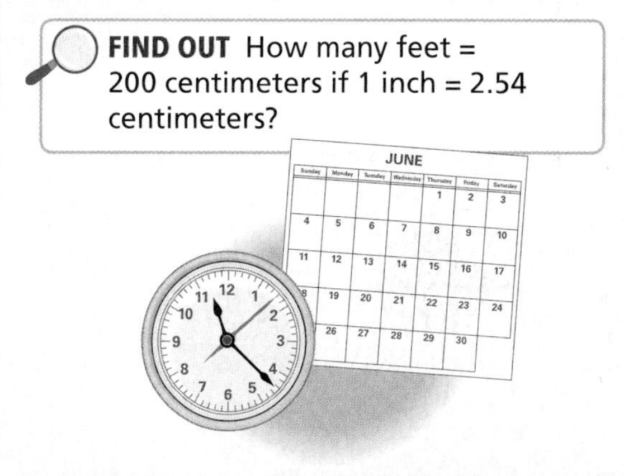

Studying technology may also help you relate to your other school subjects. Mathematics can be more relevant when you use it to calculate the velocity of a rocket. Social studies can be more interesting when you realize how advances in technology can bring about social and cultural change. Scientific principles have more meaning when you use them to build a model bridge.

Do you like working with your hands? Technology courses involve hands-on projects. Completing a successful project gives you a sense of accomplishment. It helps you develop and sharpen your skills. In this textbook, each chapter includes a design activity that will help you apply what you learn in each chapter.

How Technology Evolves

Technology has been around as long as the human race. People had to find ways of solving problems and meeting needs—in other words, to develop technology—way back in history. Technology was used to make the first tools and machines long before scientists understood the principles of how they worked. Inventions like the telescope helped scientific research make great strides forward. See **Fig. 1-3**.

Technology continually evolves. This means it grows and changes. The inventions themselves change, and so does the society from which they spring.

Changes in Society

As different cultures evolved in different parts of the world, they developed technologies to help them. Egyptians, for example, discovered architectural forms that led to their great pyramids and temple complexes. The Greeks and Romans developed the technology to build fast ships.

The technologies people create change their lives. As people began to farm the land, they developed the waterwheel to help them process grain. The waterwheel used the power of water to drive the millstones. People no longer had to grind their grain by hand. They could devote their time and energy to other things.

Waterwheel technology also brought economic and social change. Instead of everyone building his or her own waterwheel, it made sense for one person in a community to own and operate a mill. The other people could then pay the miller money or goods in exchange for grinding their grain. The economic life of the

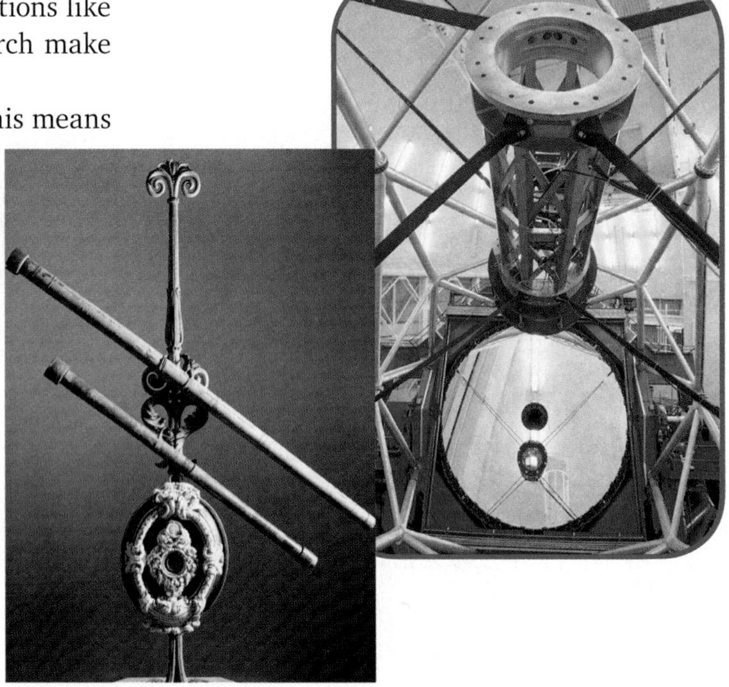

» **Fig. 1-3** Technology has evolved tremendously from the telescope that Galileo used in the 16th century to the *Gemini North telescope* in Hawaii. ***Can you think of other tools or machines that have evolved like this?***

whole community was affected. People realized technology had commercial value. This and other changes helped bring a new class of independent workers into being, which led to the growth of towns. The political landscape was altered as well. Common people had a greater voice in a country's affairs.

Changes in Inventions

We no longer use waterwheels for grinding our grain. That particular technology has changed. Inventions like the waterwheel evolve for many reasons.

New problems arise, and to solve the problems, new technologies or products are created (inventions) or existing technologies or products are altered (innovations). Did you know that most technological development is the result of changes to basic inventions? The waterwheel, for example, was eventually adapted to run large saws and machines. Someone added a different idea to the existing technology and produced something new. Changing the technology can not only improve the quality of a product, but it can also improve the efficiency of a process. A saw powered by a waterwheel could cut wood faster and in greater quantities than a saw powered by a human worker.

Ideas, knowledge, and skills may be shared by people within the same technology or across several technologies. Inventions and innovations from different areas may be combined, or a technology developed for one purpose may be used for a completely different purpose. Did you know that the design of the waterwheel evolved into the design for the giant turbines that generate electricity in many of today's power plants? See **Fig. 1-4**.

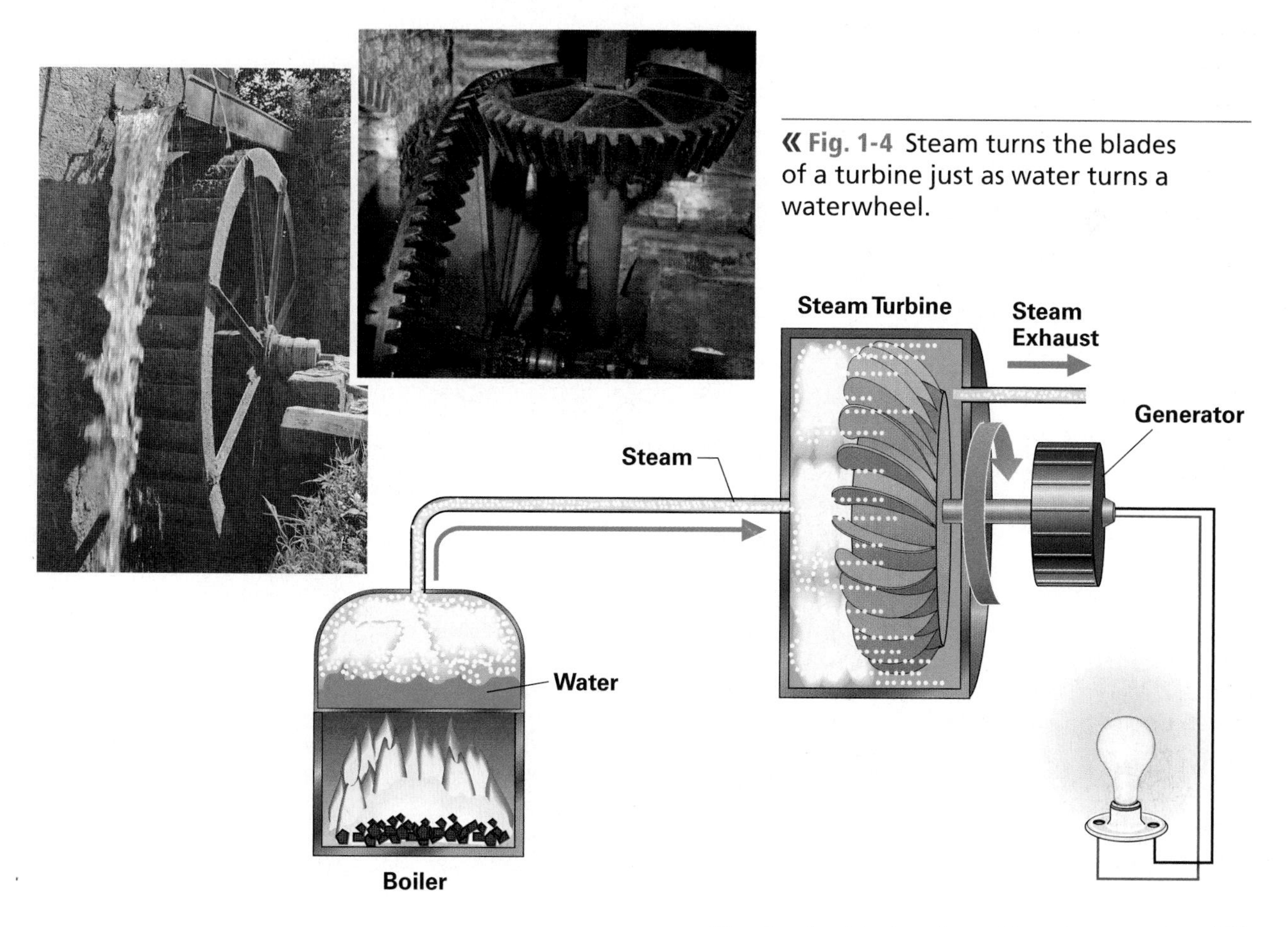

« **Fig. 1-4** Steam turns the blades of a turbine just as water turns a waterwheel.

EVOLUTION OF Technology

Technology has helped to shape and characterize historical eras. Changes occurred earlier in some parts of the world than in others, so the dates shown here are approximate.

The Stone Age was so named because stone was the most commonly used material for tools and weapons. Later, tools were made from copper, which can be easily formed.

Stone Age—
Before 3500 BCE

Iron tools, stronger than bronze, made farming easier. Permanent farming settlements developed.

Bronze Age—
About 3500 to 1500 BCE

Iron Age—
About 1500 BCE to 500

Middle Ages—
About 500 to 1300

Improved tools and weapons were made from bronze, an alloy of copper and tin. Mining, smelting, and casting methods were developed.

Improved shipbuilding aided travel and exploration. The waterwheel was used to power mills, and the increased productivity led to far-reaching social and economic changes. Papermaking techniques aided communication and the spread of ideas.

Renaissance means "rebirth," and this era saw great changes in art and science. The invention of movable type brought about an explosion in communication. Architecture blossomed. Key inventions such as the telescope, the microscope, and the counting machine led to advances in science and mathematics.

The Renaissance—
About 1300 to 1750

Industrial Revolution—
About 1750 to 1900

The development of new energy sources, machinery, and factories led to huge changes in manufacturing, construction, transportation, and communication. Education improved, and people had more leisure time.

Information Age—
About 1900 to present

Devices that can process, store, and transmit information electronically were invented and led to today's computers.

Some inventions and innovations are the result of goal-directed research. Research done to develop drugs for specific diseases is an example. Inventions and innovations can also be the unexpected result of research done for other reasons. The microwave oven was an unexpected result of research on radar to detect enemy submarines during World War II.

Economics also play a role. The United States' economy is based on capitalism. Capitalism is driven by competition. Companies must create or use technologies that help them make a profit so they are able to remain in business. For example, each year car makers must try to come out with new or improved designs to attract and satisfy customers.

Public opinion can also affect technology's development. People may demand changes for such reasons as safety. Advertising and fads may help create consumer interest in certain products. For example, clothing fads among young people may inspire adults to want similar styles.

Public opinion often influences emerging and innovative technologies. Emerging technologies are new technologies just coming into general use. Hybrid vehicles are an example. Public concern over high gas prices has inspired the invention of these cars that run on batteries as well as gasoline.

An innovative technology is a new or improved technology that is still being refined in the laboratory. However, not all innovations are universally welcomed. Cloning technology is an example. It has met with opposition from the public. Laws have been enacted in some places to limit its development. Why do you think people are concerned about cloning?

Technology Systems

Every technology can be thought of as a system. A **system** is a group of parts that work together to achieve a goal. The basic parts are input, process, and output. See **Fig. 1-5**. Thinking about a particular technology as a system helps you consider the technology logically. Its parts can be analyzed.

Input

Input includes resources that are put into the system. A resource is anything that provides support or supplies. There are seven types of

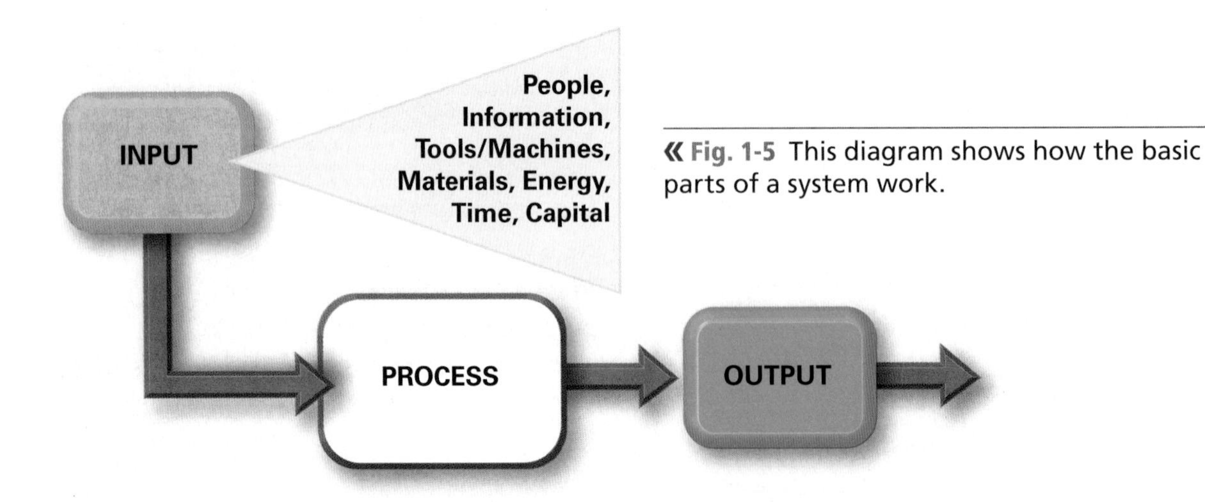

« **Fig. 1-5** This diagram shows how the basic parts of a system work.

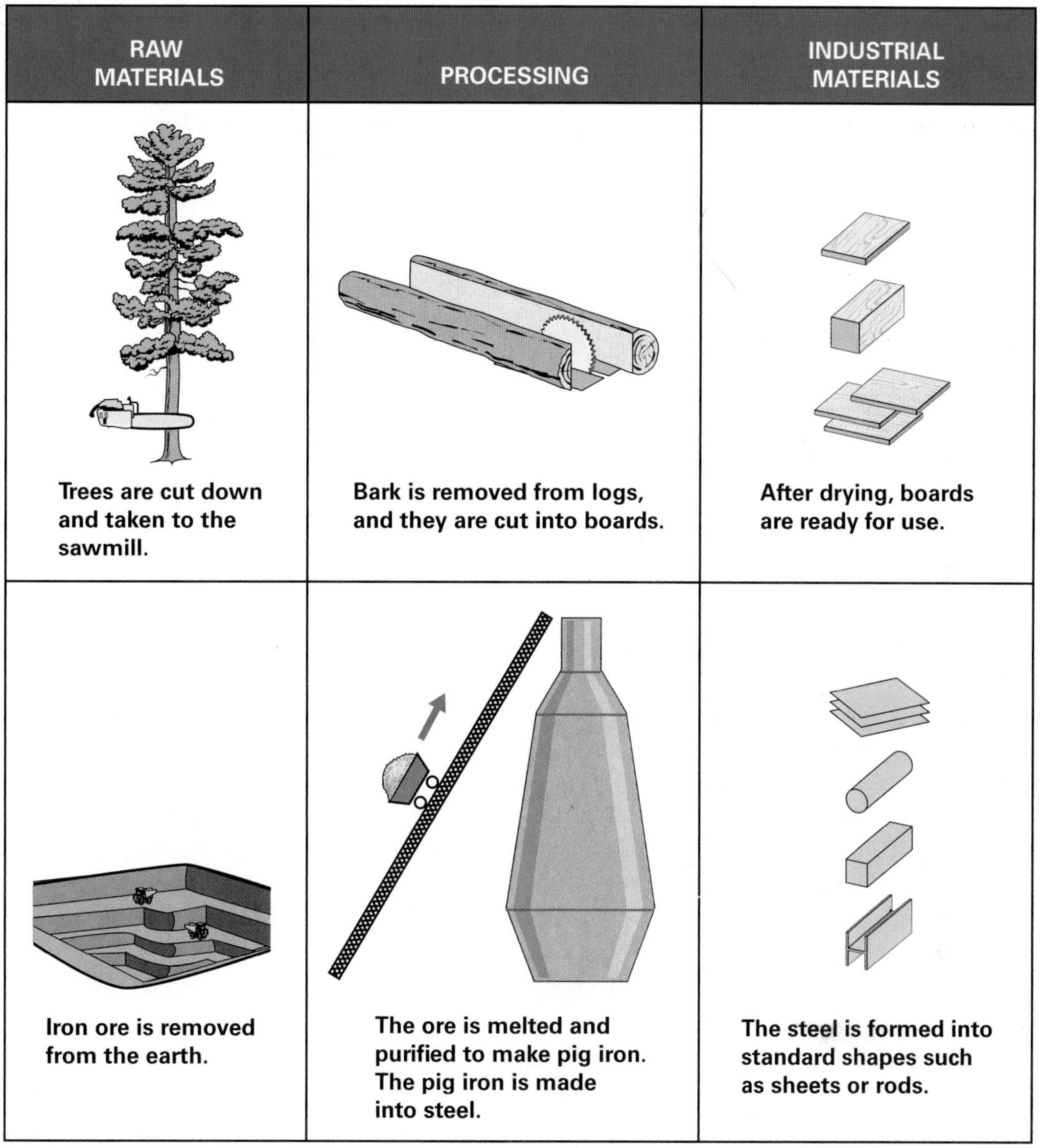

RAW MATERIALS	PROCESSING	INDUSTRIAL MATERIALS
Trees are cut down and taken to the sawmill.	Bark is removed from logs, and they are cut into boards.	After drying, boards are ready for use.
Iron ore is removed from the earth.	The ore is melted and purified to make pig iron. The pig iron is made into steel.	The steel is formed into standard shapes such as sheets or rods.

Fig. 1-6 Raw materials are natural materials obtained by such processes as harvesting and mining. To make them usable, they are turned into industrial materials.

resources that provide input for all technological systems: people, information, materials, tools and machines, energy, capital, and time.

- **People.** Perhaps the most important resource. Includes product designers, factory workers, and salespeople.
- **Information.** Anything needed to make the system work, including scientific knowledge, assembly instructions, and computer data.
- **Materials.** All the things that make up a product, including such things as wood, metal, and plastic. Materials may be classified as raw or industrial materials. See **Fig. 1-6**.

- **Tools and machines.** Anything used to make materials assume the size and form required to make the product, including measuring, layout, separating, forming, and combining tools. See **Fig. 1-7**.
- **Energy.** Includes electrical energy to power machines and light, heat energy to keep buildings warm, and human energy to do work.
- **Capital.** Includes the money, land, and equipment needed to set up and keep the system running. Working capital is the money that companies use to operate. It may be used to buy supplies or pay workers. Fixed capital includes property such as buildings and vehicles.
- **Time.** Includes the time required to build factories, design and make products, and ship products to stores.

Fig. 1-7 Common tools and machines are used to change materials into the form needed by a technology system. ***Can you identify the tools and machines used in the processes shown below?***

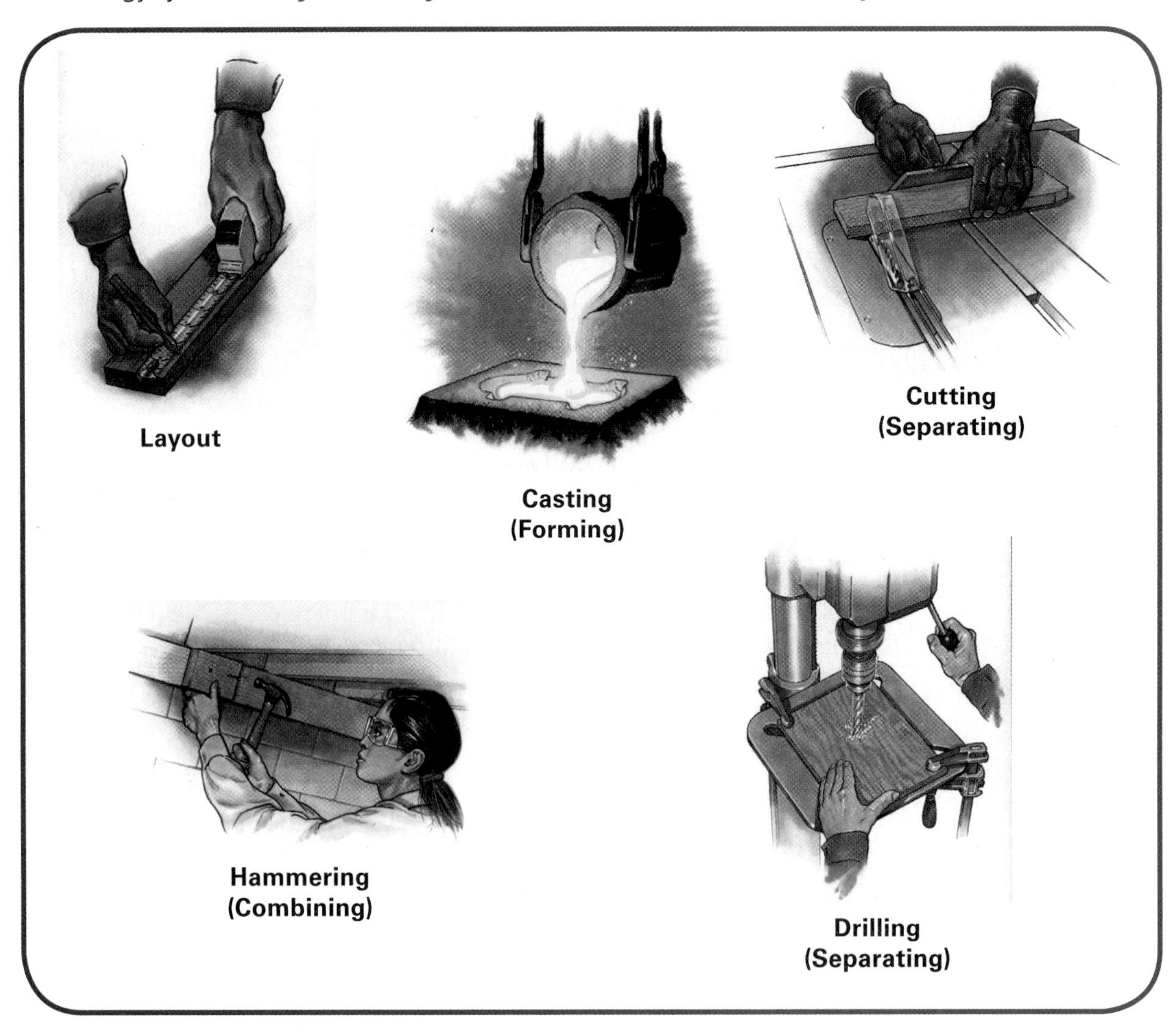

Every system uses these seven types of resources. Which specific resources are chosen depends on such things as cost, availability, appropriateness, and waste. For example, if CD players are being produced by a manufacturing system, then plastic is probably one of the material resources needed. It is an appropriate choice because plastic is lightweight and durable. It is also easily available and not very expensive.

Process

The second part of a system is the process. The **process** includes all of the activities that need to take place for the system to give the expected result. For a CD player, the process might include designing the product itself and its individual parts, making the various parts out of appropriate materials, and putting all the parts together to make the player.

The process part of a system also includes management. In order for the system to produce the expected result, the process must be well planned, organized, and controlled. However, no process works perfectly. Both people and machines are needed to identify and solve problems and to make improvements.

Output

The third part of a system is the output. The **output** includes everything that results when the input and process parts of the system go into effect. In all systems, there is, of course, the intended output, such as CD players. There are also, however, outputs that may not have been intended, such as waste that may have been created during the process or changes in society caused by the product.

Science Application

Properties of Materials The materials used in technology systems differ from one another by their properties. These properties define how they interact with other materials and with energy. To select the right material for a job, you need to know about its properties.

Mechanical properties describe how the material reacts to force. For example, when struck by a hammer, does it shatter or bend?

Sensory properties are what you can see, hear, feel, smell, or taste. Sugar is sweet and vinegar is sour.

Optical properties have to do with the way a material responds to light. For instance, window glass is transparent and steel is opaque.

Thermal properties determine how a material reacts to heat. Does it melt, burn, or change size? Does it conduct heat or insulate?

Chemical properties describe how a material reacts with other materials. Rusting, burning, and dissolving are chemical properties.

Electrical properties determine how a material conducts electricity. A copper wire is a good conductor but the plastic insulation around it is not.

Magnetic properties determine how a material reacts to a magnetic field.

FIND OUT Does a material have the same properties if it's in a different form? Obtain some aluminum foil and aluminum plating and compare their properties.

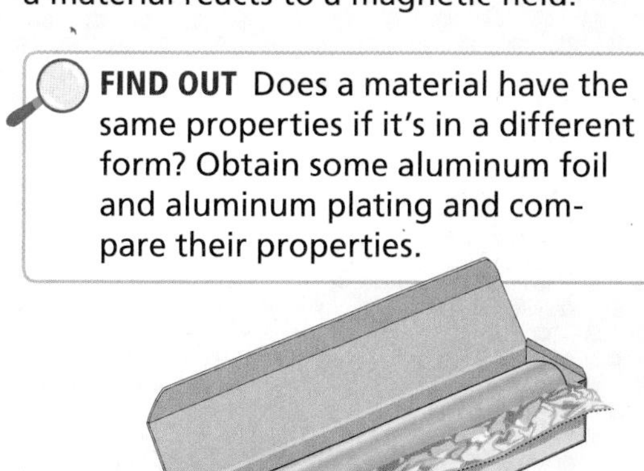

Feedback

Feedback is information about output that is sent back to the system to help determine whether the system is doing what it is supposed to do.

Not all systems include feedback. A traffic light, for example, changes from red to green whether or not cars in its lane are stalled in traffic. There is no way for that information to get back to the system and affect its cycle. This is called an open-loop system. The systems model in Fig. 1-5 on page 30 is an example of an open-loop system.

In a closed-loop system, outputs are monitored and feedback is used to make sure the system is solving the problem it was intended to solve. See Fig. 1-8. Complex systems often have many kinds of controls and ways for feedback to provide information. A system that produces CD players, for example, would receive feedback from workers checking quality within the system. It would also receive feedback from stores selling the players and from customers who bought them. Although all the parts of a system contribute to its stability, feedback is key to its success.

Subsystems

Large systems often contain many smaller subsystems. Subsystems within a manufacturing system for a CD player would include the systems that make the parts, the system that assembles them into the player unit, and the system that does the packaging.

Did you know that even big systems are actually subsystems embedded within larger technological, social, and environmental systems? A CD player manufacturing system plays a role in both the local and national economic systems. The products it produces become part of our social system when we use them to play music at social gatherings. The materials of which they are made come from the environment, and the players eventually go back into the environmental system as waste.

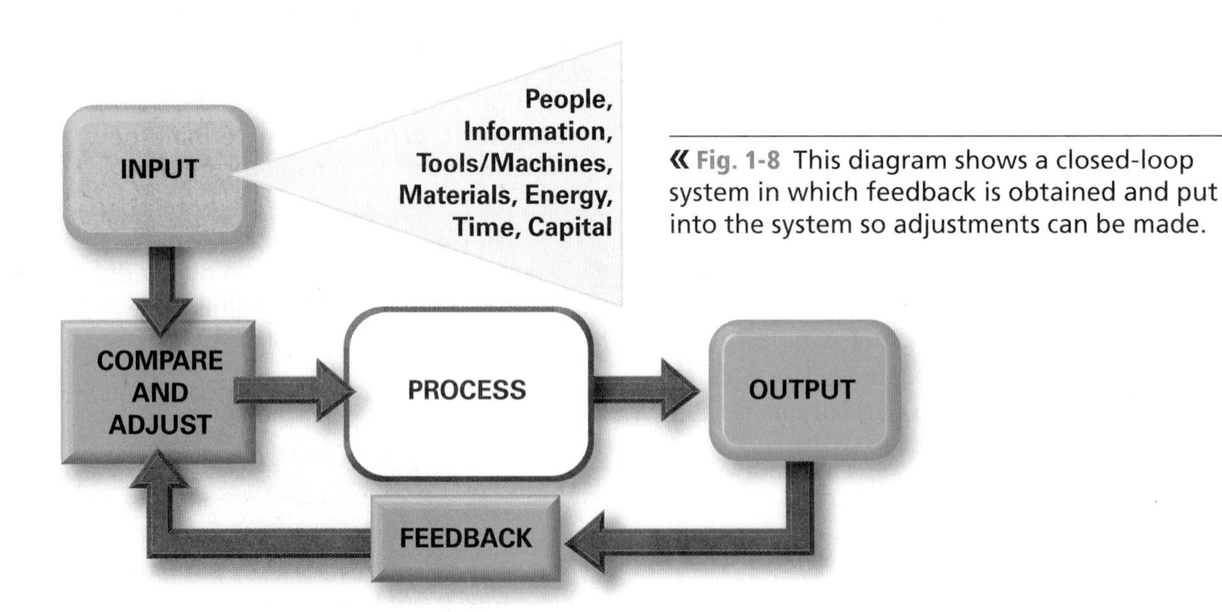

« Fig. 1-8 This diagram shows a closed-loop system in which feedback is obtained and put into the system so adjustments can be made.

Impacts and Effects of Technology

We've already looked at many benefits technology can provide. These are its desirable, or positive, **impacts** (significant effects). Sometimes, however, technologies create undesirable, or negative, impacts.

For example, for many years fossil fuels, such as coal and oil, have been used to provide heat and power. These were desirable impacts. Then it was discovered that burning these fuels causes air pollution and acid rain. Some power plants release heat into oceans, rivers, and lakes, which damages the water environment. In addition, fossil fuel sources are being used up and cannot be replaced. These are all undesirable impacts.

In the past, new technologies came into use gradually over the course of many years. It took a long time for a technology to spread from one place to another. Information traveled by handwritten letters or by word of mouth. This gave people a chance to adjust. After the invention of the printing press, however, people were able to share their ideas more easily and quickly.

Today, people can send information from one side of the globe to the other in a matter of seconds. Because of this ease of communication and the buildup of technology over centuries, we live in a time when many technologies are changing rapidly. Changes that once may have seemed subtle are now very obvious.

SAFETY FIRST

Give Your Eyes a Break. Eyestrain can be a negative impact of working on a computer for long periods of time. At least once each hour, look out a window or across the room. If you wear glasses, remove them. Let your eyes focus on a distant object. This simple exercise gives your eyes a chance to relax and protects your vision.

Fig. 1-9 Even technologies like this water-saving showerhead must be assessed to be sure they solve the problem and do not create new ones.

Technology Assessment

As you can see, technology is a mixed blessing. While solving one problem, technology can sometimes create other problems. Studying the effects of a technology is called **technology assessment**. To assess something means to determine its importance, size, or value.

Technology assessment involves looking at cost, safety, and economic and environmental impacts. Trade-offs are weighed. Some negative impacts may be considered acceptable if important benefits can be obtained at the same time. New drugs are an example. Almost every drug has side effects. How serious are they? Do the drug's benefits outweigh its negative impacts?

New technologies can sometimes be developed to reduce negative impacts of other technologies. For example, new home construction puts a strain on water supplies in some areas. Showerheads have been developed that conserve water during a shower. See **Fig. 1-9**.

However, even conservation technologies must be evaluated in terms of trade-offs. The new showerheads may save water, but does manufacturing them use energy and other resources efficiently? Saving water may prove to be less important at the time than saving energy.

Technologies can also be devised to monitor such things as the environment in order to help people make the best decisions. Methods for measuring water pollution, for example, can determine if wastewater from nearby factories threatens wildlife in a river. When technology can be aligned with natural processes, negative impacts on the environment can be reduced. Introducing special plants into waterways to act as natural filters is one example.

Forecasting Techniques

Many impacts are unexpected. Pollution, for example, was an unexpected impact of burning fossil fuels. The ability to predict, or forecast, technological events or developments is important. Both positive and negative impacts can often be identified in advance. Sometimes this involves experimentation. Water conservation methods, for example, could be tested to be sure they are not so strict that they cause hardships.

Trends can be analyzed. A **trend** is a popular movement. Researchers spot trends by looking at people's habits and preferences. Do more people seem to want to live in areas that happen to have limited water supplies? See **Fig. 1-10**. That would be a trend. The conservation technology chosen would have to address this trend.

Fig. 1-10 This housing development in the desert may overburden the water resources.

Surveys are another forecasting technique. Researchers ask people for their opinions about a particular subject. A survey of homeowners in the area might indicate whether or not current water conservation methods are practical or successful.

Ethical Use of Technology

Ethics are moral principles and values that guide conduct within a group of people. Something that is unethical may not be against the law. However, the majority of people in the group usually agree that it is wrong. For example, it is unethical for pharmaceutical companies to fail to warn doctors and patients about the side effects of a drug, especially when the effects are serious. See **Fig. 1-11**. It is also unethical for a company to make unrealistic promises about what a particular drug can do.

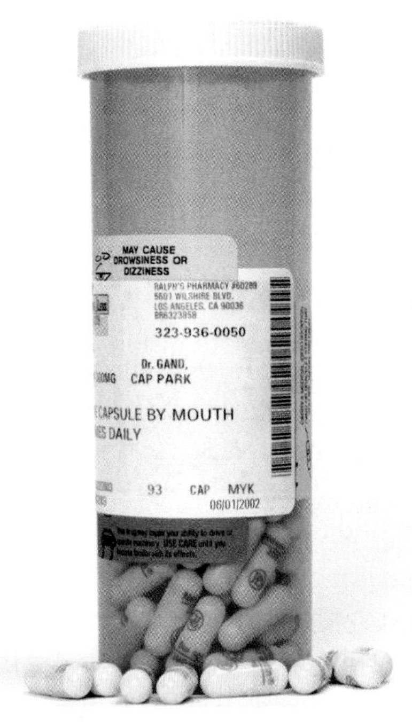

Fig. 1-11 Ethical companies show concern for the well-being of their customers. This label makes the buyer aware of possible side effects.

Ethics are important in the development, selection, and use of many technologies. For example, many people are concerned about the impacts of some types of genetic engineering on people and the environment. Researchers argue that the technology is safe, but not all impacts are obvious, and some occur gradually. Trade-offs between possible positive and negative effects must be considered. Many people would like to see a set of ethical standards established to guide this type of research.

Ethics are also important when a technology is transferred from one society to another. Cultural, social, economic, and political changes can result. During the reconstruction of Iraq, American communication technologies were distributed in greater numbers than before. More people now have access to radio and television programming. What effects do you think this is having on the Iraqi economy and political process? More information coming back from Iraq has added fuel to political battles in the United States.

Ethics come into play as well when inventors and researchers in technology base their ideas on one another's work. To use someone else's invention without permission or without giving credit to the person is unethical and often illegal. In the United States, inventors usually patent their work to protect it. A patent is a government document granting the exclusive right to produce or sell an invented object or process for a period of time. It ensures that the inventor's idea cannot be legally copied during that period.

Technology Challenges

In this chapter you've learned a little about how and why technology developed and what some of its impacts have been. What about the future? What part will technology play in your adult life?

We have recently entered the 21st century, and like past centuries, it offers special challenges. Following are several issues of world importance that engineering technology may be able to help resolve. So far, there are no easy answers. It will be up to you and other young people to find the solutions.

- **Population growth.** Although world population growth is expected to level off within your lifetime, large populations are putting pressure on the environment and dwindling resources. These large numbers of people must also be fed, housed, and given access to medical care.

Fig. 1-12 Wind can be an energy resource. Wind turbines convert the energy of moving air into electricity. ***Can you think of other ways wind energy is used?***

- **Energy.** New, sustainable (not easily depleted) energy resources must be found. See Fig. 1-12. Our use of all energy must be made more efficient, less wasteful, and less polluting.
- **Pollution.** Air and water pollution have long been a concern. What can technology do to keep air and water clean and to preserve adequate supplies of water for everyone?
- **Food production.** Although food is plentiful, better techniques for growing crops, distributing the food, and keeping food safe to eat are needed.
- **Waste disposal.** In the United States, about 1,600 pounds of trash are produced per person per year. What can be done to dispose of all that waste material? Can technology find ways to turn waste into something useful?
- **Transportation.** Vehicle exhaust pollutes the air. Roadways have become more crowded, resulting in traffic jams and short tempers. New transportation methods must be found to lessen our dependence on cars and trucks.
- **Infrastructure.** Public buildings, roads, and bridges are deteriorating and must be repaired or replaced. What should replace them, and can technology make them easier to build?
- **Medical care.** New viruses and bacteria that resist existing drugs are always emerging. Medical technologists are needed to develop new medicines. Cures for existing ills must be found.

You may one day want to work on one of these and other challenges discussed in this textbook. In the first chapter of each unit you'll find information about the many technology careers that could give you the right opportunity. What contributions might you make to the world of tomorrow?

CAREERS IN TECHNOLOGY

It is hard to think of a field in which technology does not play a part. Communication, energy and power, manufacturing, construction, transportation, and bio-related technologies touch every part of our lives. Listed below are just a few of the possible career choices that technology offers. You'll find others included in the different units in this textbook.

Drafter • Building Inspector • Steelworker • HVAC Contractor Television Camera Operator • Printer • Electrical Engineer Power Plant Technician • Oil Rig Technician • Analytical Engineer Plant Manager • Salesperson • Mechanical Engineer • Animator Quality Control Inspector • Computer Control Technician Automotive Design Engineer • Aircraft Maintenance Technician Veterinarian • Roadway Designer • Railroad Inspector Laboratory Technician • Farmer • Research Scientist

INVESTIGATE CAREERS This textbook covers six major areas of technology in Units 2–7. Place each of the above careers in one of the six major areas. Compare your answers with those of your classmates. Do you all have the same answers? Why or why not?

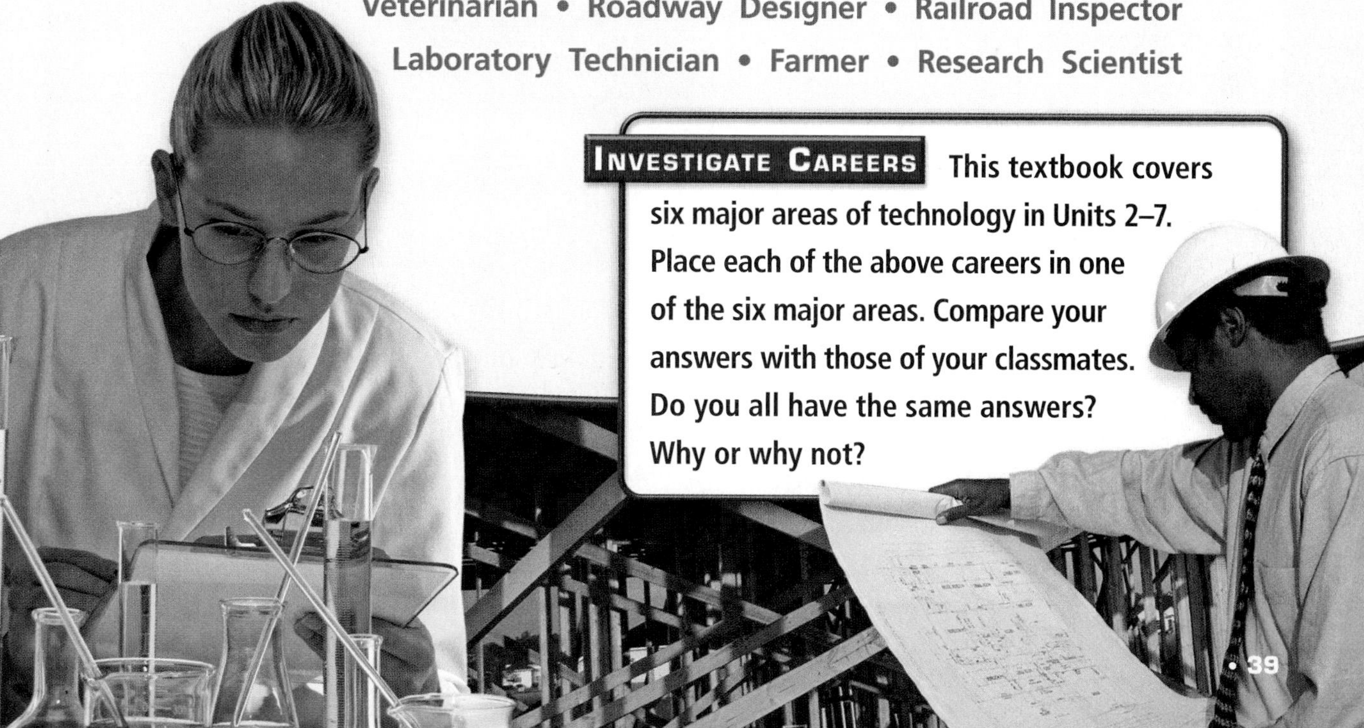

CHAP

REVIEW

Main Ideas

- Technology consists of processes and knowledge that people use to extend human abilities.
- Studying technology allows greater understanding of how the world works.
- Technology continuously evolves over time.
- Every technology is part of a system. Systems often have many subsystems.
- Technology has both positive and negative impacts on our world.
- Ethics are important in the development, selection, and use of many technologies.

Understanding Concepts

1. According to this chapter, what are three ways technology can be beneficial to you?
2. Name two ways technology evolves.
3. Define a technological system.
4. Describe one unintended impact of using fossil fuel-powered technology.
5. What are ethics?

Thinking Critically

1. What is meant by the phrase "Necessity is the mother of invention"?
2. What do sports and capitalism have in common as a driving force?
3. Describe a technology system you use frequently.
4. How might a trend get started?
5. Are unethical activities always also illegal? Explain your answer.

Applying Concepts & Solving Problems

1. **Social Studies** Report on one way in which technological progress has changed the way you live over the last 10 years.
2. **Math** Trends are often shown through percentages. In the last 100 years, the percentage of people involved in agriculture dropped from 90 percent to only 2 percent. If the U.S. has 275 million people, how many are currently involved in agriculture?
3. **Assess** Find a technological device in your home. Assess the device for cost, safety, and economic and environmental impacts. Create a presentation to communicate your findings to the class.

Activity: Design with Math and Science

Design Brief

Science, mathematics, and technology have evolved together. The earliest devices inspired people to discover the scientific or mathematical principles that enabled the devices to work. Later, science inspired technology. If a principle indicated that certain devices were possible, inventors sought to create them.

For example, the first American submarine was invented by David Bushnell in 1775. Called the American Turtle because its two halves looked like turtle shells, the submarine was designed to destroy British ships during the Revolutionary War. Bushnell used the principle of displacement to design a means of submerging the ship beneath the water.

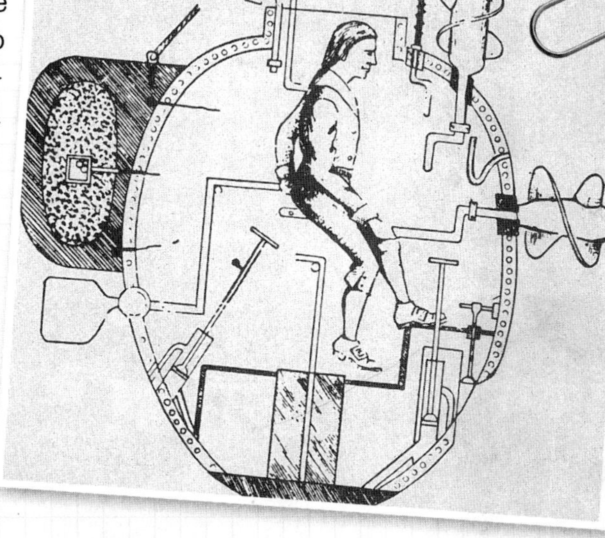

Your teacher will assign you a scientific or mathematical principle, such as reflection or buoyancy. You will then select an invention of the past and build a model of it that will prove the assigned principle.

Refer to the

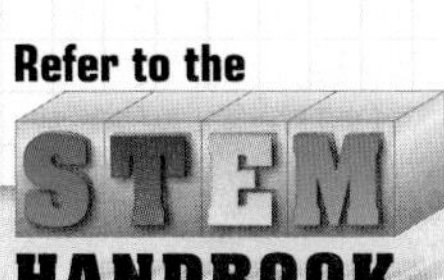

Criteria

- Your device must be an application of your assigned principle.
- You must hand in all sketches and drawings, a log sheet of your work, and a statement of the scientific and/or mathematical principles on which your device is based.

Constraints

- The device must be a scale model, not a full-size device. The model's size should not exceed three cubic feet.
- Your model may be a traditional physical model or a 3D model drawn on the computer.

Engineering Design Process

1. **Define the problem.** For the principle assigned, write a statement that describes the invention you have chosen to model.
2. **Brainstorm, research, and generate ideas.** With your team, discuss possible solutions. Hint: You may want to research several inventions. The more sources you have, the better chance you have of solving the problem.
3. **Identify criteria and specify constraints.** These are listed on page 42.
4. **Develop and propose designs and choose among alternative solutions.** Choose the invention that will best illustrate the connection to science or mathematics.
5. **Implement the proposed solution.** Decide on the process you will use for making the model. Gather any needed tools or materials.
6. **Make a model or prototype.** Create your model. Follow all safety rules.
7. **Evaluate the solution and its consequences.** Does your model work as intended? Does it demonstrate a scientific or mathematical principle?
8. **Refine the design.** After your evaluation, change the design if needed.
9. **Create the final design.** After making changes or improvements, create your final model.
10. **Communicate the processes and results.** Present your finished model to the class. Demonstrate how the model works and explain the principle involved. Be prepared to answer questions. Turn in your assignment to your teacher. Be sure to include the name of the activity, your definition of the problem, a description of how you solved the problem, and your model.

SAFETY FIRST

Before You Begin. Make sure you understand how to use the tools and materials safely. Ask your teacher to demonstrate their proper use. Follow all safety rules.

CHAP 2 TECHNOLOGY TEAMS

Objectives

- Explain the value of teamwork.
- Describe work done by engineering and design teams.
- Identify the four main teamwork skills and give examples.
- List important employability skills.
- Identify steps to take in choosing a career.

Vocabulary

- team
- engineering
- design
- criteria
- constraint
- optimization
- entrepreneur

Researching careers can help you decide on the career that fits you best. In this chapter's activity, "Investigate Technology Careers," you will have a chance to research several different careers that may be of interest to you.

Reading Focus

1. Write down the colored headings from Chapter 2 in your notebook.
2. As you read the text under each heading, visualize what you are reading.
3. Reflect on what you read by writing a few sentences under each heading to describe it.
4. Continue this process until you have finished the chapter. Reread your notes.

The Value of Teamwork

Have you ever been part of a team for sports or some other type of activity? A **team** is a group of people who work together toward a common goal. Each member of a team brings his or her own special skills, abilities, knowledge, and ideas to the mix. At their best, teams often accomplish more than individuals can do alone.

Fig. 2-1 Teamwork is important in competitions such as this one at the Technology Student Association (TSA) National Conference.

During this technology course you will probably work as a member of a team to complete different projects. See **Fig. 2-1**. If you choose a career in technology, teamwork will also become part of your work life. Because the various technologies have become so advanced, one person cannot be aware of everything having to do with a particular device or process. For this reason, 90 percent of leading manufacturers use teams. Teams provide many other benefits. Among other things, teams improve communication across a company and give workers a greater sense of job satisfaction.

Roles of Team Members

In general, team members fall into three categories: contributors, encouragers, and team leaders.

Contributors complete tasks, support other team members, ask questions, share ideas, and evaluate results. Encouragers listen and urge everyone to participate. Almost all team members alternate between being contributors and encouragers on the same project.

Team leaders keep all team members on track. They make sure everyone understands his or her role, and they set a good example. Leaders also keep an eye on the schedule.

Engineering and Design Teams

Engineering is using knowledge of science, mathematics, technology, and communication to solve technology problems. **Design** is the process of creating things by planning. Engineering and design teams are essential to most industries today. Why do you think teamwork is especially important in engineering and design?

Designing a Product Teams play a major role in designing new products or improving existing products. See Fig. 2-2. Suppose a company that makes ice cream wants to produce a new flavor based on mango. The engineering and design team first considers all the criteria and constraints. **Criteria** are the standards that a product must meet in order to be accepted. One standard might be to use fresh fruit instead of canned. **Constraints** are restrictions on a product. A possible constraint might be to produce the ice cream below a certain cost. Trade-offs may be made. What might be a trade-off for using fresh fruit in ice cream?

Fig. 2-2 Teams design products based on certain standards.

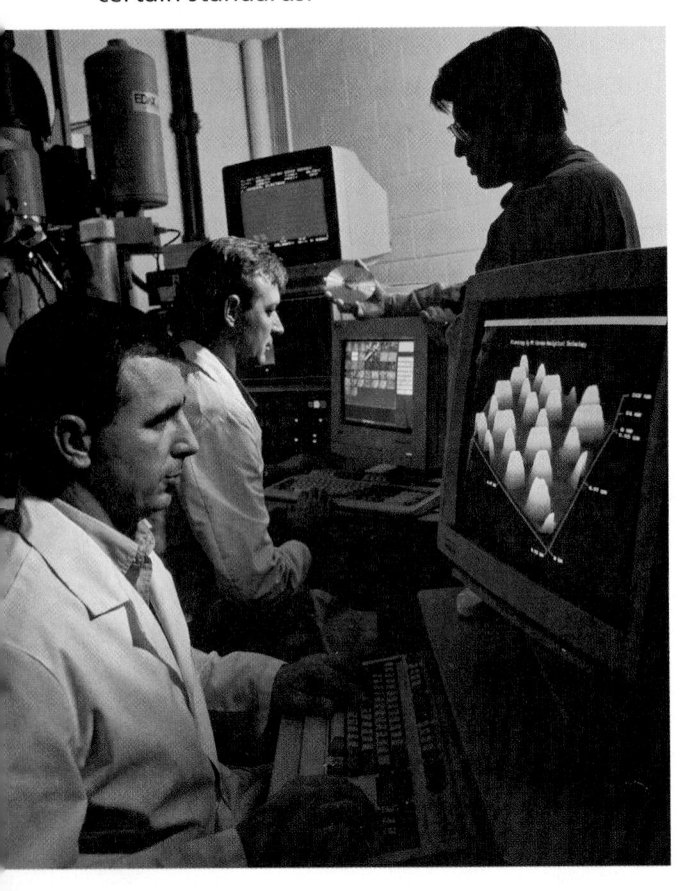

The team works together to produce the most effective and functional product or process while meeting all the criteria and constraints. This is called **optimization.** A mango ice cream would be optimized by making the most of its positive features (fresh fruit) and reducing any negative features (lower quality ingredients).

Quality Control Teams are also important to quality control. Quality control ensures that the finished product, service, or system meets the criteria and constraints. As cartons of mango ice cream rolled into the company's freezers, the team would test samples and check that costs and other factors were on track.

Management Teams may also be responsible for management functions. Management is the process of planning, organizing, and controlling work. Certain team members may be in charge of certain departments or tasks involved in producing the product.

In Chapter 3, you will find out more about how engineering and design are the foundation of many technologies. You will also learn about the engineering design process.

Teamwork Skills

As part of a team you learn several important skills. They include the ability to set goals, solve problems, reach agreements, and resolve conflicts.

Setting Goals

Without careful planning, teams may lose sight of what they are trying to achieve. Everyone on the team must understand what the goal is.

- **Be specific.** A vague goal, such as "improve safety on stairways," may mean different things to different people. "Make sure objects are not left on stairways" is more specific.

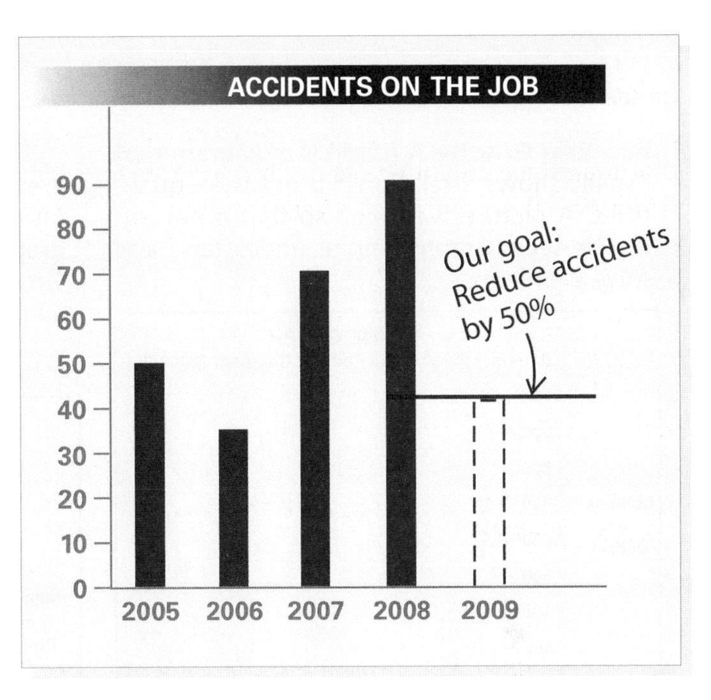

» Fig. 2-3 A goal should be stated in such a way that progress can be measured.

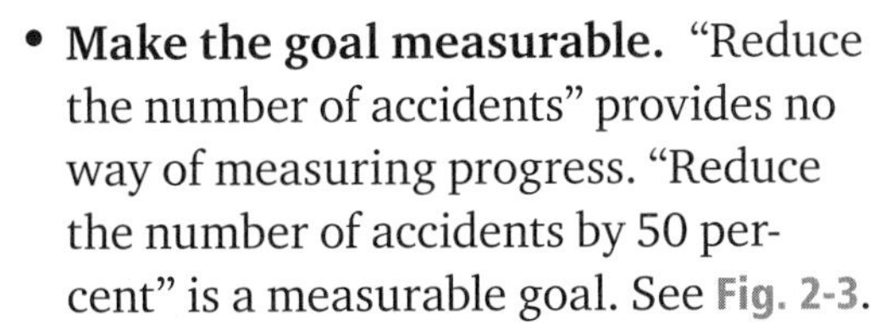

- **Make the goal measurable.** "Reduce the number of accidents" provides no way of measuring progress. "Reduce the number of accidents by 50 percent" is a measurable goal. See Fig. 2-3.
- **Be sure the goal can be achieved.** "Eliminate all accidents in two days" is probably not realistic as a goal. "Reduce the accidents by 50 percent over a period of one year" is more likely to be achieved.
- **Be sure the goal supports any larger goals.** Reducing accidents within the technology lab by getting rid of all sharp tools might work, but it defeats the purpose of the lab. How can students learn to use the tools of technology unless the tools are available? A safety goal must support the overall goals of the lab.
- **Be realistic about the time required to meet the goal.** Planning to train five technology classes in safety methods in one afternoon may not be possible. Two weeks might be more realistic.

Solving Problems

As you know, technology seeks to extend human abilities and satisfy human needs and wants. Almost all of these technological goals can be stated as problems.

There are many ways to solve problems. One way is by trial and error—trying whatever idea comes to mind and hoping something will work. However, this approach often wastes time and energy. Successful problem solving usually includes steps such as stating the problem, collecting information, developing possible solutions, selecting and implementing the best solution, and evaluating the results. These steps will be discussed in greater detail in Chapter 3.

Building Consensus

When numbers of people are involved, building consensus (agreement) can be difficult. Team members may feel strongly about a particular issue. However, decisions must be made in order for progress to continue toward the team's goal. Teams can usually reach agreement in one of the following ways.

- The team leader can decide.
- The team member with the most knowledge about the issue can decide.

Math Application

Reading Graphs A graph is a diagram that usually shows a relationship between quantities. A picture is formed so that a pattern can be seen. Graphs help us understand and solve problems.

The three most common types of graphs are bar, line, and circle graphs. Depending on the information you want to show, one graph might be better to use than others. Often the same information can be shown in all three kinds of graphs.

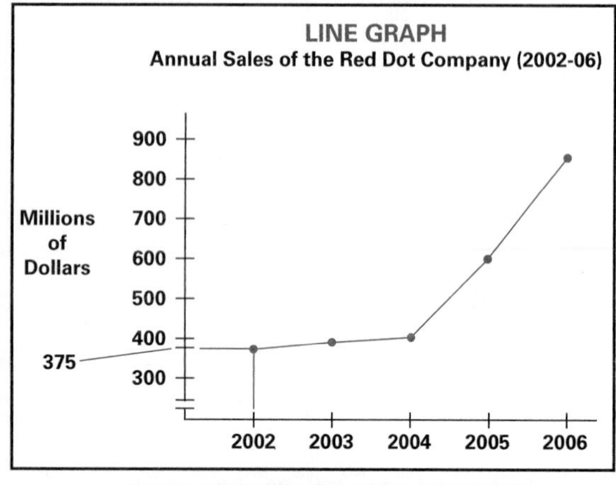

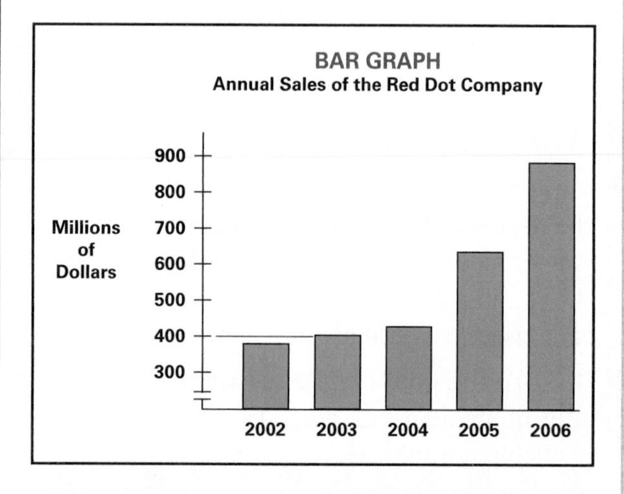

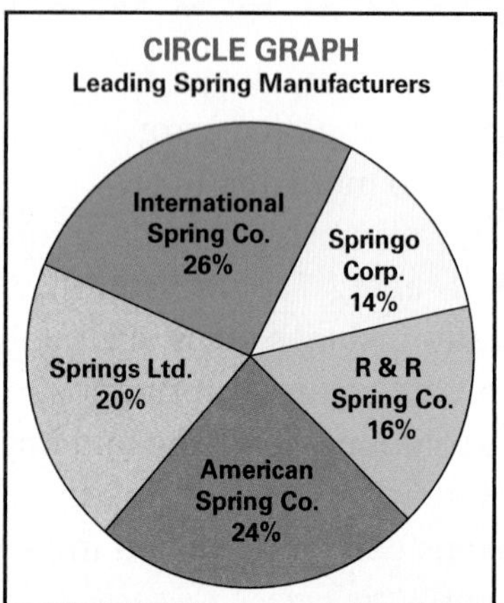

FIND OUT Create a bar graph using the information shown on the circle graph. Place the names of the five spring companies at the bottom of the graph and their percentages on the left.

- A vote can be taken and the majority allowed to rule.
- All team members' opinions can be heard, and an agreement arrived at by most of the people directly involved.

The last method is usually the most successful because every member has a voice, and most agree with the decision. However, building support for an idea may take time and negotiation. Negotiation means to reach a decision through discussion and compromise. Each party is willing to give up something in order to reach the goal. Once a decision is made, the entire team must support it.

» **Fig. 2-4** Treating one another with respect can prevent conflict between team members.

Resolving Conflict

Disagreements among team members are to be expected. They ensure that all aspects of a decision are considered so that the best decisions can be made. However, conflict is not the same as disagreement. Conflict interferes with work and may keep the team from reaching its goal.

Many conflicts can be avoided if team members treat one another with respect. See **Fig. 2-4**. Everyone should be encouraged to contribute his or her ideas. Criticism should be constructive. It should focus on the idea, not the person suggesting the idea.

Any conflict that does arise must be resolved. Resolving a conflict does not mean that one side wins and the other side loses. Often, conflicts can be settled by following these steps:

1. Have each party describe the problem in a respectful way.
2. Have each party suggest a possible solution that will be beneficial and satisfactory to all.
3. Have each party evaluate the other's suggestion and explain which things can be agreed on and which are not acceptable.
4. Attempt a compromise. Ask what each party is willing to give up in order to resolve the conflict.
5. If compromise is not possible, the parties involved should try to find a new solution.

Sometimes a mediator needs to get involved in a conflict. A mediator is an unbiased participant who helps opposing parties come to an agreement.

SAFETY FIRST

When Conflict Becomes Dangerous. Sometimes conflicts can be so upsetting for some people that they attempt to hurt themselves or others. If someone you know seems to be having trouble dealing with conflict, talk to the school counselor about it. The counselor can take steps to help the person avoid serious trouble.

Employability Skills

Have you considered a career in technology? It's not too early to think about your future. Starting now enables you to explore different areas of interest before you must make a decision. It also allows you to develop some of the skills you'll need.

You've learned about some special skills needed to be successful as part of a technology team. Other skills are also required—the same skills that make you a successful employee in any field. They include personal qualities, thinking skills, interpersonal skills, and a work ethic.

Personal Qualities

Personal qualities have to do with a person's character and personality. They are often as important to success on a job as doing the work itself. Here are the personal qualities most employers look for:

Positive Attitude People with a positive attitude try to make the best of any situation. Employers value them because they get along well with others and are willing to make an extra effort to get the job done. Increased self-esteem often comes with a positive attitude. Positive people are more likely to feel good about themselves. They realize that they make mistakes, but they believe they can improve.

Ethics People who are ethical are honest and trustworthy. They do not steal or cheat. They have integrity. They do a day's work for a day's pay. See **Fig. 2-5**.

Responsibility Employers want employees who take responsibility for the job and for their own actions. Responsible workers are dependable and reliable.

Initiative Taking initiative means doing what needs to be done without being told to do it. It also means looking for ways to help others when your own work is done.

Willingness to Learn Do you like to try new things? Are you curious about how things work? You probably have a willingness to learn. In the fast-moving field of technology, workers must keep both knowledge and skills up to date.

Commitment to Quality People who are committed to quality strive to do their best at all times. They work to meet the highest standards.

Personal Appearance Although some schools may tolerate extremely casual dress, piercings, and tattoos, most employers do not. Dressing appropriately and keeping yourself clean and well-groomed are ways of saying you are a serious person who will accomplish important things. See **Fig. 2-6**.

Fig. 2-5 Taking unscheduled breaks is cheating an employer.

» Fig. 2-6 People who work in an office dress differently than people who work on a construction site. Clothing should be appropriate for the job and acceptable to the employer.

Thinking Skills

In addition to basic academic skills, such as reading, writing, science, and mathematics, workers need thinking skills. Thinking skills include knowing how to learn and the ability to reason, set goals, solve problems, and be creative.

Someone who possesses effective study skills has already learned much about how to learn. Paying attention, correctly identifying the subject or topic, reviewing main points, and making notes are all ways in which learning is accomplished. They play a part on the job as well as in the classroom.

Reasoning is not just another term for thinking. If you have reasoning skills, you can think things through in a logical way, based on facts. It also means you are able to step back from your emotions and look at a problem objectively.

As you know, setting goals and solving problems are teamwork skills. They are also thinking skills. They involve identifying the problem or goal, determining which steps will need to be taken, and analyzing the results.

Some people have the impression that only artists, writers, and musicians are creative. This isn't true. All of us are creative to some degree. Creativity is the ability to use our imaginations to develop new ideas and products. Creativity can be applied to most fields, including technology. See Fig. 2-7.

Fig. 2-7 Creativity, combined with the practical application of technology, was used to design and build this solar race car.

Interpersonal Skills

Interpersonal skills are skills needed when interacting with other people. Some of these skills were discussed in the section on teamwork. Others include communication skills, respecting others, and teaching others.

Communication Skills These are important for several reasons. They enable you to speak or write clearly so that the information you pass along or record is accurate. They enable you to listen carefully so that you understand the message. They enable you to communicate in a courteous and respectful way so as to ensure cooperation.

You can improve communication skills by asking questions, taking notes, and giving explanations when needed. Notify others of any problems that occur. Consider your audience, because they may not be as familiar with terms used on the job as you are. To be sure everyone understands, ask if anyone has questions.

Respect for Others During your school and work life, you will meet many people from diverse backgrounds and with differing abilities. Some may be coworkers or teammates. Others may be customers. Treat them as you would want to be treated—with respect. They will be more likely to cooperate with you and make your own job go more smoothly.

Teaching Others You may be asked to teach someone on your team how to do a particular task. Approach the subject in a way that you would find interesting. Organize the information so that it's easy to follow. Take it step-by-step if possible. See Fig. 2-8. Let the learner try the different steps while you are patient and supportive. If you must criticize, do it constructively. Focus on the work, not the person.

Fig. 2-8 When you explain how to do something, it's best to take it step-by-step.

Work Ethic

People with a good work ethic believe that with hard work they will get ahead. They bring responsibility, initiative, and commitment to the job. They go the "extra mile" to get the job done.

As you might suppose, employers value workers who have a good work ethic. They are usually the workers who are given more responsibility and who advance more quickly.

Education and Training

Today, a good education is more important than ever in finding a job. Most technology jobs require at least a high school diploma. A good education also helps you learn more about the world around you. This helps you find the career in which you will be happiest and most successful.

General Education

A general education consists of courses in language arts (often called English), social studies, science, and mathematics. Some students wonder if they'll ever need what they learn. First, these courses help you learn to think and to process information. You may forget a geometry theorem, but you won't forget the logical process you used to solve geometry problems.

Second, every job requires the ability to communicate. Not only English but other courses as well ask you to listen carefully and speak and write clearly.

Third, the more you know, the more you can do. The more you can do, the easier it is to find the job that's right for you and the easier it is to advance in that job.

Career Training

Most careers require specific skills. Those skills can be learned in several ways.

- Occupational training centers offer training in trades such as carpentry. They require a high school diploma.
- Technical schools offer training for specific careers. Programs range from one to two years in length.
- Community colleges are two-year schools that offer many programs, including general education courses. Graduates receive an associate's degree.
- Four-year colleges provide a bachelor's degree, which is required for most management positions. General education classes are included.

Science Application

Interdisciplinary Engineers The interaction between engineering and other disciplines is called interdisciplinary engineering. These other disciplines include biology, business, computer science, geology, mathematics, medicine, and physics. There are also many others.

NASA is an excellent example of an organization that uses many interdisciplinary engineers. Some disciplines that are used by NASA include aerospace engineering, ceramics engineering, chemical engineering, computer engineering, electrical and electronics engineering, industrial engineering, mechanical engineering, and metallurgical engineering. Chemical engineers, for example, specialize in chemistry while working with engineering.

Students in technical schools and colleges often study interdisciplinary engineering in order to accommodate specific interests or special skills that do not fall under the traditional engineering program.

FIND OUT Research and describe the specialties of other interdisciplinary engineers who work with NASA. Can you think of projects that they might be involved in at NASA?

Choosing a Career

At this time in your life, you may not have a clear idea of the kind of work you want to do. One field may interest you today. Another field may interest you tomorrow. However, you can be thinking about some things now that will aid you later when you have to make a decision.

Self-Assessment

Most people choose a career based on their interests, abilities, and values. All should be considered.

Your interests are the things you like to do or learn about. They may include such things as school subjects, hobbies, and sports. See Fig. 2-9. It's important to choose a career in a field that interests you. Otherwise, you may be bored or unhappy.

Your abilities are those skills you have already learned. If you know how to use a hammer correctly, then that is an ability. You should also be thinking about those skills you are capable of learning or developing. For example, if you make successful sketches of house designs, then you may be capable of learning drafting skills.

Your values are your ideas and beliefs about things that you think are important or worthwhile. You may value such things as money, family relationships, and independence. Your values can affect your satisfaction with a career. For example, suppose you value family relationships. If a career requires constant travel away from home, would you be happy?

Take time to think about your interests, abilities, and values. Perhaps no job can satisfy all of them. However, keeping them in mind will help you make a wiser career choice.

Investigating Careers

You can investigate careers by doing research. One of the easiest ways to find out about different careers is to talk to people who already work at them. Ask them about jobs that you think you would like. Most people are glad to share their experiences with someone who shows an interest.

Labor unions and professional organizations provide career information. Your local library is another source. Ask to see the *Dictionary of Occupational Titles* and the *Occupational Outlook Handbook*. They provide information on thousands of different jobs, including type of work done and typical salaries. They are also accessible online.

« Fig. 2-9 Someone interested in both technology and tennis might enjoy a career designing tennis rackets and other sports equipment.

Check newspapers and the Internet for job listings. Notice what qualifications are required and what types of jobs are available in your area.

Your school counselor's office usually has many information resources about careers. Many have brochures about different industries and information about work-related topics.

Volunteer and part-time jobs can help you learn about a particular field. See Fig. 2-10. When you volunteer you offer to work without pay. What you gain is knowledge and the satisfaction of helping out.

Spending time on the job with someone is called job shadowing. It can give you an idea of the types of activities and choices a worker in the field faces.

Steps in Applying for a Job

Applying for a volunteer or part-time job is a lot like applying for a permanent job. Here are the steps to follow:

1. Prepare a résumé and/or a portfolio. A résumé is a listing of your work experience and qualifications. A portfolio is a collection of materials that exhibit your efforts and achievements.
2. Contact the organization in which you're interested by telephone or letter. If you are answering an ad, follow the instructions in the ad.
3. You will probably be asked to fill out an application. Carry important information like dates and phone numbers with you so you are prepared. Write neatly. Answer all the questions.
4. If the organization is interested in you for the job, you may be asked to come in for an interview. During the interview, your potential supervisor will ask questions about your experience and why you want the job. No one will expect you to be perfect. Just be honest and do your best. When you are finished, thank the person for the interview.

Fig. 2-10 These steps may prove helpful when applying for a job. ***Have you ever applied for a job? If so, did you use similar steps?***

Entrepreneurship

The skills and attitudes that make people successful as employees are also useful for entrepreneurs. An **entrepreneur** is someone who starts a business. The entrepreneur might have an idea for a new product or service. For example, he or she may observe a need for more housing in the community and start a construction company.

Entrepreneurs must be creative and able to think critically to solve problems. The solution they develop must be a good one, and it must fill a need.

They must also be willing to take risks. Most entrepreneurs have to borrow money to start their company. Making a company successful is not easy. Entrepreneurs must be willing to work hard.

People who are creative, ambitious, and hardworking often make good entrepreneurs. If you have these qualities, you may want to study business and marketing courses in school to help you prepare for owning and running your own business.

Free Enterprise

The economic system used in the United States is the free enterprise system. Free enterprise means individuals or businesses may buy, sell, and set prices for goods and services. The government does not interfere. However, the government does collect taxes and regulate certain activities, such as worksite safety. Our economic system is also based on capitalism. Under capitalism all or most businesses are privately owned and operated for profit.

Capitalism and free enterprise make it possible for entrepreneurs to own and operate their businesses, as well as keep most of the money earned.

Main Ideas

- Teamwork describes a group of people who work together toward a common goal.
- Engineering is using knowledge of science, mathematics, technology, and communication to solve technology problems.
- Thinking skills and employability skills are necessary for any career you choose.
- A good education is more important than ever in finding a job.

Understanding Concepts

1. What is the advantage of working in a team versus working alone?
2. Name three ways engineering and design teams are essential to most industries today.
3. Describe four teamwork skills.
4. What are four important employability skills?
5. Before you investigate careers, what should you do first?

Thinking Critically

1. How might team members have dual roles?
2. A positive attitude is important in a team setting. In what other ways is having a positive attitude or outlook important?
3. Trial and error is one way to solve problems. What are others?
4. What careers might be good for someone who is creative?
5. Can thinking be characterized as a skill? Explain your answer.

Applying Concepts & Solving Problems

1. **Research** Interview a local entrepreneur. What risks did he/she take in starting a business? Did the risks pay off? Why or why not? Report to the class.
2. **Career Activity** Select a job from your local newspaper's want ads. Create a résumé for the job. Hand in the job description and résumé to your teacher.
3. **Hypothesize** Why might an employer be as interested in a prospective employee's attendance in school as in the person's abilities?

Activity: Investigate Technology Careers

Design Brief

Technology offers many career opportunities for people of all abilities and interests. It is not too soon to start learning about them.

At your library and on the Internet are many databases for job information. They can aid you in investigating careers that might interest you in the future.

Imagine you are in charge of making a display on technology careers at a career fair. Design and create a bulletin board or some other display giving information about careers in all six areas of technology. The display should provide examples of real and current job openings, as well as general background.

Refer to the

HANDBOOK

Criteria

- Your display must cover all six areas of technology.
- Specific careers within each of the six areas must be identified.
- Each career should include the following: requirements for getting an entry-level job, requirements for advancement, expected job growth for that career over the next few years, and a general description for the kinds of work done.
- At least one "real" job opening must be located in newspapers or other sources for each career or a career closely related to it. The name of the job, name of the company, and education and experience desired must be included in your display.

Constraints

- A different area of technology should be assigned to each team member.
- Each team member must identify at least one career within the assigned area of technology that might be of interest to him or her.
- Actual job ads may be clipped from newspapers or magazines, but make sure the publications are no longer needed by someone else.

Engineering Design Process

1. **Define the problem.** Write a problem statement that describes what your objective is and what obstacles you need to overcome.
2. **Brainstorm, research, and generate ideas.** With your team, discuss possible solutions. Hint: Consider using career-information databases published by the U.S. government. They are the *Dictionary of Occupational Titles* (DOT) and the *Occupational Outlook Handbook* (OOH). You may also want to use trade magazines, such as those focused on carpentry or architecture.
3. **Identify criteria and specify constraints.** These are listed on page 58.
4. **Develop and propose designs and choose among alternative solutions.** Choose the display that will best attract students and interest them in technology careers.
5. **Implement the proposed solution.** Decide which team members will work on which field of technology. Gather any needed tools or materials.
6. **Make a model or prototype.** Design your display.
7. **Evaluate the solution and its consequences.** Do all parts of the display work toward the function of the whole? Is the overall display attractive and easily visible?
8. **Refine the design.** Based on your evaluation, change the design if needed.
9. **Create the final design.** After making any necessary changes or improvements, create your final display in your classroom.
10. **Communicate the processes and results.** Present your display to the class. Ask for your classmates' feedback on whether the display looks attractive or would hold interest at a career fair. Turn in your assignment to your teacher. Be sure to include the name of the activity, your definition of the problem, a description of how you solved the problem, and your display.

SAFETY FIRST

Before You Begin. Choose careers that are appropriate to display. Make sure you understand how to use any tools and materials safely. Ask your teacher to demonstrate their proper use. Follow all safety rules.

CHAPTER 3

THE ENGINEERING DESIGN PROCESS

These engineers are taking measurements in their attempt to design a comfortable car seat. Do you think you could be a designer? In this chapter's activity, "Design a Computer Desk," you will be given the opportunity to design and build a comfortable and useful computer desk.

Objectives

- List and discuss important concepts of engineering design.
- Define and explain the importance of human factors engineering.
- Identify the 10 steps in the engineering design process.
- Compare and contrast the problem-solving process and the scientific method.

Vocabulary

- critical thinking
- analysis
- synthesis
- evaluation
- human factors engineering
- brainstorming
- simulation

Reading Focus

1. **As you read Chapter 3, create an outline in your notebook using the colored headings.**
2. **Write a question under each heading that you can use to guide your reading.**
3. **Answer the question under each heading as you read the chapter. Record your answers.**
4. **Ask your teacher to help with answers you could not find in this chapter.**

Concepts of Engineering Design

How can a tall building be built to withstand most earthquakes? What is the most comfortable way to transport large groups of people from one place to another? What is the most efficient way to generate electricity to light homes and offices? These kinds of questions lead to the technological devices and systems we use every day and that benefit our society. They are where engineering design begins.

The Nature of Design

When you design something, you work out or create its form or structure in a skillful and creative way. You determine what the product looks like (its form) and how it works (its function). For example, suppose you have been asked to design a new toothbrush that uses a laser beam instead of bristles. Since the toothbrush will not have bristles, should it look like other toothbrushes? What if it were U-shaped to fit over the teeth? Of course, it will not work like other toothbrushes. Will it have a battery? Will safety be a concern? See **Fig. 3-1**.

A product's final form is often determined by its function. After all, the form of a laser toothbrush would ultimately depend on the machinery needed to transmit the laser beam. However, good design usually achieves a balance between the two.

Design also involves creativity. Creativity includes the ability to use resources and visualize solutions. Everyone possesses some degree of creativity. A student who ties a heavy backpack to a skateboard and pulls it along rather than carries it is being creative. He or she has developed a new way of doing something.

Fig. 3-1 These toothbrushes all have unique grips and bristles. ***How might a laser toothbrush differ in design?***

Thinking Like an Engineer

As you know, an engineer is someone who uses knowledge of science, mathematics, technology, and communication to solve technology problems. Engineers also use important thinking skills, such as critical thinking, analysis, synthesis, and evaluation, among others.

Critical thinking is usually abstract thinking. This means that it involves ideas more often than "things." When an instructor asks you to explain why a rise in gasoline prices affects the economy, he or she is asking for critical thinking.

Analysis is the act of breaking a subject into parts so that it can be understood better. An engineer designing a new car engine might take apart an existing engine to find out how the parts work.

Synthesis is the act of putting things together to form a new idea or product. The same engineer might put components from an existing engine together with experimental parts to create a new engine design. See **Fig. 3-2**.

Fig. 3-2 The design for this Chevrolet SSR Truck is a synthesis of old and new. Its engine technology is modern, but its appearance is based on that of trucks from the late 1940s.

Math Application

Engineering and Math While math is an important part of being an engineer, you should not let the amount of math an engineer uses intimidate you. As with other knowledge needed by engineers, the math required is learned gradually, step-by-step.

Think about the level of math you are proficient at now. Every math class you've taken since the first grade has built on what you learned before. If you choose now to take math courses in addition to those required, you can build a proficiency in math that will enable you to pursue a degree in engineering.

FIND OUT Not all engineers use the same amount of math. Pick three different types of engineering fields that you might be interested in and investigate colleges that offer programs for that field.

1. What are the math requirements for admission?
2. How long does it take to complete their program?

+ - π ÷

√ 3×10^{-4}

Each new design also needs to be evaluated. **Evaluation** is the act of judging the final results based on specific criteria. Does the new car engine use fuel efficiently? Does it operate smoothly and without breaking down? An engineer must be able to step back from the work and evaluate it objectively, keeping personal feelings to a minimum.

Do you use these same thinking skills in your own life? You probably do. If your bike broke down, would you consider taking it apart to find out what was wrong? That's analysis. When you make a sandwich, have you ever combined unusual ingredients, such as peanut butter and bananas? That's synthesis. Did you observe how the sandwich tasted? That's evaluation.

Considering Human Factors

Because most products and processes are made for human use, people's needs, characteristics, and habits must be kept in mind during the engineering design process. **Human factors engineering** is the design of equipment and environments to promote human health, safety, and well-being. It is also called ergonomics.

Do you use a wrist rest with your computer mouse? That's an example of human factors engineering. The wrist rest helps prevent stress and strain on your wrist. It makes the interaction between human (you) and machine (the computer) easier and more comfortable.

What other measures might make working at the computer more comfortable? How about a chair designed to relieve muscle strain? See Fig. 3-3. What about making sure the room is not too hot or too cold? These factors are all considered in human factors engineering, which includes the design of machines, work methods, and environments.

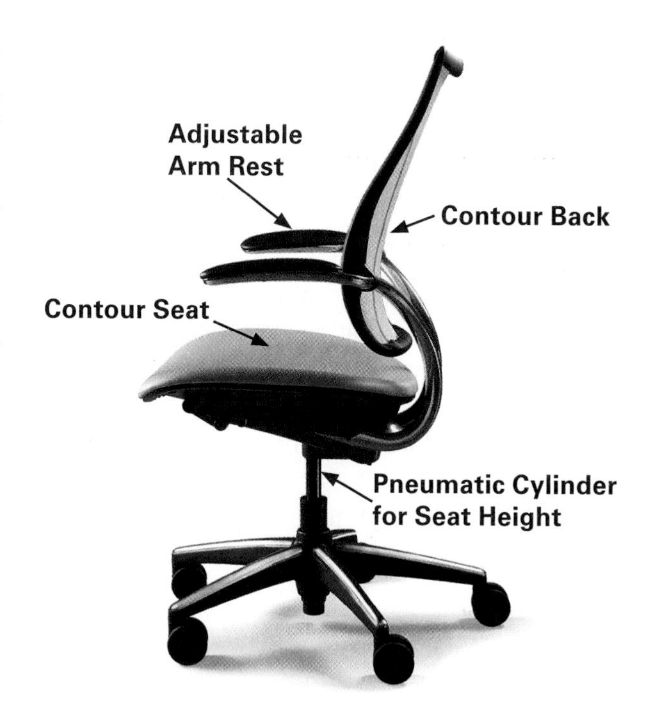

Fig. 3-3 This chair has been engineered to support the body comfortably for long periods.

Machine Design If a machine or tool is hard to use, it is inefficient, and the work takes longer. It may even be dangerous. Human factors engineering seeks to design machines and machine systems that fit the way humans move and think. Consider the way controls are placed on the dashboard of a car. Drivers can see and reach them easily. If they couldn't, it would be much harder to drive the car safely.

People come in many different sizes and shapes. They don't all have the same abilities, and some may have physical limitations. Some products therefore are made in different designs for different needs. Left-handed scissors are one example. Easy-open medicine bottles are another.

Suppose engineers are designing a new type of computer keyboard that automatically adjusts to an individual's touch. How hard does the average person press with a finger on a keyboard key? The answer may not be obvious. Engineers might conduct an experiment in which people would press keys connected to pressure sensors, which would then record the actual pressure in pounds per square inch. Engineers would then analyze this information.

Work Methods Sometimes it's not the equipment or machines that cause problems but the way we use them. For example, people who work long hours on assembly lines may experience eyestrain, shoulder pain, and other discomforts. Adjusting the position of parts as they move along the assembly line or the order in which they arrive may help prevent such problems.

The Built Environment Human factors engineering improves building design. Have you ever been in an old house where the stairway was very steep and narrow? Such stairways are hard to use, and falls are more likely. Today there are building codes that specify stair dimensions, such as the depth of the tread (the part you step on) and the height of the riser (the vertical part of the step). These standards are meant to make stairs safer and easier for people to use.

Suppose someone must use a wheelchair to get around. Homes can be built with wider doorways so that wheelchairs can pass through easily. Kitchens can be designed with lower countertops so that a person seated in a wheelchair can reach them more easily.

Science Application

Green Safety As you now know, human factors engineers are concerned with promoting human health, safety, and well-being. Even colors can be used in ergonomics.

Highway construction workers often wear green safety vests or T-shirts to be more visible to passing vehicles. Human factors engineers conducted many studies and have discovered that green is a very visible color in such situations.

FIND OUT Perform an experiment to test the effectiveness of colors. Obtain different colored T-shirts such as green, orange, red, yellow, white, and black. At marked distances (50 feet, 75 feet, etc.), compare the visibility of each color at different times of day and in sunny and cloudy conditions. Rate the visibility of the colors at different distances and conditions on a scale of poor, average, good, and best. Chart your results. Which colors were most visible? Do your results agree with those of the engineers?

Keeping the Customer in Mind

Almost every product will be used by a customer. The design must satisfy the customer if it is to be successful. Customer requirements often fall into three categories. A product that meets all three has a better chance of making a customer happy.

Basic requirements are those that the customer expects of any product in its category. Remember that laser toothbrush from earlier in the chapter? A customer's basic requirements for it would include that it clean the teeth, just as any toothbrush should do.

High-performance requirements may not be expected. However, they often inspire customers to change from one brand to another. If the laser toothbrush cleans teeth better than any other toothbrush on the market, it has met high-performance requirements.

Excitement is required when a new product is introduced to the market. Customers encounter dozens of product advertisements each day. If a new product is to stand out, something about it must be exciting. What if you could download your favorite CD into that laser toothbrush? What if using it meant you could forget about ever getting another cavity? Would the toothbrush get your attention?

Look at the ad in **Fig. 3-4**. What kinds of customer needs or expectations do you think it addresses?

Fig. 3-4 This shoe design is really new and different. ***Do you think it would excite shoppers who are looking for more comfortable shoes?***

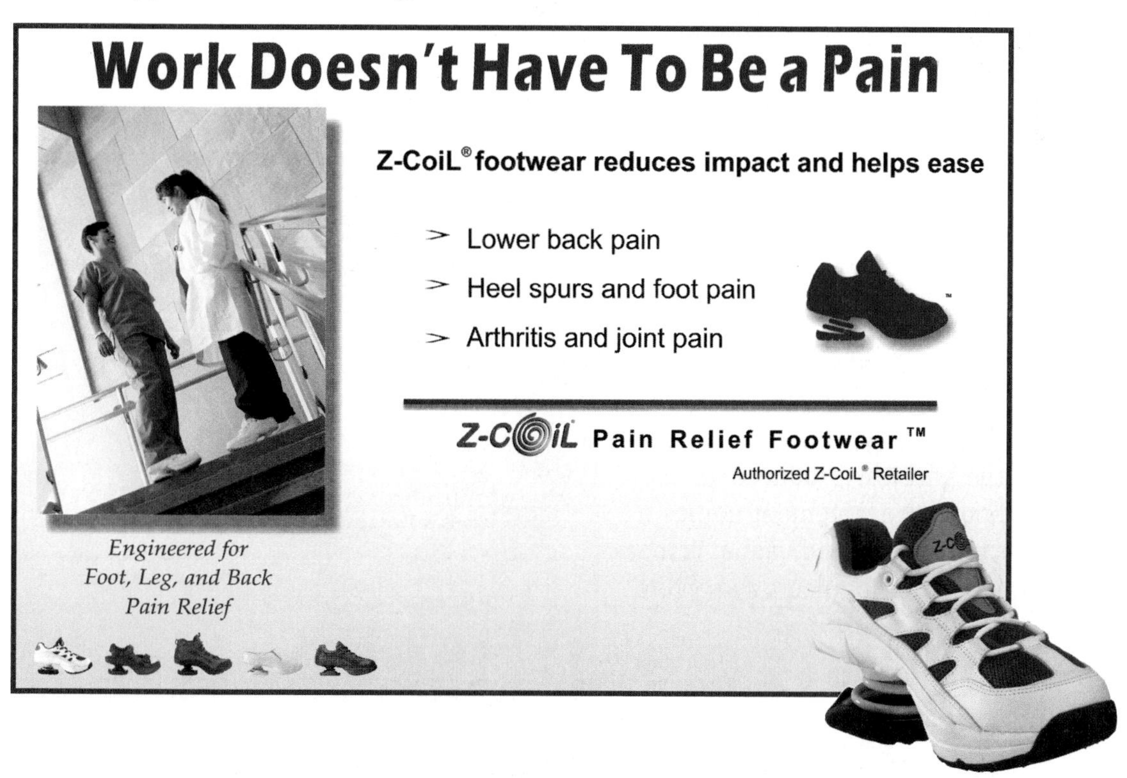

The 10 Steps in Engineering Design

Technological designs often begin as a need that must be met or a problem to be solved. The solution is the result of information available to the design team, the personalities and skills of the team members, and the decisions they make along the way.

Although not every problem can be solved with technology, many can. The process that engineers use to fulfill a need or solve a problem can be described in 10 steps that address specific technology factors. These factors include such things as cost, safety, reliability, positive and negative impacts, and ethical considerations. However, engineers do not necessarily follow the steps in order. New information may be discovered that causes them to repeat earlier steps or to skip ahead. More often, steps may be worked on simultaneously.

Step 1: Define the Problem

How do engineers know that a problem exists? Many times, someone else discovers the problem and brings it to engineers to solve. People within a company may notice a need, or a customer may ask for improvements in a particular product. Engineers are then asked to work on the problem.

Even so, not all design problems are easy to identify. Putting the problem into words helps clarify it and may suggest a possible solution. For example, suppose someone were to say to an engineering team, "You should do something about all the garbage produced in our community." Although the statement names a real problem, it is too general and vague. Is the person suggesting that the engineers design (1) ways of removing the garbage, or (2) ways to reuse the garbage, or (3) products that create less garbage, or (4) a combination of all three? Defining exactly which problem to work on avoids wasting time, money, and effort. It also helps determine which criteria the solution must meet and which constraints limit what is possible.

To give you an idea of how the steps in the design process are applied, consider the case of students in West High's technology class who were covering a unit on bio-related technology. During class they discussed the health benefits of eating organic produce (foods grown without the use of synthetic chemicals). See **Fig. 3-5**. However, they agreed that organic foods were expensive. Some students then recommended that their technology class raise organically grown fruit and vegetables in a greenhouse

Fig. 3-5 Organic foods are raised without synthetic chemicals like pesticides (insect poisons) and fertilizers. Although demand for them is growing, they are more expensive than conventionally grown food.

and provide them to the school cafeteria. Their instructor suggested students use the engineering design process to tackle the problem. First, they defined the problem: "Design a greenhouse in which to raise inexpensive, organically grown vegetables year round."

Step 2: Brainstorm, Research, and Generate Ideas

Brainstorming occurs when two or more people try to think of as many possible solutions to a problem as they can. They don't stop to evaluate or criticize each one. In fact, "silly" ideas are welcome because they may inspire someone to think of a more practical idea. Later, the list of ideas is discussed and the best ones researched.

During this step the ability to visualize, or picture something in your mind, is very helpful. Designers are often people who like to use their imaginations. They are also often resourceful when they do research. Resourceful people are willing to try many different sources for information.

The West High class brainstormed greenhouse designs, sizes, and locations. At the end of the session, they had several attractive ideas. Each team member then took responsibility for gathering information. This is one of the most valuable parts of the design process, because engineers seldom start with all the information they need to solve a problem. Research may also yield additional information about the problem that can help define it more clearly.

Some students talked to the school administration about where a greenhouse could be located. Others went to the library or searched on the Internet to find greenhouse designs that might work. One team visited local greenhouses and studied their construction. See Fig. 3-6. Still other students found out about the kinds of plants they might be able to grow and if the cafeteria would be able to use them.

Fig. 3-6 Many greenhouses are used only a few months a year. *What are a few of the design factors you should consider when building a greenhouse?*

Step 3: Identify Criteria and Specify Constraints

Criteria are the standards that the solution must meet in order to be accepted. One standard West High students found they must meet referred to the size of the greenhouse. School officials wanted some space available for other bio-related technology projects as well as science experiments.

Constraints are restrictions on a solution. One constraint was cost. The school had a limited amount of money available for the project. Another was location. The only plot of land available was shaded by overgrown bushes that might block sunlight. Trade-offs (compromises) would have to be made.

Step 4: Develop and Propose Designs and Choose Among Alternative Solutions

Teams of students began to work on plans for greenhouses based on the criteria and constraints and the research they had done. Others studied the plot of land and devised landscaping alternatives.

When all the plans had been completed, each team presented its solutions to the class and explained all the advantages. The class voted on the designs, and the two top choices were shown to school administrators for their input. Finally, one design was chosen to be implemented (carried out).

Fig. 3-7 Models help people visualize how the finished building will look.

Sometimes design teams have trouble choosing a winner among all the alternatives. They expect that a solution should be perfect and find it difficult to compromise in order to meet criteria and constraints. However, in real life many factors have to be considered. The solution finally chosen may be inferior in some respects simply because other factors limit what can be done.

Step 5: Implement the Proposed Solution

The class decided to hold a fund-raiser to increase the amount of money they had available. Some students organized the event. Others gathered plants and materials to make the greenhouse. Still others cleared the plot of land and trimmed bushes.

Step 6: Make a Model or Prototype

Meanwhile, another team built a model of the greenhouse and created a display showing the health and environmental benefits of organic farming. Models and prototypes (working models) are very useful in helping clients or customers visualize a finished design. See Fig. 3-7. The model and display were shown to school administrators for their final approval. They also took center stage at the fund-raiser.

Step 7: Evaluate the Solution and Its Consequences

During evaluation, the design is judged. It is hoped that any potential problems or negative impacts will be discovered before the solution is put into effect. School administrators had concerns about the heating unit chosen to warm the greenhouse during the winter. More research would have to be done on safety. A coach noticed that the greenhouse was in the path of potential fly balls hit from the baseball field that might damage the roof.

Step 8: Refine the Design

Any changes are now made as the design is refined. The original heating unit for the greenhouse was rejected and another chosen that had better safety features. The coach's concerns required repeating the engineering design process. A protective net was finally selected as the solution for protecting the greenhouse from fly balls.

Step 9: Create the Final Design

Finally, the plans are approved and work can begin. Over spring break, the technology class built the greenhouse. In the following weeks, a hydroponic system to grow tomatoes was installed, as well as planting beds for lettuce and other vegetables. See **Fig. 3-8**.

Step 10: Communicate the Processes and Results

Most final designs are communicated using drawings, specifications, and computer models. In a manufacturing or construction system, the plans are communicated to workers responsible for making the product. The plans must be complete and accurate. The advertising and sales departments are also informed about the product and are given information they can pass along to customers.

At West High, an article about the new greenhouse was placed in the school paper and tours given to interested faculty and students. The cafeteria staff was notified as to when the first crops could be expected. Teams of technology students were assigned greenhouse duty for such maintenance operations as watering and weeding. All that remained was for students to enjoy eating healthier foods.

Fig. 3-8 A hydroponic system grows vegetables such as tomatoes without soil. Liquid nutrients are used instead.

Other Design Processes

Two other design processes are commonly used by technologists. They are the problem-solving process and the scientific method.

The Problem-Solving Process

You may have learned about the problem-solving process in your other school classes. It is very similar to the engineering design process, except that it includes only six steps. Although the steps can be used to solve technology problems, they are not as specialized as those in the engineering design process. You may find them very useful in solving everyday problems. They are as follows:

- **State the problem clearly.** Being able to state the problem clearly may be half the job of solving it.
- **Collect information.** Sources of information depend, of course, on the nature of the problem. They could include libraries, museums, the Internet, interviews with experts, and your own lab research. Keep in mind that research may have to be done in more than one area of knowledge. See **Fig. 3-9**.
- **Develop possible solutions.** There is rarely a "perfect" solution to any problem. Most problems have more than one possible solution. Sometimes trial and error is a useful way of developing solutions. You try different solutions until you find one that works.
- **Select the best solution.** In order to choose the best one, all the solutions under consideration must be evaluated. This involves looking at the advantages and disadvantages of each one. Sometimes the criteria and constraints may interfere with a particular solution's advantages. Possible negative impacts must also be considered.
- **Implement the solution.** During this stage, models and prototypes may be made or a simulation carried out. A **simulation** imitates as closely as possible the real-life circumstances in which the solution is to be used. When testing equipment for space missions, for example, NASA may conduct a simulated space flight. See **Fig. 3-10**.

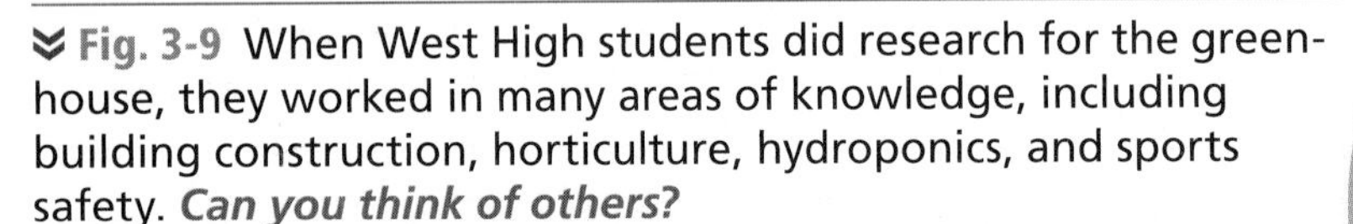

Fig. 3-9 When West High students did research for the greenhouse, they worked in many areas of knowledge, including building construction, horticulture, hydroponics, and sports safety. ***Can you think of others?***

- **Evaluate the solution.** The solution must be evaluated on the basis of established design principles. Customers often play a part in product evaluation. Does the product sell well? If not, why?

The Scientific Method

Engineers often apply scientific knowledge in developing new technologies. You may have learned about the scientific method in your science classes. The scientific method specializes in solving problems encountered in the scientific laboratory and designing solutions to them. See **Fig. 3-11**. It includes the following six steps:

- **Make an observation.** A scientist might notice, for example, that certain green plants grow better in sunlight than in shade.
- **Collect information.** Is direct sunlight required? How much sunlight do plants require in order to thrive? What kinds of plants grow best in shade?
- **Form a hypothesis.** A hypothesis is an explanation that can be tested. The scientist's hypothesis might be this: "Green plant A requires at least five hours of direct sunlight to grow well."
- **Perform an experiment to test the hypothesis.** The scientist might try to grow several specimens of plant A. Some would be placed in direct sunlight for various lengths of time. Some would be placed in shade. All of the other factors, such as water and fertilizer, would be kept the same.
- **Analyze the results.** After a set period of time, which plants did better?
- **Repeat the process to make sure the results are consistent.** Other specimens of the same plant would be tested to be sure the results were not accidental.

As with the engineering design process, observations are made, solutions are tried, and results are analyzed. When a hypothesis is confirmed, other scientists and technologists can rely on the information and use it in their own work.

Fig. 3-10 These astronauts are experiencing near weightlessness in a Neutral Buoyancy Simulator.

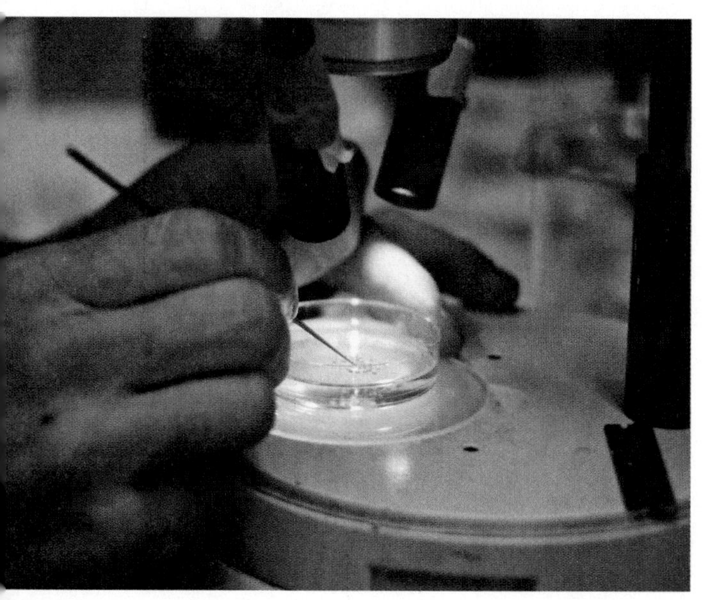

Fig. 3-11 A DNA experiment is being conducted in this lab.

CHAPTER 3 REVIEW

Main Ideas

- The nature of design is to create something in a skillful and creative way.
- Engineers use certain thinking skills when designing.
- The engineering design process includes such things as cost, safety, reliability, impacts, and ethical considerations.
- Two other design processes are the problem-solving process and the scientific method.

Understanding Concepts

1. What do you determine when you design something?
2. Discuss the importance of ergonomics in developing new product designs.
3. Identify the 10 steps in the engineering design process.
4. Why don't engineers always follow the same order of steps?
5. How does the scientific method differ from the problem-solving process?

Thinking Critically

1. How might an engineer test a computer chair for comfort?
2. Why are manufacturers concerned with fulfilling the expectations of customers?
3. Should all design problems undergo the engineering design process? Explain your answer.
4. If you were trying to design a car that would run on soda, what might the problem statement be?
5. Why should brainstorming include more than one person?

Applying Concepts & Solving Problems

1. **History of Technology** How do you think people cleaned their teeth before toothbrushes were invented? Research the topic and share your findings with the class. Explain how the nature of design relates to a toothbrush.
2. **Analyze** Determine how left-handed scissors function differently from right-handed scissors.
3. **Social Studies** Should the government protect new inventions with a patent that lasts forever (instead of a limited number of years, as is done now)? Explain your reasoning.

Activity: Design a Computer Desk

Design Brief

You learned about human factors engineering, or ergonomics, earlier in this chapter. Human factors engineers are challenged to design items that can be mass-produced but that will still be comfortable for a variety of people.

Engineers often use a group of people to test their products. For example, can a keyboard tray be within reach of someone with shorter arms and still be comfortable for someone with longer arms? Many changes may be made to a design before it can be labeled as ergonomically comfortable.

As part of an engineering team, you must create a comfortable and efficient design for a computer desk. Make a full-sized model out of cardboard or other materials for your finished design.

Refer to the STEM HANDBOOK

Criteria

- You must research and demonstrate knowledge about human anatomy and movement.
- Working drawings must be provided with your finished model.

Constraints

- Each member of your team must submit a design before a final solution is chosen.
- Your model must be approved by the majority of your classmates.
- Any tools and materials used must be approved by your teacher.

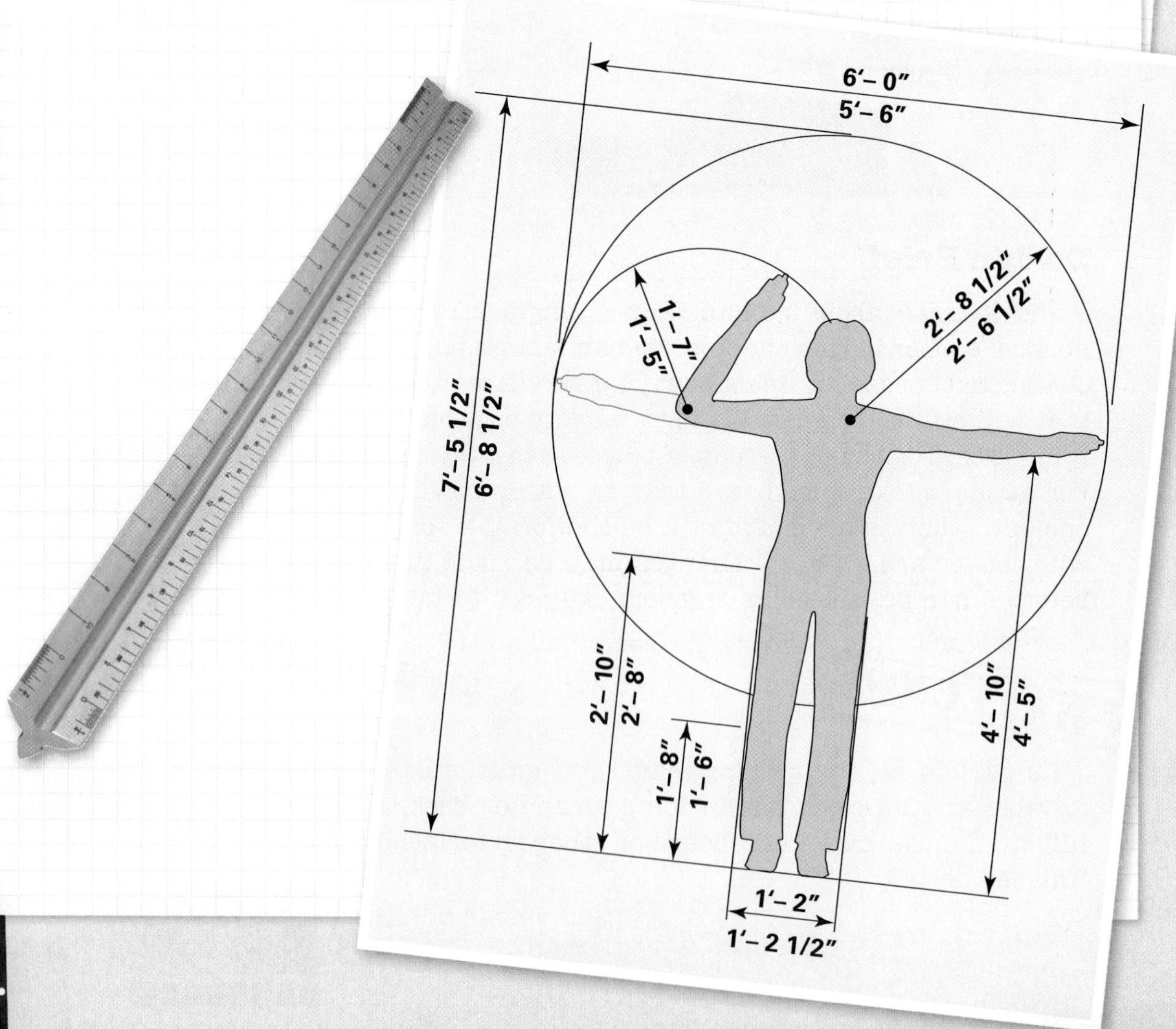

Engineering Design Process

1. **Define the problem.** What is the challenge of designing a computer desk? Write a problem statement that describes both the problem and your goal.
2. **Brainstorm, research, and generate ideas.** With your team, discuss possible solutions. Hint: Research human anatomy and motion to better understand ergonomics.
3. **Identify criteria and specify constraints.** These are listed on page 74.
4. **Develop and propose designs and choose among alternative solutions.** Choose the design that will be the most comfortable and efficient for your classmates.
5. **Implement the proposed solution.** Decide on the process you will use for making the model. Gather any needed tools or materials.
6. **Make a model or prototype.** Create your model. Follow all safety rules.
7. **Evaluate the solution and its consequences.** Does your model work as intended? Have you gotten positive feedback from your classmates?
8. **Refine the design.** Based on your evaluation, change the design if needed.
9. **Create the final design.** After any changes or improvements take place, create your final model.
10. **Communicate the processes and results.** Present your finished model to the class. Demonstrate how the model works and explain why the desk is comfortable to use. Be prepared to answer questions. Turn in your assignment to your teacher. Be sure to include the name of the activity, your definition of the problem, a description of how you solved the problem, and your model.

SAFETY FIRST

Before You Begin. Make sure you understand how to use the tools and materials safely. Ask your teacher to demonstrate their proper use. Follow all safety rules.

UNIT 2

COMMUNICATION
ENGINEERING & DESIGN FRONTIERS

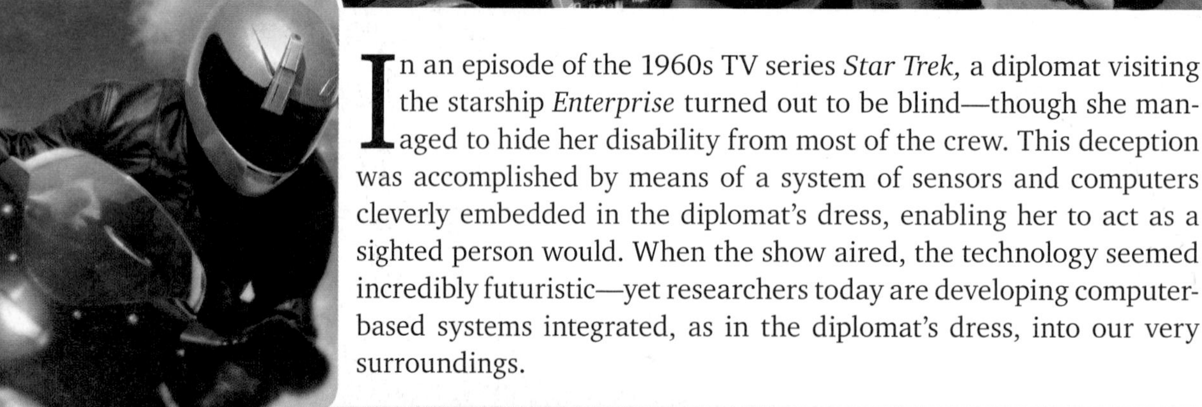

In an episode of the 1960s TV series *Star Trek,* a diplomat visiting the starship *Enterprise* turned out to be blind—though she managed to hide her disability from most of the crew. This deception was accomplished by means of a system of sensors and computers cleverly embedded in the diplomat's dress, enabling her to act as a sighted person would. When the show aired, the technology seemed incredibly futuristic—yet researchers today are developing computer-based systems integrated, as in the diplomat's dress, into our very surroundings.

Computers Everywhere

Researchers are calling this new phase of the computer era ubiquitous (all-around) computing—"ubicomp" for short. Ubicomp technology is based on miniaturization, wireless communication, speech and gesture recognition software, and other existing or emerging technologies.

Imagine walking into your house all sweaty on a hot day. At once, the air conditioner kicks in: ubicomp sensors in your wristband have sent the appropriate signal to your home's cooling system. As you head for the kitchen, you think it would be nice to hear some jazz guitar. "Kitchen ... jazz guitar," you command. The ubicomp system retrieves a sound track and sends the output to kitchen speakers. Later, when you're heading outdoors, sensors in your key ring sound a reminder that you've forgotten your key.

For some people, ubicomp capabilities will mean improved quality of life. Suppose your elderly aunt is living alone in an apartment. Ubicomp devices warn her if the stove has been left on or the bathtub left running. Wearable sensors send regular updates about her to family and professional caregivers.

These scenarios will soon become commonplace, experts say. Meanwhile, ubicomp products are showing up in stores. These include an umbrella that flashes when rain is forecast; a glowing orb (ball) that communicates various types of news by color code; helmets that show motorcycle drivers a heads-up display of speed, RPM, and gear; and garments that collect and transmit data about the wearer, allowing seriously ill persons to be monitored regularly in a nonintrusive way.

Get ready for the era of ubiquitous computing. You haven't seen anything yet!

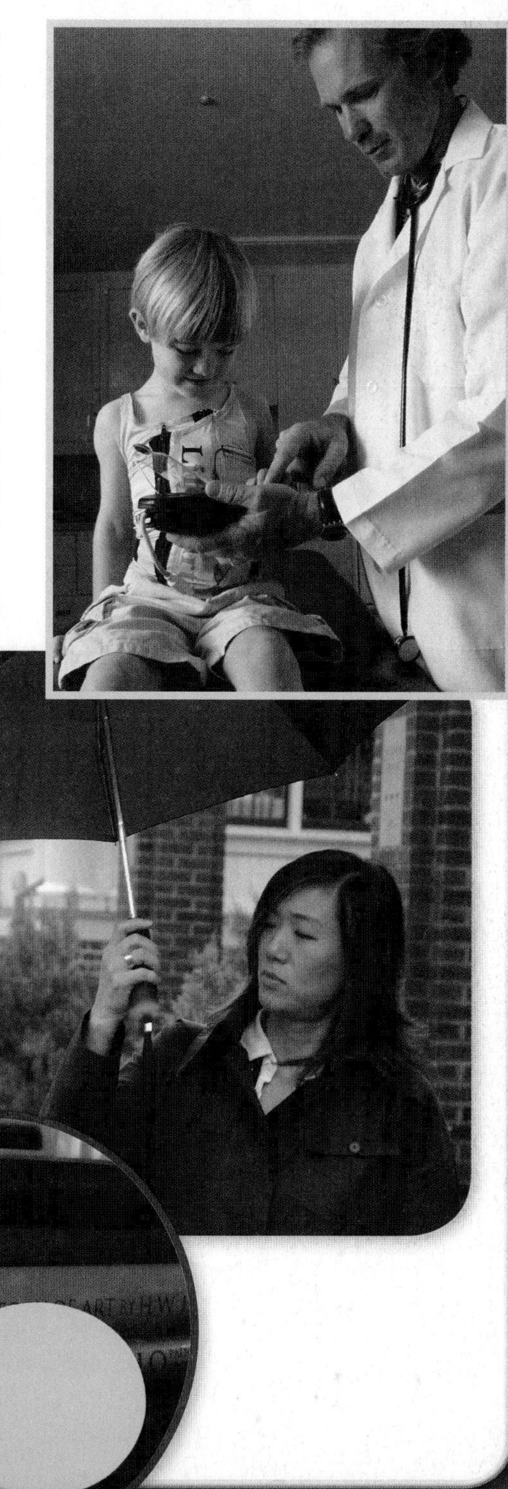

CHAPTER 4

COMMUNICATION FUNDAMENTALS

Objectives

- Identify the six purposes of communication technology.
- Describe the components and function of a communication system.
- Describe the elements of electronic communication.
- Describe the four modes of communication.

Vocabulary

- communication technology
- sensor
- communication channel
- identity theft

⮝ What affects your decision about which clothes to buy at the mall? Could it be how the clothes are displayed? **In this chapter's activity, "Design a Point-of-Purchase Display," you will have a chance to design a POP display for your favorite kind of merchandise.**

Reading Focus

1. **Read the title of this chapter and describe in writing what you expect to learn from it.**
2. **Write each term in your notebook, leaving space for definitions.**
3. **As you read Chapter 4, write the definition beside each term in your notebook.**
4. **After reading the chapter, write a paragraph describing what you learned.**

What Is Communication Technology?

Imagine a world without books, signs, computers, radios, telephones, television, or the Internet. We depend on the tools and equipment that help us to communicate—to send and receive information. We depend on communication technology.

Communication technology is all the things people make and do to send and receive information. It's the knowledge, tools, machines, and skills that go into communicating. When studying communication technology, you will discover that all areas of communication overlap. That's because advancements in electronics and computers are changing how we send and receive messages and other information.

Information comes in many different forms. It reaches its destination in various ways. For example, you often send and receive information in the form of messages by using a phone. Some companies send messages to you by advertising on television or posting ads on Web sites. A buzzer or light in a car reminds passengers to fasten their seat belts. When we study communication technology, we are exploring the ways people use their knowledge and skills to send and receive messages and other information. See **Fig. 4-1**.

Communication technology is always changing as techniques and devices are invented or improved. The invention of the telegraph in the mid-1800s gave people the opportunity to send and receive messages over long distances almost instantly. Today's methods of instant communication are of better quality and more convenient.

Fig. 4-1 This PDA (personal digital assistant) is an organizer and a calculator, will send and receive e-mails, and can access the Internet.

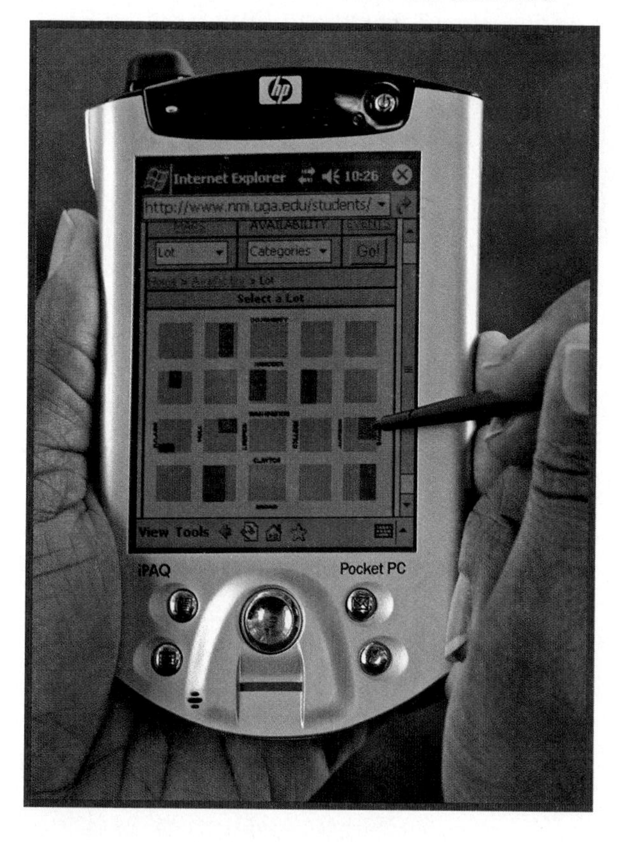

EVOLUTION OF Communication Technology

Communication technology has evolved from pictures drawn on rocks and cave walls to digital signals transmitted to distant spacecraft.

3000 BCE

The Sumerians, who lived in what is now southern Iraq, invented a system of writing called cuneiform. It used wedge-shaped characters pressed into wet clay with a sharpened reed.

1000

The Chinese made movable type by carving characters into small blocks of soft clay. The clay blocks were hardened by fire and then arranged into text.

1450s

A German metalsmith, Johannes Gutenberg, developed the movable metal type and printing press that launched the modern printing industry.

New technologies multiplied. The camera, the telephone, the phonograph, the radio, and motion pictures all were invented during this century.

The world's first commercial computer, the UNIVAC 1, was sold to the U.S. Census Bureau. It was 14.5 feet long and 7.5 feet high.

The first television sets became available. Some were kits that buyers had to assemble.

Communication technologies are converging. A single device can work as a telephone, camera, computer, and media player.

The Purpose of Communication Technology

Communication technology touches almost all parts of your life. The reasons we communicate are to inform, educate, persuade, entertain, manage, and control.

Inform

People read newspapers, watch television, listen to the radio, and surf the Internet to stay informed about a wide variety of things, such as international politics, local sports, weather, and traffic. When you are out with friends, you may use a cell phone to quickly communicate with family members. A tiny computer device with a radio transmitter can be placed under the skin of a person with a serious medical condition. In an emergency, if the person is unable to communicate, the device can inform doctors about the person's medical history.

Educate

In addition to using textbooks, teachers use video and computer programs to help you learn about many subjects. Scientists rely on other communication devices to explore and learn about nature. For example, some marine scientists use a hydrophone to detect sound waves from whales and sea lions. This device helps them learn about sea creatures. Geologists use devices called seismometers that measure vibrations within the earth in order to study earthquakes and volcanic activity. As scientists become better educated about these natural events, they will know how to predict them more accurately and be able to warn people in the area.

Math Application

Downloading and Uploading Advancements in technology increase the speed by which information is transmitted. When downloading or uploading information on the Internet, numbers such as 1,313 kbps for a download speed and 621 kbps for an upload speed indicate how fast information is being transmitted.

These numbers are measurements of the capacity (bandwidth) of the Internet connection between the user's modem and the site from which the information comes and goes.

Kbps stands for kilobits per second, or thousand-bits per second. A bit is a single binary value, meaning it can have only one of two values. In this case, the values are either zero or one.

If the download speed, for example, is 1,313 kbps, then 1,313 kilobits are being downloaded each second. Since kilo means one thousand, we can also state that 1,313,000 bits are downloaded every second, or 1.313 million bits per second.

FIND OUT If the upload speed is 621 kbps, how many bits are being uploaded each second?

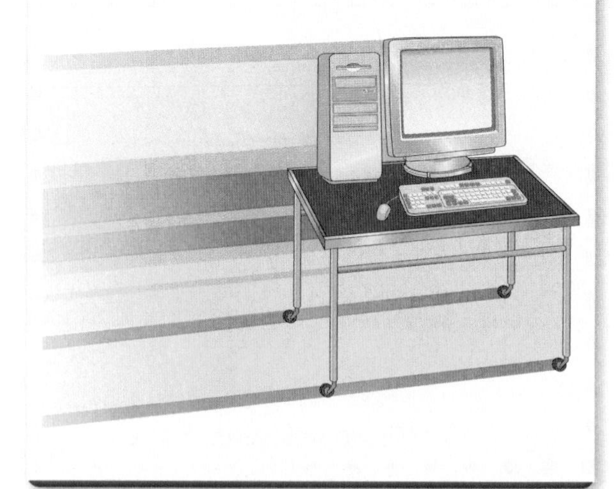

Persuade

Advertising is an example of using communication to persuade. You can probably remember seeing a commercial that made you think you wanted a certain product. Billboards along the highway may influence you to take a certain exit in order to refuel your car or satisfy your thirst. Brochures sent in the mail try to persuade voters to choose a political candidate.

Entertain

Perhaps you play computer games for entertainment or you listen to an Apple iPod™ to hear music. You may also watch television to relax and be entertained. The telephone can be entertaining because it's fun to talk with friends on the phone. Do you like to read? Books, magazines, and comics can all be forms of entertainment.

Manage

Communication devices help people manage information. Computers process data quickly and store vast amounts. Scientists and engineers use computers to keep track of data they learn about the human body. Doctors manage patient records in the same way. Law enforcement officers use data banks to manage details about crimes. See **Fig. 4-2**.

Control

Communication technology plays an important role in controlling machines and tools. Traffic signals are a common example. Computers and sensors send messages to traffic signals. **Sensors** are devices that detect (sense) such things as light or movement. They control when the light changes from red to green and back to red again. In turn, the traffic signals send a message to drivers, thus controlling the flow of traffic. Communication is also used to control where drivers go. Signs warn of one-way streets and lanes that are closed.

Fig. 4-2 Law enforcement personnel can keep vast amounts of information in computer data banks. The computer helps them pull relevant data together quickly.

The Communication System

Like all technological systems, a communication system consists of inputs, processes, outputs, and feedback. They work together to achieve the goal of sending and receiving information or messages. See **Fig. 4-3** on page 84.

Inputs are the seven resources that provide support or supplies. People, information, materials, tools and machines, energy, capital, and time are all inputs. People and equipment are two inputs in making a television show.

Processes are all the things done to or with the inputs in order to achieve the desired result. Think of your favorite movie. Planning scenes, sketching ideas, using cameras, and printing or editing images are just some of the processes used to produce it.

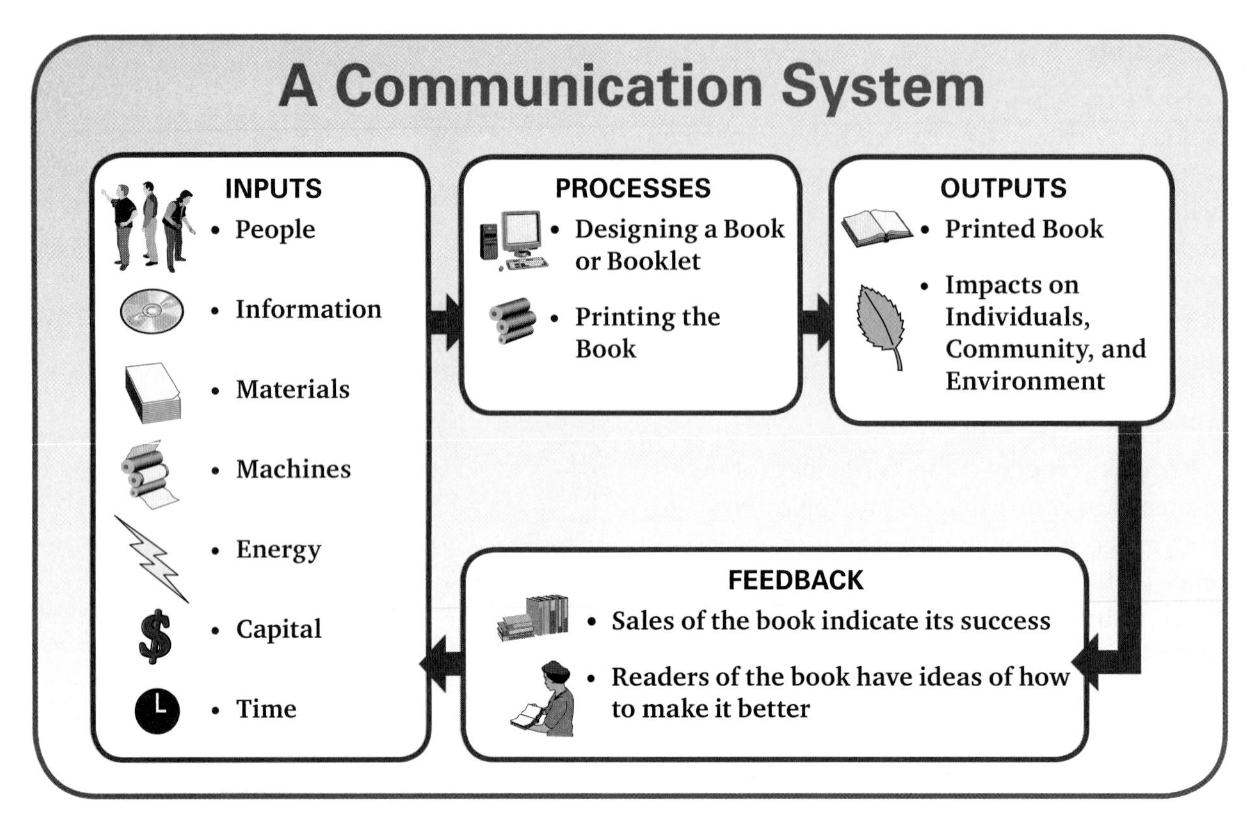

Fig. 4-3 The parts of a communication system are inputs, processes, outputs, and feedback. When they work properly, the intended communication will happen.

Outputs include all the things that result from the inputs and processes. Communication outputs come in many forms. Designed to stimulate our senses, they may be visual (what we see), auditory (what we hear), or tactile (what we can touch). See **Fig. 4-4**. Images, words, symbols, sounds, and music are all typically desirable outputs. Static, noise, or blurred images are undesirable outputs.

Feedback also occurs in a communication system. It happens when you, the sender, are told the message was received. Internet users send feedback to companies when they click on an advertisement that appears on their screen.

Elements of Electronic Communication

Many technological communication systems send and/or store information electronically. Like all communication systems, electronic communication systems consist of a source, encoder, transmitter, decoder, storage, retrieval, and destination. See **Fig. 4-5**.

All messages and information originate from a source, which is often one or more people. That message is then encoded, or changed into a signal, and transmitted (sent) via a communication channel. A **communication channel** is the path over which the message must travel to get from the sender to the receiver. This might include telephone cables or satellites that send signals around the world. Before the information is received at its destination, it must be

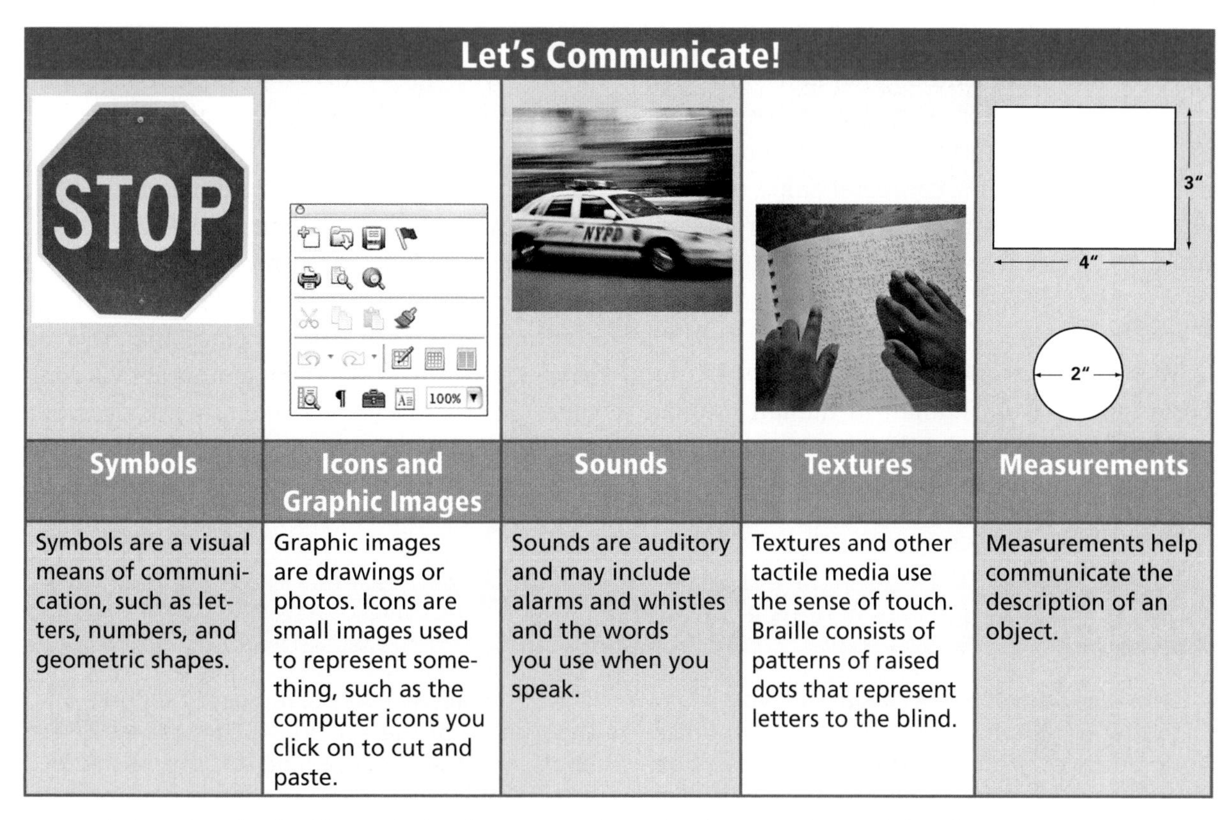

Let's Communicate!

Symbols	Icons and Graphic Images	Sounds	Textures	Measurements
Symbols are a visual means of communication, such as letters, numbers, and geometric shapes.	Graphic images are drawings or photos. Icons are small images used to represent something, such as the computer icons you click on to cut and paste.	Sounds are auditory and may include alarms and whistles and the words you use when you speak.	Textures and other tactile media use the sense of touch. Braille consists of patterns of raised dots that represent letters to the blind.	Measurements help communicate the description of an object.

Fig. 4-4 Technological knowledge is communicated using a variety of visual, auditory, and tactile means.

Fig. 4-5 Information from the sender (source) must be encoded, transmitted, and decoded. It may also be stored and retrieved.

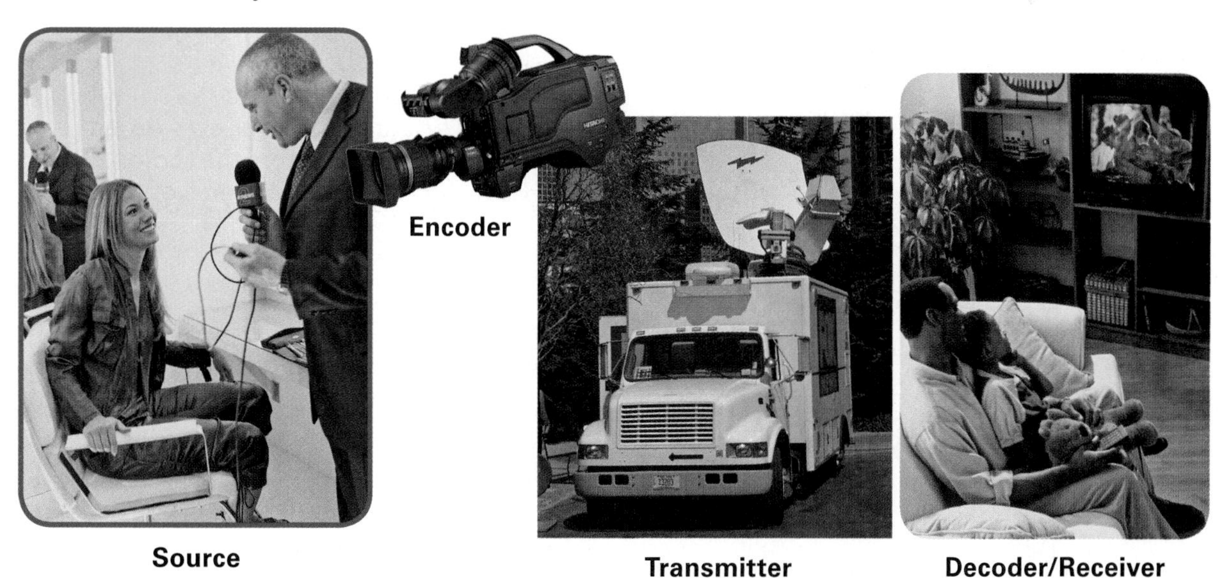

decoded, or changed back into a form that the receiver can understand. The same information may then be stored and later retrieved. For example, much information is stored in digital equipment using CDs, computer chips, or memory devices. This information is later retrieved from the device when needed.

Modes of Communication

Most messages and information travel in four basic modes, from human to human, human to machine, machine to human, and machine to machine. Many communication systems involve all of them.

When you speak with your friend on a cell phone, you are practicing human-to-human communication. When you punched in your friend's phone number, you were using human-to-machine mode. When your friend's phone rang, it was machine-to-human communication. Machine-to-machine communication occurred at many points in between. There was a large communication network in place that relayed those signals from one point to another via computers, cell phone towers, and other devices. See **Fig. 4-6**.

Science Application

Human Sound Reception Did you know that you have your own built-in communication devices? Your ears don't depend on fiber optics or electronics but, like electronic communication devices, they use mechanisms that are very complex and efficient.

Sound is transferred through the air by sound waves—motions of the particles of air. When a wave reaches your ear, it causes the ear drum to vibrate, which starts the reactions that carry information to your brain.

The ear drum is a membrane that is stretched in front of three tiny bones—the malleus, incus, and stapes. The vibrations caused by motion of the ear drum travel through the malleus and incus to the stapes. The bones act as levers to increase the force of the motion of the ear drum. The stapes then beats against the opening of the cochlea, a spiral-shaped part of the inner ear. Inside the cochlea, little hairs move in response to the vibration. These cochlea hairs transfer the vibration to the cochlear nerve. The nervous system uses electrical and chemical signals to carry the message to the brain. Your brain receives the communication and interprets it. The brain can then send messages to other parts of the body, telling them how to respond.

FIND OUT Your ears are a stereo sound system. With your eyes closed, listen to a sound across the room. How can you find the source of the sound?

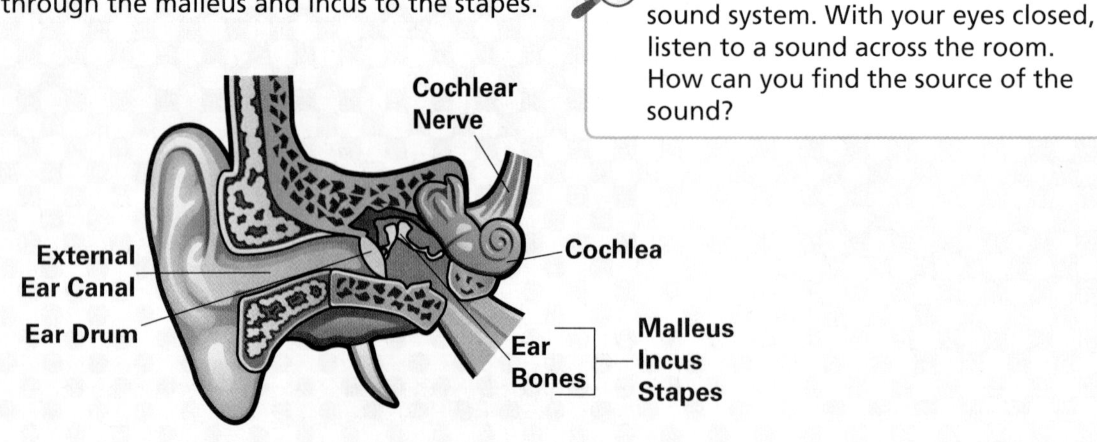

Impacts and Effects of Communication Technology

Communication technologies have many positive and negative impacts and effects on us and on our world. For example, news from Iraq affects how people feel about political leaders. As a result, those leaders may make policy changes. This is a political impact. Other impacts can include social, economic, and environmental impacts.

Disposing of old electronic equipment is an environmental issue. Disposing of these devices improperly can put hazardous materials such as lead into the waste stream. See **Fig. 4-7**. Some businesses have come into being to recycle these devices.

Many people like the convenience of cell phones and portable music players. Yet these devices may contribute to serious accidents on roadways when drivers are distracted. This drives up insurance costs. More cell phone towers may be erected to improve signals, but the structures can affect the beauty of the environment.

Communication devices affect personal privacy. Everyone benefits from the wealth of information on the Internet. However, when information is easier to access, some people will use it for wrongful purposes. The Internet has provided new opportunities for thieves. **Identity theft**, for example, is obtaining someone's personal information and using it to steal. While improvements are always being made in security systems, no system is flawless. New law enforcement methods have had to be developed to stop and find the criminals.

Fig. 4-6 This cell phone automatically relays the phone signal to a transmitting tower. ***Where does the signal go from there?***

Fig. 4-7 Electronic equipment can be recycled. This removes some hazardous materials from the waste stream.

Computer viruses are also a concern. A computer virus is a program that is created to cause harm to computer systems. It is called a virus because it is easily replicated and spread from computer to computer. Viruses can cause systems to fail, data to be lost, and time and money to be wasted on fixing the problems.

The design of computers and other electronic devices involves human factors engineering. The devices must be comfortable to use. Some people become fatigued or develop pain in their wrists after using computer keyboards for long periods. Researchers continue to study ways in which humans interact with computers that might affect physical or psychological health.

Innovations in Communication Technology

Communication technologies are constantly changing. Engineers strive to develop new and better ways of sending, receiving, storing, and retrieving information. They consider all the human factors that may be involved.

If you've done research on the Internet, you know that you cannot always easily find what you want. Computer and software engineers in the leading information technology industries are working on this problem. They are trying to figure out how to make it easier to find and use information on the Internet.

Have you used a GPS (global positioning system) device? Using signals from satellites, these navigation devices can help you find your location anywhere on Earth. More and more new cars have them installed as an aid to drivers. GPS technology is also being used to track shipments of products from place to place. It's possible for a shipper to locate the exact position of a truck carrying the merchandise from one minute to the next. You can learn more about GPS technology in Chapter 21, "Transportation Fundamentals."

Does your computer use a wireless Internet connection? Until recently, wireless hookups were available only at certain coffeehouses and other public places. Now, however, entire communities are making wireless computing available for all residents.

The Internet itself is changing. Blogs, message boards, and online communities have made it more personal. People with similar interests are finding more ways to connect.

Satellite radio is another recent innovation. The radio signal is sent to an orbiting satellite, which can then broadcast it over much greater distances than with ordinary radio.

Do you use text messaging? It has begun to replace some audio phone calls on phones that are equipped for it. When the message is short and requires no conversation, text messaging saves time.

Digital books have been around for a while, but they've never caught on with general readers. That may soon change. The reading devices have grown smaller—about the size of a paperback book—and they hold several books at a time. See **Fig. 4-8**. One day you may no longer need that heavy backpack. All your books could be carried in one hand.

Communication is one of today's fastest moving technologies. By the time you finish this course, new products and processes will already be on the horizon.

» **Fig. 4-8** This reader holds several books but is itself about the size of a single paperback.

CAREERS IN COMMUNICATION TECHNOLOGY

Communication technology offers a large variety of careers. There are communication jobs for the "hands-on" type of person (such as electrical and electronics repairers), high-tech types (computer programmers), and artistic people (graphic designers). The U.S. Department of Labor estimates that there will be nearly one million new jobs in the communication technology industry in the near future.

Graphic Artist • Computer Programmer • Computer Software Engineer

Producer • Computer Systems Analyst • Line Installer and Repairer

Computer Scientist and Database Administrator • Desktop Publisher

Commercial and Industrial Designer • News Reporter • Librarian

Computer Service Representative • Electrical and Electronics Repairer

Photographer • Public Relations Specialist • TV Camera Operator

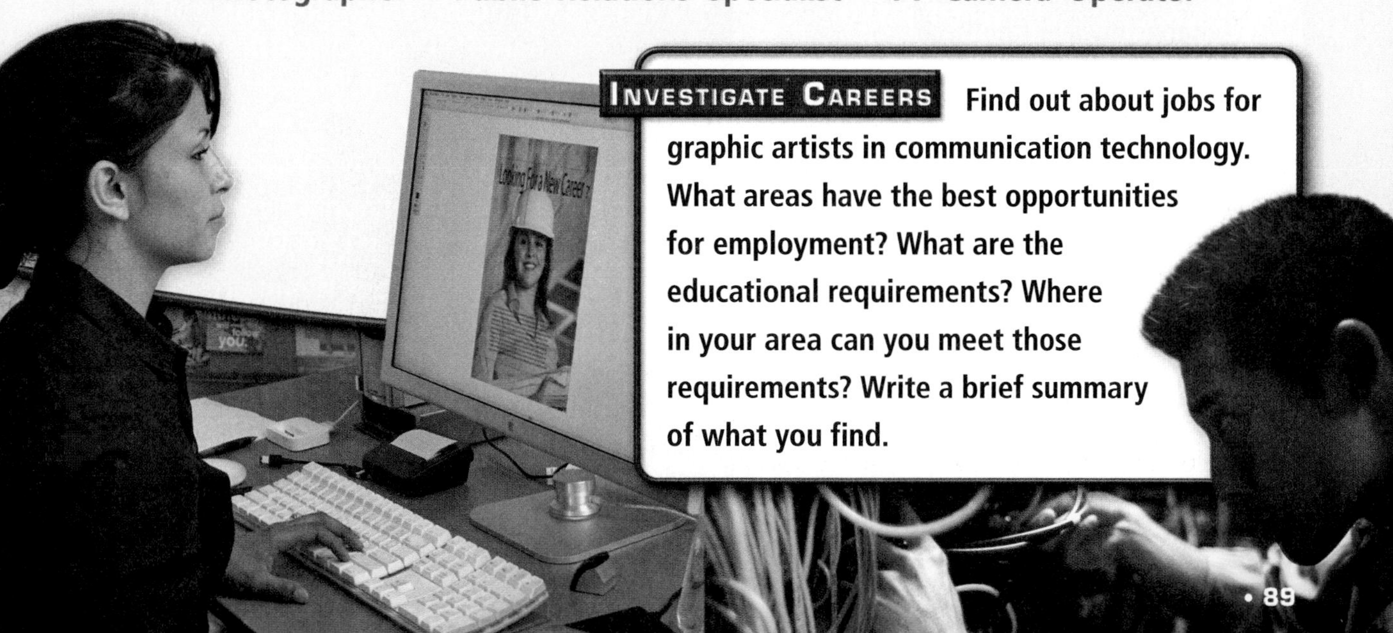

INVESTIGATE CAREERS Find out about jobs for graphic artists in communication technology. What areas have the best opportunities for employment? What are the educational requirements? Where in your area can you meet those requirements? Write a brief summary of what you find.

CHAPTER 4 REVIEW

Main Ideas

- Communication technology is all the things that people make and do to send and receive information.
- A communication system consists of inputs, processes, outputs, and feedback.
- There are four modes of communication, and they often overlap.
- Communication technologies have influenced the world politically, socially, economically, and environmentally.

Understanding Concepts

1. Why do all areas of communication technology overlap?
2. Describe the six purposes of communication technology.
3. What is the goal of a communication system?
4. Summarize the elements of electronic communication.
5. What are the four modes of communication?

Thinking Critically

1. How might the growing ease of communication over long distances change nuclear families?
2. How did the telegraph affect the United States in the 19th century?
3. What might be a negative impact of GPS technology?
4. How can you personally combat identity theft?
5. Besides not having a heavy backpack, what changes might occur if you used a digital book for all your classes?

Applying Concepts & Solving Problems

1. **Communication** Information can often be misunderstood. Whisper a message about technology to a classmate. Have that student repeat the message to someone else until the whole class has heard the whispered message. How has the message changed from the initial sender to the final receiver?
2. **Impacts of Technology** Write a page evaluating the social impacts of communication technology in your life.
3. **Design** Develop a PowerPoint presentation to show how communication has evolved over the last 100 years.

Activity: Design a Point-of-Purchase Display

Design Brief

What affects your decision about which shoes to buy at the mall? Have you ever stopped in a grocery store to pick up one item and walked out with several? How often do you drop spare change in a charity collection container at the cash register? Your decisions may have been influenced by point-of-purchase (POP) displays.

These special displays, placed where you will be most likely to respond to them favorably, come in many shapes, sizes, and forms. They may involve eye-catching graphics, creative sounds and rhythms, interesting electronic features, mechanical movements, or unique textures. However, they all communicate successfully. POP displays are effective tools for getting attention and influencing decisions.

Design and build a tabletop POP display to solicit donations for a favorite charity, nonprofit group, or student organization.

Refer to the

Criteria

- Your display must be appropriate for viewing in your school.
- Focus on a real charity, nonprofit group, or student organization.
- Use visual, auditory, and/or tactile means to entice donations from your customers.
- Incorporate graphic and/or electronic communications.
- The POP display must be self-explanatory and appropriate for the intended audience.
- You must use information that is accurate and approved by your teacher.

Constraints

- The size of the finished display must not exceed two cubic feet.
- You cannot use more than one electrical outlet.
- Your display cannot weigh more than 10 pounds (not including the donations).
- If additional materials are purchased for your display, discuss and document costs with your teacher.
- If your display is used, all proceeds must be given to the named organization.

Engineering Design Process

1. **Define the problem.** Write a statement that describes the problem, such as how to collect money for your chosen charity, group, or organization.
2. **Brainstorm, research, and generate ideas.** With your team, discuss possible solutions. Hint: Visit a mall or grocery store with the goal of identifying clever POP displays and effective techniques.
3. **Identify criteria and specify constraints.** These are listed on page 92.
4. **Develop and propose designs and choose among alternative solutions.** Choose the idea that will be the most effective for the organization you have selected.
5. **Implement the proposed solution.** Decide on the process you will use for making the display. Gather any needed tools or materials.
6. **Make a model or prototype.** Create your model. Follow all safety rules.
7. **Evaluate the solution and its consequences.** Does your model work as intended? Will it attract casual consumers in a store?
8. **Refine the design.** Based on your evaluation, change the design if needed.
9. **Create the final design.** After any changes or improvements take place, create your final model.
10. **Communicate the processes and results.** Present your finished model to the class. Demonstrate how the model works. Explain why your display will attract notice and interest. Hand in your problem statement, design notes, and final display to your teacher.

SAFETY FIRST

Before You Begin. Be sure that you understand safety guidelines about sharing information. Do not reveal personal or private information about yourself or others in a public display. Protect the privacy of individuals.

CHAPTER 5

INFORMATION TECHNOLOGY

Objectives

- Describe the basic parts of a computer system and explain how computers work.
- Explain how binary code works.
- Identify and describe ways in which computers use and manage information.
- Discuss the role of the Internet in accessing information.

A well-designed Web page is very important if you want people to use it. In this chapter's activity, "Build a Web Site About Engineering & Design," you will have the chance to create your own Web site.

Vocabulary

- circuit
- motherboard
- central processing unit (CPU)
- integrated circuit (IC)
- hard disk
- programming language

Reading Focus

1. **Write down the colored headings from Chapter 5 in your notebook.**
2. **As you read the text under each heading, visualize what you are reading.**
3. **Reflect on what you read by writing a few sentences under each heading to describe it.**
4. **Continue this process until you have finished the chapter. Reread your notes.**

Information technology is important because so many systems rely upon computers and the processing and exchange of information. When you search the Internet to find facts for your history project, then you are using information technology. Doctors may use information technologies to probe inside the human body with special computer-controlled devices. Airports rely upon information technology to schedule flights and track airplanes. Information technology has applications in nearly every field.

What Is Information Technology?

The term *information technology* is often used interchangeably with *communication technology*, but they are not exactly the same. Both refer to all of the methods, techniques, tools, and equipment that enable us to create, manage, store, send, and receive information. This information can be in the form of images, text, speech, sounds, numbers, or other data. See **Fig. 5-1**. Information technology, however, is specifically associated with computer-based communication systems. The computer is the key tool that makes information technology possible. It is an electronic device that can store, retrieve, and process data.

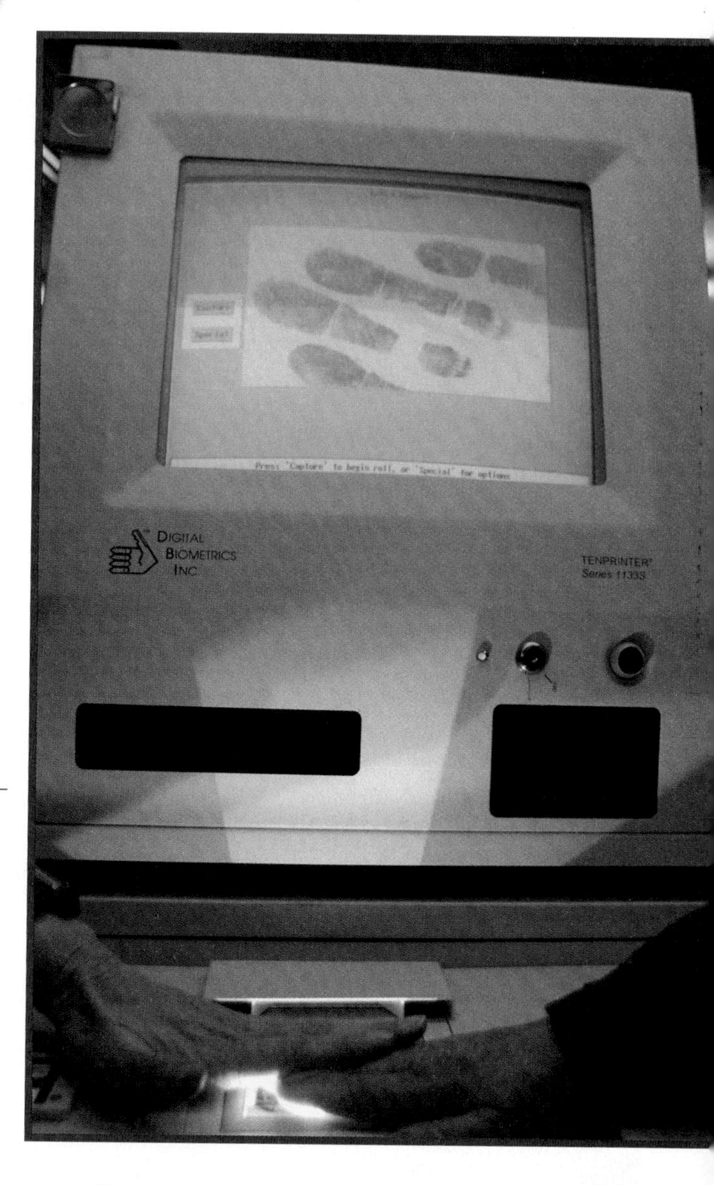

» **Fig. 5-1** Fingerprints can be stored on computer systems and accessed by law enforcement officers. ***How can this technology be used to apprehend criminals?***

Parts of a Computer

To learn more about information technology, you first need to learn more about computers. You know what a computer does, but do you know the parts of a computer and what makes it work?

A typical computer system has the following basic parts: input units, circuit boards and all the integrated circuits, storage units, and output units. See **Fig. 5-2**.

Fig. 5-2 Computers need input units, processing units, and output units in order to operate software and complete tasks.

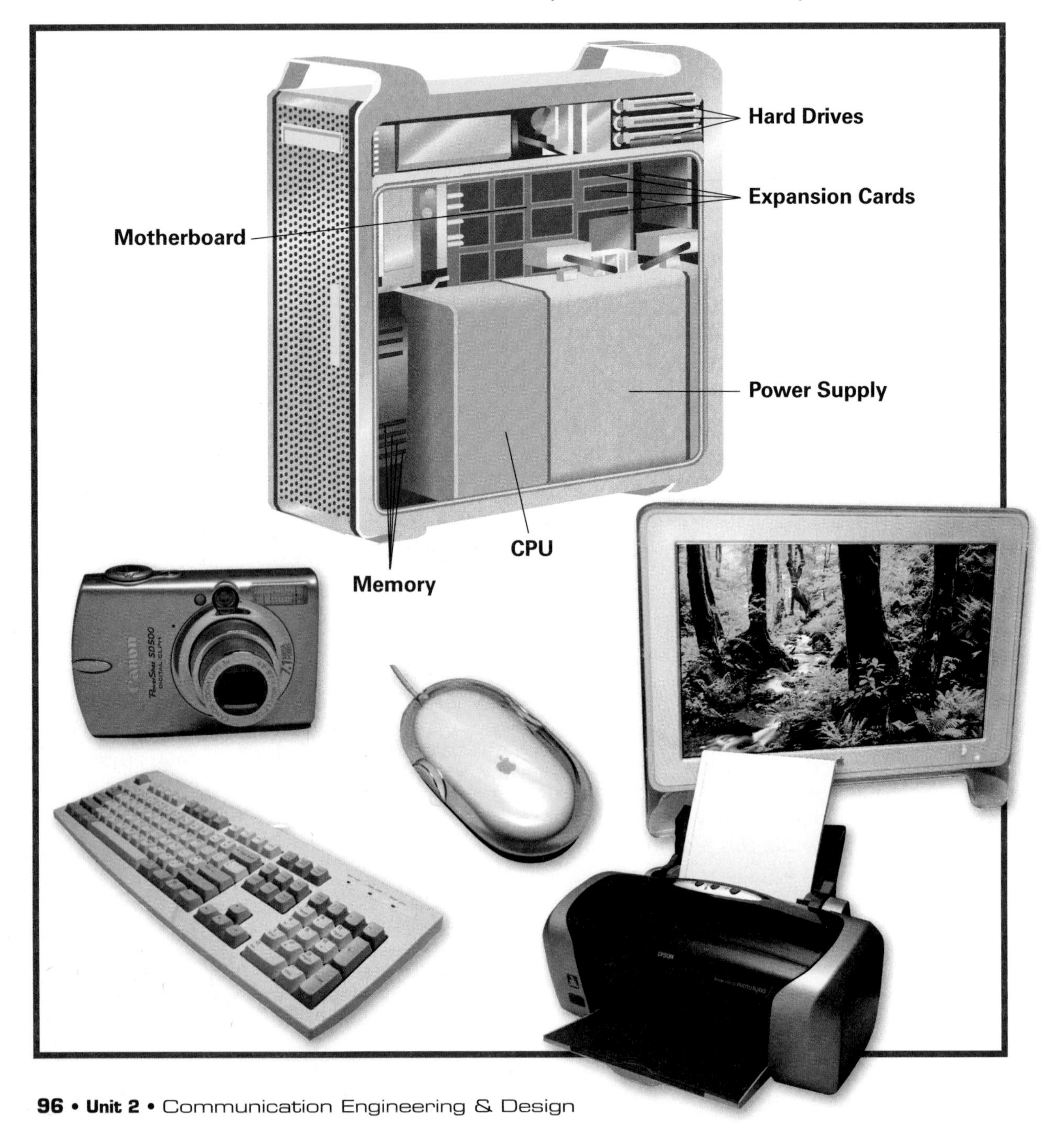

Power Supply

Like all electronic devices, the computer needs a supply of power. The power supply unit connects to the power cord that you plug into the wall to access electricity. The power supply unit also regulates the power going to other parts of your computer.

Input Units

An input unit is any device that can feed information (data, images, sounds) into the computer. Examples of input units include a keyboard, mouse, scanner, touch screen, sensor, digital camera, and joystick. A computer may have more than one input unit, but each input unit must encode (convert) the information into something that the computer can understand.

Circuit Boards

The inside of a computer is very complex. Many unique circuits work together to process the information input into the computer. A **circuit** is a path over which electric current or pulses flow. A typical circuit board connects hundreds or thousands of tiny electronic components together in a compact space. A computer contains several circuit boards. The main one is called the **motherboard** and is the backbone of the computer. All internal components are connected, either directly or indirectly, to the motherboard.

Other special circuit boards known as expansion cards also connect to the motherboard. Expansion cards are needed to enable certain output devices to work, such as the monitor (video card) and speakers (sound card). Some cards also enable the computer to connect to the Internet or to enhance its graphics.

Fig. 5-3 This is what a computer motherboard looks like.

The CPU and Other Integrated Circuits

The **central processing unit (CPU)**, also known as the brain of the computer, resides directly on the motherboard. See **Fig. 5-3**. It is also known as the microprocessor or the "computer within the computer." The CPU has three main parts:

- The control unit guides the processes and flow of information.
- The arithmetic/logic unit performs mathematical calculations with the data sent from the control unit.
- The memory unit stores the information before and after processing.

The CPU does all this work in a very orderly fashion, and it does it very quickly. In fact, it can make billions of mathematical calculations (such as multiplication, division, and addition) every second!

The CPU is the largest integrated circuit in a computer. An **integrated circuit (IC)**, often called a microchip, is a tiny piece of silicon that contains thousands of interconnected electrical circuits that work together. See Fig. 5-4.

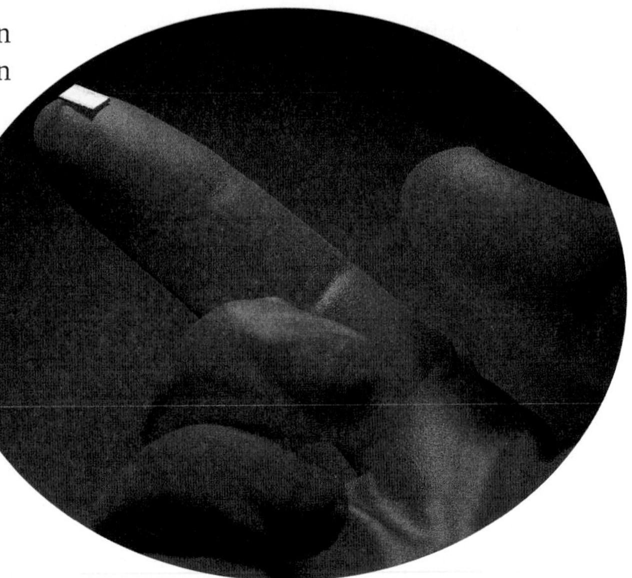

Fig. 5-4 Tiny integrated circuits like this one provide pathways for electronic signals.

Transistors are the main electronic parts found in ICs. They are like miniature electronic switches. They turn on and off in order to process information. However, transistors do not turn on and off like a light switch. Transistors have no moving parts. They control current (the flow of electricity) using one part of the transistor called the gate.

If electricity is applied to the gate, current is allowed to flow and the switch is closed (on). If the gate does not receive an electrical charge, current cannot flow and the switch is open (off). A typical CPU contains millions of these transistors. The amazing thing is that the CPU is only about the size of a dime! You would have to look through a powerful microscope to see the transistors. The CPU is one of many ICs found in a computer. However, it is the largest IC and it is the heart of the computer.

Every IC performs a certain job. Some process information while others may be used for memory only. They are combined together to perform more complex functions. Integrated circuits carry a great deal of information, yet they take up very little space. Many ICs can fit into a very small electronic device and they are used in a wide range of products. This means that products can do more, yet be smaller and lighter in weight. See Fig. 5-5. Because the electric pulses travel so fast, the products work faster. Integrated circuits cost little to operate.

Fig. 5-5 Portable digital audio devices can be made smaller and lighter thanks to integrated circuits. These circuits make it possible to store thousands of songs or other data.

Science Application

Semiconductors A semiconductor's ability to conduct electricity lies between that of a conductor (such as a metal) and that of an insulator (such as rubber or glass). Silicon is the most widely used semiconductor material. Silicon is the brittle, crystalline element found in common sand.

Semiconductors are useful in electronics because, when impurities are added to them, they can be made to carry a limited amount of electric current. Depending upon the type of impurity used, the semiconductor will carry a negative or positive charge.

When semiconductors having different charges are sandwiched together in certain ways, they can act as switches that either stop the passage of electric current or allow it to flow. Layered semiconductors can be used as amplifiers that boost weak electric signals. They can also be used as rectifiers that convert alternating current from an outlet into the direct current needed by electronic devices.

How do semiconductors relate to computers? Thousands of tiny semiconductor sandwiches (microchips) can be built in the fine layers of silicon smaller than a thumbprint. The main storage unit of a computer is made of these microchips, as are the workings of many other electronic devices.

FIND OUT Research the topic of silicon. What other products is it used for? What happens to it when it is heated?

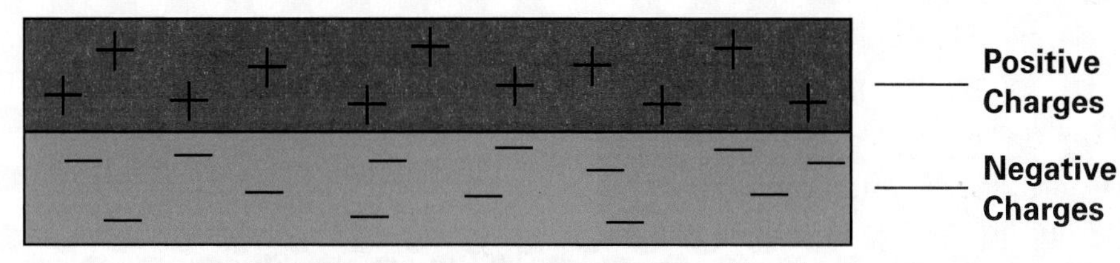

SEMICONDUCTOR SANDWICH

Because very little material is required, they can be produced cheaply in large quantities. The use of ICs has made fast communication more affordable and available to more people.

Memory and Storage Units

The memory units help the computer remember instructions and information. Every computer has several types of memory.

The internal read-only memory (ROM) in the CPU helps the computer remember which steps in the instructions come first. The information is permanently stored there and cannot be changed. The random access memory (RAM) in the CPU remembers what you tell the computer to do while the computer is on. It is a temporary form of memory.

Another type of memory unit built into the motherboard is called the basic input/output system (BIOS). BIOS is an important ROM unit that helps to start your computer. Computers also have cache memory. This is another form of RAM. Information stored in the cache memory is retrieved frequently while the computer is operating and makes processing much quicker.

Storage units are memory units that are typically used for larger amounts of data. The most important storage unit within your computer is the hard disk. See Fig. 5-6. The **hard disk** stores the computer programs that you use for doing word processing, playing games, and many other tasks. These programs are called software. The hard disk also stores all of your personal files such as reports, photographs, and correspondence. All of this information is stored based on scientific principles having to do with electromagnetism, which is magnetism caused by electricity.

Removable storage devices make it easy for you to store and carry data and software from one place to another. For many years, the floppy disk was the most common external storage device. Floppy disks are pliable pieces of plastic with a very thin coating of magnetic particles. Today, other portable magnetic storage devices can hold many times more data than the older floppy disk.

The compact disc (CD) is another popular external storage device. A laser is used to burn tiny pits into the surface of the disc. The pits form a code. When the disc is played, another laser reads the reflected light from these pits and sends that information to the computer.

Because light is used to write and read the data, CDs are called optical storage media. CDs are durable and inexpensive, and they can hold lots of data.

The digital video disc (DVD) is quite similar to a CD, but it can store about seven times as much data. The laser-made pits on the surface are smaller and more tightly packed. Although recordable DVDs are available, currently DVDs are used primarily for permanent storage, especially of movies. However, DVDs are likely to replace CDs eventually because of their quality and capacity.

Flash drives are very small storage units that plug into your computer's universal serial bus (USB) port(s). USB ports enable you to connect a large number of devices easily without having to restart your computer.

Fig. 5-6 The hard disk in your computer holds the software programs, as well as personal files, reports, and other information.

Output Units

Output units decode information they receive from the computer. The video monitor and printer are the most common ones. The monitor displays the information on a screen. The printer puts the information onto paper. There are many types of monitors and printers.

Computers can be linked to other devices, too. For example, a computer that controls woodworking machines or medical devices has output units such as switches, fans, and pumps. See Fig. 5-7.

Math Application

Kilobyte, Megabyte, or Gigabyte? Have you ever gotten confused with all the terms used to describe the amount of memory storage? For example, 1.4 megabytes (MB) and 1,400 kilobytes (KB) look different in number but are almost exactly the same size. These units of measurement are often used to represent electronic file sizes.

1,024 bytes = 1 kilobyte
1,024 kilobytes = 1 megabyte

Since computers contain more memory than individual files, larger units of memory are needed. Most commercial computer hard drives today come with memory storage measured in gigabytes.

1,024 megabytes = 1 gigabyte

The U.S. Library of Congress and other large databases currently store information in terabytes.

1,024 gigabytes = 1 terabyte

Don't be too surprised if you see computer hard drives measured in terabytes before too long. It was only a few years ago that gigabytes of information were not common in commercial computers.

FIND OUT Imagine you wanted to save files onto an 80 MB CD, and each file was 3,000 KB. How many complete files could you fit on the CD?

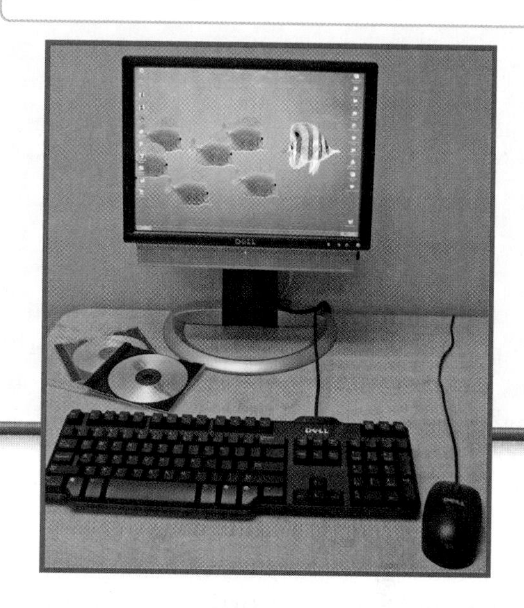

Fig. 5-7 A laser engraver is an output device that can be used to engrave wood, metal, and other materials. It can also be used to cut out materials such as those used in architectural models.

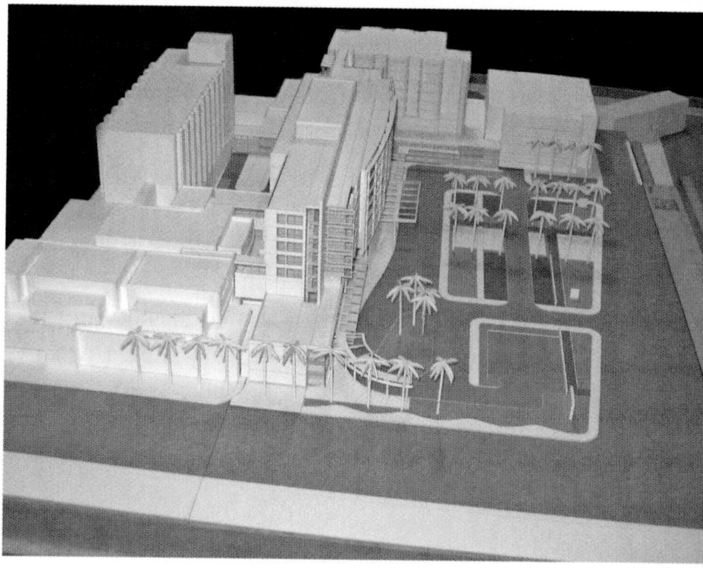

The Language of Computers

Communication can be very difficult if the people who are trying to exchange ideas don't speak the same language. Communicating with a computer can also be difficult. You have to speak its language.

Programming Languages

A computer doesn't automatically know what you want when you press a key on a keyboard, click your mouse, or provide other input. Your inputs must be translated through a programming language. A **programming language** converts data into a form that the computer can understand. For example, when you press "Z" on the keyboard, an electrical signal is created. Computer programs interpret these signals and tell the computer what to do with this information. Computer programs are written in programming languages that the computer can understand. Some of the more popular programming languages are C++, JAVA, Python, and Visual Basic. Each of these languages is used to write different types of programs.

Machine Language

Computers have their own language called machine language. It consists of only two symbols or signals: on (1) and off (0). Every computer instruction is written as a series of 1s and 0s. This sequence of 1s and 0s is digital binary code. (Binary means having two parts.) Once something has been converted into binary code, it can be sorted, retrieved, sent, or altered. Pictures, sounds, numerals, words, and letters become series of 1s and 0s. See Fig. 5-8.

Each on or off pulse of the code is called a bit. (The term comes from binary digit.) Most machines combine eight bits into a byte. Then information can be handled in larger units.

Digital Signals

All information must be changed electronically into binary code so the computer can "understand" it. The code enters the computer as electric pulses. These turn the transistors on (1) and off (0), creating different combinations of paths for the electricity to follow. Electricity flows through these paths as the computer processes the information.

The ability to convert analog data into digital information that can be used by the computer is the key to information technology systems. A digital signal is different from an analog signal. Let's clarify the difference.

Most things in nature are analog. That means they have an infinite number of levels or variations; they are continuous. Voice levels, speed, time, pressure, and light intensity are good examples. Computers cannot understand analog signals because there are too many variations. Digital signals on the other hand, are discrete, which means they have a finite number of levels. Remember that computers know two signals: on (high or 1) and off (low or

Fig. 5-8 Every time you press a key on a computer keyboard, the symbol is encoded into a form that the computer can understand—a series of on (1) and off (0) pulses.

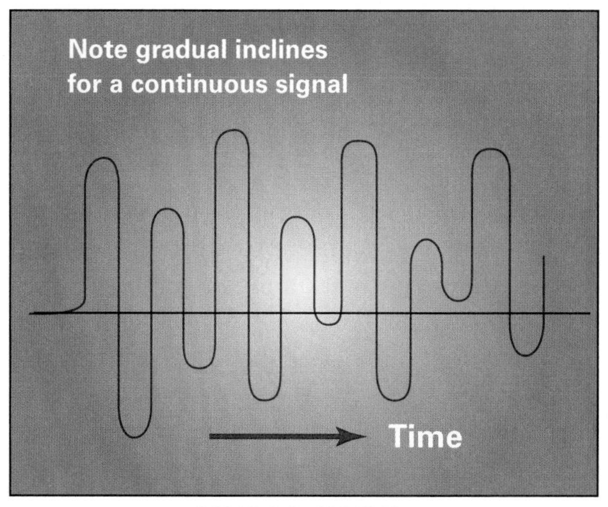

ANALOG SIGNAL

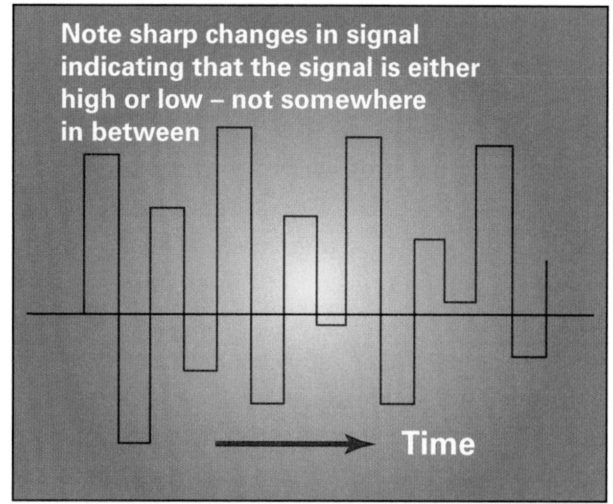

DIGITAL SIGNAL

Fig. 5-9 Notice how the analog signal reaches its upper and lower peaks gradually for a continuous signal. Note the sharp changes in the discrete digital signal. It is either high or low, not somewhere in between.

0). If you look at an analog signal and a digital signal side by side, you can see that the analog signal has many gradual changes whereas the digital signal is either completely on (high) or completely off (low). See **Fig. 5-9**.

An analog system simply transfers the information from the input to the output. For example, suppose a musician plays the guitar in front of a microphone. The microphone amplifies the sound level, but it doesn't electrically change the signal. The signal is still analog. You hear the guitar music as it is played, but it is louder. Often you may also hear noise or distortion (such as static), too.

If the guitar sounds are changed to digital signals, then you can take those signals and do other things with them. For example, you can combine the digital music signals with digital voice signals in your computer. You can filter out the distortion to make the music sound better. You could send the guitar signals through your computer modem to another computer in a South American recording studio. There the signals could be manipulated or enhanced by integrating them with the work of local Brazilian musicians.

Managing and Using Information

One of the reasons that information technology systems have become so common is that computers process information with amazing speed and accuracy. Computers are often used to store and manage massive amounts of data that can be accessed when needed for useful applications. People and companies can do things today that were difficult or impossible before.

Altering Information

When information has been reduced to binary code, it becomes fairly easy to manipulate. Pictures, for example, can be digitized and then altered. Have you wondered how you would look with a different hair style? There are computer programs that let you scan your photograph, erase your hair, select a new hairstyle from a menu, and print a picture of the new you. See **Fig. 5-10**.

Changing pictures in this way can be fun. It can also be useful. Plastic surgeons can show patients how they would look after surgery. Police departments can produce pictures of suspects based on the descriptions of witnesses. However, this type of alteration also poses ethical questions. Suppose a news photographer takes a picture of a senator surrounded by her political supporters. When the picture arrives at the news bureau, the editor notices that a tree behind the senator appears to be growing out of her head. He uses his computer to remove the tree. Is this ethical? Is the picture still true? What if it wasn't a tree behind the senator but a reputed gangster instead who appeared to be among the senator's supporters?

Biometrics

Biometrics is the name given to the measurement and analysis of a person's physical or behavioral information. Biometrics is often used to positively identify people by analyzing their unique features, such as their fingerprints, the iris of an eye, their face or hand geometry, or their voice. See **Fig. 5-11**.

Biometric systems basically involve collecting the data that is unique to the individual and storing that information in the computer until it is needed. For example, the system might record a digital image of the fingerprint. When identification needs to be made, the system detects certain traits and the data is analyzed and compared to the information on file. Then a match can be made.

There is a growing interest in finding ways to improve security in such places as schools, airports, and research laboratories. Biometric systems are one way that these needs might be met.

Artificial Intelligence

Some computers use programs to solve problems and make decisions that are commonly taken care of by humans. This is called artificial

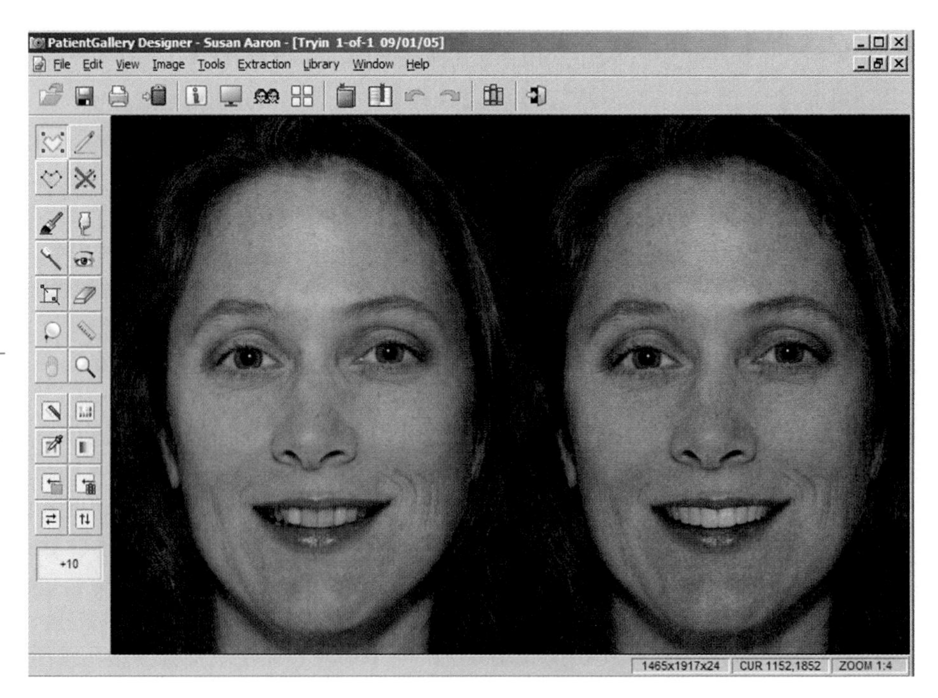

» **Fig. 5-10** This special imaging system, called PatientGallery, lets you see what your teeth could look like under a dentist's care.

» **Fig. 5-11** This customer uses her fingerprint as identification at a supermarket. ***How might this technology change the length of time spent in checkout lines?***

intelligence (AI). Computers cannot think like humans. However, some people believe it is possible to "teach" computers to imitate human thought and decision-making processes.

Computers work by following a program. Basically, these instructions are logical steps to solving a problem. However, what about solving a problem that cannot be answered by logic?

Researchers are finding that computers can solve such problems when they have instructions in the form of reasoning processes that people use. Sometimes a person makes a decision based on a good guess, or a solution may be based on the way an expert might solve the problem. In fact, one application of artificial intelligence is the development of expert systems.

When developing an expert system, developers prepare a program that has all the available facts about the topic. Experts in a particular area are interviewed. Based on the information that they provide and the facts that are available, a complex program is developed. The program instructs the computer how to solve a problem or reach a decision. This part of the program is referred to as the inference engine. It controls the order in which the experts' rules will be followed. It also infers new rules and facts when possible.

Various kinds of expert systems are in use today. Some repair shops use a system that figures out what is wrong with an engine or computer. Efforts have also been made to develop artificially intelligent pet robots. These "smart" pets can be controlled by their owners even when they are not home. Using a wireless Internet connection, you might have the intelligent dog check whether the refrigerator was left open or take a picture of a window to see if it is broken. See **Fig. 5-12**.

Fig. 5-12 These shoppers are looking at these "smart" pets, which are computer robots that look like dogs.

Research is being done to find new ways to use artificial intelligence. Scientists and engineers hope to use AI to improve medical care by making treatments tailored to individuals based on expert knowledge and biological data. Speech recognition systems are another area of research and development. These systems can recognize human speech in all its variations to make decisions or do tasks. This might be useful for improved security or to assist people with disabilities.

Multimedia Applications

Multimedia refers to the use of more than one medium to communicate. When data is presented in multiple forms such as through sound, images, text, and video, it is often referred to as multimedia.

Many information technologies involve more than one type of information. Digital animation is an example. Animation refers to the process of creating the illusion of movement through the combined sequencing of many still images. Using digital multimedia technologies that integrate music, images, and video, designers can create characters with lifelike movements and seemingly real environments. These animations are key features of many video games and movies.

Another interesting use of multimedia is in the area of virtual reality. Virtual reality (VR) uses 3D graphics to create a realistic simulation. In VR, you are part of the computer's artificial environment, called cyberspace. Instead of viewing a flat computer screen, you are "surrounded" by the image. You can interact with the computer by using special equipment such as a viewer and a glove. See **Fig. 5-13**. The computer senses your head, eye, hand, and finger movements and responds to them. You can move through the "virtual world" and control what is happening or what you are seeing by turning your head and moving your hand. Some scientists and engineers call this technology telepresence because it creates the feeling of really being there.

Virtual reality shows promise for many applications. Some researchers are experimenting with VR in the cave automatic virtual environment (CAVE). This is a special room like a small theater that is equipped with high-quality audio and video. Graphics are projected onto the walls and floor, and the VR users wear special glasses and sensors connected to the computer. Images surround the users and move with them. They get the correct perspectives and the environment seems real. One day, surgeons may be

Fig. 5-13 This virtual reality (VR) headset and data glove are used to play a variety of "Virtuality" computer games.

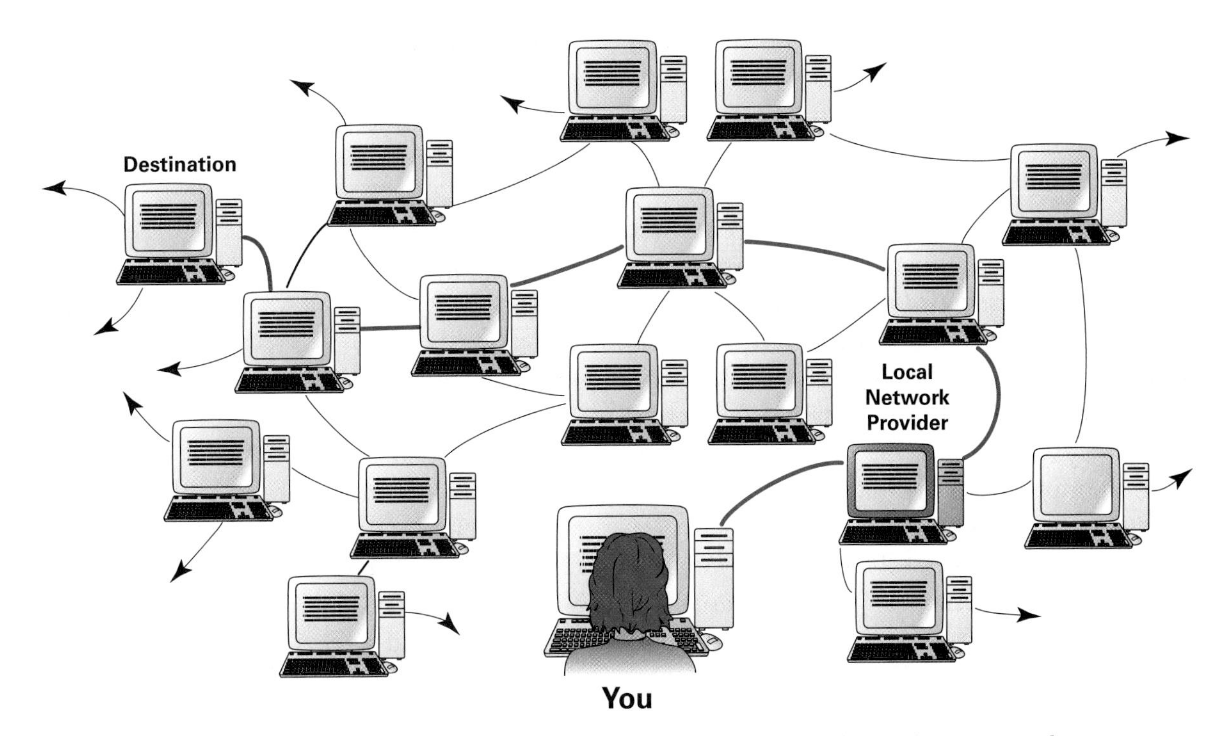

Fig. 5-14 When searching the Internet for information, you move through a maze of computers to the final destination.

trained using virtual reality. Pilots may practice maneuvers in cyberspace before using an aircraft. You may "walk through" the home you want to build before construction even begins.

A form of virtual reality called augmented reality allows people wearing special goggles to see virtual images atop those of the real world. For a tourist, for example, the goggles might show arrows and messages, such as "City Hotel: turn left, walk two blocks." The system could also be useful in construction and for military operations.

The Universal Product Code

The universal product code (UPC) is the striped code, or bar code, you see printed on most products. Each product has its own pattern of stripes. A computer reads these bar codes. The cashier at the store passes the code over a window. A laser scanner under the window reads light/no light from the stripes. The information is sent to a computer, which locates the price of the item in its memory. The computer sends the price data to the cash register. Then the computer writes the product's code number to a file that keeps track of how many units of that product have been sold.

The Internet and Online Services

As you know, you can find an immense amount of information on the Internet because it connects computers from all around the world. The Internet changes daily as new computer network sites, called nodes, are added, deleted, and updated. When you access the Internet, you have the potential to reach every other computer network that is also tapped into the Internet. See **Fig. 5-14**. You move around the

Internet electronically by jumping from one computer to another until you find the specific bit of information that you want.

What Is the Internet?

The Internet is a noncommercial computer network. That means it is not owned by any one company or operated for profit. Any person can access the Internet with the right hardware and software. One part of the Internet called the World Wide Web is written in special language called hypertext markup language (HTML). You can find information in almost any form on the Web—pictures, text, sound, and video—and on almost any topic. It provides many examples of multimedia technology. Many companies, organizations, educational institutions, and individuals have created Web sites that you can visit. Have you ever made a Web site? You will get a chance to design one in the activity at the end of this chapter.

Fig. 5-15 A good search engine is very important when you are trying to find specific information on the Internet.

To understand how the World Wide Web is organized, think about it in terms of a massive library. A Web site is like an individual book in that library. Most Web sites have several Web pages, similar to pages in a book. Each Web page has a specific address (like a page number) which is called a uniform resource locator (URL). You can tell your computer to go to a specific Web page, or URL, by just typing in that address on your computer.

Accessing the Internet

Before you can use the Internet, you must have a computer with Internet access. Two of the most common forms of access are through a cable modem or digital subscriber line (DSL). In order to find and view information on the Internet, you must use a Web browser. This is a software program that helps your computer to access and use the Internet. Netscape and Internet Explorer are two of the most popular Web browsers today. Internet service providers (ISPs), such as America Online, MSN, and Netzero, provide access to the Internet for a monthly fee.

There is so much information available on the Internet that finding what you want can be frustrating if you don't know the URL. It's like looking for a book in a large library without knowing the call number. The Internet does not have an index, card catalog, or accurate database for all the information in it. Many companies have created software programs that make it easier to search the Internet. These software programs are called search engines. Some of the more well-known include Google, Yahoo!, and Dogpile. See **Fig. 5-15**.

Using the Internet

The Internet and commercial online services make it possible for individuals to communicate with large numbers of people instantaneously. Chat rooms and instant messaging (IM) services are examples.

A chat room is a virtual place on the Internet where many people can communicate with each other in real time, usually about selected topics. For example, some online classes use chat rooms to engage students in discussions about the course topics. Instant messaging is very similar, but it's more like a private chat room where you need permission to be included in the group.

Blogs (weblogs) are also popular. A blog is like an online journal where you post information that you want others around the world to be able to read. A blog is typically a single Web page that the person continually adds to over time, just as you would add to a traditionally written journal or diary. The blog often contains links to other Web sites in which the author is interested, too. This creates what is often called the blogsphere, which is many interconnected Web sites. Today millions of people call themselves bloggers. Many blogs focus on themes, such as computer technology or cooking, while others are more freeform with bloggers writing about whatever comes to mind that day.

As you already know, you can also use online services and the Internet to send and receive e-mail. E-mail (electronic mail) refers to sending messages, letters, or documents using computers and electronic networks. You can correspond with a teen in a foreign nation or question NASA scientists about the *Hubble Space Telescope*. You can even order flowers for a special friend. Moreover, you can do any of these things 24 hours a day.

The Internet is changing the way people can access information. Students once had to visit a library in order to find information. Today, they can do research while sitting at a computer linked to the Internet. Without leaving their laboratories, scientists can use the Web to instantly share their research findings with other experts from around the world. Consumers can sell or purchase products without ever visiting a store or having a yard sale. You can even zoom in from space to a specific location on Earth and take a look around. You can tilt or rotate your view to see what's around you. The Internet can take you almost anywhere.

SAFETY FIRST

Protect Yourself. Posting personal information, such as your full name, address, phone number, school, or picture, on the Internet can be dangerous. Even naming specific places where you hang out can be unsafe. Talk to your parents and teachers about the appropriate way to communicate online.

The Future of the Internet

There is no doubt that the Internet will continue to open new doors for sharing and accessing information. A nonprofit group called Internet 2 consists of scientists, engineers, and other researchers from more than 200 universities, industries, and government agencies who are collaborating to create advanced high-speed data transfer technologies that will make it faster and easier to send and retrieve information.

CHAPTER 5 REVIEW

Main Ideas

- Information technology is specifically associated with computer-based communication systems.
- All parts of a computer work together to use and manage information.
- Computers have their own language, called machine language, which is binary in nature.
- We can do more than ever before because of computers.
- The Internet is a noncommercial computer network.

Understanding Concepts

1. What are the basic parts of the computer system?
2. In binary code, what is changed into 1s and 0s?
3. What are biometrics, artificial intelligence, and UPCs examples of?
4. Describe what a circuit is.
5. Explain how the Internet is similar to a massive library.

Thinking Critically

1. Why do you think the main circuit board is called the motherboard?
2. Why are video games some of the more complex forms of multimedia?
3. What do you think the danger is in being a blogger? How can you avoid these dangers?
4. Can output devices also be used as input devices? Explain.
5. How do you think e-mail has affected the U.S. Post Office in the past few years?

Applying Concepts & Solving Problems

1. **Writing** Research the history of the Internet and write a report using a word processing program. Be sure to include the original purpose of the Internet and how it has evolved over the years. Your report should also include conclusions regarding the effect of the Internet on the individual, society, and the environment.
2. **Evolution of Technology** Describe how computers have changed in the last 10 years. Your examples should cover all parts of the computer.
3. **Hypothesize** Why has computer technology grown so fast compared to other technologies? Identify trends in technology that might help account for the answer.

Activity: Build a Web Site About Engineering & Design

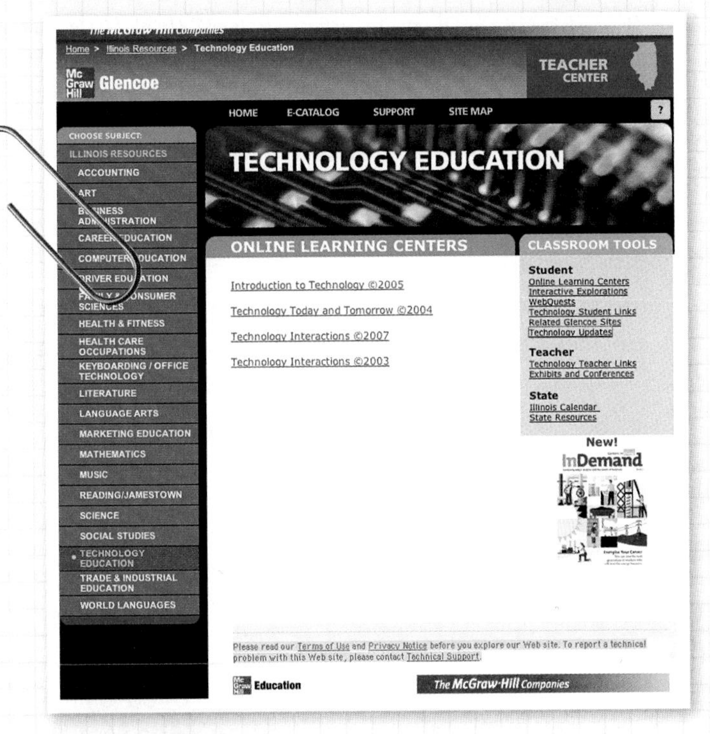

Design Brief

Given the right software and hardware, any person can create a Web page and make it available for others to access. Professional Web site designers typically spend weeks or months preparing their Web sites in order to ensure that they work well and effectively communicate their information. Then they continue to update their sites once launched to ensure that they are up-to-date and functional.

Design and build a multipage Web site focused on a specific topic of interest involving engineering and design. Your Web site should provide high school students with background information and examples with the intent to pique their interest in the topic.

Criteria

- Your site must be appropriate for posting on the Internet and for viewing in a school classroom.
- Focus on a specific aspect of engineering and design such as human factors, software engineering, or any number of topics such as those presented in this textbook.
- Effectively use text, images, sound, video, and/or data to present the content in interesting and creative ways.
- Collect information, evaluate its quality, and use only information that is approved by your teacher.
- Incorporate the following information on the main page: a clear and noticeable headline, a date, and copyright information.
- All links must be working and any problems should be diagnosed, solved, and documented.

Constraints

- Your finished Web site must include three to six pages linked together.
- Designers must use Web design software approved by the teacher and house their site at the location assigned by the teacher.
- Your working model must be tested by at least three other class members with feedback documented and answered.

Engineering Design Process

1. **Define the problem.** Write a statement that describes the task you are to accomplish and any obstacles (such as making the site attractive or functional).
2. **Brainstorm, research, and generate ideas.** With your team, discuss possible solutions. Hint: You can use this textbook or any other sources for engineering and design topics.
3. **Identify criteria and specify constraints.** These are listed on page 112.
4. **Develop and propose designs and choose among alternative solutions.** Choose the topic and identify site design elements you want to use. Look at appropriate Web sites to see what elements you may want to include. Create a map of your Web site.
5. **Implement the proposed solution.** Decide on the process you will use for making the site. Gather any needed tools or materials.
6. **Make a model or prototype.** Create your initial site. Follow all safety rules.
7. **Evaluate the solution and its consequences.** Have your site evaluated by other classmates. Have them evaluate the content as well as the design.
8. **Refine the design.** Based on the evaluation, change the design if needed.
9. **Create the final design.** After any changes or improvements take place, create your final Web site.
10. **Communicate the processes and results.** Present your Web site to the class. Use a projector or similar system so the Web site can be seen by everyone. Be prepared to answer questions. Turn in your assignment to your teacher. Be sure to include the name of the activity, your definition of the problem, a description of how you solved the problem, and any documentation you gathered during this project.

SAFETY FIRST

Before You Begin. Consider the information that you post on the Internet. It is essential that privacy and security be at the forefront of your decisions about the Web site design. Do not divulge any personal information. Do not display any photographs or use any information without the written consent of individuals, including their responsible guardian, if they are not yet 18 years of age. Strictly follow Internet safety guidelines as presented by your teacher.

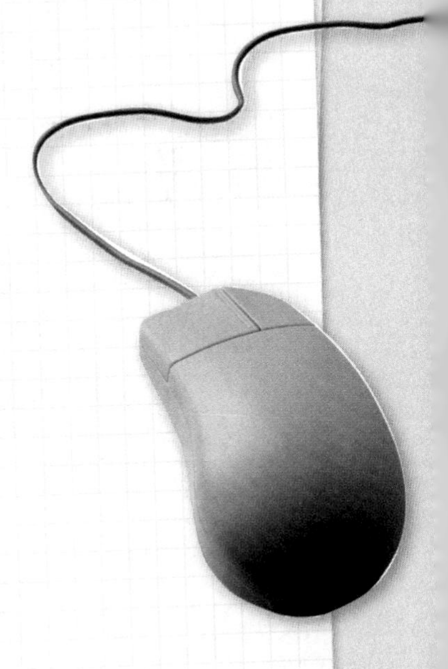

CHAPTER 6 ELECTRONIC COMMUNICATION

These students are having a good time producing a video for their technology class. In this chapter's activity, "Inform the Public," you will have a chance to produce your own public service announcement.

Objectives

- Identify and describe four types of transmission channels.
- Explain how signals are transmitted and received.
- Describe how telephones, radios, and televisions operate.
- Explain and give examples of radio frequency identification systems.

Vocabulary

- electromagnetic wave
- telecommunication
- amplitude
- frequency
- transmitter
- receiver
- modulation
- satellite
- charge-coupled device (CCD)

Reading Focus

1. As you read Chapter 6, create an outline in your notebook using the colored headings.
2. Write a question under each heading that you can use to guide your reading.
3. Answer the question under each heading as you read the chapter. Record your answers.
4. Ask your teacher to help with answers you could not find in this chapter.

What Is Electronic Communication?

Electronic communication systems involve the transmission of information using electricity. However, most modern systems use invisible **electromagnetic waves** that travel through the atmosphere. These waves are created by electric and magnetic fields. They travel through the air like ripples in a pool of water. Examples include devices such as garage door openers, baby monitors, and keyless door locks on your car that use radio waves to work. See Fig. 6-1.

Some electronic communication systems are used for **telecommunication**, which is communication over a long distance. The television, radio, and telephone are popular telecommunication systems.

In order to understand how electronic communication systems work, it's important to learn more about the ways electronic messages are transmitted and received.

Transmission Channels

Transmission channels are the paths over which messages must travel to get from the sender to the receiver. They are like highways for moving information. Information may travel over wires, cable, or optical fibers. Many devices, however, use microwaves or radio waves as a transmission channel.

« Fig. 6-1 Radio-controlled toys, such as these cars, rely upon electronic communication. You can control the cars by transmitting signals via radio waves to their receivers.

Copper Wire

Local telephone messages often travel over twisted-pair wire. This consists of two thin, insulated copper wires twisted around each other. Twisted-pair wires may also be bundled together to form large cables that stretch long distances connecting computer networks together. However, twisted-pair wire is not being used for most newly laid communication lines.

Coaxial Cable

Coaxial cable usually consists of four layers. A central conducting wire is surrounded by an insulator. Then there is another conducting layer which is surrounded by a final insulating layer. Most cable television signals are transmitted by coaxial cable.

Optical Fibers

Optical fibers (also called fiber optics) are being used increasingly to carry telephone, television, and computer data. Optical fibers are thin, flexible fibers of pure glass that carry signals in the form of pulses of light. Each optical fiber is surrounded by a reflective cladding (covering) and an outside protective coating. Electrical signals from a transmitter, such as a telephone, activate a laser. The resulting laser light pulses enter the glass fiber and bounce rapidly back and forth off the reflective surface of the cladding as they pass through. See **Fig. 6-2**. At the receiving end, the light pulses are changed back to an electrical signal. In a telephone, it is changed back into the sound of the sender's voice.

Fig. 6-2 The arrows show how light traveling through the optical fiber bounces off the reflective cladding, which keeps it inside. After leaving the glass fiber, the light is changed back into an electrical signal.

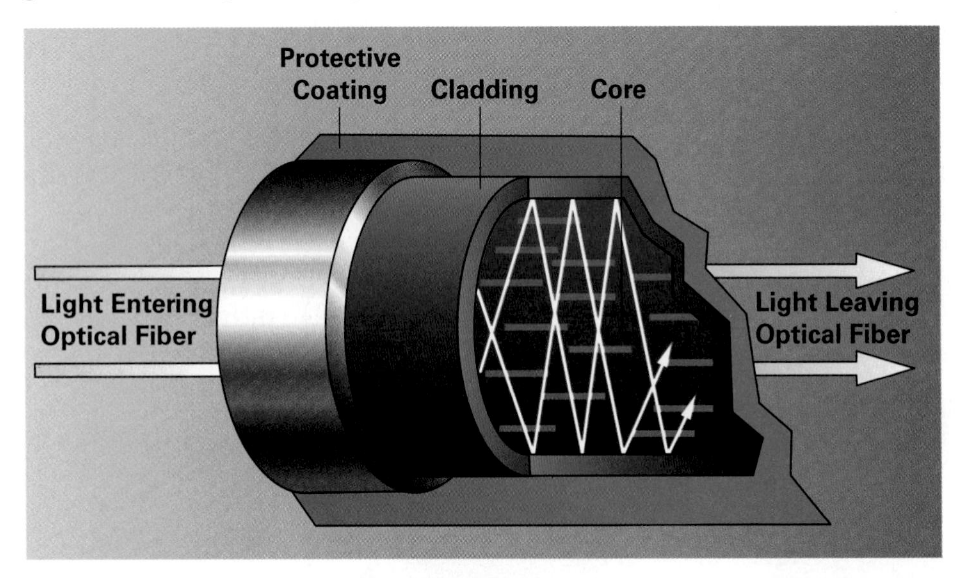

» **Fig. 6-3** Radio signals are measured in cycles per second, or hertz. ***What is the frequency of this wave?***

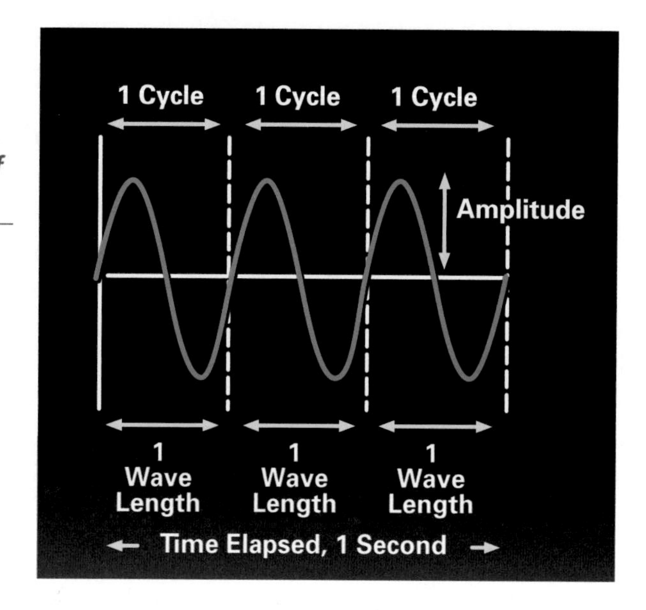

Optical fibers have many advantages over copper wire or coaxial cable. A single fiber is about the thickness of a human hair, yet it can carry much more data than a copper wire, at less cost. Optical fiber cables are much better for handling the high-speed data transmission needs of modern computer networks, including video data and the Internet. Because optical fiber cables are lighter and thinner than copper wires, they are also ideal for communication systems in places where the space is limited. In addition, the signals that travel on optical fibers do not fade as quickly as electrical signals sent on copper wire. Many of the noises you sometimes hear on the telephone, like buzzing or clicking, are also eliminated when optical fibers are used.

Electromagnetic Waves

The electromagnetic waves used by communication devices are the result of electromagnetic radiation. Visible light, for example, is a type of electromagnetic radiation that you can see. Other forms, such as radio waves and microwaves, are invisible.

All wave forms have both amplitude and frequency. **Amplitude** refers to the strength of the wave and can be seen in its height. A sound wave with greater amplitude, for example, would be stronger and louder than one with less height. **Frequency** is the number of waves that pass through a given point in one second. Frequency is commonly described in units called hertz. One hertz equals one cycle (one complete wave) per second. See **Fig. 6-3**.

Electromagnetic energy is classified according to the distance a given point on a wave travels in one cycle. This is known as the wavelength. **Figure 6-4** on page 118 shows the electromagnetic spectrum and the associated frequencies and wavelengths for common classifications.

Electromagnetic waves travel much faster than sound waves. In fact, they travel nearly 186,000 miles per second! They travel as sine waves, a shape with gradual peaks and valleys like the ones shown in **Fig. 6-4**. Since electromagnetic waves do not need wires or cables in order to be sent from one place to another, they make it much easier to transmit and receive information over great distances.

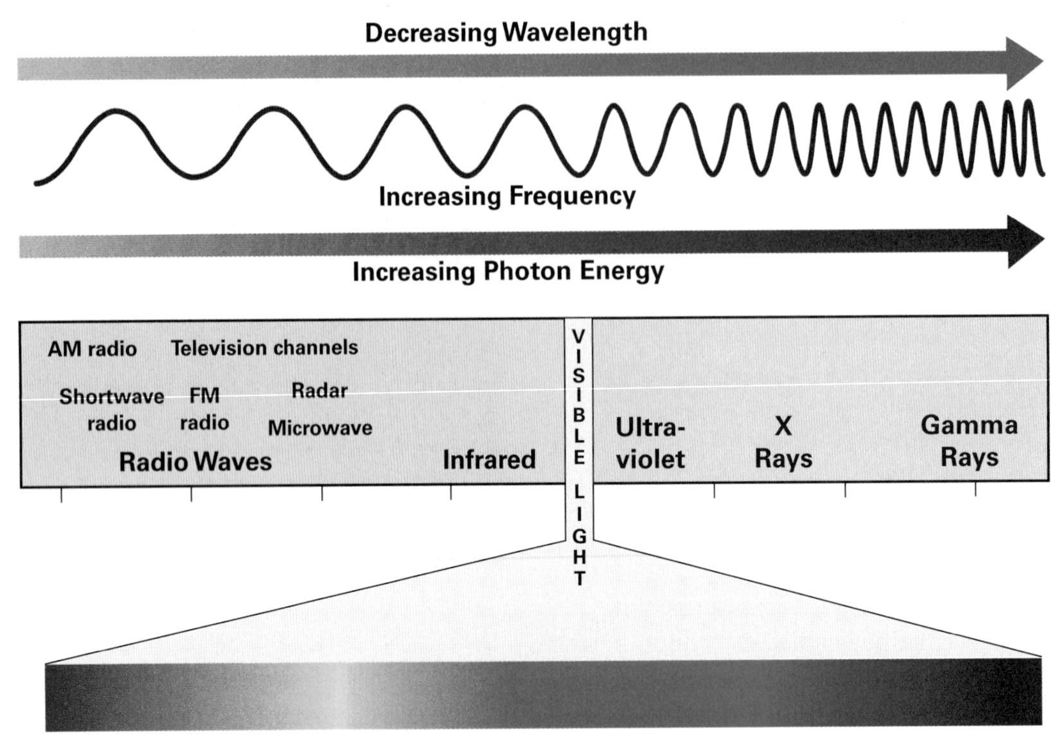

Fig. 6-4 The electromagnetic spectrum consists of bands of electromagnetic waves and their respective frequencies and wavelengths. Notice that higher frequency waves have shorter wavelengths and more energy.

Sending and Receiving Signals

In order to send and receive signals, information must be in a form that will work with the chosen transmission system. Special processes must be used if signals are going to be changed into electromagnetic waves that will be radiated into the atmosphere.

Transmitters and Receivers

Transmitters are devices that encode (change) signals into sine waves and send them out carried on radio waves or microwaves. **Receivers** collect the waves and decode the signals. The sine waves are superimposed on electromagnetic carrier waves using **modulation**.

For example, suppose a person speaks into the microphone of a cellular phone. The microphone changes the sound waves into electrical signals. The low-frequency sound signal is then combined with the transmitter's high-frequency carrier signal as shown in **Fig. 6-5 (A and B)**. These waves are modulated (altered) so that the carrier wave can transport the signal wave.

When the frequency is modulated, the waves are spread farther apart or crowded closer together as shown in **Fig. 6-5 (D)**. This type of signal is received on FM (frequency modulation) radios and cellular phones. Sometimes, the amplitude of the carrier wave is modulated as shown in **Fig. 6-5 (C)**. This type of signal can be received only by AM (amplitude modulation) radio.

» Fig. 6-5 A typical sound wave (A) is combined with a carrier wave (B). This combined wave may receive amplitude modulation (C) or frequency modulation (D).

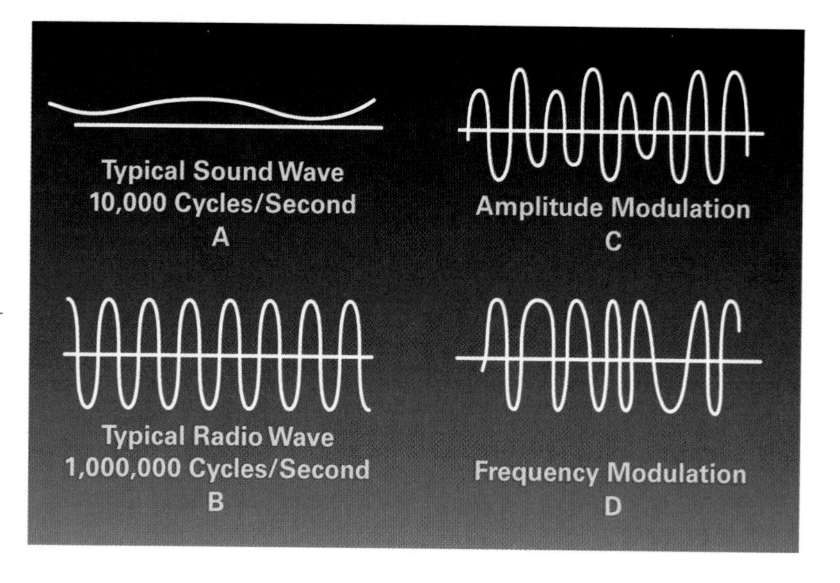

In the transmitter, the modulated signal is amplified and sent to an antenna for transmission. The waves travel through the air in all directions where they are picked up by other antennas that act as receivers. These receivers detect the signals, boost them to a usable strength, and convert them into the voice, sounds, or data originally encoded at the transmitter.

Antennas come in many different forms. Their design is dependent on the types of signals they will send and receive. See Fig. 6-6.

» Fig. 6-6 Antennas transmit and receive signals that travel as electromagnetic waves. The shape and size of antennas vary depending on the types of waves and the distance they must travel. Shown here are a radar antenna, cellular tower, and baby monitor set.

Satellites

A **satellite** is a device placed into orbit above the earth to receive messages from one location and transmit them to another. The satellite serves as a transmitter and a receiver. A satellite contains many transponders. Transponders are radios that receive signals at one frequency, amplify them, and then transmit them at a different frequency. The satellite is a relay station for signals. For that reason, a satellite is often called a "mirror in the sky."

Satellites are placed in orbit more than 22,000 miles above the earth. They travel at the same speed as the earth rotates. Thus, they remain above the same part of the earth at all times. This is called a geostationary orbit.

Satellites have greatly influenced our modern communication systems. For example, satellites help transmit numerous types of messages, including telephone and television signals. Satellites are used to transmit printed information as well, such as stories and pictures for the *Wall Street Journal* and *USA Today* newspapers. What other ways are satellites used, even in your own neighborhood?

Satellites make it possible to communicate instantly. Live broadcasts depend on satellites to transmit messages about events as they are happening. People all over the world can watch the World Series games while they are being played. Millions of people from around the world were able to witness the World Trade Center crisis on September 11, 2001, by means of satellite TV.

How do satellite systems work? Signals are sent to orbiting satellites through earth stations. An earth station (sometimes called a ground station) is a large, pie-shaped antenna. It receives signals and transmits them to the satellite. This is called the uplink. The satellite receives the signals and transmits them to another location back on Earth. This is called the downlink. Receiving earth stations capture the signals and send them to the desired receivers. See **Fig. 6-7**.

Many signals are passing through the air constantly! When a signal is transmitted, the sender puts a certain code at the beginning of the message. The code directs the signal to the intended receiver. In addition, messages can be scrambled. Then, only certain earth stations can pick up the message and decode the information.

Fig. 6-7 Satellites receive and retransmit electromagnetic signals from one place to another using a system of uplinks and downlinks.

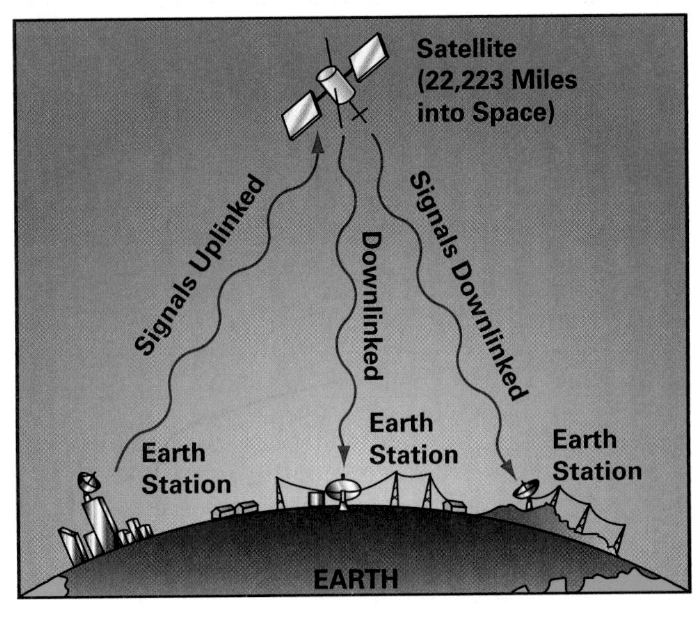

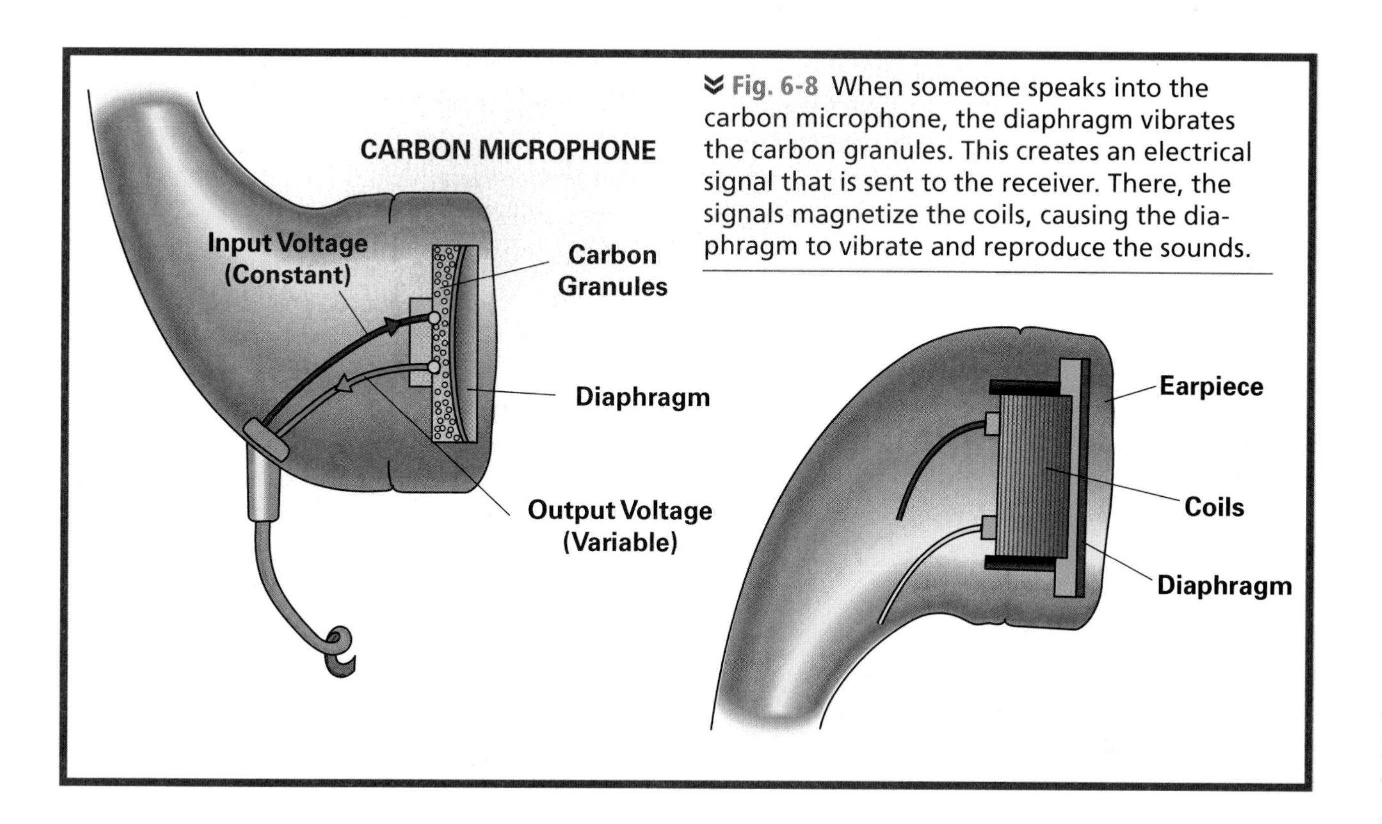

Fig. 6-8 When someone speaks into the carbon microphone, the diaphragm vibrates the carbon granules. This creates an electrical signal that is sent to the receiver. There, the signals magnetize the coils, causing the diaphragm to vibrate and reproduce the sounds.

How the Telephone Works

The three most common electronic communication technologies are the telephone, radio, and television. Let's take a closer look at them.

The telephone is based on the principle that sound waves cause vibrations that can be encoded into signals. The telephone includes a transmitter, located in the mouthpiece, and a receiver, located in the earpiece.

In older phones, the sound waves from a person's voice cause a flexible piece of metal—the diaphragm—to vibrate. The diaphragm presses against carbon granules and produces a varying electrical current. When the speaker pauses, the current stops. See Fig. 6-8.

The telephone receiver works much like the microphone (transmitter) in reverse. The receiver contains a wire-wrapped iron core. Connected to the iron core is a flexible metal diaphragm. When the transmitted electrical signal enters the receiver, it travels through the coil, magnetizing the iron core. This forms an electromagnet. The magnetic field pulls on the metal diaphragm. The diaphragm vibrates and reproduces the sound.

Electronic Telephones

Modern telephones are mostly digital electronic devices. Integrated circuits (ICs) have replaced many of the parts in the older telephones, such as the bulky wire coils.

In the microphone, other devices that produce electrical currents in response to sound waves have replaced the carbon granules. For example, the microphone uses metallic plates with opposite (positive/negative) charges to produce the current. The varying sound waves created by the speaker's voice alter the space between the oppositely charged plates, and this produces a varying current. See **Fig. 6-9**. Integrated circuitry has expanded the features available, such as last number redial, speed dialing, caller identification voice mail, call forwarding, and more. It has also made portable phones possible and economical.

Cellular Telephones

The most significant development in telephone technology in the past few decades has been wireless communication with lightweight cellular telephones. These devices use radio waves to transmit conversations, thereby omitting the need for connecting wires and cables.

As you move about with your mobile telephone, your call is transferred from one operating area to another in order to maintain good signal transmission. Each operating area, or cell, may range from a few hundred meters to several miles in diameter. Every cell has its own transmission tower which is linked to an electronic switching office.

A cellular phone contains a device called a transceiver, which is a transmitter and receiver combined into a single unit. The transmitter portion converts the signals into radio waves, which are transmitted to an antenna. The receiver portion of the transceiver receives radio waves from the antenna, changes them back into electrical signals, and sends them to the phone's control unit.

Today, most cellular calls use digital technology, as discussed in the previous chapter. It converts the voice into a computer language before transmitting it via radio waves. Digital technology makes it easy to encrypt electronic signals so that no one can listen in on the conversations. In addition, digital cellular technology makes available new options called personal communications services (PCS). They include messaging and many other services linked to your cellular phone. You need only one phone number to have access to all the services.

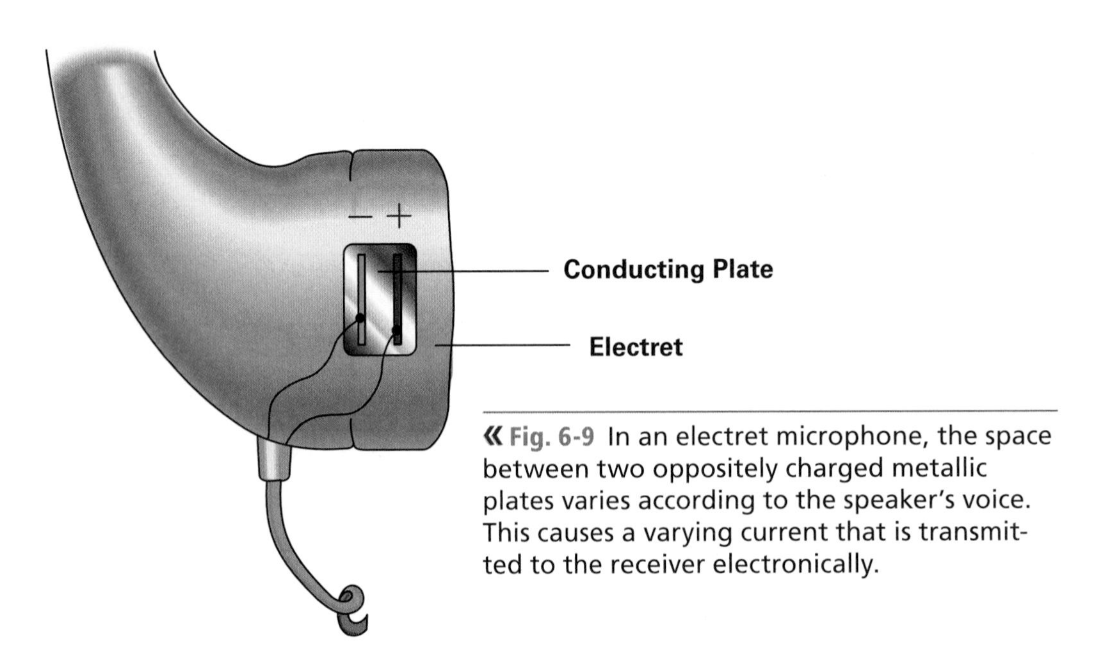

« Fig. 6-9 In an electret microphone, the space between two oppositely charged metallic plates varies according to the speaker's voice. This causes a varying current that is transmitted to the receiver electronically.

Getting a Message to the Right Place

No matter what transmission channel is used to send the message over the telephone, another major problem still exists. With millions of telephones in the world, how can you get a message to the right one? When telephones were first introduced, the number of connections possible was very limited. Today, however, almost all homes and other buildings have at least one phone.

Can you imagine a live person connecting every single call you make? At one time, all telephone exchange systems required human operators to connect senders to receivers. Now we have electronic switching. It makes connections faster and provides better signal transmission. These telephone exchange systems use computers to route a call through a series of telephone exchange centers to the intended destination.

How Radio Works

Radio is a form of mass communication—simultaneous communication with large groups of people. Like other forms of electronic communication, a radio system uses signals. As with a cell phone call, the radio signal's amplitude or frequency is electronically modulated and amplified. Then an antenna transmits the signal through the air where other antennas can pick up the signal.

Each radio station and device that works on radio waves (such as CB radios, military radios, and pagers) is assigned a certain frequency. The numbers on your radio dial indicate frequencies of different radio stations. When you set your dial at a number, you are tuning in the frequency of a particular station. The antenna on the radio picks up the waves that have that frequency. The circuitry in the radio separates the carrier wave from the sound signal. This

Science Application

Wave Frequencies As you know, when waves are described by their frequency, the unit used is the hertz (Hz). One hertz is one wave per second. Most radio broadcasts use waves that have frequencies in the kilohertz (kHz) and megahertz (MHz) ranges—thousands or million of waves per second. A specific group of frequencies is available for each type of broadcast, including AM radio, shortwave radio, FM radio, and television.

Many other wireless technologies use electromagnetic waves. The Federal Communications Commission (FCC) has assigned frequencies of the electromagnetic spectrum for cell phone systems, air traffic control, global positioning systems, television remote controls, and even baby monitors and garage door openers.

FIND OUT Devices that broadcast electromagnetic waves in communication ranges must be approved by the FCC. Make a list of items in your home that might broadcast signals. Check their labels to see if they have FCC approval.

Examples of Common Frequencies

Frequencies	Uses
30 kHz to 300 kHz (Low)	Cordless telephones
300 kHz to 3 MHz (Medium)	AM radio
30 MHz to 144 MHz (Very High)	Local TV stations and FM radio
806 MHz to 960 MHz (Ultra High)	Cell phones

process is called demodulation. The circuitry also boosts the electrical signal and sends it to the speaker, which changes the signal back into sounds.

Satellite Radio

Have you noticed that you are unable to listen to your favorite local radio station when you travel too far from home? Radio stations usually broadcast their signals in specific geographical areas, perhaps 30-40 miles from the station's transmitters. As you move farther from the station's transmitter, static interferes and the signal may eventually be lost completely.

With satellite radio, also called digital audio radio service (DARS), you can listen to a specific station as you move from one location to another. This is made possible because the digital signals are transmitted via satellites and radio receivers on the ground. See **Fig. 6-10**. You are able to pick up the station you desire because the satellite signals are broadcast over large areas and can be captured by receivers in many locations. Plus, digital technology makes it possible to encode additional information with the signals. Your satellite radio receiver can also display information, such as the name of the song and artist.

Although conventional radio is free, satellite radio is not. As with cable television, subscribers pay for the service. However, the service typically provides more than 100 different stations, and much of the programming is commercial-free.

How Television Works

Television is an electronic system of transmitting pictures and sounds over a wire or through the air. Today, television is a popular telecommunication tool with a variety of program options. Nearly every household in the United States has at least one television set—some have one in every room and vehicle. The picture is clear and sharp, and so is the sound. A complex broadcasting network is set up to transmit television programs around the world 24 hours a day.

The Television Camera

A television camera is most often referred to as a camcorder. Can you guess which two devices combine to make a camcorder? The camera converts images and light (video) and sounds (audio) into electrical signals that can be recorded. The recorder puts this information

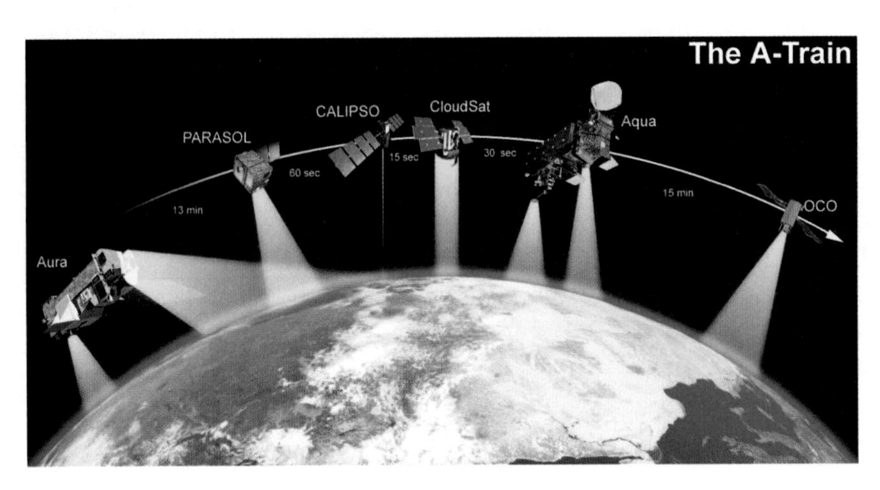

« **Fig. 6-10** In a satellite radio system, communication satellites orbiting Earth relay signals back to receivers on Earth.

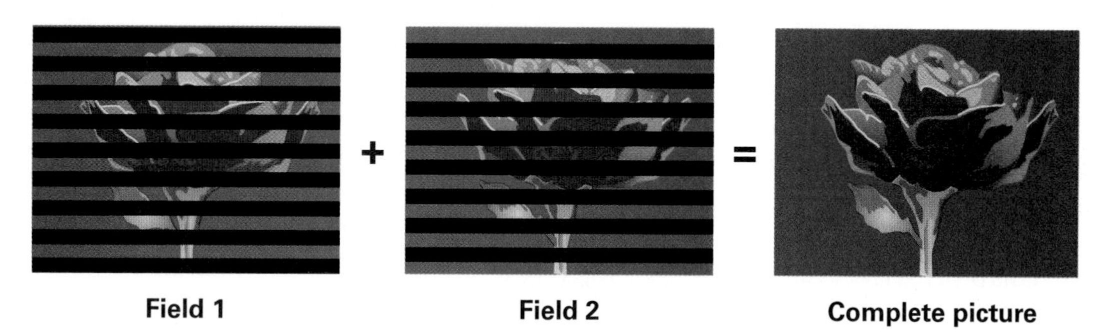

Fig. 6-11 The camcorder's CCD has thousands of tiny cells that convert light from the image into electrical signals. The camera's electronics turn this pattern of charges into a picture. The process of interlacing divides the frame into two alternating fields.

Light

Photosites

onto magnetic tape, discs, or memory cards. It also allows you to play back the recorded images and sounds.

Camcorders record audio and video information separately. A microphone is used for sound, while images and light are converted into electrical signals using a **charge-coupled device (CCD)**. The CCD is a tiny (less than one centimeter across) panel having about 500,000 light-sensitive cells called photosites, or photodiodes. See **Fig. 6-11**.

Light reflected from the image enters the camera's lens. The lens focuses the light on the CCD. The amount and color of light hitting each photosite on the CCD determines the electrical charge produced. Electronic circuits in the camera combine these thousands of separate signals into a complete picture called a frame. In a process called interlacing, each frame is divided electronically into two fields of alternating lines. Typical camcorders record 30 complete frames per second, creating 60 separate fields. When these series of images are played back on your television, your brain "sees" them as a continuously moving scene.

Different mixtures of red, green, and blue light can produce every color in nature. In order to produce accurate color images, the CCD determines levels of red, green, and blue light for every frame. Most basic camcorders use red, green, and blue filters over individual photosites to do this. Electronic circuitry in the camera interprets this information in order to re-create images that accurately reflect the true colors of the original scene.

Did you know that CCDs are also used in digital cameras? To learn more, read Chapter 7, "Graphic Communications."

Transmission

Broadcast television signals (audio and video) are first sent to an antenna. The signals are amplified and modulated before they are transmitted via antenna, satellite, or cable. This process is very similar to the way radio signals are sent.

However, television signals use higher frequencies. The channel selector on the television receiver tunes in the frequency of the channel, just as the dial does on the radio.

Your television receives the transmitted signals and converts them to electrical signals. The video and audio signals are then separated. The audio signals are sent to the television's speaker, which converts the signals back into sound. The video signals are sent to the display unit, which is usually a cathode ray tube (CRT).

SAFETY FIRST

TV Safety. Television components often retain a dangerous electrical charge for a while even after the set has been turned off or unplugged. Leave repairs to professionals who are trained in working with electronic devices.

The Television Receiver

The television screen is the flat part of the CRT. In a color television, the surface of the CRT is covered by groups of red, green, and blue phosphors. A phosphor is a substance that emits light when given energy. Each dot of light produced is called a pixel. The pixels are arranged in a pattern of dots or stripes that make up the color image.

Three electron guns in the back of the CRT move back and forth across the screen from left to right 525 times for each screen image. One gun is used for each of the primary colors of light—red, green, and blue. See **Fig. 6-12**. The beams excite the phosphors, making them glow. The intensity and color of the phosphors varies according to the electrical signals recorded by the camera at the scene.

The sweeping action occurs 60 times every second. Because of the interlacing technique used by the camera, alternating lines are swept each time, putting only half the lines on the screen with each pass.

It is very important that the proper electron gun hits the correct phosphors according to the way the image was recorded on the original camcorder. Electronic controls help to synchronize this process, as does a shadow mask on the CRT. The mask contains many holes that direct the beams to the correct phosphors. Your brain interprets these rapid changes as a continuously moving image.

In flat-panel plasma displays, pixels are made from cells filled with electrically charged gas atoms (plasma). Electrodes extending across the screen intersect these cells. Current from the electrodes excites the gas atoms, and they release energy. This energy interacts with the phosphors coating the inside of the screen and producing the color image.

Digital Television

New guidelines from the Federal Communications Commission are requiring eventual adoption of digital technologies. This brings significant changes in television broadcasting. During the transition phase from analog to digital, analog transmission television sets will still function. When the changeover is complete and analog transmission is discontinued altogether, a digital television set (DTV) will be required in order to receive television broadcasts.

The initial expense of this technological change will be high for consumers and broadcasters, but the long-term advantages should be worth it. Digital systems are more like televisions combined with computers. You can manipulate video images, store images and

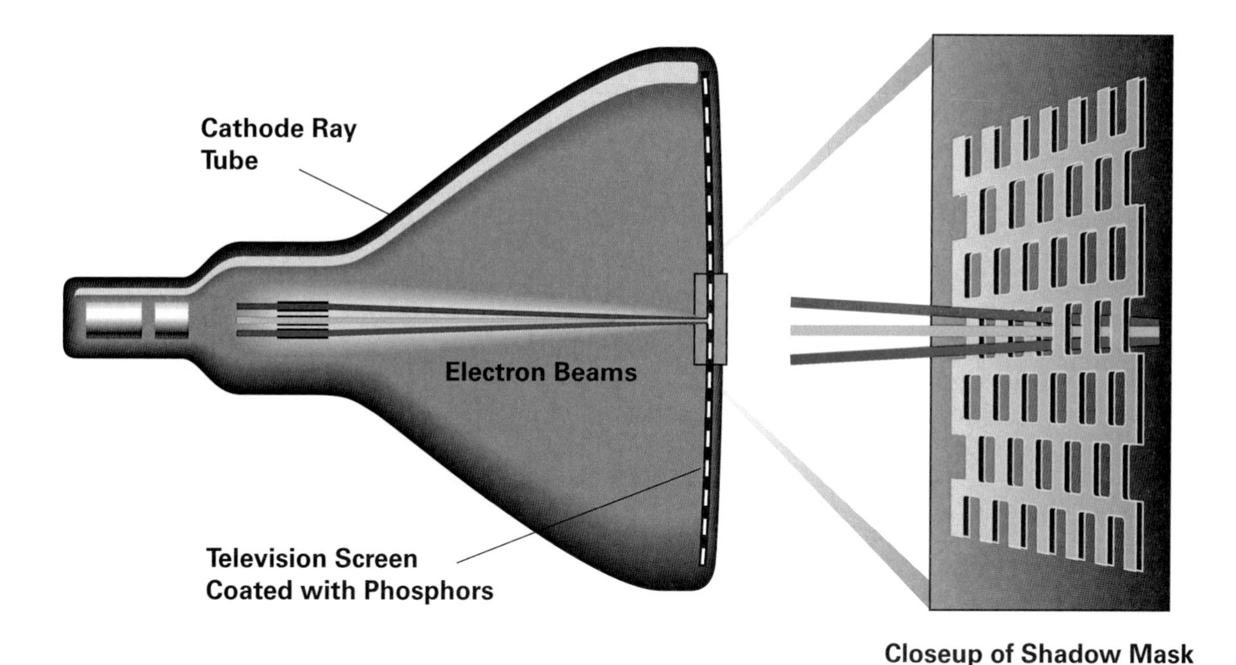

Fig. 6-12 Each of the three electron guns inside the cathode ray tube scans the screen for one of the primary colors of light. The phosphors glow according to the electrical signals received. The shadow mask ensures that the beam reaches the correct color target.

sounds, electronically search for programs of interest, and more. You can modify colors and experiment with sounds. You can even divide your TV display into sections and watch several programs at the same time. Who knows what your TV will be capable of in the future? The possibilities seem unlimited.

High-Definition Television

High-definition television (HDTV) is a special class of DTV. It makes cinema-quality video and CD-quality sound available on home television sets. Picture clarity is improved because the number of scanning lines is doubled. More scanning lines mean that the HDTV screen can be made larger without losing the sharpness of the picture. In the United States, the number of lines scanned will be increased from 525 to 1,080. In addition, HDTV has a different aspect ratio, which is the comparison of width to height. Conventional screens have a 4:3 aspect ratio. This means the screen is 4 units wide and 3 units high. HD screens have a 16:9 aspect ratio, or 16 units wide and 9 units high. This makes the screen look more like a movie screen. See Fig. 6-13 on page 128. Some HDTV screens offer 10 times more picture detail than analog TV screens. Viewers have remarked that the picture is so clear they can see wrinkles on actors' faces, textures in fabrics, and individual blades of grass on a ball field.

HDTV technology is being sold today. The cost of these new sets is expected to be high at first. This is common for new electronic products until the demand increases and the cost of producing them decreases.

» **Fig. 6-13** High-definition plasma television sets have a different aspect ratio, making them look more like movie screens. These sets have great clarity, and their flat design makes it convenient to mount them on a wall.

Advances in Electronic Communication

Some engineers are designing innovative ways to apply electronic communication technologies to new products and systems. Others are finding ways to improve our comfort and safety when using electronic communication devices.

RFID Technology

Radio frequency identification (RFID) technology is a wireless system that can be used to track goods or vehicles.

An RFID tag is basically a microprocessor (computer) and antenna packaged together in an extremely small space. Each tag can carry a unique product number and other information. A special radio-emitting scanner or transceiver reads the information when the tag is within range. An RFID tag can be made the size of a grain of sand and embedded in paper or other products. It has also been approved for implantation in humans as a medical identification tag.

RFID technology has many applications. One day it may completely replace the universal bar codes that are currently in place on nearly every product you purchase. Instead of scanning each product at a checkout station, you would simply walk out of the store. A scanner placed in the doorway would automatically "read" the tags as you exit. A computer would then total your bill and charge your account directly.

Electronic toll collection systems also use RFID technology. A special transponder is mounted on the windshield of your vehicle. As you pass through the toll lane, an antenna activates the transponder and detects your account information so that you can be charged the proper amount. This speeds up the toll collection process and alleviates traffic backups. See **Fig. 6-14**.

RFID technology has many benefits. However, some people are concerned about how access to personal information will affect privacy. Opponents of RFID technology sometimes refer to tags as "spy chips." Using this technology will require trade-offs between the positive and negative effects.

Reducing EMI

Did you know that any device with an electrical circuit emits some electromagnetic radiation? Human factors engineers must design products and systems so that people will not be harmed by this radiation.

Electromagnetic radiation can also cause unwanted electromagnetic interference (EMI) that can disrupt how other electronic devices work. For example, some medical researchers are concerned about the effects of EMI on patient care. Signals from nearby radio transmitters, police radios, or cellular phones have caused patient monitors to not work properly. Electrical devices for maintaining the heartbeat, called pacemakers, and other medical appliances have also malfunctioned.

Engineers and product developers can address these problems by designing electrical circuits to reduce EMI. They can also shield the devices. Shielding involves adding coverings that block unwanted signals.

Fig. 6-14 Radio frequency identification technology makes it possible for some motorists to pay their tolls without stopping to drop coins in a basket or to hand cash to an attendant.

Math Application

Binary Bar Codes Bar codes are used to assign a unique identification code to products. Bar codes store data, such as alphabetic and numeric characters, in different width and spacing patterns of printed lines. They also come in patterns of dots, concentric circles, and hidden images.

Bar codes are encoded with the use of the binary number system. It is based on only two symbols: zero (0) and one (1). Zeros generally correspond to spaces, while ones correspond to bars. The binary system is simple to design, which is why it is used.

We normally use the decimal number system of 0, 1, 2, 3, 4, 5, 6, 7, 8, 9, etc. The first two numbers in decimal and binary systems are represented the same. Zero is represented as 0 and one is represented as 1. However, in the binary system, the decimal 2 is represented as 10 (one-zero) and 3 as 11 (one-one). In converting from binary to decimal, for example, 10 (one-zero) is equal to $(1 \times 2^1) + (0 \times 2^0) = 2$.

FIND OUT Write the equation for turning 11 (one-one) into the decimal 3. Research to learn more about the binary numbering system.

CHAPTER 6 REVIEW

Main Ideas

- Electronic communication systems involve the transmission of information using electricity.
- Transmission channels are the paths over which messages must travel to get from sender to receiver.
- The three most common electronic communication technologies are the telephone, radio, and television.
- Advances in electronic communication include RFID tags and technologies used to reduce electromagnetic interference.

Understanding Concepts

1. Name the four types of transmission channels.
2. How are signals transmitted and received?
3. What principle does a telephone use?
4. How is frequency used in both radio and television?
5. What is an RFID?

Thinking Critically

1. How have cell phones changed social interaction?
2. How many mediums can a sound wave pass through?
3. List five ways your school would change without the use of electronic communication.
4. With your fellow students, discuss the positive and negative impacts of RFID technology.
5. How is electronic communication used in distance learning?

Applying Concepts & Solving Problems

1. **Math** A geostationary orbit is about 22,000 miles above the earth. How many kilometers is that?
2. **Science** What slight adjustments need to be made in order for satellites to remain at the same place in the sky relative to the earth?
3. **Writing** As this chapter discussed, electromagnetic signals are linked to lasers. Research and write a report on lasers, explaining such concepts as "critical angle" and "total internal reflection."

Activity: Inform the Public

Design Brief

Most commercials are brief advertisements designed to persuade viewers to purchase goods or services or support certain causes. Public service announcements (PSAs) are noncommercial advertisements that inform and educate viewers about important issues such as drug abuse.

Broadcasting a PSA on local or national stations may be very expensive. However, it does not usually take a lot of money to plan and produce a PSA for personal or school use and the results can still be amazingly effective!

Plan and produce an attention-grabbing 60-second public service announcement to inform, educate, and/or persuade a specific audience about a meaningful issue or concern.

Refer to the

Criteria

- Your topic should concern an issue of interest and importance to young people (e.g., bullying, staying in school, being tobacco-free, enrolling in technology education classes).
- Your PSA must be appropriate for airing on the school cable television network or viewing in a school classroom.
- You must identify your target audience (somewhere between the ages of 5-18).

Constraints

- The final video must be submitted with documentation of the steps your team followed.
- The length of your PSA must be 60 seconds (plus or minus five seconds).
- If available, use digital editing software (e.g., Apple iMovie®, Windows Movie Maker®, Adobe Premiere®) to combine your audio and video clips together to create a seamless program.

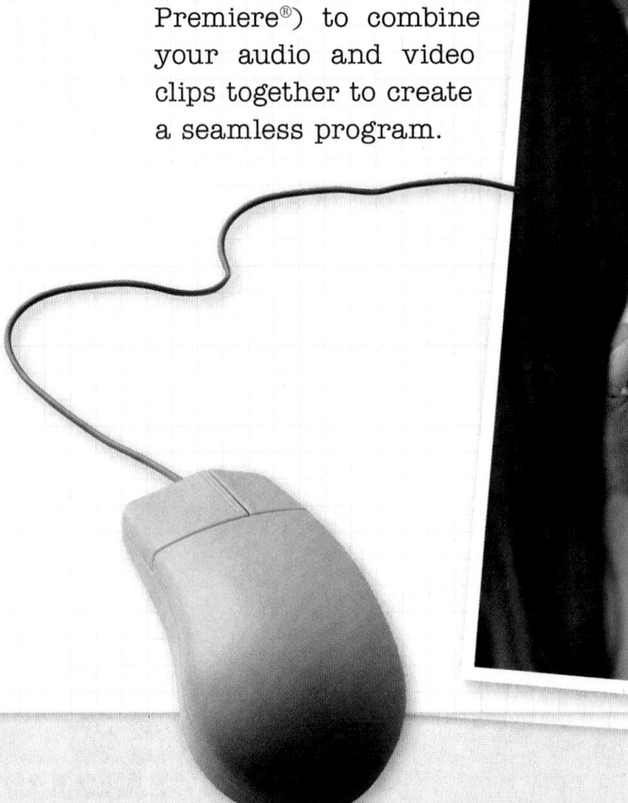

Engineering Design Process

1. **Define the problem.** Write a statement that describes the problem you are going to solve. For example, which topic will you choose? Be sure to choose a subject you find interesting.
2. **Brainstorm, research, and generate ideas.** With your team, discuss possible solutions. Hint: You may want to review PSAs that have already been developed.
3. **Identify criteria and specify constraints.** These are listed on page 132.
4. **Develop and propose designs and choose among alternative solutions.** Choose the best design that will solve your problem. For example, you might want to use real actors instead of just graphic images. Investigate different software to see what kind will best help you create your video.
5. **Implement the proposed solution.** Decide on the design you will use. Gather any needed information or software tools.
6. **Make a model or prototype.** Create a storyboard for your PSA.
7. **Evaluate the solution and its consequences.** Compare the storyboard to your original design. Will your solution solve the problem? What impacts and effects will result from this solution? Ask at least three classmates to review your design and provide written feedback.
8. **Refine the design.** Based on your evaluation, change the design if needed.
9. **Create the final design.** After you have your teacher's approval, create your video.
10. **Communicate the processes and results.** Present your finished PSA to the class. Be prepared to answer questions about possible impacts. Collect evidence of their impressions of the program. Turn in your video to your teacher. Be sure to include the name of the activity, your definition of the problem, a description of how you solved the problem, and your storyboard.

SAFETY FIRST

Before You Begin. When creating your video, be sensitive to the needs of your audience. For example, a video with repetitive flashes may prove harmful to those with certain medical conditions.

CHAPTER 7 GRAPHIC COMMUNICATIONS

Objectives

- List the principles of design.
- Describe the different printing processes.
- Compare and contrast film and digital photography.
- Explain why drafting is referred to as the universal language.

Vocabulary

- graphic communications
- principles of design
- digital workflow
- holography
- computer-aided drafting (CAD)
- rapid prototyping

A well-designed board game can be fun for everyone. In this chapter's activity, "Design a Board Game," you will have a chance to design your own creative board game.

Reading Focus

1. **Read the title of this chapter and describe in writing what you expect to learn from it.**
2. **Write each term in your notebook, leaving space for definitions.**
3. **As you read Chapter 7, write the definition beside each term in your notebook.**
4. **After reading the chapter, write a paragraph describing what you learned.**

What Is Graphic Communications?

Graphic communications is the field of technology that involves the sending of messages and other information using visual means. Photographs, newspapers, magazines, books, packaging, and game boards are just a few examples of graphic communications media. Media is a means of communication. How many forms of graphic media can you identify where you are right now?

A variety of processes and techniques are involved in graphic communications systems. They include printing, photography, and drafting. All require an understanding of design.

Principles of Design

Certain designs or pictures seem to catch our attention better than others. For example, many organizations develop their own symbols with which they are quickly and easily identified. These symbols are called logos. Look at the logos in Fig. 7-1. What makes them appealing? A good design captures the attention of

Fig. 7-1 Here are examples of logos that you might see on an average day. ***How might logos be used to influence what stores you visit at this shopping mall?***

the intended reader or viewer. It effectively communicates a message and leaves a lasting impression.

Designers must consider how a design will be used. They must think about the function of the design. Does it need to appeal to young people or older people? Will it be used in a magazine or on a billboard? Will it be read and thrown away, as with a newspaper, or will people look at it for many years, as they do a photograph?

Many factors can have an impact on the effectiveness of a design. When designing a graphic message, the **principles of design** can be used to help determine the effectiveness of the design. As you read about each principle, look at the pages in front of you. Try to find examples of each principle.

Balance The visual weight of images in a design is referred to as balance. Some designs are symmetrical, or formally balanced. Formal balance is achieved when a line drawn down the center of the design creates two halves that are very similar to one another, or symmetrical. Others are asymmetrical, or informally balanced. The objects in the design may look different but have equal weight to the eye. See Fig. 7-2. The viewer must gain a sense of balance

Fig. 7-2 A design with about the same amount of illustration and type on both sides of an imaginary centerline is formally balanced. If there is unequal amount on either side, but both sides appear to have equal "weight," the design is informally balanced.

FORMAL BALANCE

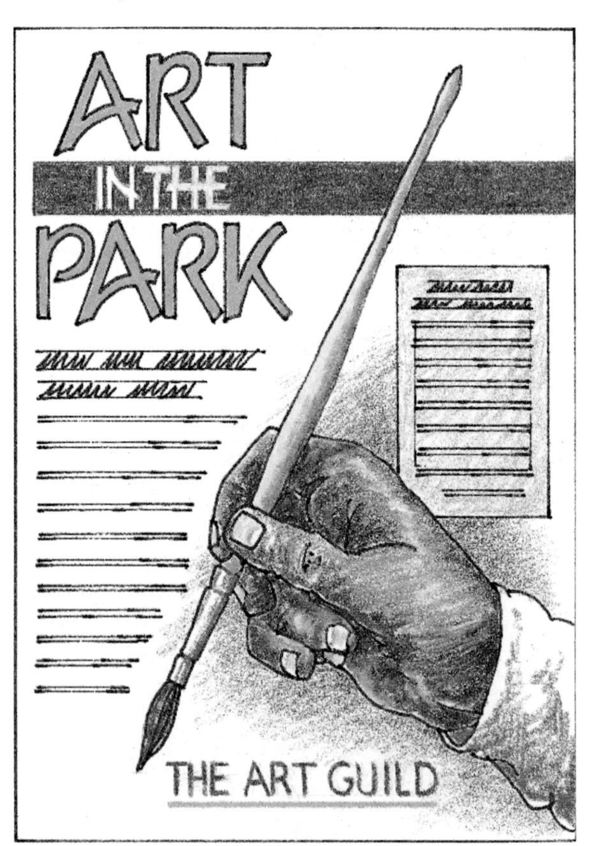

INFORMAL BALANCE

Math Application

How to Measure Type Printers, typographers, and other people who work with material set in type have their own measuring system. This system accounts for sizes in the type itself and in the space between one line of type and another. The basic units of type measurement are the point and the pica.

A point is equal to 1⁄72 inch. Point size is measured from the top of a capital letter to the bottom of the descender on a lowercase letter. (The descender is the part of some lowercase letters—g, j, p, q, and y—that goes below the main body of the letter.) In general, capital letters are about two-thirds the height of the overall point size. The type shown measures 18 points. To find its size you would measure from the top of the T to the bottom of the y, as shown.

A pica is equal to 12 points (or 1⁄6 inch). Printers use picas to measure larger sizes, such as the length of a line or column of type.

A typographer's rule is used to measure points and picas. The rule has several other scales, including those for inches and rule widths.

FIND OUT Use a typographer's rule to measure the width and depth of a full column of main text in this book.

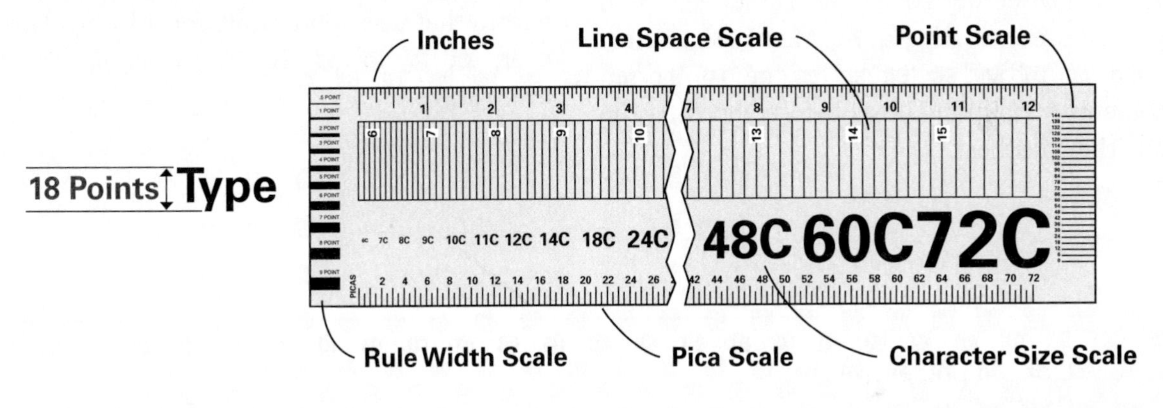

from the design—the space, type, and artwork should be positioned to make the viewer feel comfortable. If the design is not balanced, the viewer may become confused or lose interest.

Proportion The size relationship of the various parts of a design is referred to as proportion. The sizes of type, drawings, and photographs are all considered carefully. The designer wants to give the right amount of attention and space to each part of the message. A design that is out of proportion will appear awkward and displeasing to the viewer.

Emphasis refers to techniques used to call attention to certain parts of the message. The designer wants the viewer to especially notice important parts of the message. Emphasis is often accomplished through simple techniques such as underlining, *italicizing*, **bold-facing**, adding color, changing the size of the type, or using arrows →.

Variety Adding different elements to a design creates variety. It's not all the same. It's not boring. Variety adds interest.

Rhythm addresses eye movement and flow. It can be accomplished by repeating elements. The placement of type and graphics on the page can also guide the viewer's eye.

Unity/Harmony refers to the overall effect of the design. If all aspects of the design appear to belong together and work well together, the design has unity. Unity is achieved when the designer carefully plans the kinds of type, colors, and shapes used. If even one element doesn't "fit," the effect of the design can be ruined.

The Creative Design Process

How do designers begin creating a design? The first three steps of a typical creative design process are shown in **Fig. 7-3**. Most designs start out as thumbnail sketches. Designers will often come up with many ideas before choosing the one they like best.

Some designers then make rough layouts of the design that are most appealing or functional. The rough layout provides more detail and it is drawn full-size. They may share these layouts with clients to get their feedback as well. After making a more comprehensive layout, the design's final form must be prepared. This final form will vary according to what kind of graphic product is being created and how it will be reproduced or distributed.

For some printed products, designers create a mechanical, which is also called camera-ready art or a pasteup. This is a layout that combines the type, art, photographs, and other features together in place. Traditionally, the mechanical was photographed and then converted to a form that was appropriate for printing. However, today most designers use computers at every stage.

Fig. 7-3 Here you can see the creative design process from thumbnail sketches, to rough layout, to full-color comprehensive layout.

» Fig. 7-4 Books, money, and CDs are just a few examples of printed materials you might see each day.

Digital workflow is the use of computers for each stage of the design process. Sometimes this process is called desktop publishing or electronic page composition. Many types of page layout software programs make it easy to create graphic designs on the computer. Complex, full-color layouts can be assembled and printed or distributed electronically.

Using computers for page layout makes it much easier to combine text with photos and illustrations. Changes can be made quickly and accurately. Text can be enlarged or a picture moved from the top of the page to the bottom. A headline in red can be changed to green. A designer doing the work by hand would need to redraw the entire layout. With a computer, these changes take only seconds to complete.

Photos entered into the computer's memory can be altered and inserted into the layout. Usually designers use photos taken with digital cameras. However, a scanner can turn a regular photo into digital information that the computer can understand. Then the designer can alter that information. With a few mouse clicks, the designer can remove a tattoo, add a missing tooth, or change a red car to black. (While this capability is often useful, designers must be careful to avoid altering key factual information, which may cause ethical issues.)

The design process still relies on the creative and artistic skills of designers. However, digital technology has greatly affected how these designers produce their ideas and the speed at which they work.

Printed Communications

How many printed materials do you see each day? Cereal boxes, candy wrappers, wallpaper, CD labels, soda cans, dollar bills, books, and many flags are all printed. See Fig. 7-4.

Sometimes graphic messages are distributed electronically via e-mail, Web pages, cell phones, or other means. However, many graphic messages are converted into a form that can be printed. Several stages are involved in the printing process. Because digital workflow has streamlined many of these stages, it is now much easier to turn your idea into a printed product.

Before printing was invented, people recorded information by hand. If several copies were needed, someone would handwrite the information many times over. Printing processes make it possible for people to easily duplicate many copies of a message. Printed products can be customized so that customers receive unique versions at the same speed as mass-produced versions.

Printing can be done in a number of ways. The type of printing process used depends on what is being printed, the quality of reproduction desired, costs, and the speed at which the job must be done.

Generally, printing techniques are grouped according to the type of printing surface used. Major groups of printing processes include relief, porous, planographic, gravure, and electrostatic. Other printing processes include photography, ink-jet printing, laser printing, and 3D printing.

The processes used to reproduce messages onto a surface have changed considerably over the years. Different processes are used for different reasons. However, the basic idea is still the same. No matter how the message is printed, it must be visually effective. As computers and other devices have changed our printing machines and methods, the message being communicated is still what's most important.

Fig. 7-5 Relief printing is printing from a raised surface.

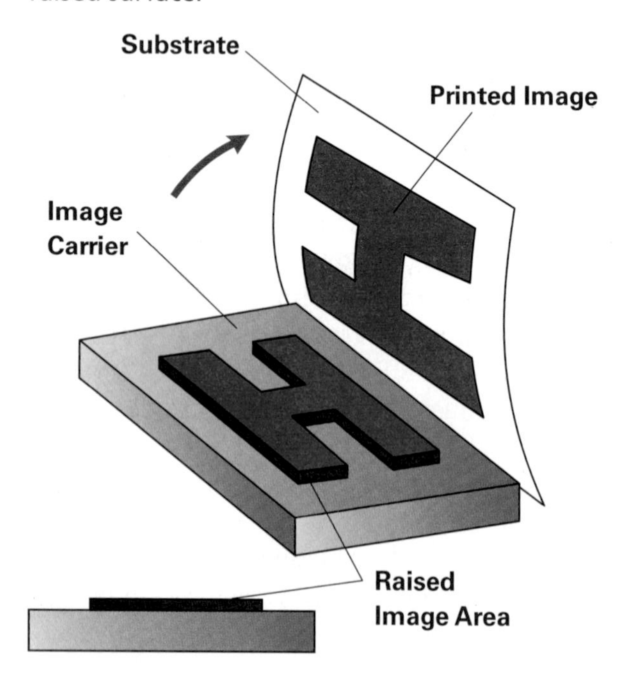

Relief Printing

Relief printing processes are methods that print from a raised surface. See Fig. 7-5. Parts of the image are raised above the printing surface. These pick up ink and transfer it to the substrate (print material), which is often paper.

Letterpress is a very old form of relief printing in which reverse images of all the characters (letters, numbers, and symbols) are formed into metal type. Ink is rolled over the raised type. Then paper is pressed on the type to transfer the image. This process is repeated for as many copies as desired.

Johannes Gutenberg, the "Father of Printing," is given credit for inventing movable metal type and making letterpress possible. Today, however, letterpress is seldom used. Flexography is the most popular relief printing process.

In flexography, images are formed in plastic or rubber sheets that resemble ordinary rubber stamps. These flexible plates are mounted on rotary presses. Plates are usually made using photographic methods or computer-driven lasers that engrave the surface.

Flexography is used extensively in the packaging industry, especially for food packaging. This is because water-based inks can be used that are more environmentally friendly than solvent-based inks.

Porous Printing

A porous material has many small openings (pores) that allow liquids to pass through. In porous printing processes, ink or dye is passed through an image plate or stencil and trans-

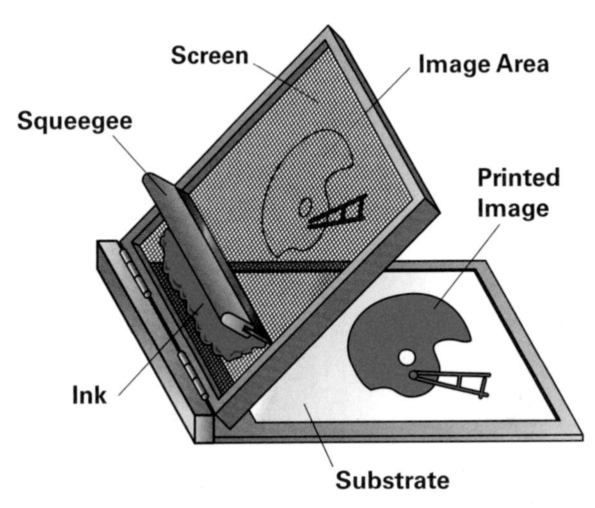

Fig. 7-6 Screen printing is a porous printing process. Ink passes through a stencil attached to a porous screen onto the surface below.

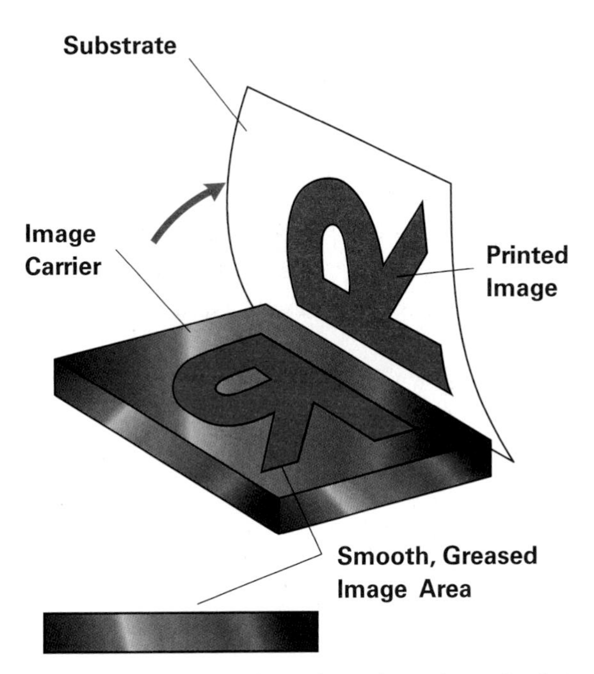

Fig. 7-7 Lithography is based on the principle that grease and water do not mix. Because the printing is done from a flat (plane) surface, this is a planographic process.

ferred onto the substrate. A stencil is a thin piece of material with holes cut out or etched in the shape of the desired design. Stencils can be made in many different ways. The material being printed on may be paper, cloth, glass, or another material. The most common porous printing process is serigraphy. It is more commonly referred to as screen printing or silk-screen printing.

In serigraphy, a stencil is adhered (made to stick) to a porous screen. All areas of the screen are blocked out except for the area to be printed. A special tool called a squeegee is used to force the ink through the stencil and onto the substrate. See Fig. 7-6.

Screen printing is a simple and flexible printing process. It can be used to print on almost any surface. It is commonly used on fabrics, such as T-shirts and sweatshirts. Even products like drinking glasses, mugs, or shampoo bottles are commonly screen printed.

Planographic Printing

Any process that involves the transfer of a message from a flat surface is called planographic printing. Offset lithography is the most popular planographic printing method. It is also the most common printing process used today. Many books, newspapers, brochures, and stationery products are created this way.

Lithography is based on the principle that grease and water do not mix. An image is created on a flat plane using a material that has a grease base. The image attracts ink because ink, too, has a grease base. All areas of the printing surface that do not contain the image are covered with a water mixture, which repels the ink. See Fig. 7-7. Multiple copies of the image can be made by repeating the process using a printing press.

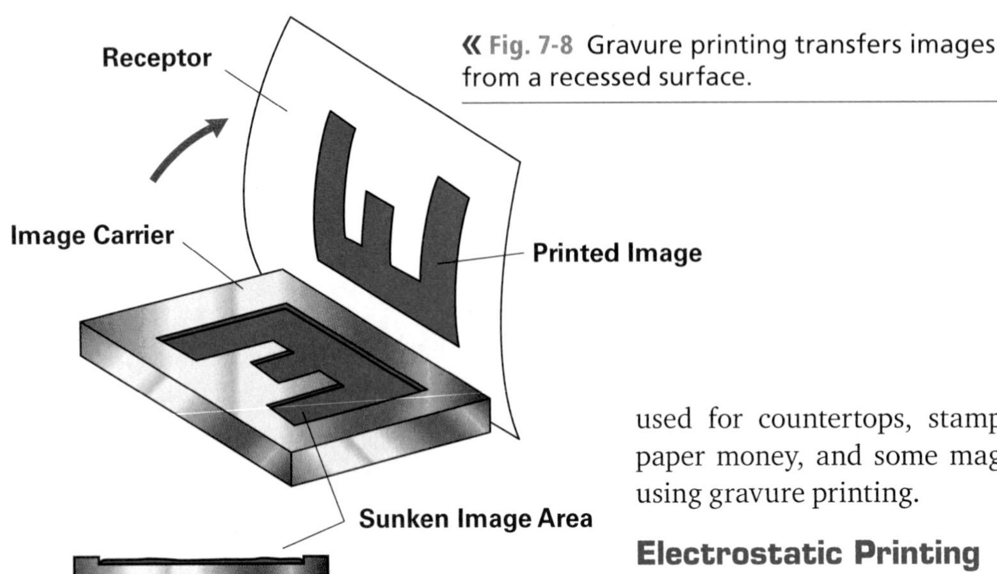

« Fig. 7-8 Gravure printing transfers images from a recessed surface.

Gravure Printing

Gravure, or intaglio, printing is the exact opposite of relief printing. In gravure printing, images are transferred from plates that have sunken areas. The images are etched or carved into the surface with computer-controlled devices or lasers. Each plate has many tiny holes, called cells, that when combined, form the shape of the letters, symbols, and other design elements. The cells are filled with ink and the surface is wiped to remove ink from non-image areas. Then, paper is forced against the plate. The paper absorbs the ink. The image transferred to the paper is the identical form of the engraved image on the plate. See Fig. 7-8.

Making gravure plates can be expensive because the large copper plates must be carefully machined and engraved. If a plate becomes damaged or a mistake is made, re-engraving can be time consuming and costly.

Gravure printing is often used for long press runs or jobs that require very high quality. The plates are extremely durable and the reproduced image is sharp. Art books, laminates used for countertops, stamps, United States paper money, and some magazines are made using gravure printing.

Electrostatic Printing

Many offices and quick-printing companies use electrostatic printing processes. In fact, you have probably used this technique a few times yourself at the library or local copy center. Copier machines that you use to make quick duplicates of your original fall into this printing category. If you use a laser printer to output from your computer, then you are also using an electrostatic process.

Electrostatic printing is based on the principle that opposite electrical charges attract, whereas like charges repel. Basically, when an original is placed on the glass window, it is exposed to a plate inside the copier. The plate has been given a positive charge. See Fig. 7-9. A light removes the charge from the nonimage areas of the plate. The image area remains positively charged. A toner material is given a negative charge. It sticks to the positive image areas of the plate. A piece of paper is given a positive charge and passed over the toner plate. Because opposite charges attract and the positive charge is stronger than the negative charge, the toner is transferred to the paper. A heating element sets the toner on the paper permanently.

Electrostatic printing makes it possible to output or duplicate messages quickly. New digital printing systems that rely on this tech-

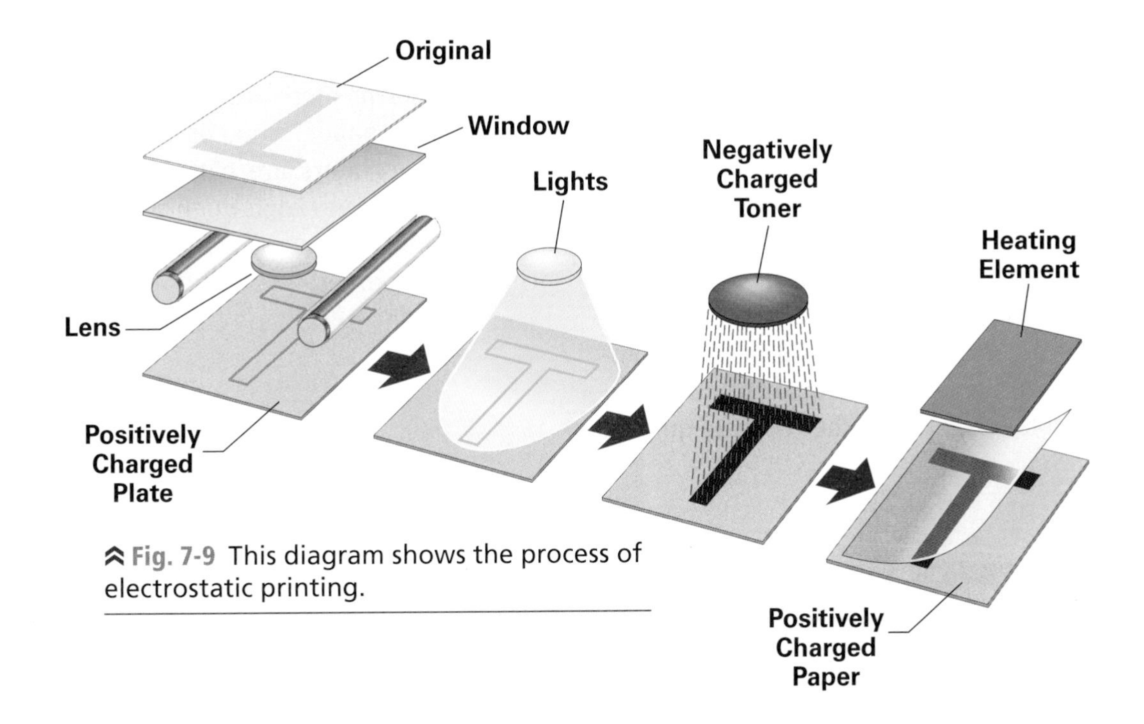

Fig. 7-9 This diagram shows the process of electrostatic printing.

nology are already transforming the industry. They enable four-color printing jobs to be completed in minutes instead of hours or weeks using high-speed, specialized printers.

Photographic Printing

Photography is often considered to be a printing technique. Basically, in photographic printing, light is projected through a plate (usually called a negative) onto a light-sensitive material. After processing, the image appears. (Photography is discussed in greater detail later in this chapter.)

Ink-Jet Printing

Another technique that does not easily fit into the other categories is ink-jet printing. During this process, ink jets spray ink onto the substrate. Digital computer data controls the tiny nozzles that spray the ink droplets. Because there is no contact between the image carrier and the substrate, this process can be used to print on a wide variety of materials and in full color. In fact, some fine artists use large-format ink-jet printers to reproduce their work. Companies can use massive ink-jet printers to print billboards, banners, and vehicle wraps more than 16 feet wide and almost any length. See Fig. 7-10.

Fig. 7-10 This large vehicle wrap was printed on an electrostatic or ink-jet printer and then applied to the truck.

Laser Printing

Laser printing is an increasingly popular form of printing. As you may know, lasers are devices that strengthen and direct light to produce a narrow, high-energy beam. The most common laser printers work much like electrostatic copiers. They print onto regular paper. Many homes, schools, and businesses have these printers hooked up to their desktop computers. They provide good-quality black and white or color output, and they operate quietly.

Other laser printers are designed to expose photographic paper, film, or printing plates. These laser printers are called image setters and they make very high-quality images. They are commonly found in art departments or printing companies.

Science Application

Laser Printers As you may know, static electricity is what makes a lightning bolt touch the ground or why you sometimes get a slight shock when touching a door knob. Static electricity is simply an electrical charge built up on an insulated object. Since oppositely charged atoms are attracted to each other, objects with opposite static electricity fields cling together. This principle is also what makes laser printers work.

After a file is sent to the printer, a drum that has a positive charge begins to turn. A laser shines light and creates an image of the information on the drum as a pattern of electrical charges—also called an electrostatic image. As the drum is exposed to the laser, it takes on a negative charge.

Once the electrostatic image is completed, the printer releases positively charged toner that sticks to the negatively charged drum. Then paper, passing through a charged roller, gets a negative charge. Since the negative charge is greater than that of the drum, the toner sticks to the paper as it passes over the drum. The electrostatic image is thus transferred to the paper. The paper then passes through two heated rollers, called the fuser, which melts any loose toner. This is why paper is warm when it comes out of the printer.

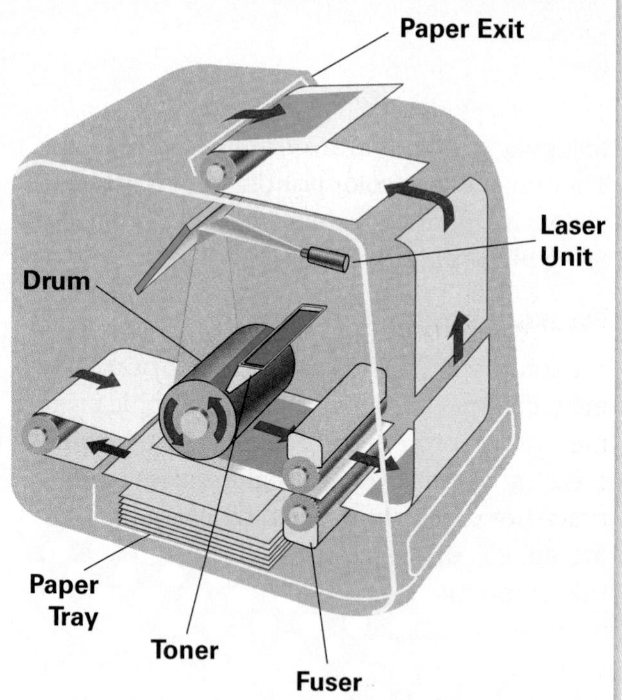

FIND OUT Write your name with glue on a piece of paper. Before the glue dries, pour glitter or sand onto the page and shake off any excess. How does doing this imitate how a laser printer works?

Engineering the Printing Process

New technological advancements continue to be made in printing. Engineers work to solve problems and improve printing operations from start to finish.

Computers and integrated circuits are an important part of nearly all modern printing processes. They make the digital workflow possible. This improves speed, ease, and accuracy. Computers ensure accurate ink coverage. They help to determine how cyan (blue), magenta (red), yellow, and black inks can be combined to form any color needed. Computer-controlled printing operations convert plain white paper or plastic film into vibrant graphic media.

At one time, printed products were designed and printed in the same place, then shipped to where they were sold. Today, the design can be transmitted electronically to a printer closer to where a product may be sold. A book designed in Chicago, for example, may be printed and sold in China. This reduces shipping costs and enables the product to be received by the customer more quickly.

Digital technologies are changing how and when products are printed, too. Printing-on-demand technologies allow products to be printed on an as-needed basis. If changes in the products must be made, there are few in the warehouse that must be discarded. Storage costs are lower, and overproduction seldom occurs.

Variable data printing (VDP) is a type of printing-on-demand. It is often referred to as personalized publishing. The same basic document layout might be used for thousands of customers, but parts of the document are customized for individuals. Insurance companies use this system to send policy statements to customers with the customer's unique information. Credit card companies use VDP to send customized discount coupons to cardholders based on their interests and buying habits.

Communicating Through Photography

Often, photographs can relate so many ideas and feelings that words are hardly necessary. Most graphic communications media depend on photography to send a message. Few magazines and newspapers use only words. Photographs help us to capture feelings, expressions, concepts, images, and information. In science and technology, photos enable researchers to study microscopic forms and to explore the failure or success of experiments. Teachers and authors use them to educate and to explain complex information. Most of us use photos to record times and events in our lives.

How Cameras Work

The photographic process is similar to how your eyes work. Eyes sense reflected light and transmit images to the brain. Cameras, whether film or digital, have basic parts that allow a similar process to occur.

When you take a picture, you usually use a viewfinder or display screen to locate and position your subject. Light is reflected from a subject and focused through the lens of a camera onto film or a sensor. The lens directs the light rays so that they are correctly positioned and do not appear blurred.

An aperture (opening) lets light from your subject into the camera. An aperture diaphragm is required to increase or decrease the amount of light that can reach the film or sensors. See **Fig. 7-11** on page 146.

Every camera must also have a shutter. A shutter is a device that controls the amount of time that light is allowed to reach the film or sensor. When you take a picture and hear a click, you are hearing the shutter open and close.

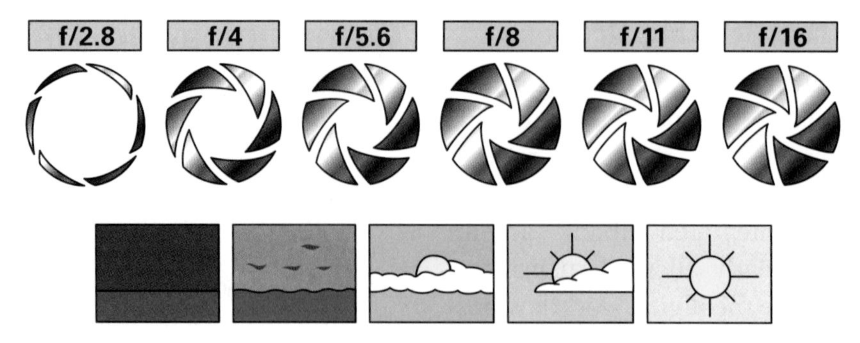

Fig. 7-11 The size of the aperture is identified by numbers called f-stops. Notice that the larger the f-stop number, the smaller the opening. On cloudy days, the opening must be large; on bright sunny days, it should be small.

When the film or data is developed, you see a reproduction of what the camera "saw." The biggest difference between film and digital cameras is how the image is stored and developed. See Fig. 7-12.

Using a Film Camera

Photographic film comes in a variety of formats and sizes to fit different cameras and applications. Rolled film comes in canisters or cartridges that fit inside your camera. Sheet film is most often used by professional photographers and printers.

Despite the size or shape, all film contains one or more thin layers of a light-sensitive material called an emulsion. The emulsion is the part of the film that captures the image. It must be chemically processed before the image can be seen.

Films are rated according to how sensitive their emulsion is to light. This is often called the speed of the film. A high-speed film, such as ASA/ISO 400, is very sensitive to light. It does not need much light for a good exposure. Photographers choose films according to the lighting conditions where they will be shooting the film.

You have already learned that a lens, aperture, and shutter are all necessary to get the correct amount of light onto the film. Every film camera must also have a completely light-tight space in which to store the film. A filmholder is also needed. This holds the film in position so that it will not move during an exposure.

Using a Digital Camera

Digital cameras record images electronically instead of on film. You can see the images instantly by looking at a small display screen on the camera. You can also view them on a television screen or computer monitor. Digital cameras are quickly replacing film cameras for personal and professional uses because of their convenience.

Digital cameras store images as digital information. The light is focused onto a sensor that converts the images to electrical signals. One sensor commonly used is the charge-coupled device (CCD), which you may have already learned about in Chapter 6, "Electronic Communication."

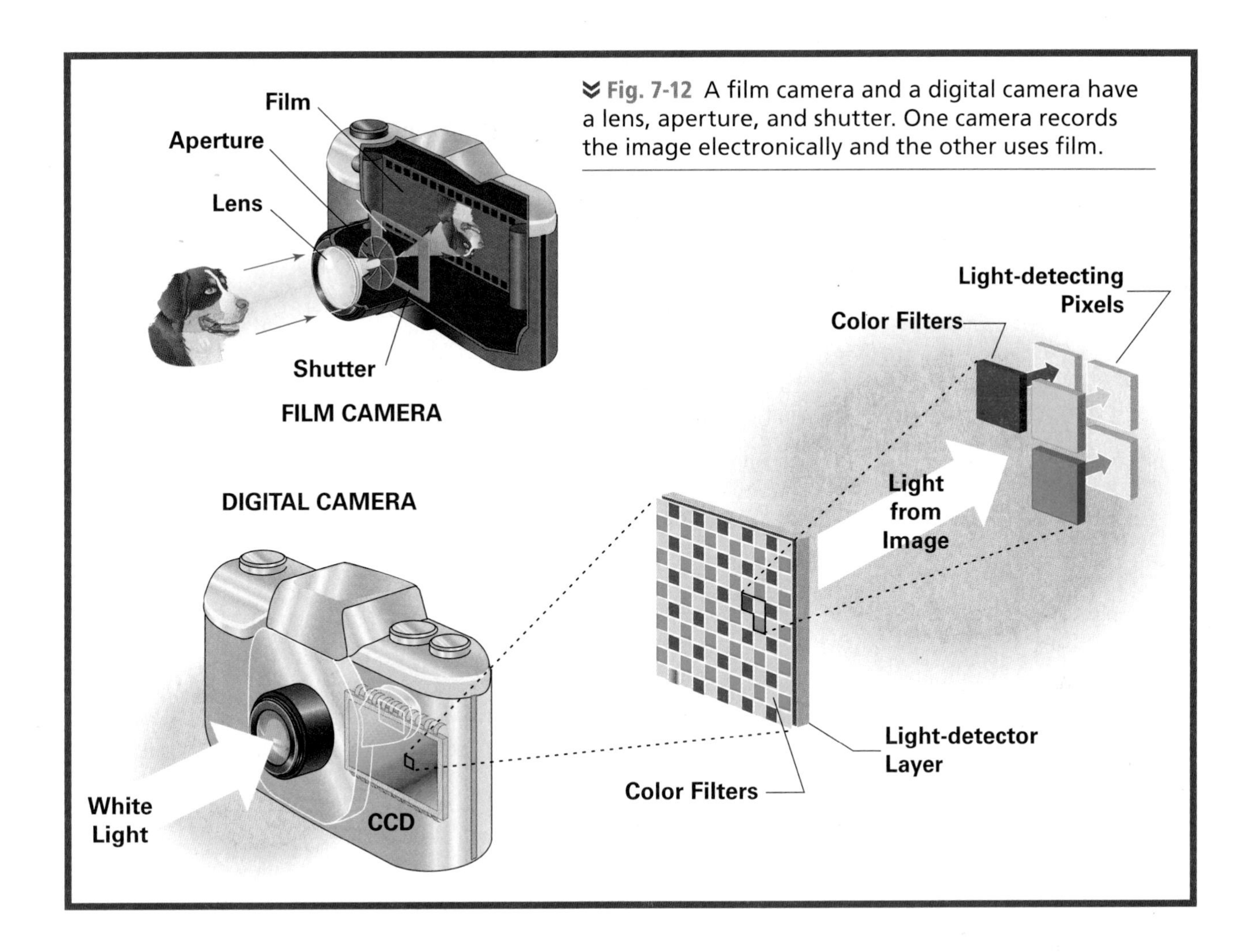

Fig. 7-12 A film camera and a digital camera have a lens, aperture, and shutter. One camera records the image electronically and the other uses film.

A CCD is usually divided into millions of tiny light-sensitive cells, or photosites, arranged in rows and columns. The signal from each of these cells produces a tiny dot of light called a pixel. These dots make up the picture. See Fig. 7-13 on page 148. The CCD in a high-quality camera might produce 3,072 pixels horizontally by 2,304 pixels vertically. Since the images produced have 7,077,888 pixels, it is called a 7-megapixel device. Lower-quality cameras produce fewer pixels. The more pixels a photo has, the higher the resolution, or sharpness of detail.

The information collected by the CCD is digital data. This data is easily stored in a camera's internal memory or transmitted via the Internet or e-mail. The images are also easily modified. You can crop out parts that you do not like, alter colors, and add special effects with ease using your computer. You can print these photos at home or send them to professional photofinishers for output on high-quality papers or other products, such as address labels, mugs, and T-shirts.

Fig. 7-13 When you enlarge a digital image, it is easy to see the many pixels that make up the image. Higher-quality images usually have more pixels than those of lesser quality.

Photographic Composition

Composition is the way in which all the elements in a photograph are arranged. When you compose a photograph, you plan what will be in the picture and how it will be positioned. To communicate effectively and to create a visually pleasing photo, you will want to compose it carefully.

Balance Just as layouts should appear balanced, so too should the subjects in a photograph. When composing a photograph, consider the position of the subjects. Ask yourself questions like, "Is there too much on one side of the picture?" or "Is there appropriate space at the top and bottom of the picture?"

Photographers often follow the rule of thirds to position a subject in a photograph. The rule of thirds divides the image area into thirds both vertically and horizontally. A composition is more interesting when the subject is located at the intersections of these imaginary lines, rather than in the exact center. See Fig. 7-14. By positioning the subject off-center at one of these locations, the viewer gains a better sense of balance. The picture is more visually pleasing.

Framing A photo is sometimes more effective if the photographer uses a foreground image (part of the scene nearest the viewer) to frame a background image. This is a good technique to use when photographing scenic landscapes. Other times a background image can be used to frame a subject. For example, the trunk and branches of a tree can be used to frame a person standing in front of the tree.

Simplicity Photographers should usually keep photos simple. This is done by concentrating on the center of interest in a picture. Background can be distracting to viewers. Sometimes, such as in portraits, no background is needed.

Leading Lines A photographer often needs to "lead" the viewer into a picture. By positioning a subject appropriately, leading lines can be used to pull the attention into the scene. This technique is especially important when photographing subjects that have apparent line

» Fig. 7-14 This photo clearly shows the rule of thirds. Note that the main subject is placed off-center where the lines intersect.

structure. For example, the "lines" created by roads, bridges, fences, and buildings can be seen easily. See Fig. 7-15.

Holography

Holography is the use of lasers to record realistic images of three-dimensional objects. In a photograph, you see only one side of the subject. In a hologram (the 3D image), you see the subject from different angles as you change your view. You can see the front, back, and sides.

Holograms are a good way to record many realistic details about an object. It seems like the object is right there in front of you and that you can pick it up and hold it. Holograms are not used as often as photographs. They are more difficult and expensive to reproduce. For this reason, some organizations use them for security purposes. Credit card companies and software makers have used holograms for many years to discourage counterfeiters. The technology to make and reproduce holograms is improving all the time. One day, holograms may be used in places where you now see photographs.

» Fig. 7-15 The lines created by the fence and trellis lead the viewer into the courtyard.

Young Innovators

Through the Lens

Have you ever heard that "a picture is worth a thousand words"? Since its invention, photography has become the medium of choice for communicating a large amount of information in a small amount of space.

Photography contests provide the opportunity to get involved and challenge yourself to find the best way to present an idea. Choosing a photography contest to enter is as easy as choosing your favorite subject. Do you like landscapes? Perhaps you prefer portraits or maybe sports photography. Whatever your interest may be, chances are there is a photography contest in that subject area.

As with any contest, photography contests have their own guidelines that must be followed. These rules will vary by contest, so it is always a good idea to review the rules before submitting a work. The following guidelines will give you a general idea of which common elements judges may look for in a photograph.

Composition refers to the way the elements of the photograph come together to create a visually pleasing image. You may hear such terms as "rule of thirds" and "leading lines" when discussing composition.

Exposure is the degree to which your subject is rendered in the photograph. An ideally exposed image will have a range of values (lightness or darkness) that will provide enough contrast to make the image pleasing to the eye.

Presentation includes the size, overall quality of the image, and the physical manner in which the photograph is presented (matted vs. unmatted). The entire presentation has an effect on the image.

All of the elements combined harmoniously can produce stunning images. If you would like to try your hand at photography, there are many contests in a variety of subject areas available to you. Ask your teacher for more details on photography contests and how to enter.

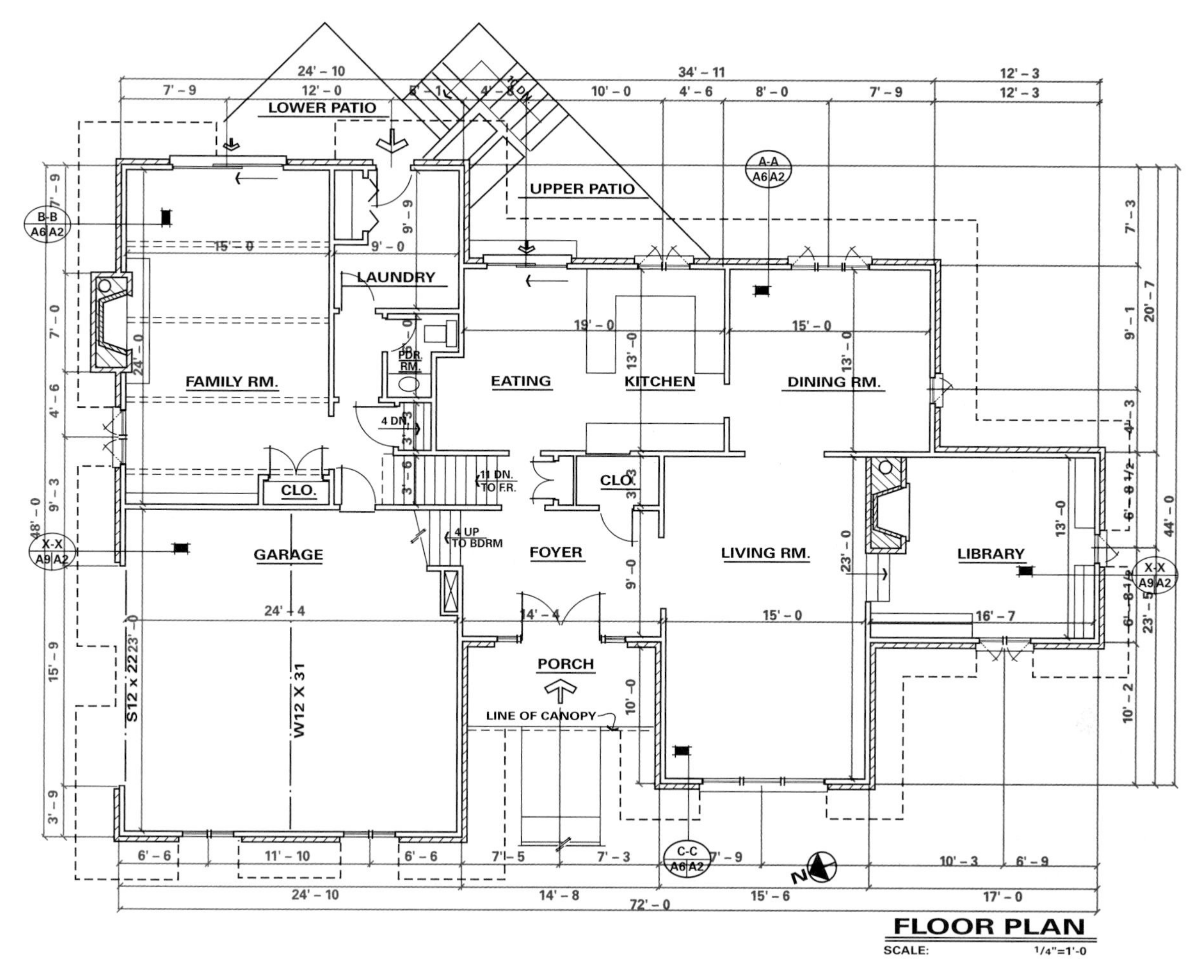

Fig. 7-16 This house floor plan is an example of a drafted drawing. It accurately shows the placement and size of each room and other structural features, such as fireplaces, stairs, windows, doors, and closets.

Drafting: The Universal Language

Drafting is the process of creating technical drawings that describe the size, shape, and structure of an object. It is an accurate drawing process used for nearly every product or structure made today—large or small. Integrated circuits, shoes, tools, cars, bridges, and skyscrapers are just a few examples of things that start as drafted drawings.

Drawings can communicate ideas effectively and accurately. For example, all the design details of a building could not be verbally described to construction workers. Detailed plans like those shown in **Fig. 7-16** are required.

Drafting is often referred to as the universal language. A person who understands the basic symbols, lines, and rules can understand the message regardless of who communicates it. You don't need to understand the Japanese language in order to understand a drawing made by a Japanese-speaking person.

People who are skilled at recording and understanding drafted messages are called drafters. Many other careers also require a thorough understanding of drafting. Architects, engineers, electricians, and plumbers, for example, regularly make or use drafted drawings.

Drawings can be made with computers or with simple tools like pencils, rulers, T-squares, triangles, and compasses. Regardless of the materials and tools used to communicate the message, the thinking process is the same.

Designers often begin by making sketches of their own ideas. Sketches show the ideas roughly but neatly. They show the proper size relationship between parts, but they don't include much detail. See Fig. 7-17.

After the designer chooses the sketch that best represents the idea, detailed drawings are made. The drawings accurately describe the shape, size, dimensions, and details of the product and all its parts.

Many different types of drawings are used to describe things accurately. Each drawing has a different purpose and use.

Multiview Drawings

Multiview drawings show two or more different views of an object. The views are drawn at right angles (perpendicular) to one another. The technique used is called orthographic projection. The best way to think about this process is to picture the object that you want to draw inside of a glass box. Each view of the object is projected on a different side of the imaginary box. When the box is unfolded, you can see each view as it would appear on a piece of paper. See Fig. 7-18.

There are six possible views: top, bottom, right side, left side, front, and rear. Many objects can be described completely in only three views: top, front, and right side. The object's

Fig. 7-17 Most projects, such as this birdhouse, start with a sketch. Sketching allows you to put your design ideas on paper in a quick and easy way. After that, a more finished drawing can be made.

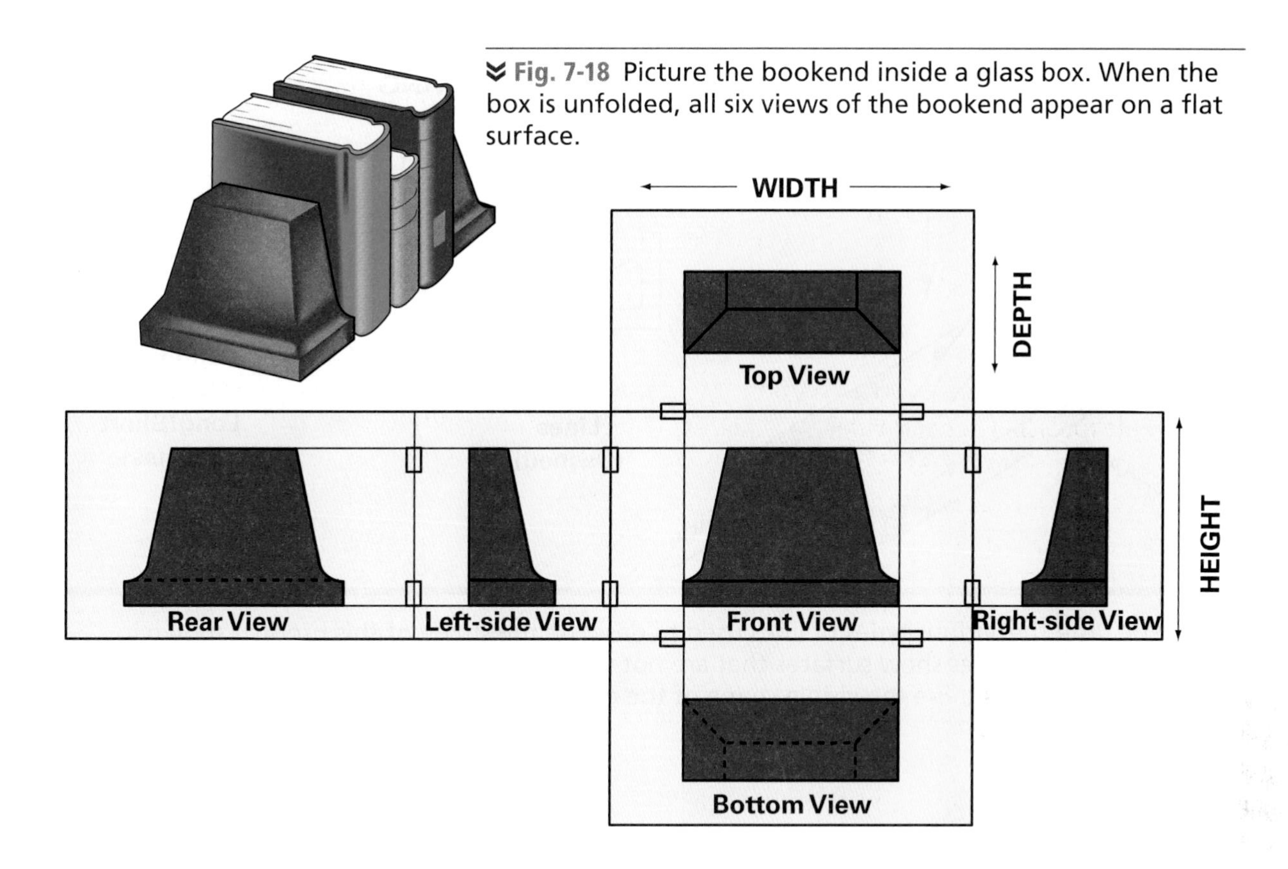

Fig. 7-18 Picture the bookend inside a glass box. When the box is unfolded, all six views of the bookend appear on a flat surface.

height can be shown in the front and right-side views. The object's width can be shown in the front and top views. The object's depth is seen in the top and right-side views. Cylinders can be described in only two views. How many views are needed to describe a sphere?

In order to create accurate multiview drawings, drafters must understand and follow standard drafting practices. For example, every detail of the object must be shown in each view. If a certain feature cannot be seen from a particular view, it must still be shown with special lines or symbols. Study Fig. 7-19 on page 154. Notice that the holes in this object are shown as dashed lines in some views. This means that the holes are hidden in that view. Notice that some lines are curved to show the curved surfaces of the object. Centerlines are shown as alternating long and short dashed lines. They show the centers of holes and arched shapes. Object lines are the most prominent lines on the drawing. They represent all surfaces and details that are visible from that view.

Often multiview drawings are called working drawings. This is because they provide information for making the product. Working drawings are fully dimensioned to show the exact size, shape, and location of every detail. Symbols are used to represent specific types of materials, fasteners, electrical components, plumbing fixtures, and structural parts.

Most drawings are not made to the actual size of the product. A full-size drawing of a house would be much too large to be usable. A full-size drawing of an integrated circuit would be too small to read. For these reasons, drawings are made "to scale." This means they are enlarged or reduced by a certain proportion.

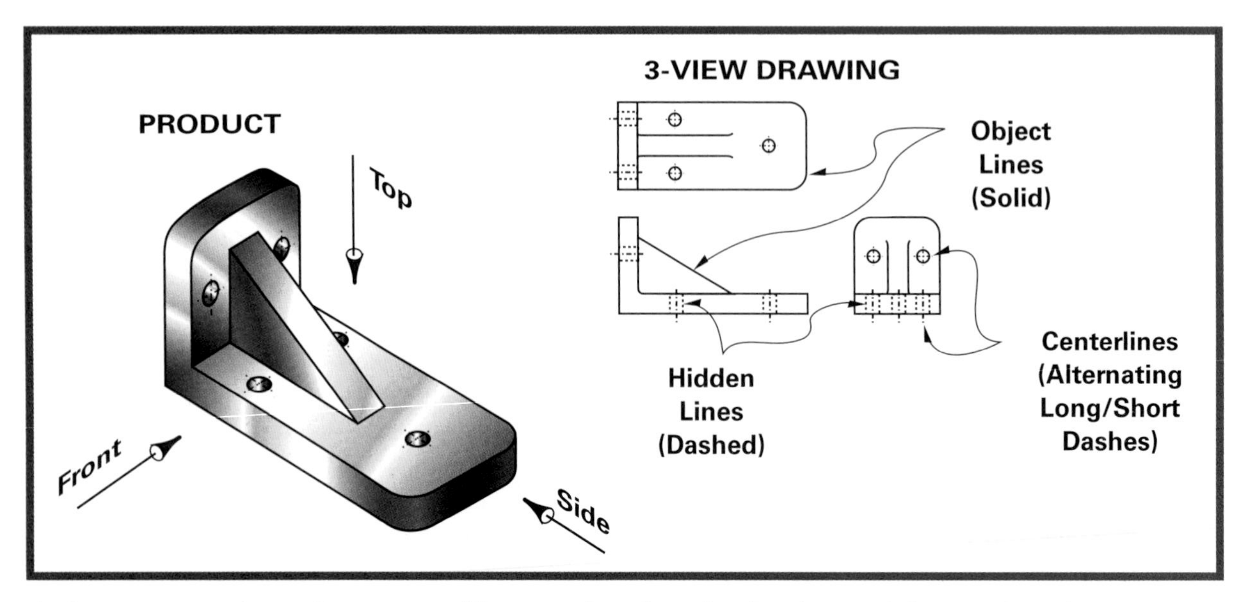

Fig. 7-19 Note the various types of lines used to describe the shape of this product shown in three views. Hidden lines show surfaces that are not visible in that view. Centerlines show centers of holes. Object lines define the visible shape of the object.

For instance, a drafter may use the scale ¼ in. = 1 ft. or 1 cm = 1 m. By using scales, drafters can create a drawing that is in proportion to the actual object but has a practical size.

Pictorial Drawings

Pictorial drawings show objects as they appear to the human eye. Most people can easily understand a pictorial drawing because it looks quite realistic. There are three types of pictorial drawings: isometric, oblique, and perspective.

- **Isometric.** In an isometric drawing, the object is tilted forward 30 degrees and rotated 30 degrees so that its edges form equal angles. (Isometric means of equal measure.) In isometric drawings, one corner always appears closest to you. See Fig. 7-20.
- **Oblique.** One perfect, undistorted view of an object is shown in an oblique drawing. The other sides of the object are shown at an angle. If you draw a simple cube, you have just made an oblique drawing. See again Fig. 7-20.
- **Perspective.** These drawings help people visualize how an object would appear in real life. Parts of the object farther away from you appear smaller. Parallel lines disappear into the distance at the drawing's vanishing point. See Fig. 7-21.

Technical illustrations may be based on any type of pictorial drawing. They provide technical information in a visual way. They often include shading to help improve clarity or add emphasis. Technical illustrations are commonly found in repair manuals and assembly instructions. They are often accompanied by graphs, charts, or diagrams. Exploded views are one type of technical illustration. They show how parts of a product fit together.

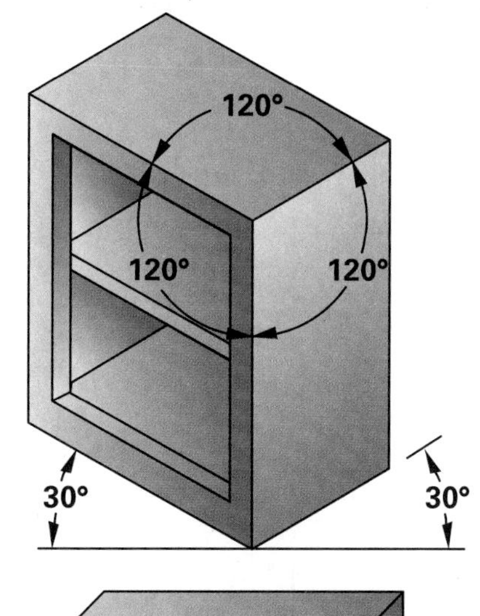

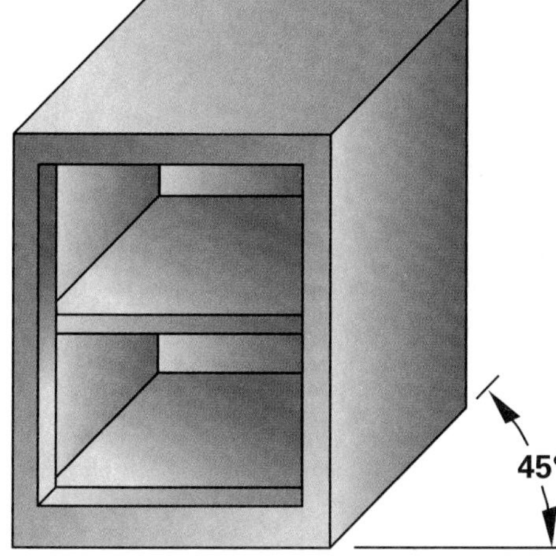

≈ Fig. 7-20 The cabinet drawing at the top is an isometric drawing, and the drawing at the bottom is an oblique drawing of the same cabinet.

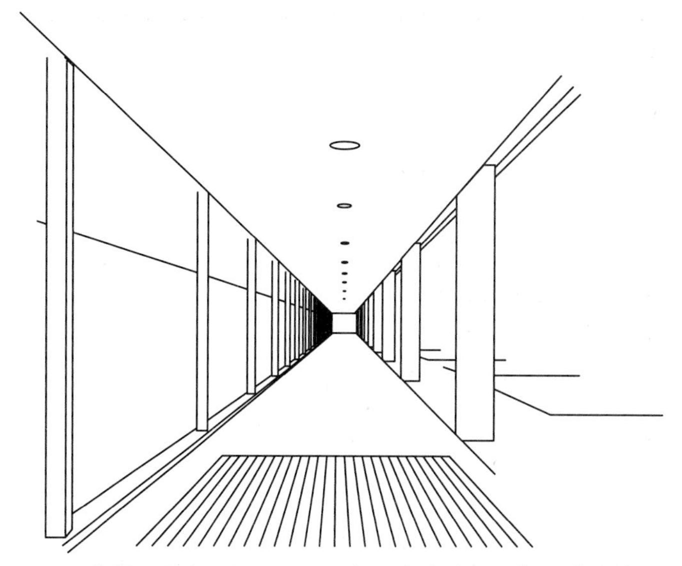

≈ Fig. 7-21 A perspective drawing often looks more realistic. Note how the receding lines appear to meet at a distant vanishing point.

Pattern Development

A pattern is a full-size outline of an object. Patterns are created for many products made from flat materials that are folded, shaped, or combined into three-dimensional products. Leather wallets, pizza boxes, clothing, hot air balloons, and metal heating ducts and pipes all started as patterns. See Fig. 7-22 on page 156.

Computer-Aided Drafting

While the previously mentioned drawings can be done by hand, these same drawings can be done on the computer much more easily. **Computer-aided drafting (CAD)** is the use of a computer system to produce technical drawings and/or design a product. Today, most drafters use CAD to make drawings. The CAD software provides all the electronic tools needed, as well as large data banks of common drafting symbols. Rather than draw the symbol from scratch, the drafter simply selects it from the data bank and places it in the drawing.

CAD users can create objects of any shape to any scale desired. Dimensions can be added and color used to enhance details and improve clarity. Drawings can be easily modified using commands such as "copy" or "move." New copies can be output to a printer or plotter.

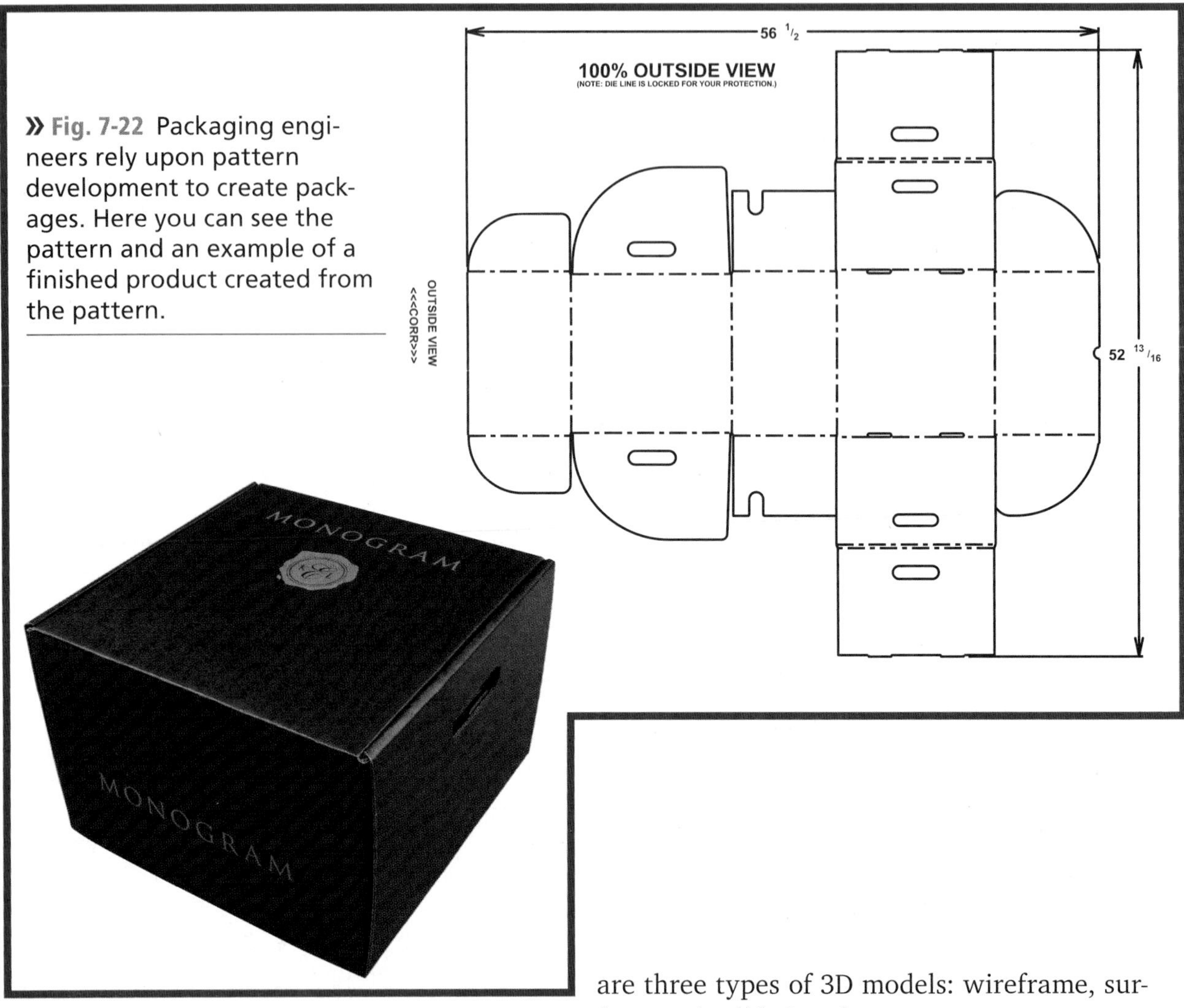

» **Fig. 7-22** Packaging engineers rely upon pattern development to create packages. Here you can see the pattern and an example of a finished product created from the pattern.

CAD can also be used to produce more accurate drawings than manual drafting. Users can pick specific points of an object when drafting. The computer can then pick the exact same point each time, which is harder to do when drafting by hand.

All CAD drawings are saved as digital data that can be sent easily via computer networks or the Internet. They may also be inserted electronically into reports and other documents.

In addition to two-dimensional drawings, CAD programs enable drafters to create three-dimensional representations of objects. There are three types of 3D models: wireframe, surface, and solid. See **Fig. 7-23**.

- **Wireframe.** This type of modeling can also be thought of as "stick-figure" modeling. The model only shows lines connected together. Wireframe modeling is usually used to show the amount of space between objects.
- **Surface.** A surface model can be compared to a piece of paper that is shaped into an object. Surface models have no thickness. They may look like the object one way but may be seen as only a thin line or curve when viewed at another angle. These models can be used to display the area of an object.

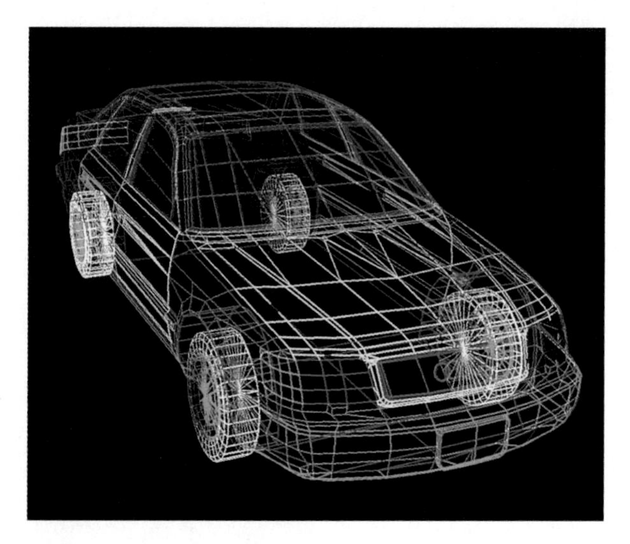

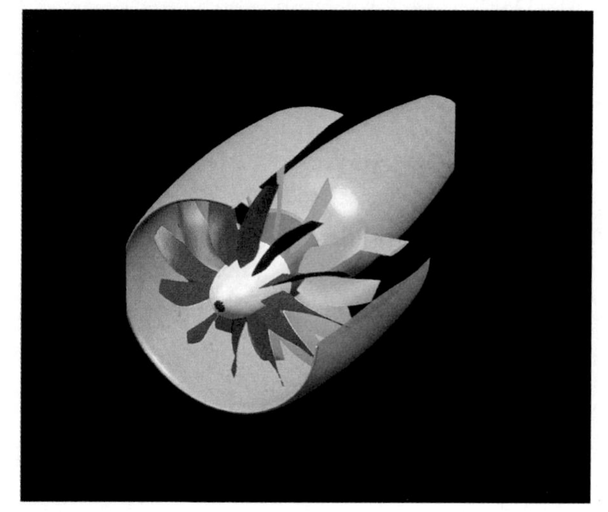

Fig. 7-23 Two ways of modeling on a CAD system are wireframe and solid modeling.

- **Solid.** Models that include the shape, area, and volume of an object are solid models. Engineers often use solid modeling to assess properties and conduct analyses of products before they are built. For example, solid modeling can help the engineer determine whether the parts will fit together correctly. When appropriate, models can be animated, which may make analysis easier. Animation is also useful when trying to explain to someone else how the product works.

CAD can also be used to create a physical model. In the past, all such models had to be made by hand. Now 3D CAD drawings can be used to create real models using a process called **rapid prototyping**. Other names for this process are stereolithography and 3D printing.

In rapid prototyping, 3D CAD drawings are read into a 3D printer. A laser traces the product's shape into a special liquid plastic. The plastic then hardens. The shape of the product is formed layer by layer until the model is complete. See Fig. 7-24. Variations of rapid prototyping involve using other materials such as plastic powder, metal, or paper. In some cases, the result is not just a model but a finished, working product. For example, researchers are even studying ways to create prosthetic arms and legs using rapid prototyping technology. Each prosthetic device is unique because it has to fit the individual precisely. Rapid prototyping technology is well suited to such custom manufacturing.

Fig. 7-24 After a CAD drawing of this funnel was sent to a 3D printer, a physical prototype was created layer by layer.

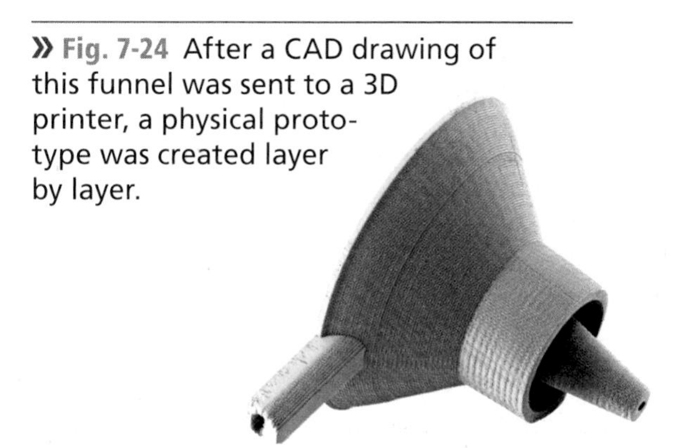

CHAPTER 7 REVIEW

Main Ideas

- Graphic communications technology involves the sending of messages and other information using visual means.
- Principles of design help to determine a design's effectiveness.
- Steps in the creative design process include making thumbnail sketches, rough layout drawings, and a final form.
- Printing, photography, and drafting are all forms of graphic communications.

Understanding Concepts

1. List the principles of design.
2. What is digital workflow?
3. What principle is electrostatic printing based on?
4. What is the main difference between film cameras and digital cameras?
5. Why is drafting referred to as the universal language?

Thinking Critically

1. Do you think the principles of design were used successfully in this textbook? Why or why not?
2. Discuss the role of graphic communications in a free enterprise system. For example, how does advertising help businesses compete?
3. Photos can now be digitally altered. What ethical situations might arise?
4. You learned about the first three steps in the creative design process. What steps might come next?
5. Identify the type of printing processes used for various classroom items.

Applying Concepts & Solving Problems

1. **Math** Drafting requires an understanding of scale measurement. Make a scale drawing of one room in your home. Use a scale of ½" = 1'. Do a second drawing of a different room using a scale of ¼" = 1'.
2. **Design** Select a simple square or rectangular object and make a multiview drawing of it. Show at least three views in their proper locations.
3. **Language Arts** Prove that "a picture is worth a thousand words." Show a picture to a classmate and ask him or her to describe it as completely as possible. How many words were actually used?

Activity: Design a Board Game

Design Brief

Do you know which board game is played more than any other in the world? According to Hasbro Corporation, it's Monopoly®! More than 200 million copies have been sold in 80 countries and in 26 different languages. Charles Darrow patented the game in 1935, but variations of it already existed. Don't you wish you had invented Monopoly? For this design brief you'll put your creative energies to work and try to invent a new game. Who knows? It just might become the most played game in the world one day!

Design and build a prototype of an exciting new board game, including all game components and packaging.

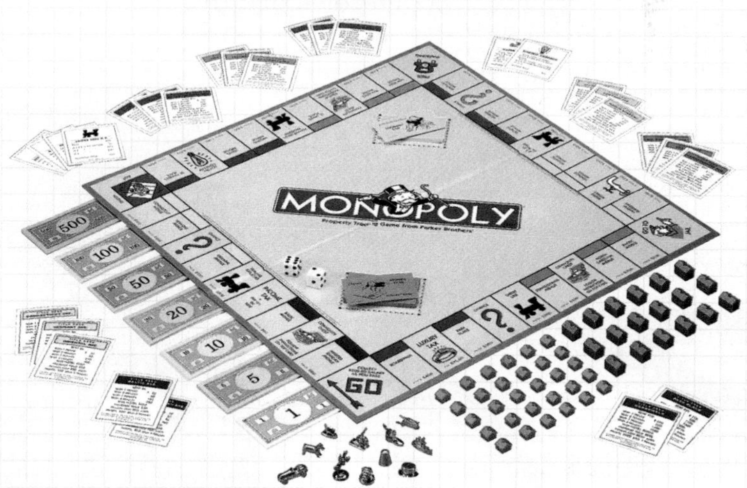

Criteria

- Your game must be suitable for play by ages 10 to adult.
- Some type of playing board must be used. It cannot be a simple card or dice game.
- All game components (e.g., dice, spinner, tokens, cards, etc.) should be included.
- Your game must be neatly enclosed in a suitable and marketable package.
- Instructions should clearly describe the rules for someone unfamiliar with the game.
- Your game must be made of durable materials or mounted securely on a sturdy material such as foam core board or heavy chip board.
- You must effectively use the principles of design.

Constraints

- The packaged game cannot be larger than 18" × 12" × 3".
- Use digital imaging software, graphic layout software, and/or computer-aided drafting software.
- Your package must be generated using pattern development techniques and following procedures for either manual or computer-aided drafting.

Engineering Design Process

1. **Define the problem.** Write a statement that describes the problem you are going to solve. For example, is the main objective of your game to entertain, or do you need to meet a certain profit?
2. **Brainstorm, research, and generate ideas.** With your team, discuss possible solutions. Hint: Gather an assortment of board games and examine them with a critical eye to determine appealing or unappealing features.
3. **Identify criteria and specify constraints.** These are listed on page 160.
4. **Develop and propose designs and choose among alternative solutions.** Choose the best design that will solve your problem. For example, if you are targeting a specific age group, get their opinions on what kinds of games they like.
5. **Implement the proposed solution.** Decide on the design you will use. Gather any needed information or software tools.
6. **Make a model or prototype.** Create a model of your game.
7. **Evaluate the solution and its consequences.** Does your game work as intended? Does it include all principles of design?
8. **Refine the design.** Based on your evaluation, change the design if needed.
9. **Create the final design.** After you have your teacher's approval, create your final prototype.
10. **Communicate the processes and results.** Present your finished game to the class. Then hand in the assignment to your teacher. Your finished product must be accompanied by complete documentation of the design and problem-solving process, including evidence of having considered multiple solutions to the problem.

SAFETY FIRST

Before You Begin. Use caution when cutting heavy cardboard, chipboard, or foam core board. Be sure you use a sharp knife and a metal straightedge. Keep a protective surface under your board. Position hands and fingers so they will not be harmed if the knife slips. Always cut away from your body. Have your teacher demonstrate the proper technique. Follow all safety rules.

UNIT 3

ENERGY & POWER
ENGINEERING & DESIGN FRONTIERS

In 1859 near Titusville, Pennsylvania, Edwin Drake drilled the first oil well in the United States. That was the beginning of the world's love affair with petroleum. In Drake's day, consumers used a petroleum product called kerosene to light their lamps, but gasoline and diesel were soon powering our transportation systems.

Our world economy today hums on oil—but the love affair has gone sour. We are all too aware that petroleum is polluting our atmosphere and making our economies hostage to oil politics and prices.

Enter Biofuels

It's welcome news, therefore, that energy industries are intensively developing and improving biofuels. A biofuel is any fuel produced from plant or animal products or their wastes. More plants and animals can always be grown, so biofuels—unlike oil, coal, and other fossil fuels—are renewable. Though biofuels release potentially harmful greenhouse gases into the atmosphere, they participate in a balanced carbon cycle. As biofuel plants grow, they trap atmospheric carbon. When they are burned as fuel, they release the carbon back into the air. Fossil fuels, on the other hand, transfer carbon from Earth's interior to the atmosphere.

One biofuel, E10, is already quite common at the pump. It is a blend of 90 percent gasoline (from petroleum) and 10 percent ethanol (from crops). Fuels with higher ethanol percentages are also becoming available.

Biobutanol, also made from crops, is now coming on the market. This fuel offers some advantages over ethanol. It is more stable and thus easier to store and move, and it yields higher fuel efficiency.

Another fuel, biodiesel, is made by chemically combining fat or oil, such as used cooking oil or soybean oil, with alcohol. Biodiesel is usually mixed with petroleum-based diesel fuel, though some engines can run on pure biodiesel.

At the "cutting edge" of biofuel technology today are techniques for converting almost any plant material—even grasses, cornstalks, and wood—into fuel. When this technology becomes practical and cost-effective, biofuels may well eclipse petroleum in the world's energy economy.

CHAPTER 8 ENERGY & POWER FUNDAMENTALS

≈The energy generated in this bike race is tremendous! In this chapter's activity, "Measure Energy and Power," you will have a chance to study and measure forms of energy and power.

Objectives

- Identify the forms of energy and power.
- Explain why energy must be controlled.
- List impacts of energy and power technology.
- Explain how to measure different forms of energy and power.

Vocabulary

- energy
- work
- potential energy
- kinetic energy
- power
- Ohm's law

Reading Focus

1. Write down the colored headings from Chapter 8 in your notebook.
2. As you read the text under each heading, visualize what you are reading.
3. Reflect on what you read by writing a few sentences under each heading to describe it.
4. Continue this process until you have finished the chapter. Reread your notes.

What Is Energy?

We use energy every day in countless ways. Our bodies use it to carry out life processes. It powers our vehicles and other machines, heats and cools our buildings, and gives us light.

Although energy is found in nature, technology has enabled us to pull it out of nature and put it to use. It was by creating sails, for example, that ancient peoples were able to tap into the energy of the wind and use it to propel their boats. See **Fig. 8-1**.

» **Fig. 8-1** Sail technology enables us to harness the speed of the wind.

Energy and Work

Energy is the capacity to do work or to make an effort. **Work** is using a force to act on an object in order to move that object in the same direction as the force. When you use your energy to lift a heavy backpack, you have done work. Work also involves useful motion. If you had tried to lift the backpack but failed, no work would have been accomplished. As work is done, one kind of energy is converted into another. The food (chemical) energy in the pizza you ate for lunch was turned into mechanical energy as your muscles lifted that backpack.

EVOLUTION OF
Energy & Power Technology

All other areas of technology rely on energy and power technology.

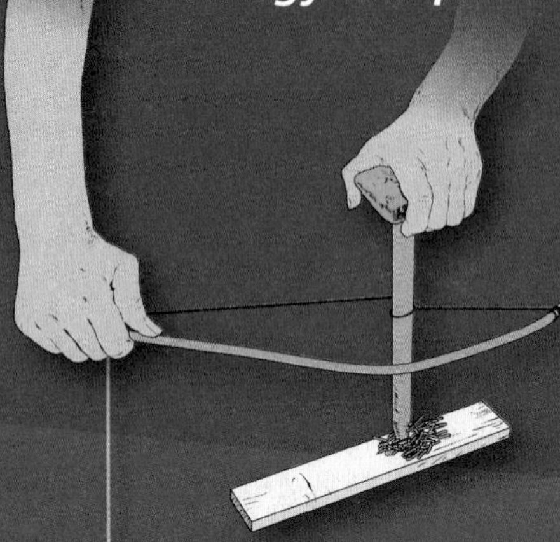

Making fire was one of the first great technological breakthroughs. One early device for making fire was the bow drill. Fire technology made other technologies possible. Fire supplied the heat energy for processing pottery (6500 BCE) and smelting metal (4000 BCE).

7000 BCE

200 BCE

Developed to supply power for grinding grain, waterwheels were later used to pump water, operate bellows for smelting metal, and power sawmills and textile mills. Waterwheels were the chief source of power for industry until the end of the 18th century.

late 1700s

Improved steam engines eventually replaced waterwheels as the main source of power for industry. The steam engines also greatly increased the demand for coal, which was used to heat the water to make steam.

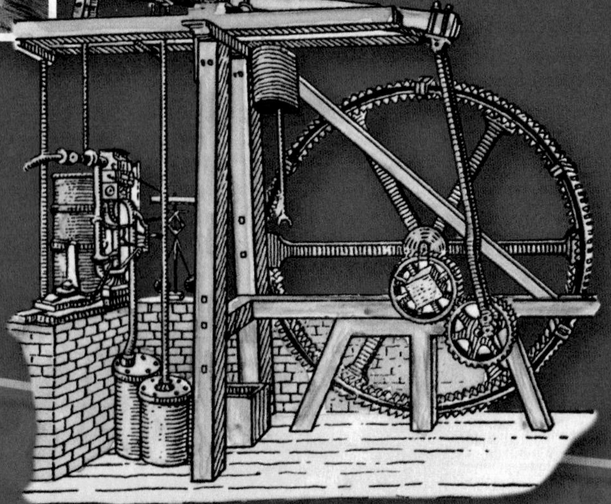

The first oil well in the United States was drilled in 1859 in Titusville, PA. Early petroleum refineries produced mostly kerosene, which replaced whale oil as a fuel for lamps. Gasoline was considered a waste product.

Improvements in generators made large-scale use of electric power practical for industrial, commercial, and residential applications.

THE WESTINGHOUSE ELECTRIC COMPANY

OF

PITTSBURGH,

Manufacturers of Isolated Incandescent Plants, and Contractors for Central Stations.

It is believed that the advantages of our System place us beyond competition.

Capital investing for dividends will do well to close no contracts until our proposals are considered.

Address
THE WESTINGHOUSE ELECTRIC CO.,
Pittsburgh, Pa.,
or
WESTINGHOUSE, CHURCH, KERR, & CO.,
17 Cortlandt Street,
New York.

16 Candle-power Lamp.

1850s

1870s

1880s

The world's first nuclear power plant went into operation at Obninsk, USSR, in 1954. The first U.S. nuclear plant opened in 1957 in Shippingport, PA.

1950s

2000+

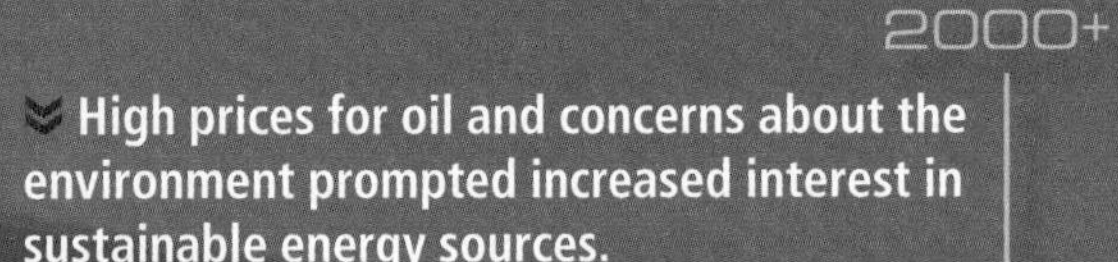

High prices for oil and concerns about the environment prompted increased interest in sustainable energy sources.

Although designs for an internal-combustion engine appeared as early as the 1600s, the first successful internal combustion engines were developed in the late 19th century. These engines spurred the production of automobiles, which in turn created a demand for gasoline.

Types of Energy

Energy can take many forms, but there are only two basic types: potential and kinetic. **Potential energy** is stored energy, or energy at rest. A piece of coal is just a black rock until its potential energy is tapped. A bowstring has only potential energy until it is released. See Fig. 8-2. **Kinetic energy** is energy in motion. Its power has been released. The heat radiating outward as the coal burns is kinetic energy. Ocean currents, wind, and electricity are other examples of kinetic energy.

Technology takes forms of potential energy, such as oil, natural gas, and water stored behind a dam, and turns them into kinetic energy. Whenever we use energy, it is as kinetic energy.

Forms of Energy

All energy can be classified into six basic forms: mechanical, thermal, radiant, chemical, electrical, and nuclear. See Fig. 8-3. Each form can be changed into all other forms.

Mechanical Energy is the energy of motion. It is usually visible. Every moving object has mechanical energy. It is the energy in a hammer that strikes a nail and the energy in your legs as you pump the pedals on a bike.

Thermal Energy is heat. Heat is given off by movement of the atoms and molecules in a substance. The faster they move, the more heat energy is produced. Heat is often the result of energy conversions.

Radiant (Light) Energy radiates outward in all directions. Visible light occurs as photons (tiny particles within atoms) escape into the surroundings. Other forms of radiant energy used in technology include X rays and gamma rays. Most light energy is produced by the sun.

» Fig. 8-2 The bowstring has potential energy. As the archer releases it, the potential energy changes into kinetic energy.

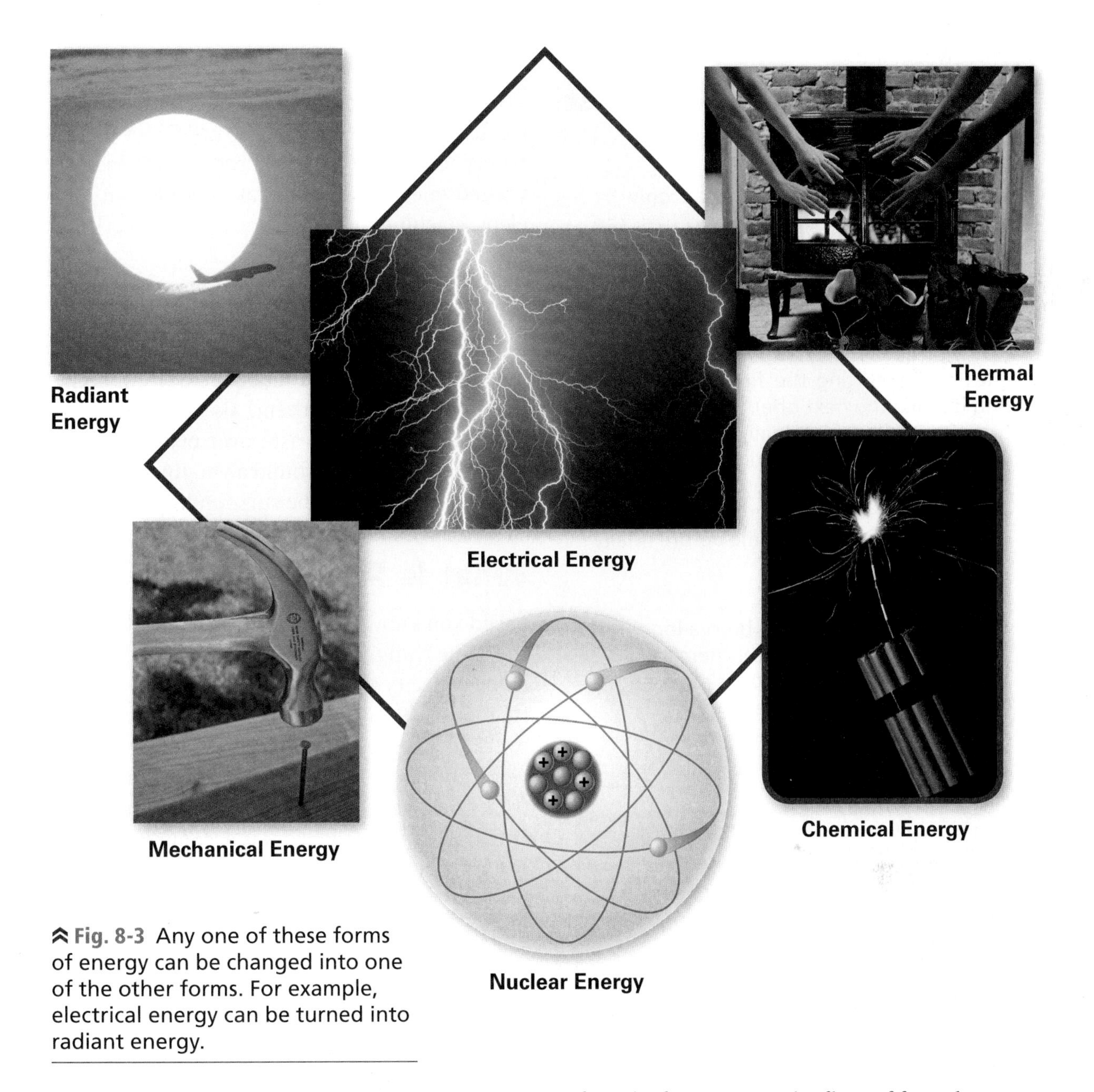

Fig. 8-3 Any one of these forms of energy can be changed into one of the other forms. For example, electrical energy can be turned into radiant energy.

Chemical Energy is released when the bonds between atoms in molecules are broken and the atoms rearranged. Chemical energy is stored in the foods you eat and released by body processes. Chemical energy is also found in fuels and explosives.

Electrical Energy is the flow of free electrons (tiny particles within atoms) from molecule to molecule within conductors. In nature it's found in a bolt of lightning.

Nuclear Energy is contained within the nucleus of atoms. It is released when atoms of certain substances, such as uranium, are split apart in a nuclear reaction. The most common use of nuclear energy is in power plants.

Science Application

Law of Conservation of Energy Did you know that energy can be neither created nor destroyed? Energy can only be changed from one form to another. The amount of energy in a system always remains the same.

Have you ever seen a toy called Newton's cradle? It demonstrates the energy conservation law. Once kinetic energy is put into motion (lifting the first ball and letting it strike the next one), it is transferred from ball to ball. Gradually, the energy is converted into sound and heat energy. Because little energy is lost as sound and heat, the balls at either end can continue back and forth for some time before finally stopping.

FIND OUT When you burn a log of wood in a fireplace, chemical energy from the wood is released. What two forms of energy is it converted to and where does that energy go?

Energy Efficiency

One of the factors to be considered when we use energy is efficiency. Efficiency is the ability to achieve a desired result with as little wasted energy and effort as possible. An efficient machine accomplishes a lot of work for the amount of energy used.

When technology systems use energy, some is always wasted. However, some systems are more efficient than others. Electrical motors, for example, waste only about 10 percent of the energy used to run them. Gasoline-powered automobile engines waste over 60 percent. In fact, it is impossible to build an engine that does not release heat into the surrounding area.

What Is Power?

Did you know that energy and power are not the same thing? You already know that energy is the capacity to do work and that work is useful motion. **Power** is the measure of work done over a certain period of time as energy is converted from one form to another or transferred from one place to another. Power helps us measure how quickly the work is accomplished.

For example, you can cross a lake in a motorboat or you can swim. In both cases, the work done—crossing the lake— is the same, but the motorboat gets you across faster. It supplies more power.

Power is commonly measured in horsepower (hp). This measurement is based on the amount of work a horse can accomplish in a certain period of time. If the motor on your boat puts out 6 hp, it produces as much power as six horses.

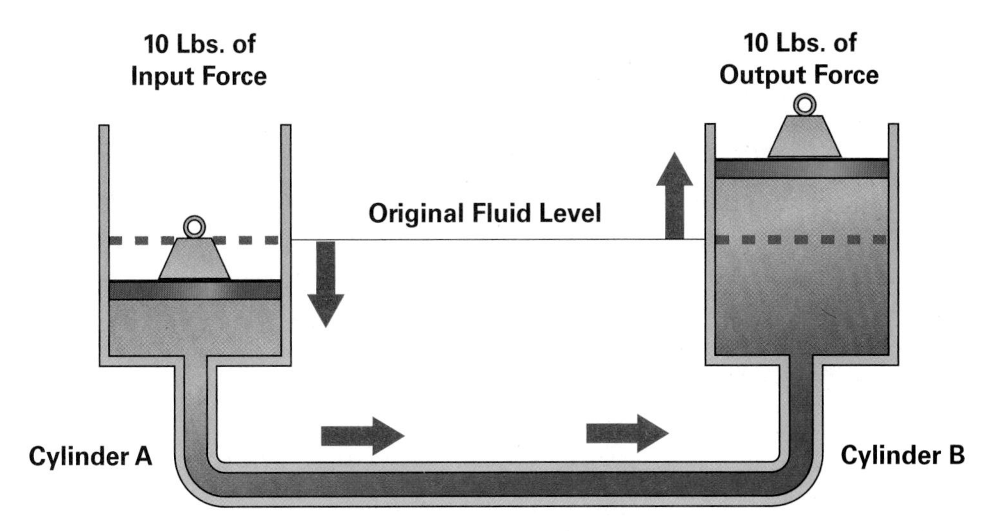

Fig. 8-4 Pressure applied to the liquid in cylinder A is transmitted to the liquid in cylinder B, causing it to rise.

Forms of Power

Three forms of power are commonly used in technology: mechanical, electrical, and fluid power.

Mechanical Power is a measure of the work done by means of mechanical energy over a certain period. You already know that mechanical energy is the energy of motion. When an engine moves the wheels of a car, it is producing mechanical energy (the motion of the turning wheels). Mechanical power is a measure of how much work the engine accomplished over time to get the car where it was going.

Electrical Power is a measure of the work done by electrical energy over a period of time. A 6-hp motor can push a boat across a lake faster than a 4-hp motor, so it is more powerful.

Fluid Power is produced by outside energy sources, such as a motor. The fluids transmit the energy. It is important to remember that fluids themselves are not the source of fluid power.

Pneumatic power is a measure of work produced using pressurized gases. Some automatic nailers are powered by pressurized gas. Hydraulic power is a measure of the work produced by putting liquids under pressure. It is often used to power heavy construction equipment. See **Fig. 8-4**.

Energy and Power Systems

Like all technologies, energy and power systems require the same seven resources as inputs—people, information, materials, tools and machines, energy, capital, and time. See **Fig. 8-5** on page 172. For example, suppose a car manufacturer wants to develop and market a new model of truck. Automotive engineers and researchers are needed to develop the design. Metal, plastic, and other materials are needed to make the truck. Energy is needed to run the machinery in the factory. Capital is needed to pay the workers.

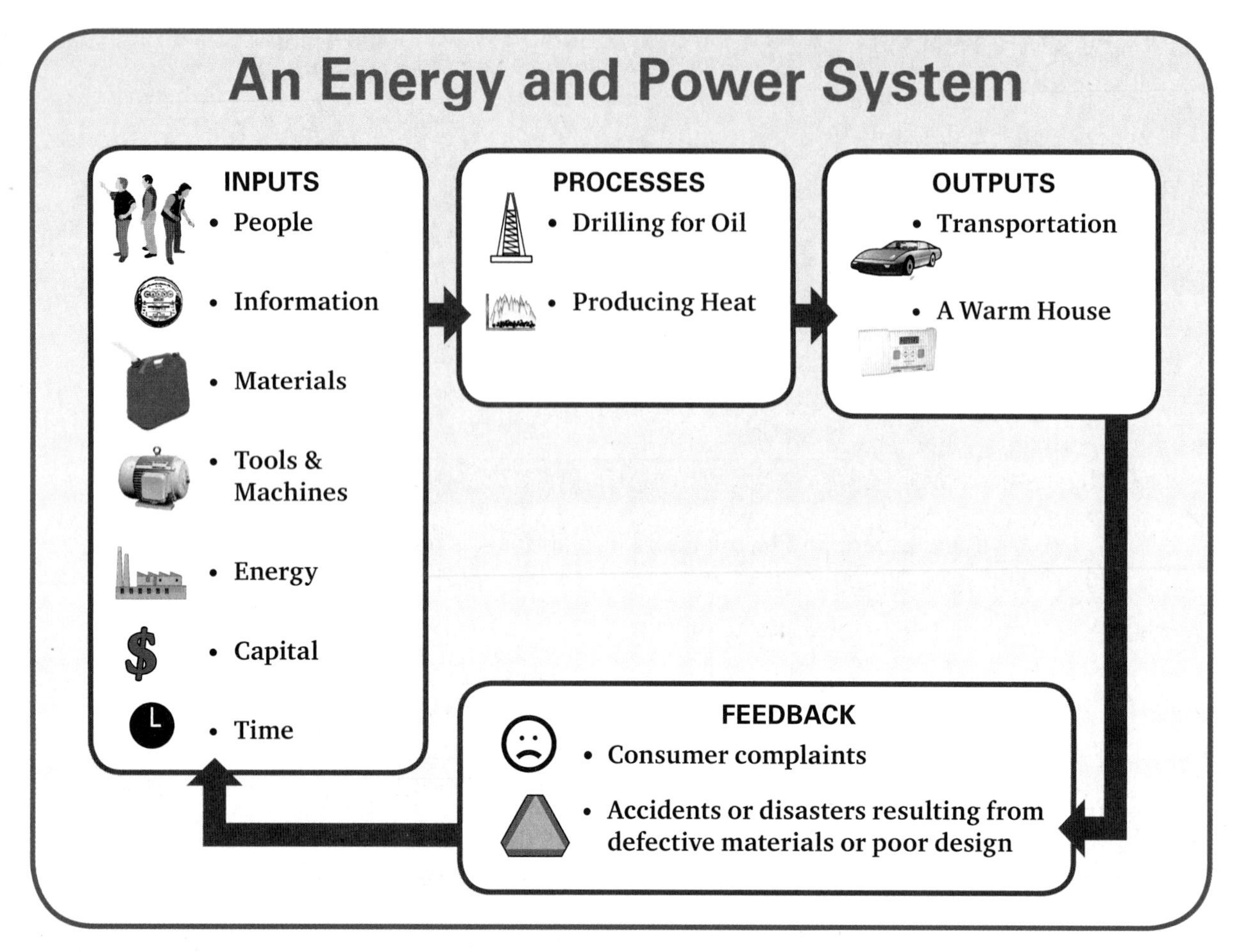

Fig. 8-5 Like all technology systems, energy and power systems have inputs, processes, outputs, and feedback.

Processes would include all the tasks as far back as creating the new truck design. Other processes, such as forming the metal, would take place inside the manufacturing plant. Still other processes would include distributing the new trucks to car dealers and selling them.

As you know, outputs are the results of the system. In our example, the finished trucks would be the primary desired output. However, not all outputs are positive, and some may be unexpected. Suppose plastic used for the bumpers did not perform well in safety tests. High damage costs might then be a negative output.

Feedback occurs when information about the outputs is put into the system. In our truck example, the company might be told about the weak bumpers after a certain number had failed during accidents. However, sometimes this information is not acquired until much later.

Controlling Energy and Power

Energy and power are used by all technology systems. Transportation uses energy to move people and goods. Industry uses it to power factories and manufacture goods. Residences and commercial buildings use energy for light, for heat and cooling, and to power machines.

In order to use energy effectively, we must control it. Early humans, for example, may have first learned what fire could do as they watched lightning strike a tree. The fire would not have been useful to them, however, until they learned to control the process, such as by using flint to start their own fires.

Before building a system to control energy and power, technologists must consider three main questions:

- What is the original source of the energy?
- How will it be changed and/or moved from one place to another?
- How will it be eventually used?

Suppose a community decides it needs to control and use electrical energy. The original source might be nuclear energy, which produces heat energy. The heat is used to boil water, which produces steam. The steam turns the blades of a turbine (mechanical energy). The turbine produces the electrical energy, which is sent to homes and office buildings.

Other factors that must be considered include where the energy source is located, the amount of energy that must be produced, and the length of time it must be controlled. These affect how the energy is stored and transported.

Storing Energy

Energy can be stored only as potential energy. Wood, for example, contains potential thermal and light energy and is easy to store. It is impossible to store the fire (kinetic energy) created as the wood is burning.

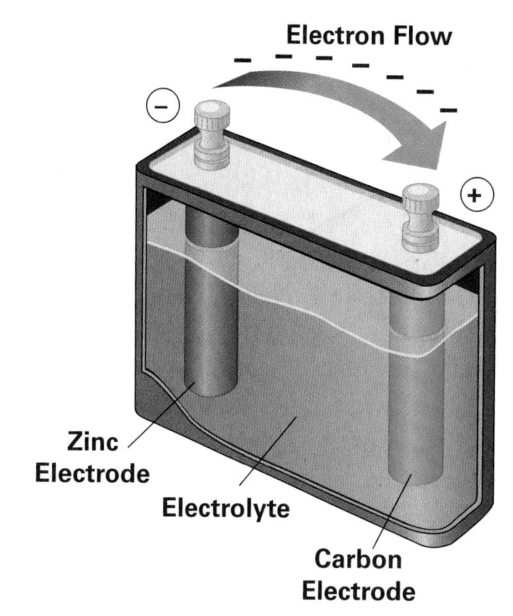

Fig. 8-6 Mixing the chemicals in this battery causes the release of free electrons. The electrons remain stored as potential energy until the battery is hooked up to a conductor.

The most common way of storing potential energy is in batteries. Batteries store potential chemical energy that, when released, produces electrical energy. See Fig. 8-6. Although battery storage is limited, the capacity is increasing through advanced battery technology.

Water is sometimes used to store potential mechanical energy. One method involves pumping water into a high reservoir, like a water tower tank. When the water is released, gravity pulls it down. As it moves, it produces kinetic mechanical energy and is used to power turbines. The water is then recaptured and pumped back into the tower. There it becomes potential energy to be released again and again.

Transporting Energy

Potential energy is often easy to transport, such as in the form of fuels. Gasoline, for example, is refined from oil and then stored in tanks. From the refinery, trucks transport the gasoline to gas stations where it is stored again. It is then pumped into the tanks of cars where it is again stored until it is burned in the car engine. During movement, the gasoline remains as potential energy.

We can also transport kinetic energy. As electricity travels along power lines long distances from generating plants, for example, it is in a kinetic form. The electrons are constantly moving around the wire. Mechanical energy can also be transferred over short distances by means of belts, gears, and shafts. Pneumatic and hydraulic systems can be used to transmit energy as well.

Impacts and Effects of Energy and Power Technology

As you know, all technology systems produce impacts on us and on our environment. Energy and power systems are no different. However, because all other systems use energy and power, those impacts are multiplied.

Environmental Impacts

Most of our energy for transportation and industry comes from burning fossil fuels like gasoline and coal. This causes air pollution, the presence of unwanted elements in the air. Examples include carbon monoxide, carbon dioxide, and sulfur dioxide.

Carbon monoxide is a colorless, odorless, poisonous gas. Breathing it reduces the ability of your lungs to carry oxygen. Automobile exhaust fumes mixed with fog create smog, which may lie in a yellowish layer over cities. See Fig. 8-7. When people breathe in smog contaminants over a long period, lung tissues can be damaged.

« Fig. 8-7 Fog and exhaust fumes combine to form smog, which is harmful to breathe.

Coal contains sulfur. When coal burns, it releases sulfur dioxide into the air. Combined with oxygen and water vapor, sulfur dioxide creates a weak sulfuric acid. When sulfuric acid mixes with nitric acid, another pollutant, it forms acid rain. Acid rain destroys trees and crops, kills fish, and damages buildings and monuments. Industries that burn coal can install special equipment that removes most of the sulfur, reducing the damage.

When too much carbon dioxide builds up in the air from burning fossil fuels, it can cause the greenhouse effect. The carbon dioxide prevents heat from escaping the atmosphere, which can raise temperatures. You have probably heard of global warming, a gradual warming of the earth's climate. Some researchers think that the greenhouse effect contributes to global warming.

Uranium is not a fossil fuel, but when it is used to release nuclear energy, the waste products are so dangerous they can cause serious health problems. Radioactive nuclear waste remains dangerous for a long time and is difficult to dispose of safely. Currently, it is stored in concrete containers. In the future it may be buried in a remote location. See **Fig. 8-8**.

Fig. 8-8 This heavily guarded site is used to store waste from nuclear plants. ***Where do you think sites such as this one would be located?***

Mining coal close to the surface can cause another type of pollution—areas of unattractive wasteland where no trees or grass grow. After the coal is removed, most surface mines are now covered with topsoil and replanted to restore the landscape.

Depletion of Resources

The supplies of some of our energy resources, like fossil fuels, are limited. As Earth's population grows, however, these resources are being rapidly used up. Also, as more and more countries are becoming technologically advanced, their need for energy increases.

Energy conservation is one way of extending our supply of resources. Conservation involves careful management and efficient use of energy supplies. People can "dial down" thermostats in the winter to save on energy used for heat in their homes and offices. Cars and trucks can be engineered for better fuel efficiency, and vehicles that use alternative fuels can be developed. Industry can practice conservation and switch to more efficient equipment.

Measuring Energy and Power

When planning a task, it is useful to know how much energy is needed. Will 10 gallons of gas in your car, for example, get you from Los Angeles to San Francisco? However, not all energy needs can be measured in terms of gallons as gasoline can. In order to measure energy, we use concepts like work, power, force, torque, pressure, and heat. Both customary and metric measurements may be used. See **Fig. 8-9**. (More metric conversions can be found in the STEM Handbook.)

Measuring Work

As you know, work is using a force to move an object. Work involves useful motion. This motion may involve the movement of electrons in an electrical wire or the soaring of a rocket through the atmosphere.

When measuring work, it is helpful to think of it as mechanical energy. Using the customary system, mechanical energy is measured in foot-pounds (ft.-lbs.). One foot-pound of work is equal to the lifting of 1 pound a distance of 1 foot. The formula reads as follows:

$$\text{work} = \text{weight (pounds)} \times \text{distance (feet)}$$

For example, how much work does a 150-pound man accomplish by climbing a 25-foot flight of stairs?

$$\text{work} = 150 \text{ lbs.} \times 25 \text{ ft.}$$
$$\text{work} = 3{,}750 \text{ ft.-lbs.}$$

Measuring Power

If you recall, power is a measure of the work done over a certain period of time. Power is commonly measured in units of horsepower. One horsepower is equal to the energy needed to lift 550 foot-pounds in 1 second. To calculate horsepower, the foot-pounds of work are divided by the time in seconds and multiplied by 550. The formula is written like this:

$$hp = \frac{\text{weight(lbs.)} \times \text{distance (ft.)}}{\text{seconds} \times 550}$$

Going back to our example of the 150-pound man climbing a 25-foot stairway, how many horsepower would he develop over 15 seconds?

$$hp = \frac{150 \text{ lbs.} \times 25 \text{ ft.}}{15 \text{ seconds} \times 550}$$
$$hp = 0.45$$

Fig. 8-9 Customary and Metric Units Used in Measuring Energy and Power

Measurement	Customary Unit	Metric Equivalent
Mechanical energy	1.000 foot-pound	1.356 joules
Power (mechanical)	1 horsepower	746 watts
Force	1.000 pound	4.448 newtons
Torque	1.000 pound-foot	1.356 newton-meters
Pressure	1.000 pound per sq. in.	6,895 pascals
Heat	1.000 Btu per second	1,054 joules
Electricity	1.000 watt	1.000 watt

Instead of horsepower, the watt (*W*) is used to measure electrical power. It equals one joule of electrical energy per second. A microwave oven, for example, may operate using 1,000 watts (1,000 joules) of power at the highest setting.

Measuring Force

Force is any push or pull on an object. Gravity is an example of a force. When you lift a book, you must exert a force on the book that exceeds the pull of gravity or the book will not be moved. The weight of an object is a measure of the force of gravity acting in a vertical direction. Force is also measured in terms of weight when an object is pushed or pulled horizontally, or in a back-and-forth direction.

Force can be used instead of weight to calculate the amount of work done. The following formula is used:

$$\text{work} = \text{force} \times \text{distance}$$

For example, suppose a man pushes a 200-pound barrel of nails a distance of 10 feet along a floor. He uses 55 pounds of force. How much work does he accomplish?

$$\text{work} = 55 \text{ lbs.} \times 10 \text{ ft.}$$
$$\text{work} = 550 \text{ ft.-lbs.}$$

Force can also be used instead of weight to calculate horsepower. This is the formula used:

$$hp = \frac{\text{force} \times \text{distance}}{\text{seconds} \times 550}$$

For example, how much horsepower does the man produce if it takes him 15 seconds to slide the barrel along the floor?

$$hp = \frac{55 \text{ lbs.} \times 10 \text{ ft.}}{15 \text{ seconds} \times 550}$$
$$hp = 0.067$$

In the metric system, force is measured in newtons (N) instead of pounds. Two other types of force used in measuring energy and power are torque and pressure.

Math Application

Horsepower in the Fast Lane As you now know, horsepower is the measure of work done over a certain amount of time. The term was first used by the Scottish engineer James Watt in describing the power of his steam engines. He compared his engines to the number of horses they could replace for pumping water out of coal mines.

The work performed by a car's engine can also be measured in horsepower. Have you ever watched a NASCAR race? Today's NASCAR racers can produce upward of 800 horsepower. Engineers create special engines for these race cars that allow them to produce the power needed for a typical race. After the engine is built, it runs on a dynamometer that measures the engine's power. Once the engine passes all tests, it's ready to be put into a car for a race.

FIND OUT NASCAR mandates that all cup cars must weigh about 3,400 pounds at the time of a race. If a race track is 10,560 feet (2 miles), and a car completes a lap in 86 seconds, how much horsepower is produced?

Measuring Torque

Torque is a turning or twisting force. When you open a jar of peanut butter or pedal the wheels of a bike, you are using torque.

Torque is applied with a circular motion, and measuring it involves using the radius (distance from the center) of the circle. Torque is calculated in pound-feet by multiplying the force used by the radius of the object being turned. See **Fig. 8-10**. (Note that the terms *pound-feet* and *foot-pounds* are not the same.)

$$\text{torque} = \text{force (lbs.)} \times \text{radius (ft.)}$$

For example, suppose a mechanic uses 30 lbs. of force to turn a wheel that has a radius of 1.5 ft. What is the torque?

$$\text{torque} = 30 \text{ lbs.} \times 1.5 \text{ ft.}$$
$$\text{torque} = 45 \text{ lb.-ft.}$$

In the metric system, torque is measured in newton-meters (N-m). That is the force in newtons multiplied by the radius in meters.

Measuring Pressure

Pressure is force spread out over a certain area. The area is measured in square inches, and its size (length × width) must be calculated first. The answer is given in pounds per square inch. This formula is used:

$$\text{pressure} = \text{force} \div \text{area}$$

For example, suppose you want to store a 60-pound weight on a shelf. Its base measures 10 inches by 8 inches. How much pressure does it exert on the shelf?

$$\text{area} = 10 \text{ in.} \times 8 \text{ in.}$$
$$\text{area} = 80 \text{ sq. in.}$$

$$\text{pressure} = 60 \text{ lbs.} \div 80 \text{ sq. in.}$$
$$\text{pressure} = 0.75 \text{ lb. per sq. in.}$$

This means that each square inch of the shelf must support 0.75 lb. Keep in mind that pressure is calculated as force per unit of area. The total amount of force depends on the total amount of area. See **Fig. 8-11**.

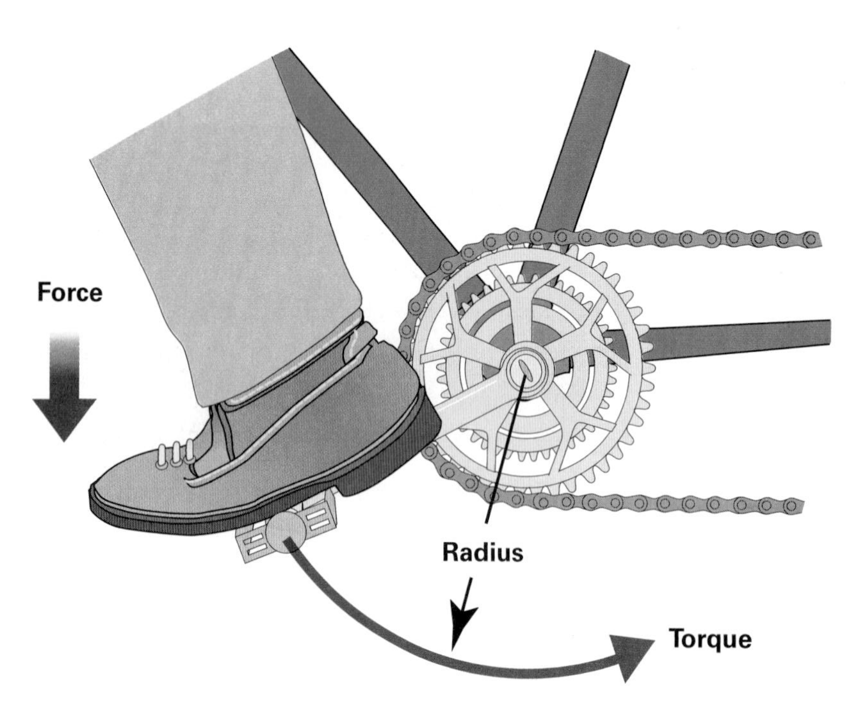

Fig. 8-10 Torque is calculated by multiplying the force by the radius.

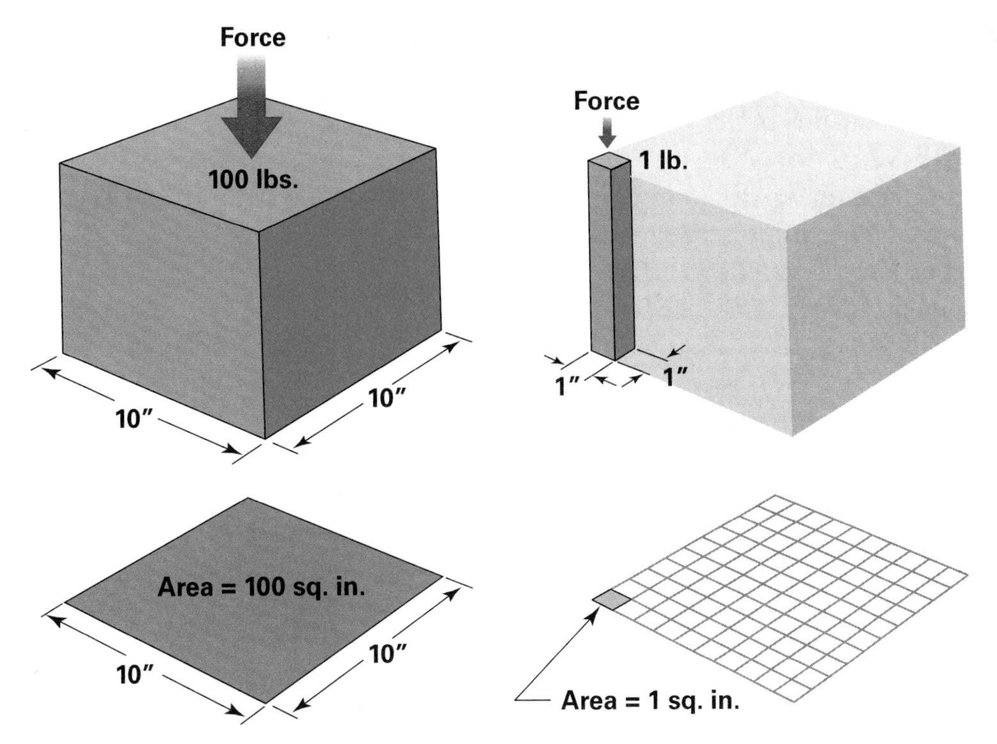

Fig. 8-11 The block shown here on the top left applies pressure to each square inch of supporting area.

In the metric system, pressure is measured in pascals, or the force of one newton acting on an area of one square meter.

Rather than such things as weights on shelves, units of pressure are commonly used to measure the force of liquids and gases, as in pneumatic and hydraulic systems. This is the formula used:

$$\text{force} = \text{pressure} \times \text{area}$$

For example, suppose you want to calculate the force of the air compressed inside a balloon. The air pressure is ¼ psi and the inside surface area of the balloon is 80 square inches.

$$\text{force} = \tfrac{1}{4}\ \text{psi} \times 80\ \text{sq. in.}$$
$$\text{force} = 20\ \text{lbs.}$$

Measuring Electricity

The basic unit of measurement for electricity is the coulomb. One coulomb is enough to light an average lightbulb for one second. Other measures are made of amperage, voltage, resistance, and wattage.

- Amperage (I) is the rate at which current flows through a conductor. The single unit is the ampere.
- Voltage (E) is the pressure that pushes current through a conductor. The single unit is the volt.

- Resistance (R) is the opposition to the flow of current. A single unit is the ohm. Resistance is affected by outside factors, such as temperature.

Ohm's law is a mathematical formula using amps, volts, and ohms. It states that it takes one volt to force one amp of current through a resistance of one ohm. See Fig. 8-12. The formula is written as:

$$\text{voltage } (E) = \text{amperage } (I) \times \text{resistance } (R)$$

If you have any two of the measurements, you can calculate the third. For example, if a circuit has a resistance of 20 ohms and a voltage of 120, what is the amperage?

$$I = E \div R$$
$$I = 120 \div 20$$
$$I = 6$$

Electrical power is measured in watts. Watts are calculated by multiplying amps times volts. The formula is written as:

$$P = I \times E$$

If a hair dryer produces 1,000 watts of power and the voltage is 120, how many amps flow through it?

$$I = P \div E$$
$$I = 1{,}000 \div 120$$
$$I = 8.3$$

Measuring Heat

Heat energy is measured in Btu's (British thermal units). One Btu is the heat needed to raise the temperature of one pound of water one degree Fahrenheit. Heat energy is often measured in Btu's per hour. In the metric system, it is measured in joules.

Heat energy can be changed into power with engines. However, this is not an efficient process. Much of the heat is lost into the air or absorbed into the metal of the engine. By learning more about energy now, we can develop more efficient ways to use energy in the future.

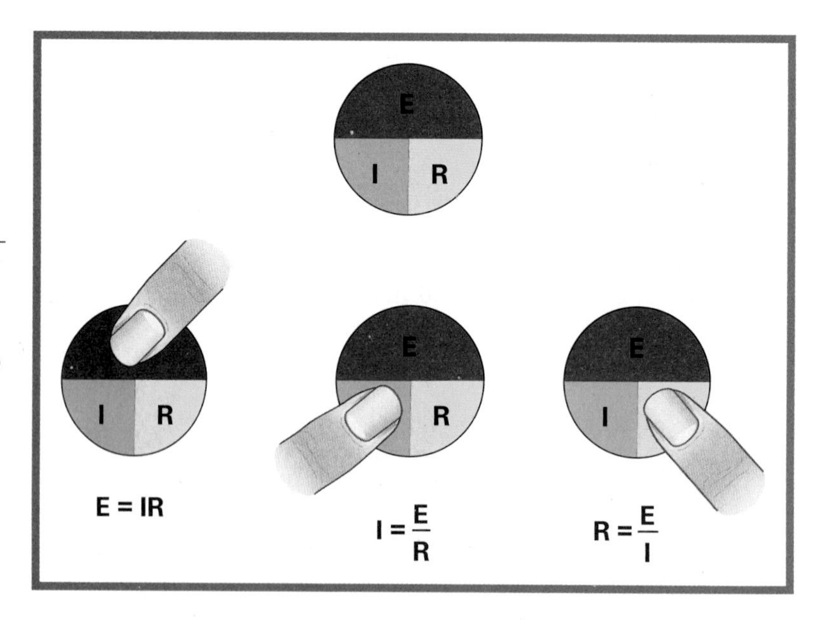

» Fig. 8-12 To use this Ohm's law diagram, put your finger over the value you want to find. The two remaining variables will give you the correct equation to use.

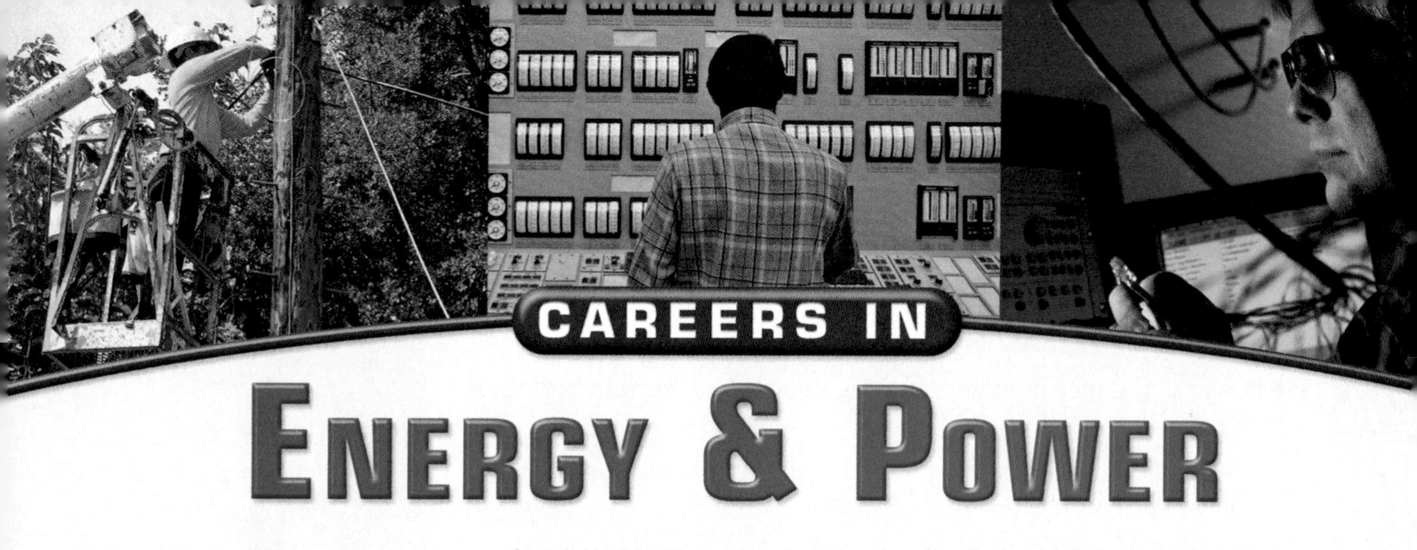

CAREERS IN ENERGY & POWER

Energy and power technology includes careers in many areas. There are energy and power jobs for the "hands-on" type of person (electrical power line installer and repairer, for example), high-tech types (computer specialist), science-minded people (geoscientist), and people persons (customer service representative). The U.S. Department of Labor estimates that there will be about three-quarters of a million new jobs in the energy and power industry by the year 2012.

Electrical Engineer • Nuclear Engineer • Geoscientist • Geophysicist Biological Scientist • Petroleum Geologist • Computer Specialist Customer Service Representative • General Office Clerk • Accountant Auditing Clerk • Electrical Power Line Installer and Repairer Industrial Machinery Mechanic • Bookkeeper • Gas Plant Operator Electrical and Electronic Engineering Technician • Mineralogist Power Plant Operator • Power Distributors and Dispatchers Nuclear Technician • Water and Liquid Waste Treatment Plant Operator Utilities Meter Reader

INVESTIGATE CAREERS Interview someone about his or her job in the field of energy and power technology. Ask questions related to job duties, education, years of experience, and opportunities for advancement.

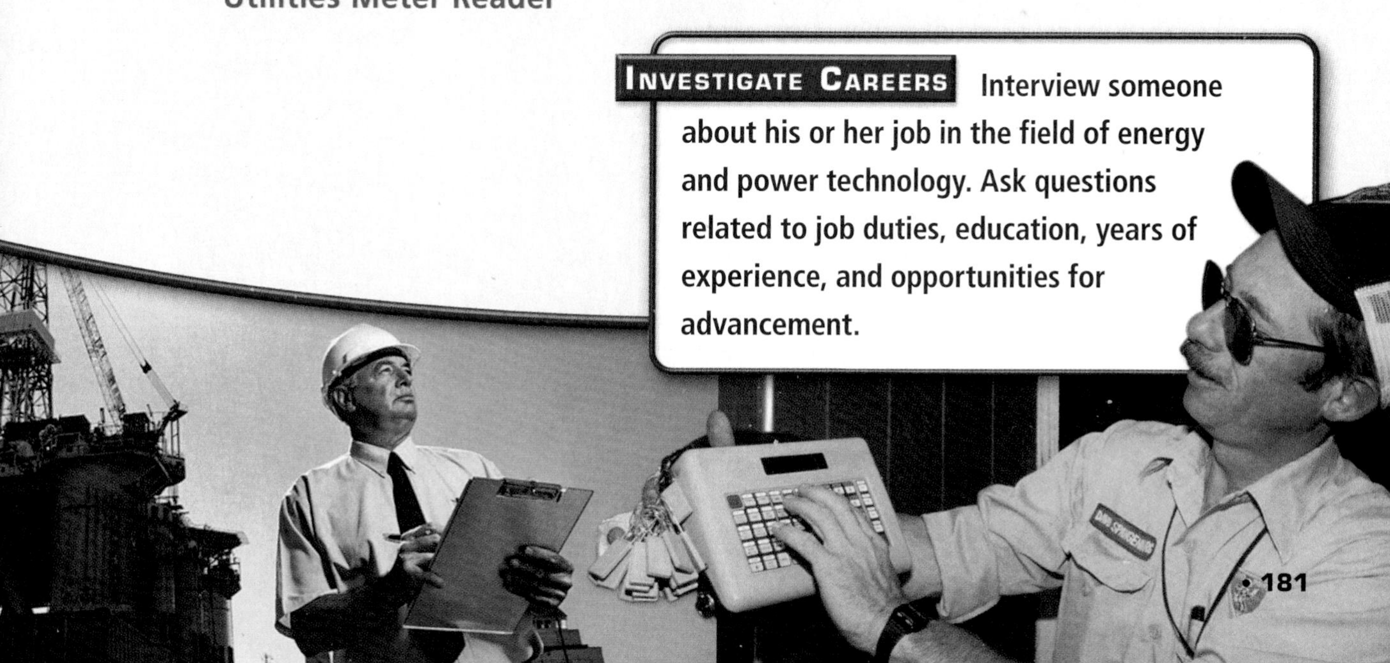

CHAPTER 8 REVIEW

Main Ideas

- Energy is the capacity to do work or make an effort.
- Power is the measurement of work done over a certain period of time.
- The impacts of energy and power are significant because all technology systems use energy and power.
- When planning any task, it is useful to know how much energy and power is needed.

Understanding Concepts

1. What are the six forms of energy?
2. Identify the three forms of power.
3. Why must energy be controlled?
4. Name two major impacts of energy and power technology.
5. How do you measure the amount of work done?

Thinking Critically

1. We gain energy from food we eat. How do we measure this energy?
2. Why do you think that air conditioners are rated in Btu's, a unit of heat measurement?
3. Is turning a light on always an example of electrical energy? Explain.
4. Does it take less force to drive nails in with a pneumatic nailer than with a person swinging a hammer? Explain your answer.
5. What two forms of energy do you use most in a typical day?

Applying Concepts & Solving Problems

1. **Social Studies** Before the 1950s, many homes did not have air conditioning. Where did people go to cool off?
2. **Design** Create a device that uses one of the six forms of energy.
3. **Formulate** Ask your parents for copies of their energy bills for up to one year. Compare how much energy was used each month. Why was more energy used some months compared to others?

Activity: Measure Energy and Power

Design Brief

In this chapter you have learned the formulas for measuring energy and power. Engineers use such formulas during the design process to make products that meet certain standards. Later, they may use them again to test product performance.

Select at least three measures of energy and/or power discussed in this chapter. Design an experiment for each one in which you use the formulas provided. Then create a display or poster explaining one of the processes you used. For example, you might measure the work accomplished by five different team members who must carry a backpack filled with books up stairs. You might then create a poster illustrating how you worked out the answers.

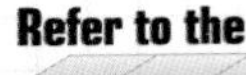

Criteria

- Design experiments to measure at least three examples of energy or power from the following: work, power, force, torque, pressure, heat, and electricity.
- Show your calculations using the different formulas on paper and turn them in at the end of the activity.
- Your display may be designed for a tabletop or it may be created using some kind of computer software, such as animation software.
- You should vary your experiments from the problems described in the text. Be creative.

Constraints

- If you design a tabletop display, it must measure no more than 12 inches by 18 inches.
- Your calculations must be accurate and done correctly.

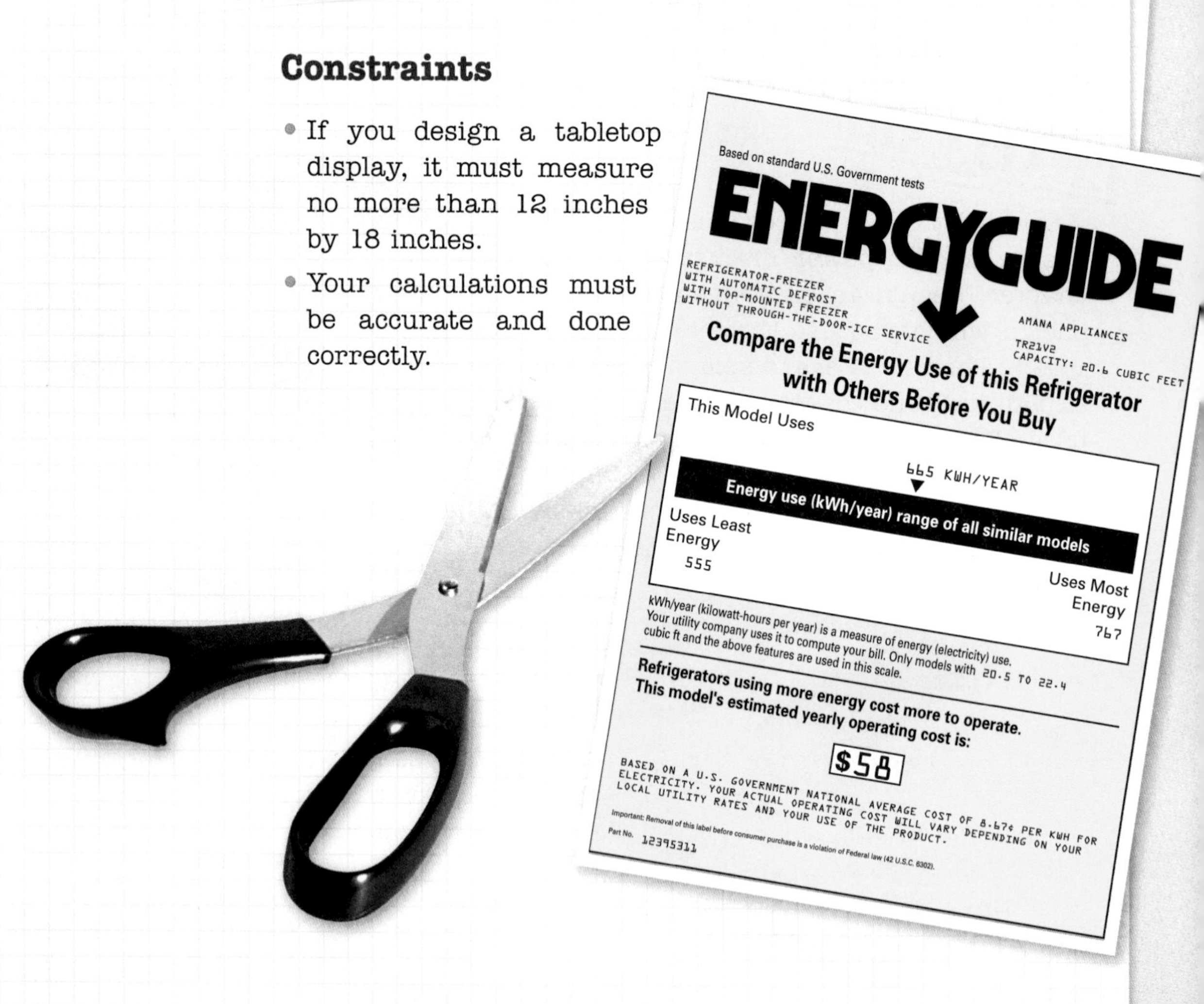

Engineering Design Process

1. **Define the problem.** For the measures you have selected, write a statement that describes the problem you are going to solve. For example, how much force would a person have to use to move a 100-lb. bag of cement 30 ft. across a floor?
2. **Brainstorm, research, and generate ideas.** With your team, discuss possible solutions. Hint: You may want to review the formulas and how they are worked to be sure you understand them.
3. **Identify criteria and specify constraints.** These are listed on page 184.
4. **Develop and propose designs and choose among alternative solutions.** Choose the best design that will solve your problems. For example, avoid designs that require materials that are hard to acquire or tasks that are too difficult for a high school student to accomplish.
5. **Implement the proposed solution.** Choose the processes you will use for your experiments. Gather any needed tools or materials. Then carry out the experiments.
6. **Make a model or prototype.** Create a rough design for your display or poster.
7. **Evaluate the solution and its consequences.** With your teammates, check your calculations to be sure they're accurate. Ask classmates to view your design to see if it explains the measuring process clearly.
8. **Refine the design.** Based on your evaluation, change the design if needed.
9. **Create the final design.** After you have an approved design, create your final display or poster.
10. **Communicate the processes and results.** Present your finished display to the class. Be prepared to answer questions. Turn in your assignment to your teacher. Be sure to include the name of the activity, your definition of the problem, a description of how you solved the problem, and your calculations.

SAFETY FIRST

Before You Begin. Make sure you understand how to use any tools and materials safely. Have your teacher demonstrate their proper use. Follow all safety rules.

CHAPTER 9 ENERGY SOURCES & CONVERSIONS

≈ These solar collectors are used to harness energy from the sun. In this chapter's activity, "Convert Energy," you will have a chance to design a device that performs a simple task by using different forms of energy conversion.

Objectives

- Identify and discuss exhaustible, renewable, and inexhaustible energy sources.
- Describe the methods of conversion for several energy sources.
- Discuss the negative and positive effects of several energy sources.

Vocabulary

- fossil fuel
- nuclear fission
- bioconversion
- photovoltaic cell
- geothermal energy
- electrolysis

Reading Focus

1. As you read Chapter 9, create an outline in your notebook using the colored headings.
2. Write a question under each heading that you can use to guide your reading.
3. Answer the question under each heading as you read the chapter. Record your answers.
4. Ask your teacher to help with answers you could not find in this chapter.

Types of Energy Sources

The sun is the fundamental source of almost all the earth's other sources of energy. Plants absorb the sun's energy and use it to grow. We obtain that energy when we eat plant and animal foods. Over many ages, the remains of plants and animals are turned into coal and oil. Warmth from the sun creates wind, another source of energy. The sun's energy may also be used directly to power such devices as solar cells or batteries.

All sources of energy can be classified into three main types: exhaustible, renewable, and inexhaustible.

Exhaustible Sources of Energy

Exhaustible energy sources are those that cannot be replaced. They include uranium and fossil fuels, such as coal, oil, and natural gas.

Coal

Coal is a black or brown rock that can be burned. It is obtained through both underground and surface (strip) mining operations. See Fig. 9-1. Once washed of dust, coal can be used immediately as a fuel.

Coal is our most abundant fossil fuel. **Fossil fuels** come from plants and animals that died millions of years ago. Covered by sediments and subjected to pressure and heat, they eventually formed coal, oil, and natural gas deposits.

Coal is generally found in two forms: soft and hard. Soft coal contains more sulfur, a known pollutant, but is usually found closer to the earth's surface and is cheaper and safer to obtain. Hard coal is less polluting and produces more Btu's. However, it is much more dangerous and costly to mine as it is generally found in veins deep within the earth.

« Fig. 9-1 Miners use powerful machines to extract coal underground. The work is very dirty and loud.

At our present rate of use, it is estimated that we have enough untapped coal reserves to last 500 or more years. As research continues, coal's major problems may be solved and we will use the reserves faster. For example, gasoline and diesel fuel can be made from coal using heat and pressure.

Oil

During the last 60 years, oil and natural gas have replaced coal for home and transportation use. Oil and gas are more easily stored and moved and are less polluting than coal. Oil, in its refined state, is the most common energy source for transportation. Fuel oil is used to heat our houses. Oil is also used to make plastics. These uses have made oil the world's most important fuel.

There are three sources of oil: crude oil, shale oil, and tar sands.

- Crude oil is a thick, dark liquid trapped in pockets in the earth. Workers drill into the oil pockets—some over four miles below the earth's crust—and pump the oil out. See **Fig. 9-2**.
- Shale oil is oil trapped inside shale rock. Over three-quarters of the oil reserves in the United States are in the form of shale oil. The shale must be mined and the rocks crushed and heated to extract the oil, a time-consuming and costly process.
- Tar sands are the third source of oil, which is trapped within the sand. Heat is used to extract the liquid oil.

Fig. 9-2 Drilling rigs such as these are used to bring crude oil to the surface. Some rigs resemble small communities where workers stay for weeks at a time.

Fig. 9-3 Refineries convert crude oil into useful products such as gasoline, diesel fuel, and motor oil.

After the oil is obtained, it must be refined and separated into useful substances. See **Fig. 9-3**. During refining, the oil is heated until it changes into a gas. As the vapors rise, they condense into different liquids. Heavy oil products, such as diesel fuel, do not rise very high and are drained off. Lighter oil products, such as gasoline and kerosene, rise higher. This process is called fractioning. The oil that remains after fractioning is further refined by cracking. Cracking breaks down the oil into smaller gasoline molecules.

Unfortunately, oil has two major problems as an energy source. One of these is limited supply, especially of crude oil. Our reserves of oil are very low.

The other major problem with oil is that it creates pollution. Crude oil spills damage shorelines and kill fish and other wildlife. Burning oil in any form creates air pollution. Technology has reduced these problems, but the cost makes oil an increasingly expensive energy source.

Math Application

Calculating Miles per Gallon Oil is a valued commodity. When the price per barrel of oil increases, the price per gallon of gasoline also increases. You or your parents may be directly affected by elevated gas prices. These prices have changed how the consumer chooses a vehicle. While SUVs are better for larger families and for storage, they cost more to fuel.

Engineers are currently designing vehicles that are more fuel efficient. You may see labels on new vehicles that say "30 mpg." This means that the vehicle can travel 30 miles by the time it uses up one gallon of gasoline.

FIND OUT Today's consumers are considering how many miles they can travel before refueling. Knowing these amounts can help when deciding on the appropriate vehicle to purchase.

1. If a car can travel 35 miles per gallon, and it has a 15-gallon tank, how many miles can the car travel before running out of gas in ideal conditions?
2. If the same driver travels 238 miles every week, approximately how often does he or she need to get gas?

CITY MPG 30 — Fuel Economy Information — DOE EPA — HIGHWAY MPG 40

« Fig. 9-4 Large tankers such as this one can carry enormous amounts of liquified natural gas.

Natural Gas

Natural gas is most commonly found in underground pools or with crude oil deposits and is obtained by drilling. Not all deposits contain the same type of gas. The most common gas is methane. Others include ethane, propane, and butane. Each has a different Btu rating, and propane is the hottest. The mixture of the natural gases used as fuel is generally based upon availability and cost.

Natural gas is transported or stored under pressure. See Fig. 9-4. The pressure is increased to a high point, and then the gas is cooled to a liquid state. Although pressurized containers of gas can be explosive, design and careful handling of containers makes them safe.

Did you know natural gas is our cleanest fuel? It creates fewer pollutants. It has also been the cheapest fossil fuel for many years. However, as reserves have decreased in size and as the cost of locating and processing natural gas has increased, the cost has risen. Not enough natural gas exists in our current reserves to meet demands. Fortunately, researchers have uncovered other sources.

Geopressure reserves can be found in high-pressure brine (salt water) pools deep within the earth. When the pools are tapped, the gas is released. Today, this process is very expensive, but as methods of separating the gas from the brine are improved, these new reserves will increase our supplies.

Tight sand reserves consist of natural gas trapped in a type of hard, dense sandstone found in the Rocky Mountain region. The sandstone is broken apart by injecting high pressure fluid into it. Natural gas is released as the stone breaks apart. The gas must be processed to remove impurities such as dirt, water, and sulfur. Specialty gases are drawn off to be used as is or later mixed to create higher Btu gas.

Uranium

Uranium is a radioactive substance found in many metal ores and is used to produce nuclear energy. Nuclear energy is most often used to produce electrical energy. Although the United States' supply of uranium is limited, there is enough to last many years, based on our current numbers of only about 100 nuclear power plants. See Fig. 9-5.

Mined from the earth, uranium is found in two basic types, or isotopes. One of these isotopes, uranium-235 (U-235), is rare and is used as the basic fuel in nuclear reactors. The other isotope, uranium-238 (U-238), is less valuable as an energy source. Although generally found together, the two isotopes are separated during the processing operation.

Nuclear Fission Nuclear power is generated in a nuclear power plant during controlled nuclear fission. **Nuclear fission** is the splitting apart of the nucleus of an atom. See Fig. 9-6. The atoms are bombarded with free neutrons (small particles that have escaped the nucleus of other atoms). When an atom splits, it forms two or more smaller nuclei, more free neutrons, and gamma rays. It also releases lots of energy. Under the right conditions, the new free neutrons strike other uranium atoms and a chain reaction takes place. If the reaction is maintained, it produces a steady flow of energy.

Fig. 9-5 Nuclear power plants convert nuclear energy into electrical energy.

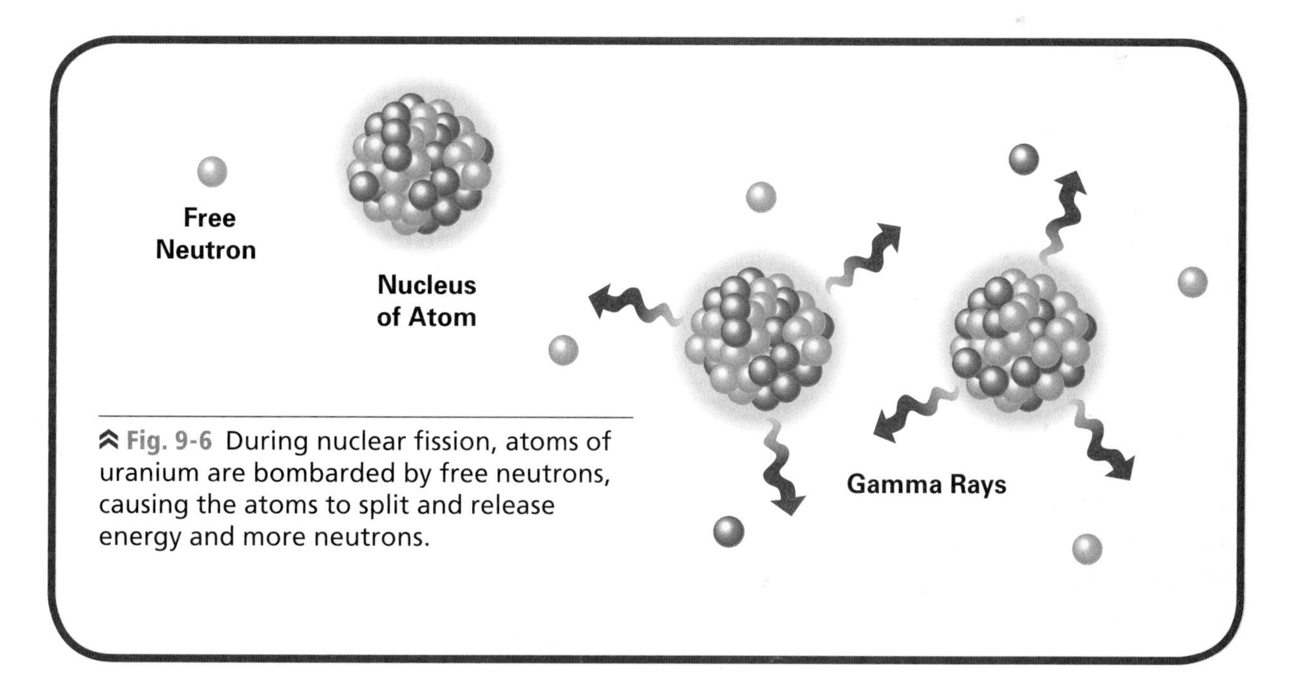

Fig. 9-6 During nuclear fission, atoms of uranium are bombarded by free neutrons, causing the atoms to split and release energy and more neutrons.

Young Innovators

Competing at the INTEL ISEF

The Intel International Science and Engineering Fair (ISEF) is a gathering of exceptional talent in science, mathematics, and engineering from around the world. The competition is an annual event, and each May almost 1,500 students from 40 nations come together to share ideas, show cutting-edge science projects, and compete for over $4 million in awards and scholarships.

Students must first compete in an Intel ISEF-affiliated science fair at the local, regional, state, or national level. After competition, the fair may name up to two individual project finalists and one team project to compete in the Intel ISEF. Students are encouraged to submit their own, original research projects for exhibition at their local, regional, or national fair.

Students who make it all the way to the Intel ISEF will have the opportunity to discuss their research and its contributions to scientific knowledge and our understanding of the world. They also have the chance to discover the different types of research in which their peers are engaged and perhaps even collaborate on future projects. Past projects have included the design and construction of a magnetohydrodynamic (MHD) generator using a hybrid rocket motor, experimentation with bacterial fuel cells, and methane production from waste material as a source of renewable energy.

The Intel International Science and Engineering Fair is a competition for students with an interest and dedication to excellence in science, mathematics, and engineering. Have you been tinkering with a science project lately? Have you discovered new and exciting information that could enhance our understanding of the world in which we live? Perhaps you have a novel concept for a new invention or innovation. If you have a contribution that you would like to make to the world of science or technology, then the Intel International Science and Engineering Fair may be just the incentive you have been looking for to become involved.

During the fission reaction, heat energy is released. In the most common type of nuclear plant, a pressurized-water reactor, the core water is heated to about 580 degrees Fahrenheit. It is kept under pressure so it will not boil. The water is then piped through a steam generator where it heats a second supply of water, which turns to steam. Pressure from this steam turns the blades of a turbine. The turbine turns a generator that produces electricity.

Pressurized-water reactors are a type of light-water reactor. The water is used to help transfer the heat produced. Experimental models of another type of reactor called a breeder reactor have been built. This type uses U-238 and plutonium-239 as fuel. During fission, more plutonium is created, increasing the supply of fuel. Liquid sodium instead of water is used to absorb heat. Breeder reactors are more difficult to operate and more costly to build.

Exposure to radiation from a nuclear reaction can damage human cells. Special precautions must be taken to prevent radiation leaks. Although nuclear power plants built in the U.S. have proved very safe over the years, some people still have safety concerns. Disposal of radioactive nuclear wastes, substances formed during fission, is another problem. Currently, wastes are stored at the power plant for future disposal.

Fig. 9-7 Corn is one source for ethanol.

Renewable Energy Sources

Renewable sources of energy are those that can be used indefinitely if they are properly managed and maintained. Wood, plants, and waste products are all renewable sources. Most renewable energy is processed from these easily obtained sources. As the supply and cost of fossil fuels change, renewable sources of energy will grow in importance. The most promising renewable sources include ethanol, methanol, biodiesel fuel, waste products, and wood.

Ethanol

Ethanol is ethyl alcohol, a compound derived from plants. A mixture of nine-tenths unleaded gasoline and one-tenth ethanol can be used as an alternative fuel in automobiles and trucks. Called E10, it saves about 10 percent of the oil needed to produce pure gasoline. Some new "flex-fuel" cars and trucks are able to run on E85, a mixture containing 85 percent ethanol, or even E95 (95 percent ethanol). Ethanol mixtures produce less pollution than pure gasoline.

The sources of ethanol include sweet sorghum, sugar beets, and grains, such as corn. See Fig. 9-7. Through distilling, ethanol can be produced in large quantities at a reasonable cost. During distilling, the source material is heated and the vapors cooled and collected as a liquid.

Methanol

Methanol, also called methyl alcohol, is a clean-burning liquid fuel that can be used in vehicles. It can be made from natural gas and coal. Renewable methanol is made from wood, plants, or waste.

Methanol produces more Btu's than ethanol, so it does not need to be mixed with gasoline. It can be used as an alternative fuel in flex-fuel vehicles.

Using methanol as a transportation fuel produces much less pollution, less dependence on crude oil, and a more stable fuel cost. Its disadvantage of slower engine starting has been overcome with advances in engine design.

Biodiesel Fuel

Biodiesel fuel is produced from vegetable oil. The oil is heated to a temperature of 130 degrees, filtered, and mixed with additives. Many people produce their own biodiesel from cooking oil waste. Commercially, biodiesel is made from soybeans and other oil producing bean plants.

Biodiesel fuel is cleaner burning than regular petroleum-based diesel. Exhaust fumes contain fewer pollutants, and the smell of the fuel is more pleasant. When mixed properly, the power level of the fuel is as high as that of regular diesel.

However, because it is such a clean fuel, biodiesel loosens dirt and rust from surfaces, which can sometimes clog vehicle fuel systems. Also, straight vegetable oil fuels become very thick at temperatures below 45 degrees and require a heated fuel system.

Waste Products

Bioconversion is a method of obtaining energy from waste products. These include food product wastes, animal wastes, paper, cardboard, and wood wastes. All of these can be either burned or converted into fuels such as alcohol, oil, and methane gas. One process, called thermal conversion, or thermal depolymerization, produces a high-quality fuel oil from waste products such as turkey parts and plastic bottles. See Fig. 9-8.

Bioconversion disposes of unwanted waste, and production facilities can be located almost anywhere. The resulting fuels are low in pollutants. Using waste as fuel also helps limit the waste that must be disposed of in landfills.

Wood

Wood and wood waste are two of the oldest forms of heat energy. People have long used wood for cooking and heating. In pioneer days, wood was a major source of energy. Today, millions of homes are again heated by burning wood. Some cities and businesses burn wood and wood waste, together with other forms of waste, to provide electricity.

As an energy source, wood has one major disadvantage. It is not clean burning and creates air pollution.

Researchers are studying other uses for wood. For example, wood can be converted to liquid and gaseous fuels, such as methanol.

Although wood is a renewable energy source, trees take a long time to grow. Its use as a fuel competes with its use in the construction industry. Other industries competing for the existing supply of wood include paper production and furniture manufacturing. For wood to become an important energy source, the number of trees now grown on tree farms and in managed forests will need to quickly increase.

» Fig. 9-8 These used plastic bottles are compacted into bales. They are then shredded and eventually used to make new bottles.

Inexhaustible Energy Sources

Inexhaustible energy sources are those that will always be available. It does not matter how much of them we use. They include the sun, flowing water, wind, tides, ocean heat, salt ponds, earth heat, and hydrogen.

The Sun

Solar energy is energy directly from the sun. We can collect it in the form of heat and light and put it to work.

If all the energy arriving from the sun each day were collected and controlled, we would have all the energy we need. Unfortunately, our devices for collecting and controlling the sun's energy in large quantities are inefficient and expensive. Also, the actual amount of time each day when we can collect the sun's energy is limited, especially in winter months and on cloudy days. In spite of recent advances, the technology for storing solar energy is it still not efficient. Today, we use the sun's energy directly for thermal (heat) energy and for electricity.

Thermal Energy As you know, thermal energy, or heat, is produced by the movement of molecules. The transfer of heat through a solid substance is called conduction. The closer the molecules are packed together, the faster the heat is transferred. When heat is transferred through a fluid, such as air or water, the process is called convection. Like warm air rising toward the ceiling in a room, the warmest fluid molecules always move toward those that are cooler. You have probably felt heat coming from the sun when sunlight touched your skin. This is heat transfer by radiation, or electromagnetic waves. The waves do not heat the atmosphere as they travel through it. The heat is transferred to comparatively dark surfaces. Hot water can be produced by collecting the sun's thermal energy through solar panels. The hot water is then used to heat homes, businesses, and factories. Passive solar heating systems use water or other materials to collect the heat and move it naturally through the building. Active solar heating systems collect and store the heat. Another system moves it through the building.

In larger systems, with the use of mirrors, the water can be heated to steam and used to operate electricity-generating turbines. Back-up systems are needed to supply heat on cloudy days.

≈ Fig. 9-9 Each photovoltaic cell on this panel collects solar energy and converts it into electricity.

Electricity Devices that convert sunlight directly into electricity are **photovoltaic cells.** Light energy travels as tiny energy particles called photons. When photons strike the materials contained in a photovoltaic cell, they release electrons, particles within the atoms of the materials. The flow of these free electrons is electricity. See **Fig. 9-9**.

When coupled with a battery storage system, photovoltaic cell technology is more efficient than solar panels. The cost of producing photovoltaic cells is decreasing. Many hand-held calculators contain them. They are also used to power traffic signals and signs, to send river flow information, and to power communication equipment. In many parts of the world where there are no large power plants, portable computers, televisions, and lights operate on solar cells.

Flowing Water

We gather the potential energy of water by trapping it behind dams. See **Fig. 9-10**. Much of this water is used to produce electrical power in a hydroelectric power plant.

As water drains from the reservoir created by the dam, it flows through a sluice (a special waterway tunnel). Rotated by this directed flow of water, a turbine near the end of the

≈ Fig. 9-10 The Hoover Dam, located in the Black Canyon of the Colorado River, holds back a body of water 115 miles long. Near the dam, the water is 589 feet deep.

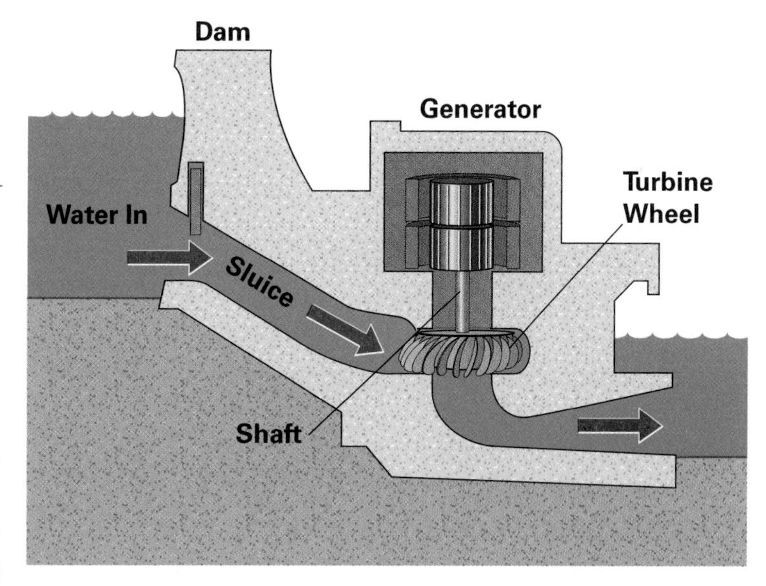

» **Fig. 9-11** Flowing water drives the blades of the water turbine wheel. The turbine rotates the generator shaft, producing electricity.

sluice drives a generator that produces electricity. See **Fig. 9-11**.

About 20 percent of all electricity produced in the United States is from hydroelectric power. While hydroelectric power is clean, power plants are costly to build. Not all of our rivers can be dammed. Some are needed for water transportation routes. Also, dams require that large areas of land be abandoned for reservoirs. The natural environment is altered and wildlife may be destroyed.

Wind

Wind energy was a very important energy source before the rural parts of the United States were developed. Where the wind blew much of the time, people built windmills. They used this direct mechanical energy to pump water, grind grain, and perform other useful tasks. When the wind did not blow, however, the work did not get done.

Today windmills have been engineered to produce electricity. The wind turns a specially designed propeller that is connected to an electrical generator. See **Fig. 9-12**. The electricity is stored in banks of batteries. Unfortunately, large amounts of electricity cannot be stored economically, and there is a limit to the economic use of windmills. However, large farms of wind-powered electrical generators are being installed near cities and industries as a backup source of electricity.

Fig. 9-12 Windmills are being used as backup sources of electricity in some cities. ***How might geography play a role in which cities have windmills near them?***

Ocean Tides, Waves, and Currents

Much is unknown about energy sources that lie below the surface of oceans. However, some, such as tides and currents, have proven to be useful.

Tides are the rising and lowering of water levels caused by the moon's gravitational pull and the rotation of the earth. High tide is when the water has risen to its highest level. Low tide is the opposite. Changes leading to high and low tide last about six hours each.

In some parts of the world the difference between high and low tide is over 40 feet, but differences much less than that are more common. Strong currents develop as the water moves in and then out. If the water is channeled through a turbine, electricity can be produced. See **Fig. 9-13**. Because the flow of the water reverses twice each day, the engineering required for the turbine generator is more complex. The cost of operating and maintaining tidal-controlled hydroelectric plants is high, but tidal energy is very reliable.

Ocean waves and currents are other sources of energy. Devices like the sea snake, a 450-foot-long red steel snake that floats on the ocean's surface, use the ocean's motion to power generators that provide electricity. This technology is already being used in countries like Scotland and Portugal. Researchers estimate that it could eventually provide up to 13 percent of the world's energy.

Ocean Heat

Tropical oceans collect and store heat from the sun. Water near the surface is warm. Water deep under the surface remains very cold. This temperature difference is used to produce electricity, a process called ocean thermal energy conversion (OTEC).

An OTEC system is much like a refrigerator. A liquid, generally ammonia, is cycled through a closed system of pipes. At one end of the system, where the ocean water is very cold, the ammonia is a liquid. As the ammonia cycles through the hot ocean water, its pressure increases and it changes into a gas. The pressurized gas turns a turbine that is connected to an electrical generator. See **Fig. 9-14**. After the ammonia gas has done its work, it is cooled by the cold water into a liquid and the cycle starts over again.

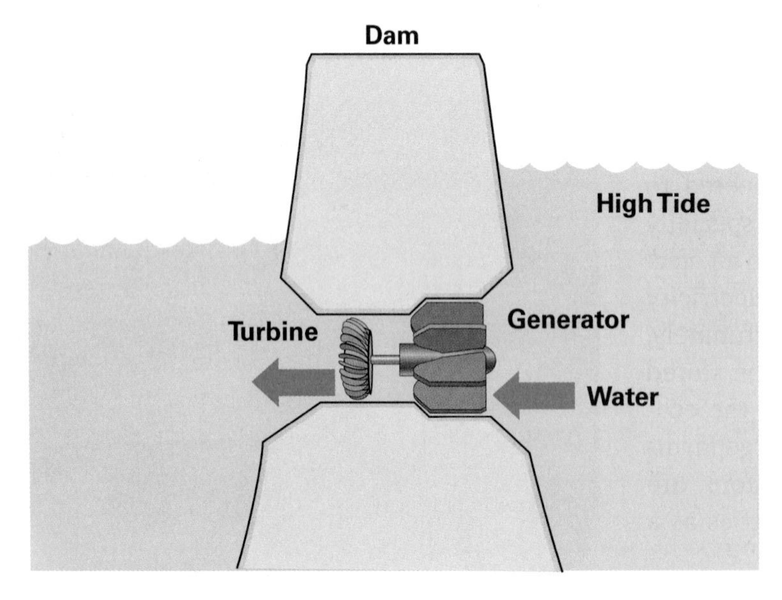

« **Fig. 9-13** The tide moves in and out and turns the turbines on this electric generator. Many megawatts of electricity can be produced.

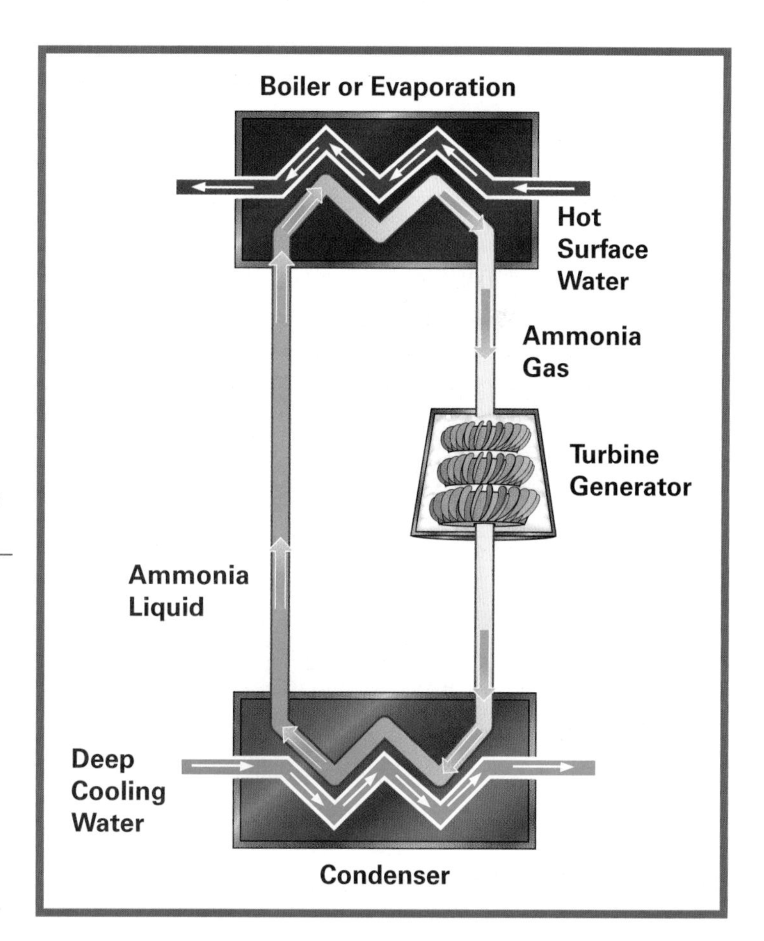

» Fig. 9-14 A simplified drawing of an OTEC plant. Hot surface water vaporizes the liquid ammonia, which produces pressure to drive the turbine generator.

The technology involved in an OTEC system is very simple. However, the plumbing required to control the cold and hot ocean waters and the sealed ammonia system is not simple. The advantages of OTEC systems mean that research will continue.

Solar Salt Ponds

Solar salt ponds, natural ponds generally found in the western U.S., are another source of electrical energy. The sun's rays pass through the top layer of salt water and heat the water on the bottom. Since there is very little movement in heavy salt water, the bottom water gets hotter and hotter, often reaching 250 degrees. A system similar to that used for OTEC heats and cools ammonia, which operates a turbine generator. However, there are a limited number of salt ponds, and there is little interest in further developing this energy source.

Earth Heat

Geothermal energy is heat generated within the earth. It is the result of the decay of radioactive materials beneath the earth's crust. This heat is trapped unless it escapes through geysers and volcanoes.

The most efficient use of geothermal energy is to produce or capture steam in order to power electricity-generating turbines. See Fig. 9-15 on page 200. Usable geothermal energy can be found in hot, dry rock fields; dry steam fields; hot water fields; and fields of lesser heat. The engineering and technology available to use these potential energy sources vary.

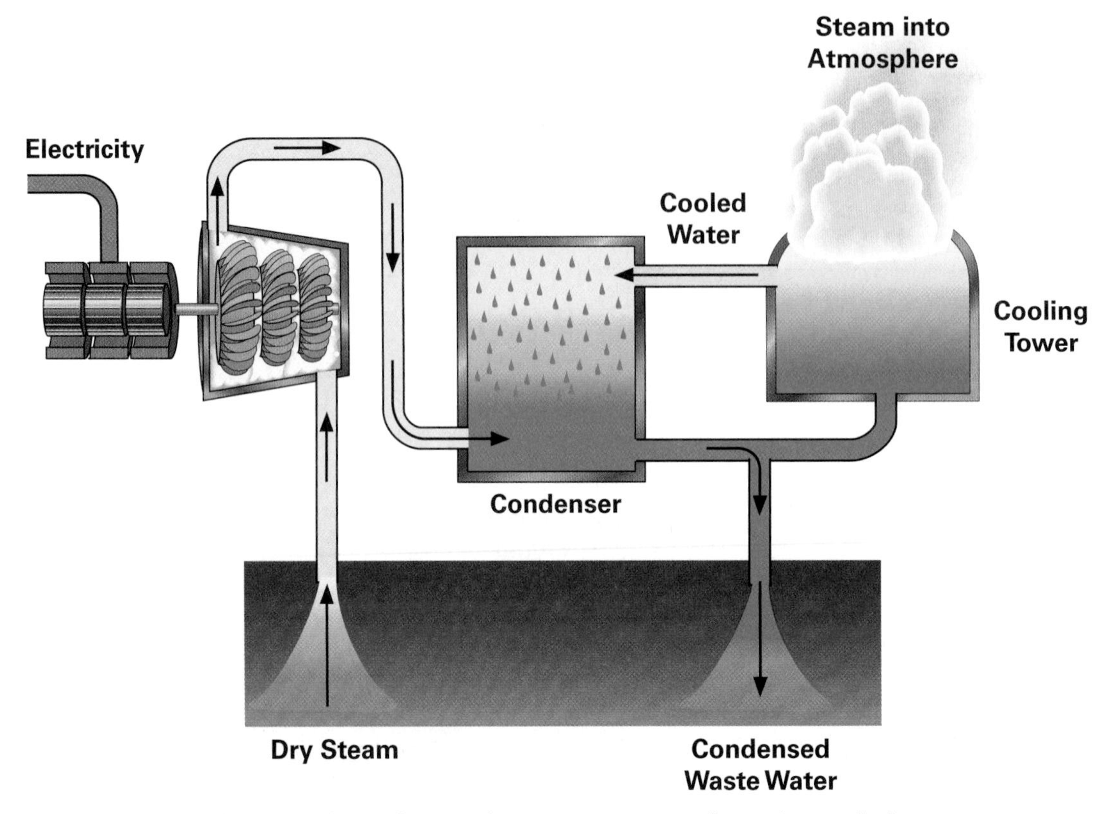

Fig. 9-15 This diagram shows how a dry steam-powered geothermal plant operates.

Where hot, dry rocks are close to the surface, water is injected to produce steam. The steam is piped to the earth's surface and used to power turbines.

Dry steam fields are easily tapped to gain pressurized steam to produce electricity. Hot water fields contain water ranging from 350 to 700 degrees. While inside the earth, water does not boil because of the surrounding pressure. As the hot water comes to the surface, pressure is released and it begins to boil. Boiling water makes steam, which can be harnessed to drive turbines.

Fields of lesser heat also contain hot water but at a lower temperature. With these lower temperatures, steam is not created, but the heat energy can be used directly to heat structures or even to operate a land-based system used to drive a turbine.

Hydrogen

Pure hydrogen is one of the most common elements on Earth and burns easily, causing no pollution. There are few problems using hydrogen as a fuel. However, it is hard to produce.

Science Application

Using Boiling Points The temperature at which a liquid boils depends on the pressure of air (or atmosphere) around it. Areas of low or high pressure will have different boiling points than those areas with standard air pressure. For example, the standard boiling point of water is 212 degrees Fahrenheit, or 100 degrees Celsius. However, if you are up on a mountain where the air pressure is much lower, the boiling point will also be much lower.

How might something as simple as an adjustable boiling point affect you? Imagine you are making soup and you live in an area of typical elevation. The directions on the soup container may tell you to heat the water at a specific temperature for a specific time. Now imagine if you lived somewhere where the air pressure is much lower. The directions would be different, wouldn't they? Food packages today often have disclaimers about boiling water according to the user's elevation.

FIND OUT Research the boiling points at different altitudes in North America. Create a map showing a variety of boiling points. You've now created a map that shows changes in elevation.

Water consists of two atoms of hydrogen (H) bonded to one atom of oxygen (O). Hydrogen bonds in water are very strong and require a great amount of energy to be broken. Although there are several ways of doing this, electrolysis seems to be the most promising.

During **electrolysis,** electricity passes through water. This separates the hydrogen and oxygen atoms. So far, the cost of the electricity used for the electrolysis is greater than the value of the produced hydrogen. As the cost of electricity decreases and the cost of other transportation fuels increase, the use of hydrogen-fueled vehicles may increase.

Hydrogen can also be used to produce electricity in a fuel cell. Fuel cells are like batteries. Hydrogen is combined with oxygen, and electrons are released.

CHAPTER 9 REVIEW

Main Ideas

- All sources of energy can be classified into three types: exhaustible, renewable, and inexhaustible.
- Exhaustible energy sources are those that cannot be replaced.
- Renewable sources of energy are those that can be used indefinitely if they are properly managed and maintained.
- Inexhaustible energy sources will always be available.

Understanding Concepts

1. What is the fundamental source of almost all the earth's other sources of energy?
2. How can electricity be produced from flowing water?
3. What is the one major disadvantage of using wood as an energy source?
4. Define *bioconversion*.
5. How can electricity be used to release hydrogen?

Thinking Critically

1. What problems might occur if different nations placed sea snakes throughout the earth's oceans?
2. In what ways could the government encourage energy conservation?
3. How would society be different if wind was a primary source of energy today?
4. What factors might be considered when planning the location of a nuclear power plant?
5. If a nearby forest was cut down to provide energy, what would change in the amount of visible wildlife?

Applying Concepts & Solving Problems

1. **Energy and Power** Monitor either your own or someone else's driving habits. How can gas be conserved? Create an analysis to show your findings.
2. **Science** Design an experiment that will allow water to be heated past 212 degrees Fahrenheit without boiling.
3. **Assess** Research wind farms and report on how efficiently they use energy. If possible, visit a wind farm near your area.

Activity: Convert Energy

Design Brief

How much energy do you use in a typical day? You may only think of tasks that require a lot of work, such as participating in a physical education class or working at a job after school. However, even the simplest tasks use various forms of energy and energy conversions.

Design a device that will accomplish a simple daily task using different forms of energy and several energy conversions. Be creative—your device does not have to be practical as long as it performs energy conversions.

Refer to the STEM HANDBOOK

Criteria

- Your device must use an odd assortment of materials found around the house or school. For example, a skateboard, toothbrush, rope, and candle might be used to create an automatic tooth brusher.
- Your device must make use of at least three forms of energy.
- Your device must involve at least four operations in which energy conversion takes place.
- The energy conversions must be labeled with tags or signs.

Constraints

- The work achieved by the device must simulate a task you complete on a daily basis.
- Your device must be a real model.
- You must demonstrate your device for the class, and it must accomplish the task for which it has been designed.

Engineering Design Process

1. **Define the problem.** Write a problem statement that describes what must be accomplished in order to complete your chosen task.
2. **Brainstorm, research, and generate ideas.** With your team, discuss possible solutions. If necessary, review the forms of energy in Chapter 8. Hint: Rube Goldberg is a cartoonist who designed complex energy conversion devices using random materials. Research his designs. Maybe your device will be even more creative than his cartoons!
3. **Identify criteria and specify constraints.** These are listed on page 204.
4. **Develop and propose designs and choose among alternative solutions.** Choose the best design that will solve your problem. For example, avoid designs that require complex machinery or other equipment.
5. **Implement the proposed solution.** Decide on the design you will use. Gather any needed tools or materials. Then make a drawing.
6. **Make a model or prototype.** Create a working model of your device.
7. **Evaluate the solution and its consequences.** With your teammates, check your device to make sure it works properly.
8. **Refine the design.** Based on your evaluation, change the design if needed.
9. **Create the final design.** After you have an approved design, put on the finishing touches. Label your energy conversions.
10. **Communicate the processes and results.** Demonstrate your device for the class. Be prepared to answer questions. Turn in your device to your teacher. Be sure to include the name of the activity, your definition of the problem, and a description of how you solved the problem.

SAFETY FIRST

Before You Begin. Make sure you understand how to use the tools and materials safely. Have your teacher demonstrate their proper use. Follow all safety rules.

CHAPTER 10 POWER SYSTEMS

Objectives

- Name and discuss six simple machines.
- Describe several devices used to transmit mechanical power.
- Compare and contrast hydraulic and pneumatic systems.
- Explain the difference between alternating and direct current.

Vocabulary

- load
- mechanical advantage
- simple machine
- fluid power
- Boyle's law
- direct current (DC)
- alternating current (AC)
- series circuit
- parallel circuit

The wheel and axle of this dragster is a powerful application of a simple machine. In this chapter's activity, "Inventory Simple Machines," you will identify simple machines that you use in your daily tasks.

Reading Focus

1. **Read the title of this chapter and describe in writing what you expect to learn from it.**
2. **Write each term in your notebook, leaving space for definitions.**
3. **As you read Chapter 10, write the definition beside each term in your notebook.**
4. **After reading the chapter, write a paragraph describing what you learned.**

Types of Power Systems

The three most common types of power systems include mechanical, fluid, and electrical. Each of these systems usually converts one type of energy into another. See **Fig. 10-1**.

As you know, power is a measure of work done over a period of time. Power systems are used to drive other technological devices and systems or to provide force.

Power systems must have a source of energy, a process, and a load. Energy is one of the inputs power systems use. During the process, that energy is converted into a form that can accomplish work. The **load** is defined in two ways: (1) The load is the amount of resistance the power system must overcome to do the desired work. The power system of a car, for example, must overcome the inertia and weight of the vehicle to get it moving and keep it moving. (2) The load is also the amount of force output by the power system. The load is the force a car engine must exert to move the car.

Fig. 10-1 This backhoe uses all three types of power. ***Can you name other machines that use all three power systems?***

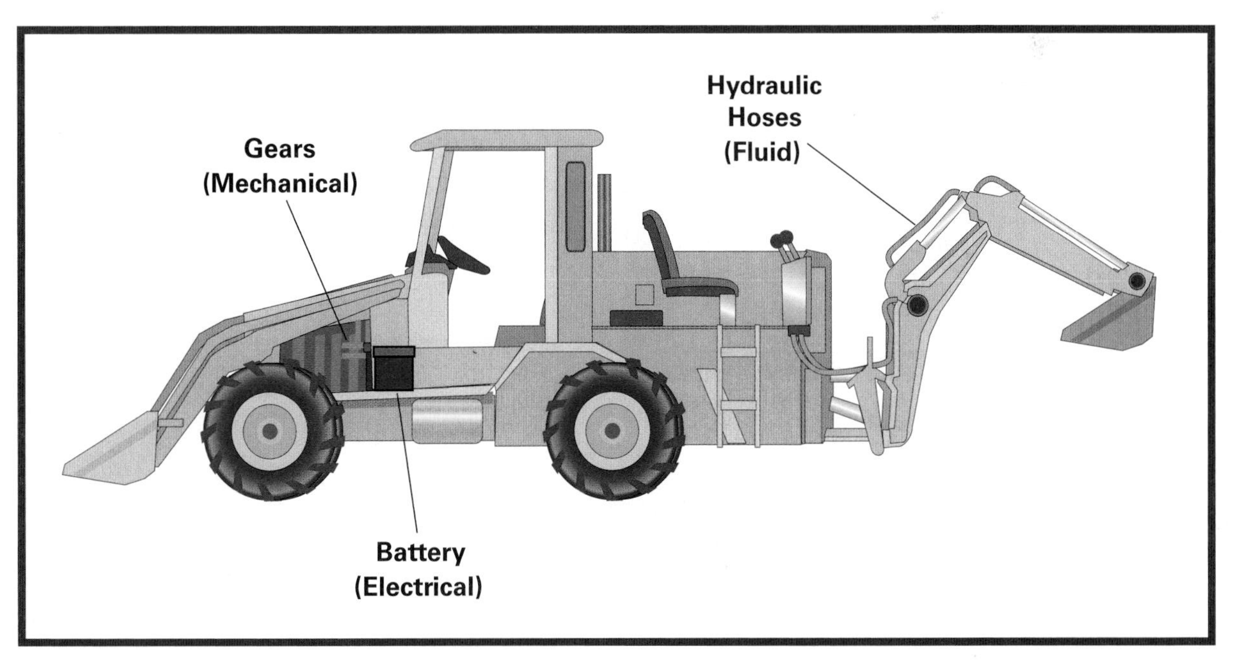

As we study mechanical, fluid, and electrical systems, you will learn about the principles underlying each one and the basic methods of controlling the operations.

Mechanical Power

As you know, mechanical energy is the energy of motion. When it is timed, it becomes mechanical power. Machines are the devices we use to convert and control mechanical power. All machines work on the same basic principles, which include mechanical advantage and the principles of the six simple machines.

Mechanical Advantage

Using a machine multiplies human strength or force. This is called **mechanical advantage**. For example, you could not pull 200 pounds along a sidewalk using just a rope. Pushing the weight in a wheelbarrow, however, would make it possible.

One way to calculate mechanical advantage is to compare output force to input force or input distance to output distance. The difference is written as a ratio.

For example, in Fig. 10-2, a man is lifting a load (output force) of 30 pounds using an input force of 6 pounds. The lever provides a mechanical advantage of 30 to 6, or 5:1. Also, the distance the man must move his end of the lever is five times as far as the 30-pound load is moved up.

With any machine, the distance ratio and the force ratio will equal each other, so the only information needed to calculate mechanical advantage is the distances or forces involved.

Simple Machines

Simple machines are devices that create mechanical advantage. All complex machines are based on these six machines—the lever, the wheel and axle, the pulley, the inclined plane, the wedge, and the screw.

Lever The lever is a bar that rests on a pivot point called a fulcrum. See Fig. 10-3. The lever is classified according to where the load, force, and fulcrum are located. For example, a first-class lever works like a pry bar while a second-class lever works like a wheelbarrow. Any change in the location of the fulcrum or the distances of the load or force makes a difference in the mechanical advantage ratio and the effort it takes to complete the work.

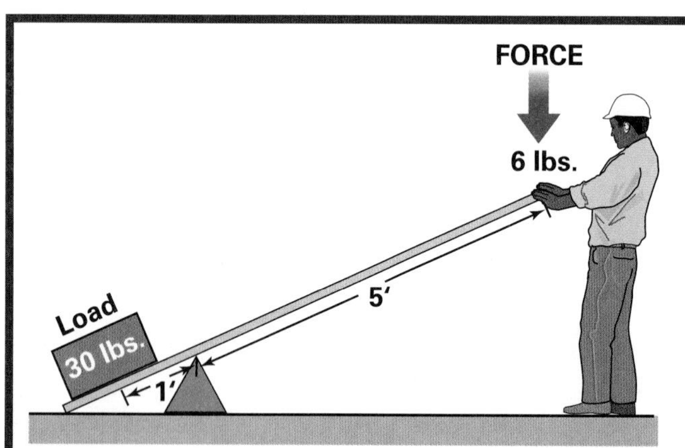

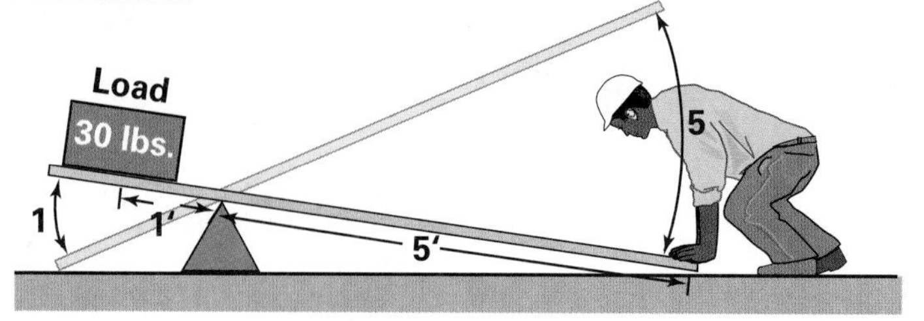

» Fig. 10-2 If 6 pounds of force move a 30-pound load, the mechanical advantage is 30:6, or 5:1, or 5. However, the 6 pounds of force must move five times as far as the 30-pound load.

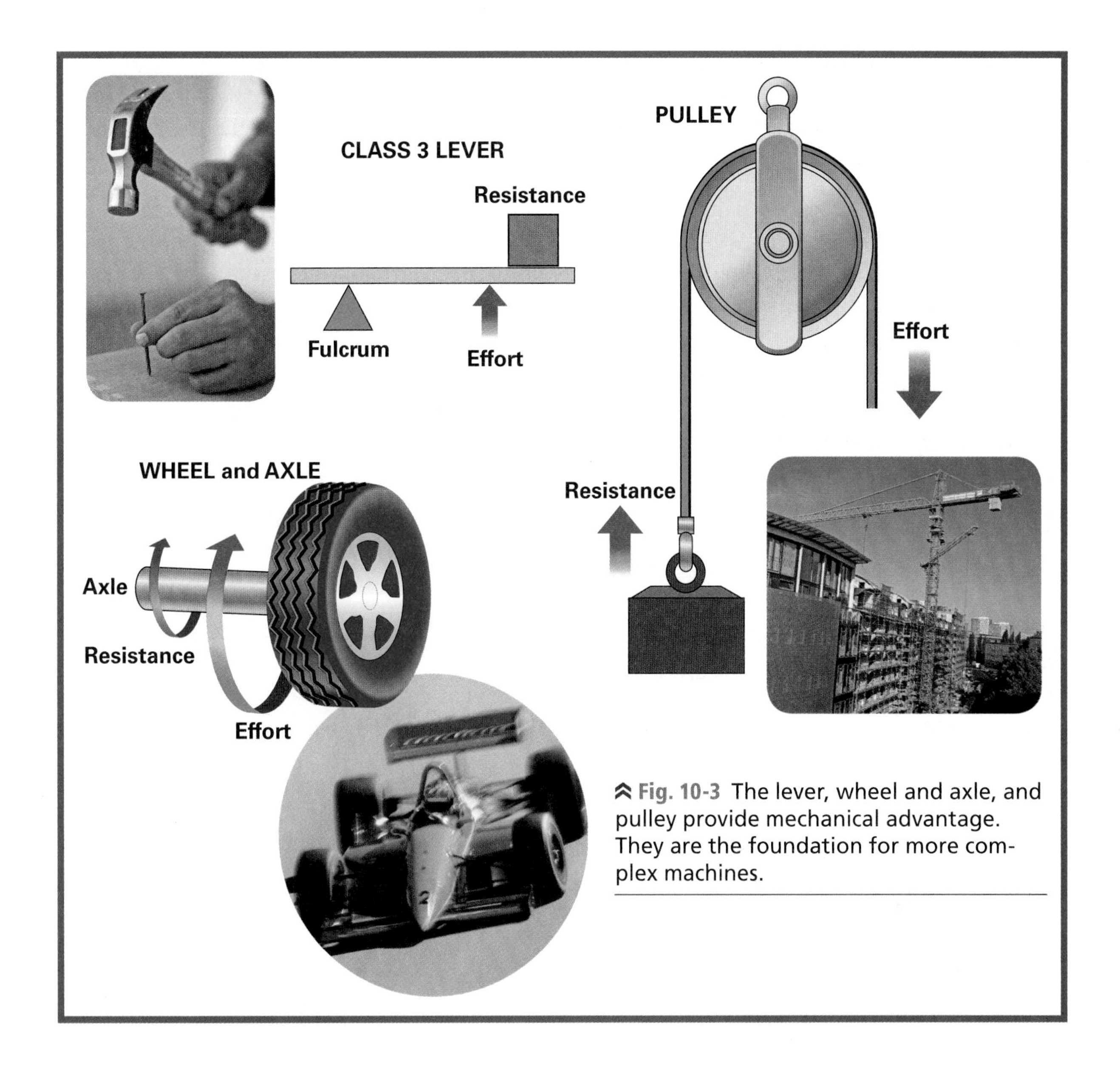

Fig. 10-3 The lever, wheel and axle, and pulley provide mechanical advantage. They are the foundation for more complex machines.

Wheel and Axle A rod or shaft attached to a wheel is called a wheel and axle. It is much like a lever except the fulcrum is at the center and the lever rotates around it. See Fig. 10-3. The mechanical advantage is gained by the location of the force or the load. (The load is attached to the axle.) The greater the wheel radius (distance), the greater advantage you have gained. The closer to the center (fulcrum) the force is applied, the faster the load rotates. For example, suppose you use a hand winch to pull a boat onto a boat trailer after a day of fishing. The longer the handle on the winch, the larger the diameter of circular motion and the less arm muscle you must use to load the boat.

Pulley Like the wheel, a pulley turns around an axis. Most pulleys are grooved to carry ropes or belts. See Fig. 10-3. The direction of the load can be changed, as well as the mechanical advantage of either force or distance. Like the wheel and axle, the pulley works on the principle of the lever in that the fulcrum is centered between the force and the load.

Inclined Plane The inclined plane makes use of a sloping surface. See Fig. 10-4. You can calculate its mechanical advantage by comparing the height the load is raised by the length (distance) of the ramp.

Suppose you want to raise an oil drum that weighs 150 pounds to the height of 2 feet. Would you rather lift it straight up or roll it up a ramp?

Wedge A wedge is made from two or more inclined planes all meeting at one edge or point. See Fig. 10-4. Axes and nails are examples of wedges. The mechanical advantage is measured by comparing the thickness with the

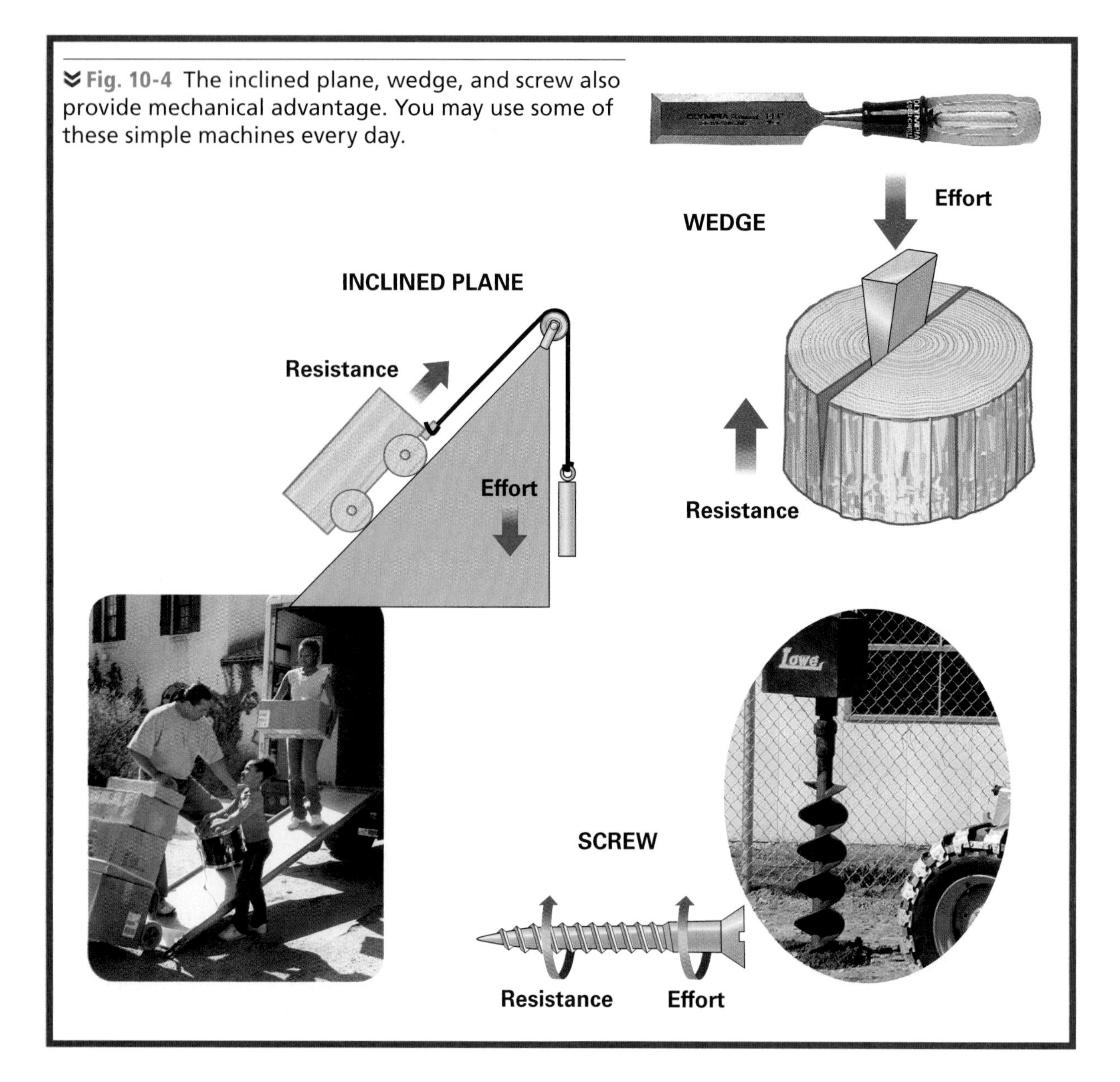

Fig. 10-4 The inclined plane, wedge, and screw also provide mechanical advantage. You may use some of these simple machines every day.

length. When using a wedge to split wood, for example, the wedge multiplies the force of the hammer used to drive it.

Screw An inclined plane cut in a spiral around a shaft is called a screw. See Fig. 10-4. This arrangement provides a long and gradual slope around the shaft and produces a high mechanical advantage. For example, a standard ½" bolt that has 20 threads per inch creates a mechanical advantage ratio of over 30:1.

Controlling Mechanical Power Systems

Mechanical power can be used directly without changing it. This is called direct drive. An example is a standard lawn mower. The blade is connected directly to the crankshaft of the engine.

However, usually mechanical power is changed in some way before we put it to use. In that case, control devices like gears, pulleys, sprockets and chains, clutches, and couplings transmit the power. Typically these devices change the input power in three ways:

- Switch it on and off. For example, the clutch in a car disengages the power.
- Change its direction. For example, the gears in a car may cause it to go backward.
- Change its force and speed. For example, control devices supply more power to the car to get it moving than when it is already in motion.

Gears are basically wheels with teeth cut around their outer edge. The teeth of one gear mesh with the teeth of another. See Fig. 10-5. The force is transferred from a shaft through a hole in the center of the gears. Changing the number or size of gears increases speed or power. The gear connected to the power source is called the driver gear. The last gear in a gear system is called the driven gear.

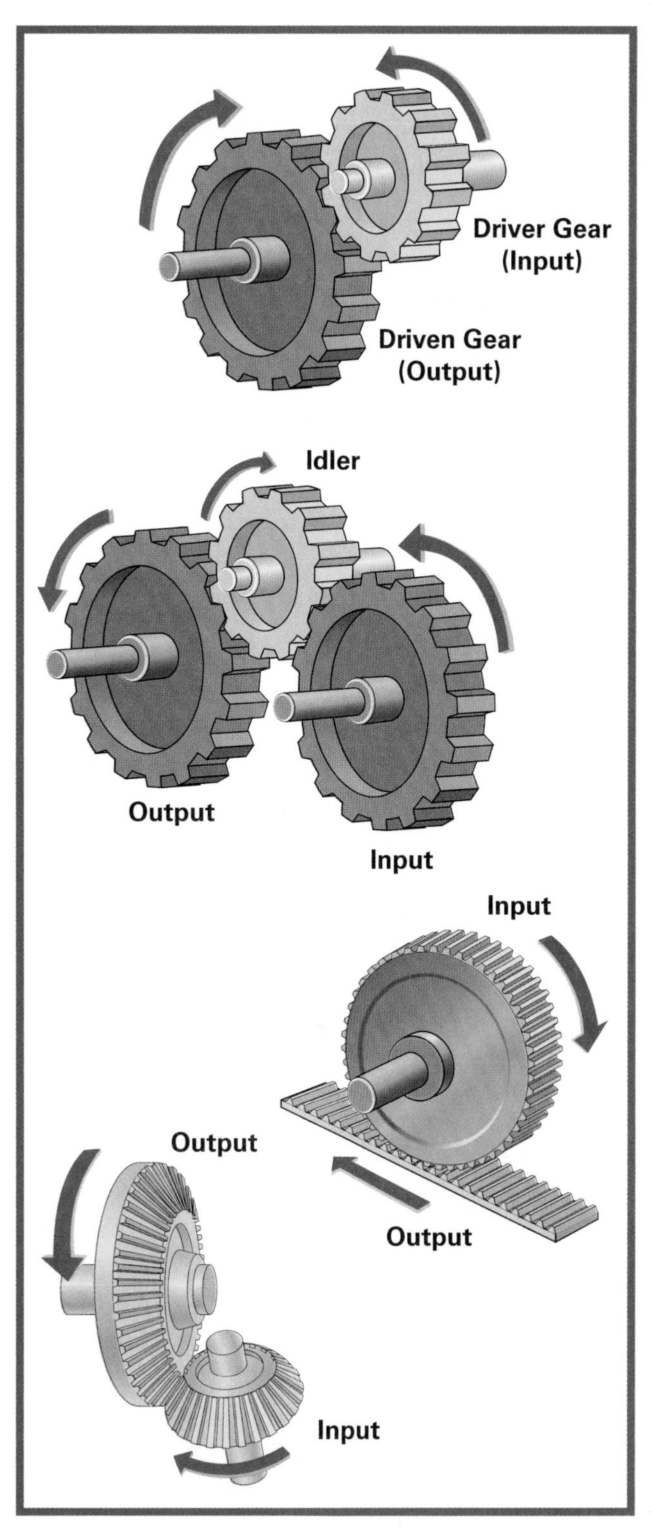

Fig. 10-5 The types of gears shown here transmit speed or power in different ways to accomplish different tasks.

Changes in force and speed are calculated in terms of a gear ratio. This is the number of teeth on the driver gear compared to the number of teeth on the driven gear.

Pulleys and Belts A pulley and belt power system is not the same as the simple machine pulley discussed earlier. See **Fig. 10-6**. A pulley is a grooved metal disk connected to a shaft. A flexible belt fits into the groove and transfers power from one pulley to another.

Two pulleys connected by a belt can either increase or decrease the speed or power of the shafts. Some tools, such as drill presses, have multiple-size pulleys or adjustable-size pulleys enabling quick speed and/or power changes.

Sprockets and Chains are driven by a pulley/belt instead of another gear. The teeth on the sprocket mesh with the chain. The timing of the rotation is controlled. Bicycles operate using a sprocket and chain system.

Clutches control on and off switching. There are generally two types: friction and centrifugal. Friction clutches operate using pressure to transfer power flow. They are found in vehicle transmission systems. When the driver steps on the clutch pedal, the pressure is released and no power is transmitted. A centrifugal clutch works on the principle of centrifugal force. The faster an object is rotated around an axis, the more outward force that object develops and the more pressure it exerts. Examples of centrifugal clutches can be found in wood-cutting chain saws and some clothes washing machines.

Couplings are permanent connections used to transmit power. Rigid couplings offer a stable connection between a power source and a driven device. The advantage of couplings is that they can be disassembled to change power sources or appliances. A special coupling called a universal joint allows shafts connected to it to be out of alignment yet still operational. There are many different universal joints in cars and trucks. A constant velocity joint (CV) is used in vehicles.

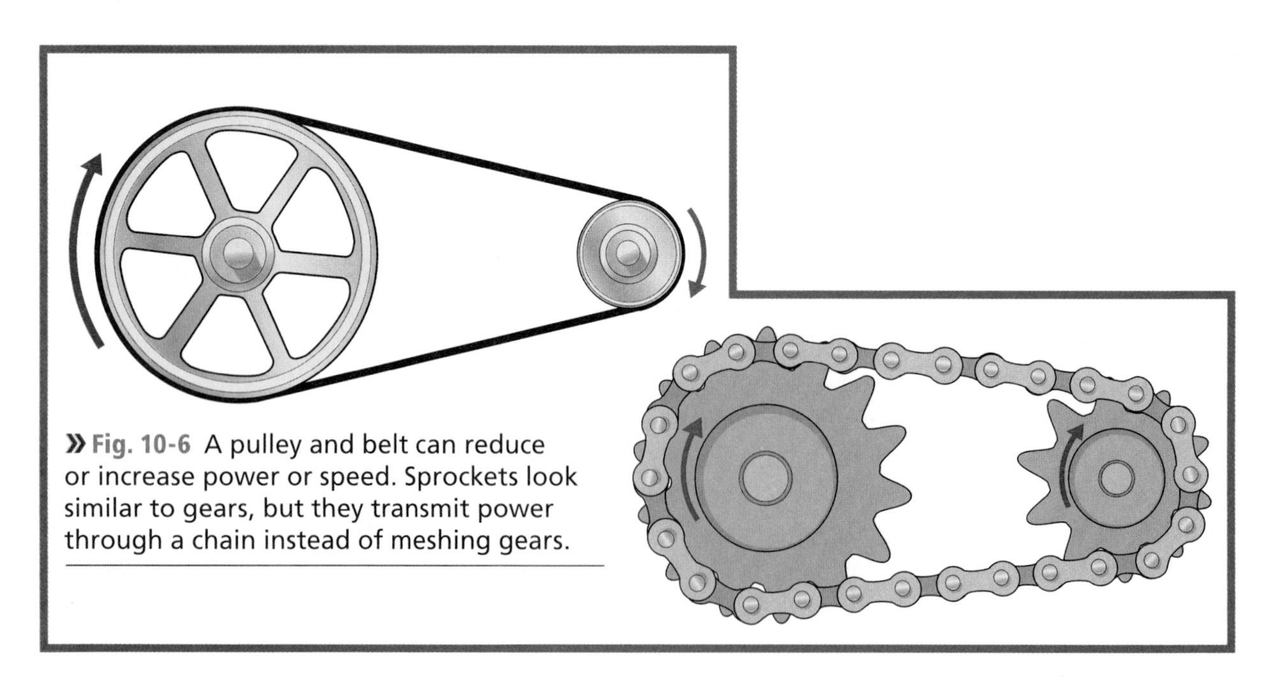

» **Fig. 10-6** A pulley and belt can reduce or increase power or speed. Sprockets look similar to gears, but they transmit power through a chain instead of meshing gears.

Fluid Power

Fluid power is the use of pressurized liquids or gases to control and transmit power. The use of fluid power systems can be traced back over two thousand years.

The liquids and gases in a fluid power system receive pressure from an outside source, such as a compressor. Fluid pressure can be either a positive or negative number based on atmospheric pressure. A gauge pointing to 0 is actually reading 14.7 pounds per square inch (psi). This is the force of the atmosphere on everything. Any reading above 0 is a positive pressure. Readings below 0 indicate a vacuum.

Work can be accomplished by a fluid power system at positive psi readings or in a vacuum but not at 0 psi. For example, when you suck on a straw, you create a vacuum and draw the liquid to your mouth. When you blow into the straw, you create pressure and the liquid bubbles in the glass.

The control of fluid systems often depends on volume. A fluid's volume is the amount of space it takes up. You cannot compress a liquid. Gas, on the other hand, will compress as pressure is applied. See Fig. 10-7.

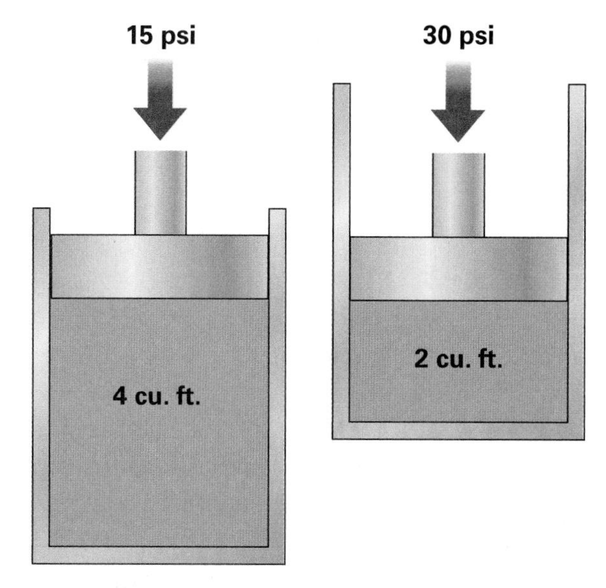

Fig. 10-7 Shown here is an example of fluid power. ***If you increased the pressure to 60 psi, how would the amount of cubic feet change?***

Boyle's Law

Compression of gases is described in Boyle's law, named for Robert Boyle, an English scientist. **Boyle's law** states that if the pressure of a gas increases, the volume decreases (provided that the temperature is constant). For example, if you double the pressure on an enclosed gas, its volume is cut in half. This volume-pressure relationship also works in reverse. A change in volume creates an equal but opposite change in pressure.

Temperature also affects the volume of a gas. An increase in temperature increases the volume. A decrease in temperature decreases the volume.

Think of a tire as an example. When you add air to a flat tire, you add volume until the tire comes to full size. The pressure of the gas increases. During driving, if the tire hits a curb, the change in the shape of the tire instantly decreases the volume, and thus the pressure goes up. Sometimes the tire will blow. That same tire, provided it does not have a leak, will decrease in pressure during cold weather to an unsafe level.

Pressurized Liquids

Hydraulics works because the liquid will not compress. By increasing the pressure of the liquids, a very high mechanical advantage is possible. In a closed leak-proof system filled with liquid, the same pressure applied on one end of the system will be transferred to the other

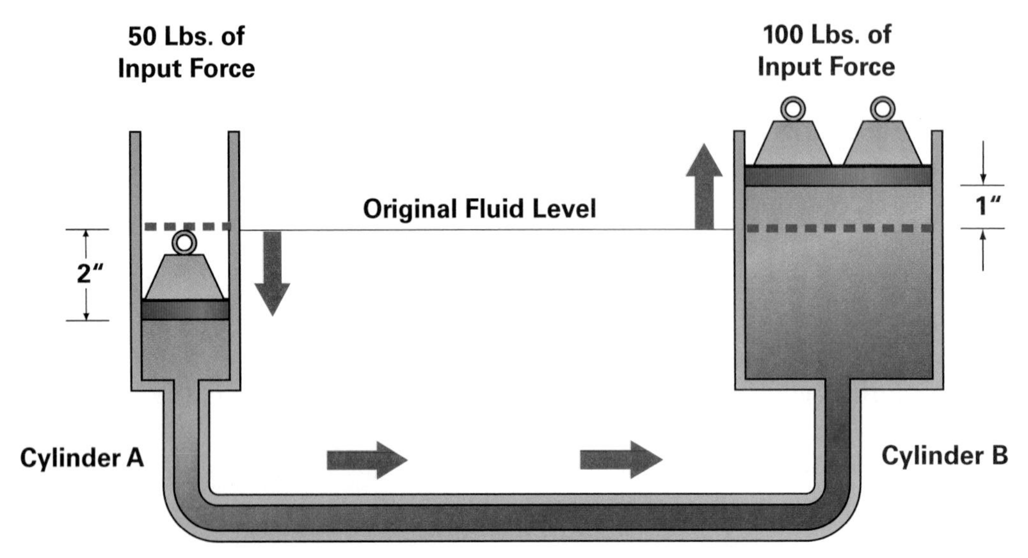

Fig. 10-8 Pressure applied at one end of the hydraulic system is transferred to do work at the other end.

end of the system. See Fig. 10-8. Difference in advantage is gained by changing the distance, pressure, and/or piston size of the different devices that make up the system.

Controlling Fluid Power Systems

Gases are compressed by a compressor and liquids are pressurized by a pump. The process of transmission and control through pipes and hoses is the key to an effective fluid power system.

Pneumatics Modern uses of pneumatic power include tool operation, assembly line processing, and high-speed drilling. High-speed dentist drills operate on pneumatic power as do most food processing assembly lines. One of the major reasons these operations use pneumatic power systems is cleanliness. Filtered compressed air or other inert gases do not contaminate like oil-operated power systems.

Most of these operations also need low horsepower for high-speed tool operation. The exception is a power nailer, which operates on the explosive power of quickly released compressed air. See Fig. 10-9.

Hydraulics The components of a hydraulic system include a pump to pressurize the oil that comes from a reservoir, through a filter, into transmission lines. The oil then passes through a hand-operated shut-off valve, by a pressure relief valve, to a control valve. From the control valve, the pressurized oil goes to an actuator, which is usually a cylinder. The actuator converts the fluid power into mechanical power. See Fig. 10-10.

Notice how the pressure relief valve allows the pump to remain operational even though the actuator is not operating. The pump keeps a constant oil pressure, but when there is no work to be done, the oil bypasses the system through the pressure valve and returns to the reservoir. The pressure relief valve is also a safety feature in the event that the mechanical arm of the piston is overworked or becomes lodged in one position.

Combination Systems When greater force is required, a system that combines the advantages of both hydraulic and pneumatic systems is used. This system is called an air-over-oil power system. Automobile lifts are an example. See Fig. 10-10 again.

When the system needs to be used, an operator opens the control valve to a tank of pressurized air. The pressurized air is piped into a pressure cylinder. There is oil in the bottom of the cylinder. The pressurized air forces the oil from the air-over-oil pressure cylinder into the hydraulic cylinder. The hydraulic cylinder raises the lift and the car.

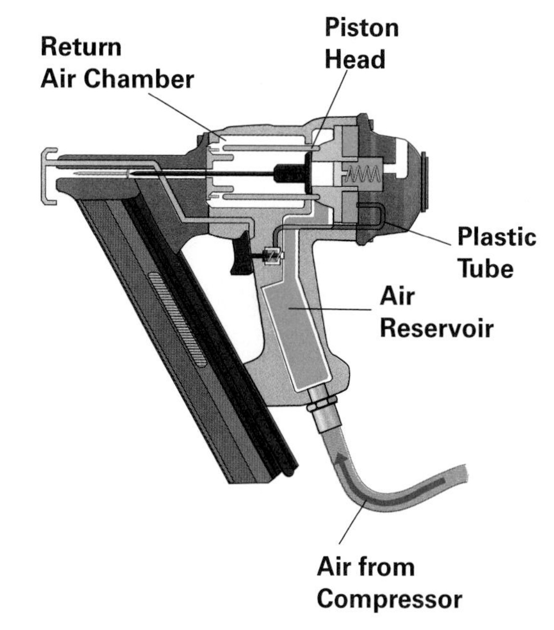

Fig. 10-9 This power nailer is a pneumatic tool. It uses compressed air to perform work much quicker than using a hammer.

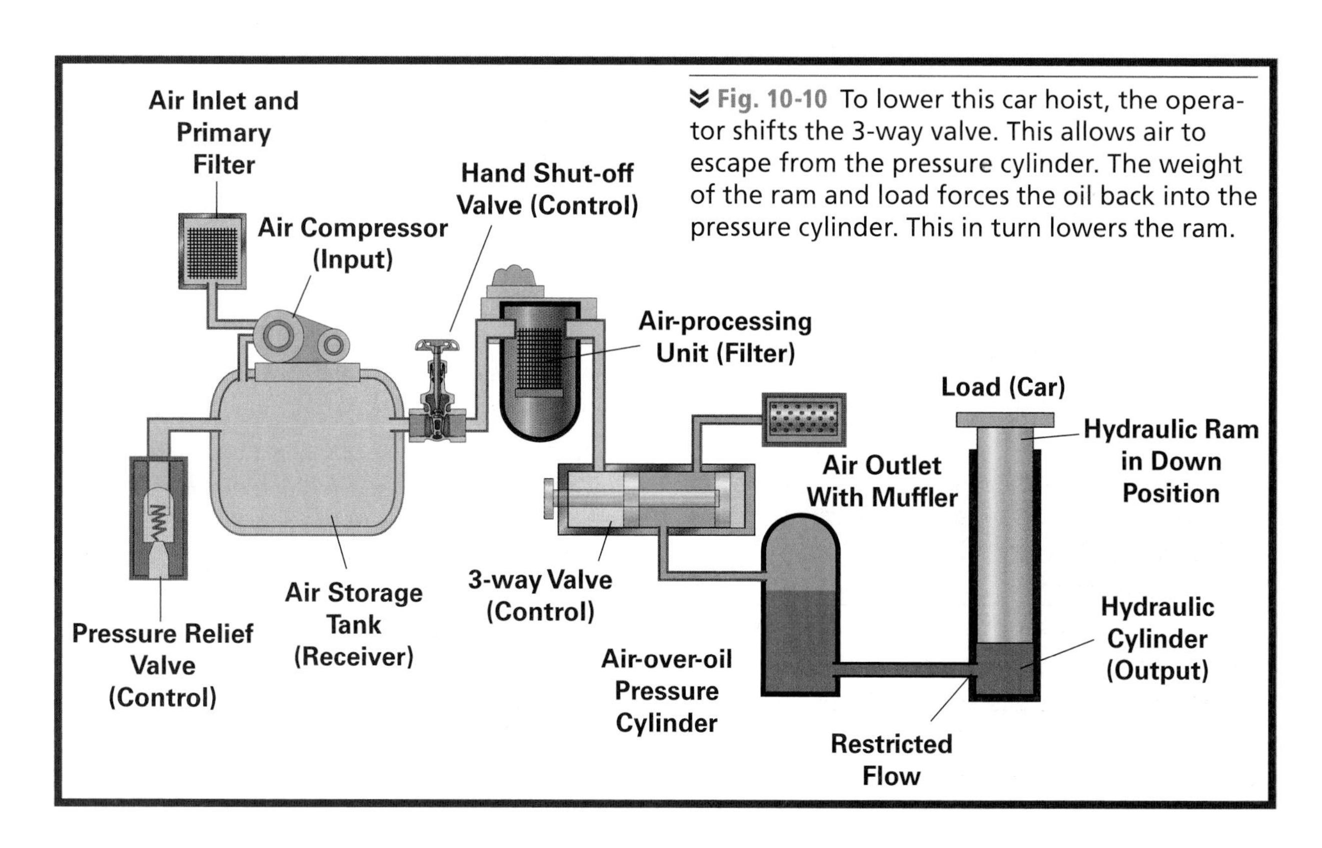

Fig. 10-10 To lower this car hoist, the operator shifts the 3-way valve. This allows air to escape from the pressure cylinder. The weight of the ram and load forces the oil back into the pressure cylinder. This in turn lowers the ram.

To lower the car, the operator releases the pressurized air in the air-over-oil pressure cylinder. As the pressure is released, the oil goes back into the air-over-oil pressure cylinder and the hydraulic cylinder lowers.

Science Application

Animatronics The construction of robots to look and act like living things is called animatronics. The word comes from Audio-Animatronics™, which is a form of robotics created for Disney theme parks. This technology has been used to create anything from pirates having sword fights to dinosaurs threatening to attack. How do these robots simulate movement? The answer is by using fluid power systems.

Pneumatic "muscles" control small actions such as blinking and mouth movements. For larger movements, such as limbs or a turning head, hydraulic systems are often used. Engineers who design animatronics use fluid power systems, combined with the latest electronic technologies, to make these robots seem as realistic as possible.

FIND OUT What field of science would engineers need knowledge of in order to simulate human movement in a robot?

Electrical Power

We use electrical power in countless ways. It provides light, heat, and motion. To understand how it works, you must understand the principles on which it is based.

Atoms consist of three types of particles: protons, neutrons, and electrons. Protons and neutrons form the center of an atom, which is called the nucleus. The electrons revolve around the nucleus in paths called shells.

Electrons have a negative charge, and protons have a positive charge. Particles with opposite charges are attracted to one another. An atom seeks to be in a balanced state. The positive charge of the protons in the nucleus keeps the negatively charged electrons in balance. However, electrons in the outer orbits sometimes break free and move to other atoms that are out of balance and have a positive charge. See Fig. 10-11. This movement of electrons is called electrical current.

The movement of electrons occurs more readily in some elements than in others. Copper, with 29 electrons, seems to be one of the best conductors of these positive/negative forces. Other materials resist the movement.

Types of Current

Alternating current and direct current are the two types of electrical flow available from our energy sources. In **direct current (DC)**, the electrons flow in only one direction. DC power can be stored in batteries, and AC cannot. Also, DC current is used to operate most electronic equipment because the flow of electrons is constant and lower voltages accomplish greater amounts of work.

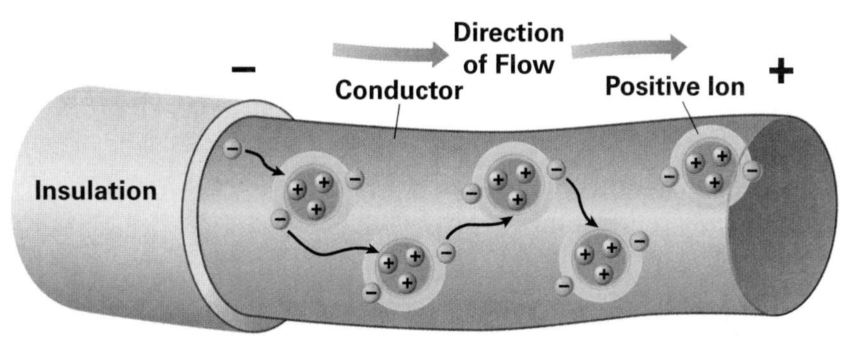

Fig. 10-11 Atoms have protons, neutrons, and electrons. An atom is balanced if there are the same number of electrons as protons. If there is an imbalance, the electrons will jump to another atom. This movement of electrons creates electricity.

Math Application

Calculating LED Power A light-emitting diode (LED) is a semiconductor device that emits ultraviolet, visible, or infrared radiation when electricity is applied to it. LEDs were originally used just as power indicators. Now they are used in any number of electronic devices, from watch displays to traffic signals.

A typical LED requires about 1.5 volts of direct current at 10 milliamps to begin emitting light. Visible LEDs usually produce red light (but blue, yellow, and green colored versions are also made).

An LED's peak (output) drive power is emitted at one particular wavelength, which gives it its characteristic color. The most common peak wavelengths are 780, 850, and 1,310 nanometers (where one nanometer equals one-billionth of a meter). The peak drive current is usually 50 to 100 milliamps.

Most typical LEDs operate with no more than 30 to 60 milliwatts of electrical power (where one watt equals 1,000 milliwatts). Power is defined as $P = EI$, where P is power (milliwatts), E is voltage (volts), and I is current (milliamps).

For example, if an LED requires 1.5 volts at 20 milliamps of drive current, its power rating ($P = EI$) is $P = 1.5$ volts × 0.02 amps = 0.030 watts, or 30 milliwatts.

FIND OUT If an LED has a power rating of 150 milliwatts at 1.5 volts, how much current is used?

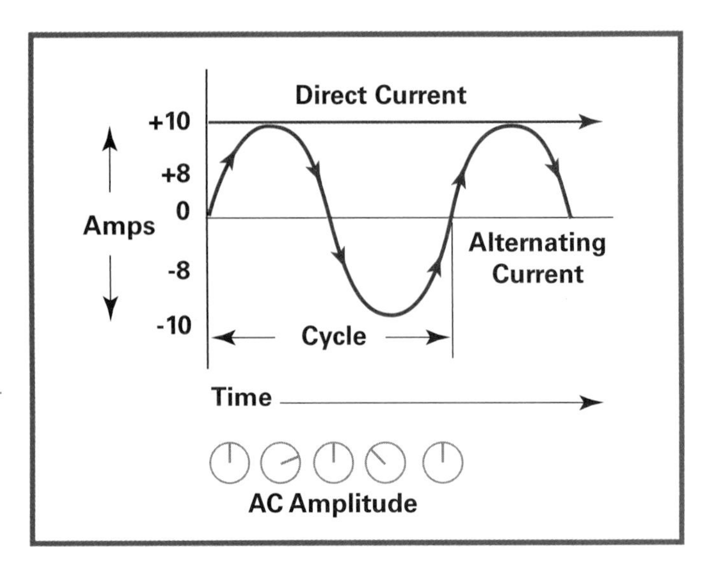

» Fig. 10-12 In a 10-amp circuit, AC current and DC current will conduct the same amount of current. However, the AC current is a back-and-forth movement of electrons. The DC current is a steady flow in one direction.

Alternating current (AC) is the most common type of current used in household and industrial circuits. AC is used to send electrical power over long distances. In AC, the direction of electron flow is first in one direction, then in the opposite direction, and then in the original direction. This sequence makes up one AC cycle, or hertz. See Fig. 10-12. In many countries, including the United States, AC flows at 60 cycles per second.

Types of Circuits

Now that you've learned about the types of current, you can begin to understand how an electrical circuit works. Electrical circuits both begin and end at the same power source. A circuit must also contain a device that uses electricity, such as a buzzer or lightbulb. If electrons do not keep flowing, the circuit isn't complete and electricity will be shut off. An electronic switch is used to open and close the circuit, turning things on and off.

In a **series circuit**, electricity flows along a single path to more than one electrical device. If one device in the path stops working, they all stop.

In **parallel circuits**, electricity flows along more than one path. If one device stops working, the other devices are unaffected. In your home, what happens when a lightbulb goes out? Do all the lightbulbs on that circuit go out? Probably not. Most of the circuits in your home are parallel circuits.

All the outlets in your home are connected to a fuse box or circuit breaker. These are used to shut off electricity in case of a power overload. If the circuit system is set up properly, an electrical fire caused by a power overload can be easily avoided.

SAFETY FIRST

Electrical Safety. Although a basic understanding of how electricity works is useful when studying technology, it does not make you an electrician. Always treat electricity with respect. Leave electrical repairs to professionals.

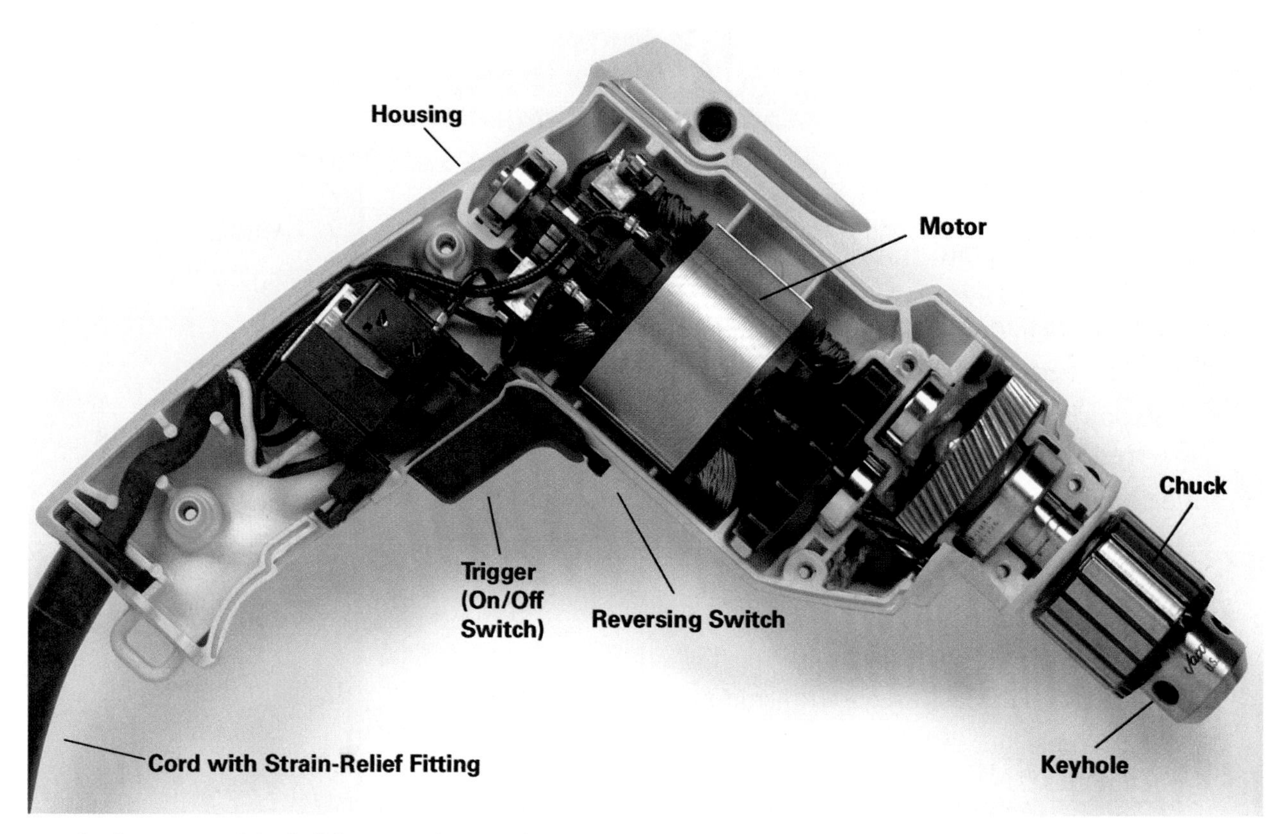

Fig. 10-13 This drill has an electrical motor that changes electricity into rotary motion.

Controlling Electrical Power Systems

Electrical power systems are grouped into four major work output categories: light, motion, heat, and communication. Electricity is converted into other forms of energy that do work. An example is the electrical motor, which changes electricity into rotary motion to operate such devices as drills. See **Fig. 10-13**.

Solenoids change electricity into linear motion. A solenoid is a coil of wires with a movable plunger core. The plunger may be attached to water valves on a clothes washing machine, switches on circuit controls, damper doors on heating systems, or the power locks on your car.

Electrical heat systems are used to dry clothes and heat air and water. In industry, electrical heat is used to liquefy metals and plastics during manufacturing.

Countless innovations in communication use electricity. From simple door bells to highly complex microwave and telecommunication operations, many systems are powered by electrical energy.

CHAPTER

10 REVIEW

Main Ideas

- The three most common types of power systems are mechanical, fluid, and electrical.
- Mechanical energy is the energy of motion. When it is timed, it becomes mechanical power.
- Fluid power is the use of pressurized liquids and gases to control and transmit power.
- Electrical power systems provide lift, heat, and motion.

Understanding Concepts

1. What are the six simple machines?
2. Name several devices used to transmit mechanical power.
3. What is the major difference between hydraulics and pneumatics?
4. What is the primary difference between AC and DC?
5. What creates electrical current?

Thinking Critically

1. How might hydraulics and pneumatics be used in other fields of technology discussed in this textbook?
2. What is the difference between a fluid and a liquid? Explain why gas isn't a liquid but is used for fluid power.
3. Identify electronic communication devices in your school.
4. Bicycles have sprockets and chains. How many other simple machines does a bicycle have?
5. What lights up a power button on a computer?

Applying Concepts & Solving Problems

1. **Design** Create working drawings of a device that uses at least three simple machines.
2. **Math** Who would exert more force on a see-saw: a person weighing 100 pounds at 2.5 feet from the fulcrum, or a person weighing 75 pounds at 3.5 feet away?
3. **Science** The Periodic Table of Elements shows the elements and how many electrons are in the atoms of each one. Study this table and make lists of good and bad conductors of electricity.

Activity: Inventory Simple Machines

Design Brief

In this chapter you have learned about the six simple machines and how more complex machines are based on them. Engineers recognize the simple machines as they design and use the many tools needed to produce products. Do you recognize them? Let's find out.

Find devices in your home and school environments that use the principle of a simple machine. List each device and explain how it provides a mechanical advantage. Calculate the mechanical advantage of at least one device.

Criteria

- Find at least one example of each of the six simple machines. Although some complex devices may incorporate more than one of the simple machines, each device can represent only one in your inventory. For example, if a power tool is based on both a lever and a pulley, you can use it as representative of only one of them.
- You must include a brief paragraph explaining how the device provides mechanical advantage.
- You must calculate the mechanical advantage for at least one device by comparing input distance to output distance.

Constraints

- You must include at least six different devices in your inventory.
- Your calculations of mechanical advantage should be shown on paper and turned in to your teacher with your inventory.

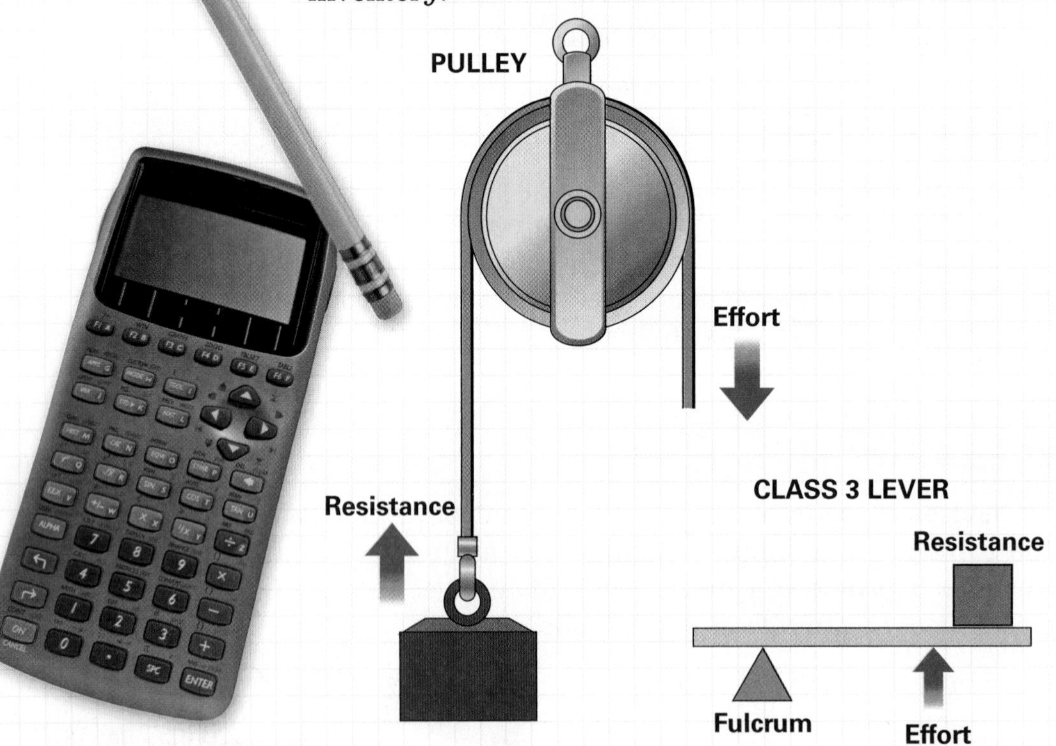

Engineering Design Process

1. **Define the problem.** Write a statement that describes the problem you are going to solve. For example, you might describe the inventory task in detail.
2. **Brainstorm, research, and generate ideas.** With your team, discuss possible devices that might appear in your inventory. Hint: Devices you might consider include pliers, wrenches, chisels, can openers, knives, dishwashers, trowels, wheelbarrows, and playground equipment. Do research to find out how certain devices work.
3. **Identify criteria and specify constraints.** These are listed on page 222.
4. **Develop and propose designs and choose among alternative solutions.** Choose the best devices and inventory design that will solve your problems. For example, avoid devices that are so complex that you're not sure which simple machines they include.
5. **Implement the proposed solution.** Decide on the devices you will include in your inventory. Then study the devices and determine how mechanical advantage is obtained.
6. **Make a model or prototype.** Create a rough draft of your inventory.
7. **Evaluate the solution and its consequences.** With your teammates, check your inventory to be sure it's accurate. Double-check your calculations for mechanical advantage.
8. **Refine the design.** Based on your evaluation, change the inventory if needed.
9. **Create the final design.** After you have an approved inventory, copy it neatly.
10. **Communicate the processes and results.** Present your finished inventory to the class. Be prepared to answer questions. Turn in your assignment to your teacher. Be sure to include the name of the activity, your definition of the problem, a description of how you solved the problem, and your calculations.

SAFETY FIRST

Before You Begin. Even though you are only taking an inventory, make sure you handle any tools or machines safely. Be sure any power tools are unplugged before examining them. Follow all safety rules.

UNIT 4

MANUFACTURING
ENGINEERING & DESIGN FRONTIERS

Think about the cell phone, the PDA, or even the drinking glass you use. Do you ever consider how these manufactured products were made? If you're like most people, you use them without much thought beyond "what it will do for me."

If you were a manufacturer, though, you'd probably be very concerned with how the products you market are designed and made. For just a moment, think like a manufacturer.

Rapid Manufacturing

Today's global economy generates intense competition to produce the best and the cheapest goods. So manufacturers are keenly interested in streamlining the manufacturing process to save time and cut costs, without sacrificing quality. Rapid manufacturing (RM) is a group of emerging technologies that enable manufacturers to go directly from a computer-based 3D description to a manufactured part or product.

Many RM technologies involve a technique called layering fabrication. It starts with a 3D image of the object in CAD (computer-aided design) software that is then virtually "sliced" into very thin cyberlayers. The RM system instructs machines to deposit the thin layers, one at a time, to build up the real-life, 3D object. Ink-jet printers spray layers of liquid plastic to build up the object, or laser-equipped devices fuse powder to make the layers.

RM technologies can produce complex objects that are more durable than parts made in conventional ways, such as die-casting or molding. RM also opens the door to a whole new generation of superstrong composite materials. Aircraft manufacturers use RM to turn out nearly indestructible jet and airplane components. Medical manufacturers produce better, longer-lasting replacement parts such as artificial hip sockets. The potential applications of RM are almost infinite.

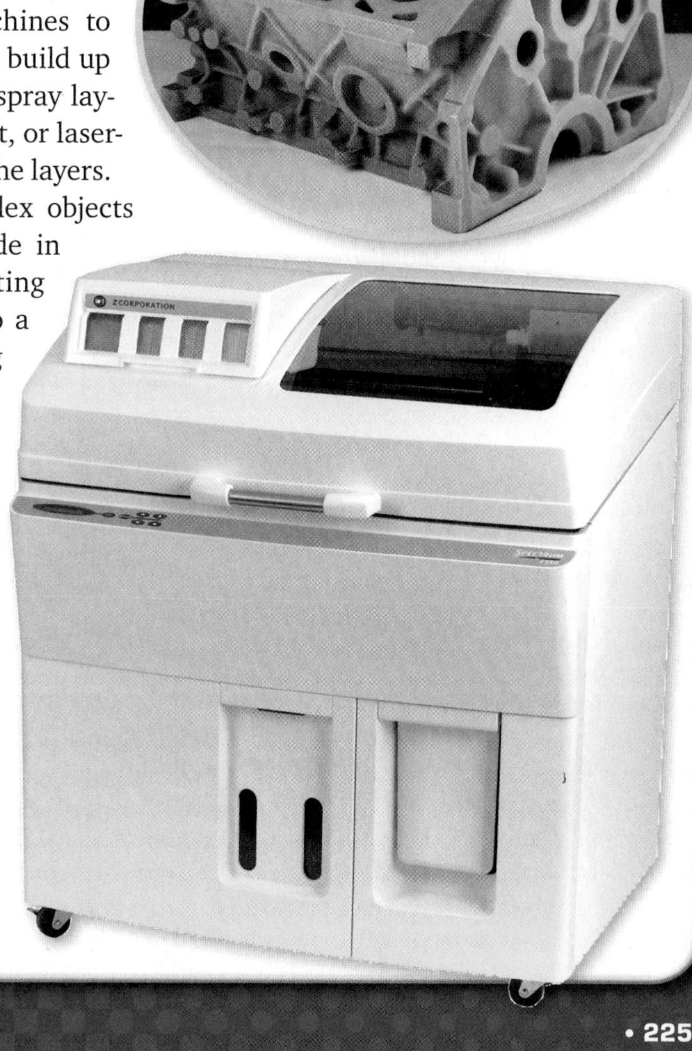

Highly customized manufactured products may be in the RM future. Imagine sitting for a scan of your head—as input for a bike helmet. The RM software creates a cybermodel of your head, slices up the model, and prompts the RM machine to layer a lightweight, sturdy helmet that fits your head perfectly!

CHAPTER 11

MANUFACTURING FUNDAMENTALS

This athlete is using a special wheelchair built with titanium and carbon fiber. In this chapter's activity, "Create Composite Materials," you will have a chance to create and test several samples of composites.

Objectives

- Explain the importance of manufacturing.
- Name the three basic types of production.
- Explain how companies compete in the global market.
- Discuss several manufacturing processes.
- Name some of the areas on which manufacturing has an impact.

Vocabulary

- nanotechnology
- global market
- NAFTA
- profit
- productivity
- custom production
- intermittent production
- continuous production
- assembly line

Reading Focus

1. Write down the colored headings from Chapter 11 in your notebook.
2. As you read the text under each heading, visualize what you are reading.
3. Reflect on what you read by writing a few sentences under each heading to describe it.
4. Continue this process until you have finished the chapter. Reread your notes.

What Is Manufacturing?

Manufacturing is the use of technology to make the things people want and need. Can you imagine your life without manufacturing? You would have no bicycle to ride, no television to watch, and no athletic shoes to wear. There would be no clothes, no furniture, and no airplanes. Lifesaving devices such as artificial hearts could not even be imagined. Manufacturing creates tools that make tasks easier, industrial materials that are stronger than natural materials, and devices that enable us to send messages into outer space. With manufacturing, we extend our reach and improve our quality of life.

If you build a bookcase from lumber, you are manufacturing a product. However, when most people think of manufacturing today, they think of the manufacturing industry. In factories, parts are made and then put together to create products. See **Fig. 11-1**.

On the next two pages, you can learn how factories evolved. Today's manufacturing plants are very different from the factories of the past, and more changes are coming rapidly. Computers are used in every aspect of manufacturing, from designing the product to storing and shipping it. New materials and methods are making entirely new products possible. One of the biggest changes on the horizon involves very small things: atoms and molecules. **Nanotechnology** involves building machines or materials on an atomic or molecular level to give them new properties. For example, nanotech fibers have been used to make a "smart" fabric that resists stains and wrinkles.

» **Fig. 11-1** In this chocolate factory, a worker is assembling very large pieces of chocolate for commercial candy bars.

EVOLUTION OF Manufacturing Technology

To manufacture originally meant "to make by hand." In that sense, manufacturing began in the Stone Age, when people started making tools, weapons, clothing, and pottery. Over time, manufacturing evolved from individuals producing small numbers of custom-made goods to factories mass-producing large quantities.

1200s

In Europe, demand for more and better wool cloth spurred the development of cottage industry. Merchants bought wool and sent it out to peasants who used their own machinery and worked in their homes to spin and weave the cloth. Cottage industry prevailed until the 18th century.

1800s

1913

With inventions such as the power loom, production became more mechanized. A steam engine or waterwheel provided the power for these large machines. People went to work in a special building that housed these machines. The building was called a factory. A company owned the factory and paid the workers. These changes were the beginning of the Industrial Revolution.

Adapting processes he had observed in meat-packing plants, Henry Ford implemented assembly line production in his automobile factories. Jobs were broken into systematic steps. Workers stayed in one place, and the work came to them on a moving assembly line. These methods were widely adopted by other manufacturers.

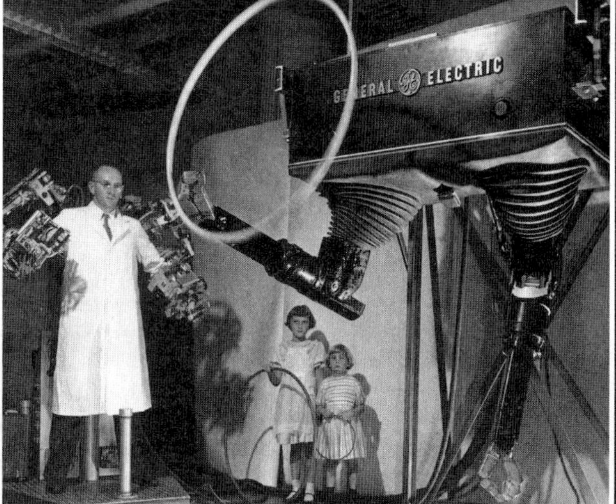

Industrial robots were developed to do heavy lifting or dangerous tasks. Some, such as GE's "Handyman," were directly operated by a human controller.

1960s

1970s

During the 1970s, manufacturers began using computer-aided design (CAD) and computer-aided manufacturing (CAM) to increase productivity. In 1994, the Boeing 777 became the first commercial aircraft to be designed entirely on a computer.

2000+

Innovations such as nanotechnology and rapid manufacturing continue to change manufacturing processes. Demand for low-skilled workers continues to decline, but the need for highly skilled technicians is increasing.

Manufacturing Systems

As you know, a system is a group of parts that work together to achieve a goal. A system is needed to produce manufactured goods. To work efficiently, a manufacturing system needs to be well managed. Like other systems, a managed production system has inputs, processes, outputs, and feedback. See **Fig. 11-2**.

Input includes any of the seven resources put into the system. People in manufacturing may design products, purchase materials, run machines, assemble parts, inspect products, or sweep floors. Metals, plastics, wood, glass and other ceramics, textiles, and rubber are all materials commonly used in manufactured products. The tools and machines used in manufacturing may be hand tools, portable power tools, or large machines and equipment. Energy provides power, heat, and light for manufactur-

Fig. 11-2 The parts of a manufacturing system are inputs, processes, outputs, and feedback. When they work properly, the result is a finished product, such as a new car.

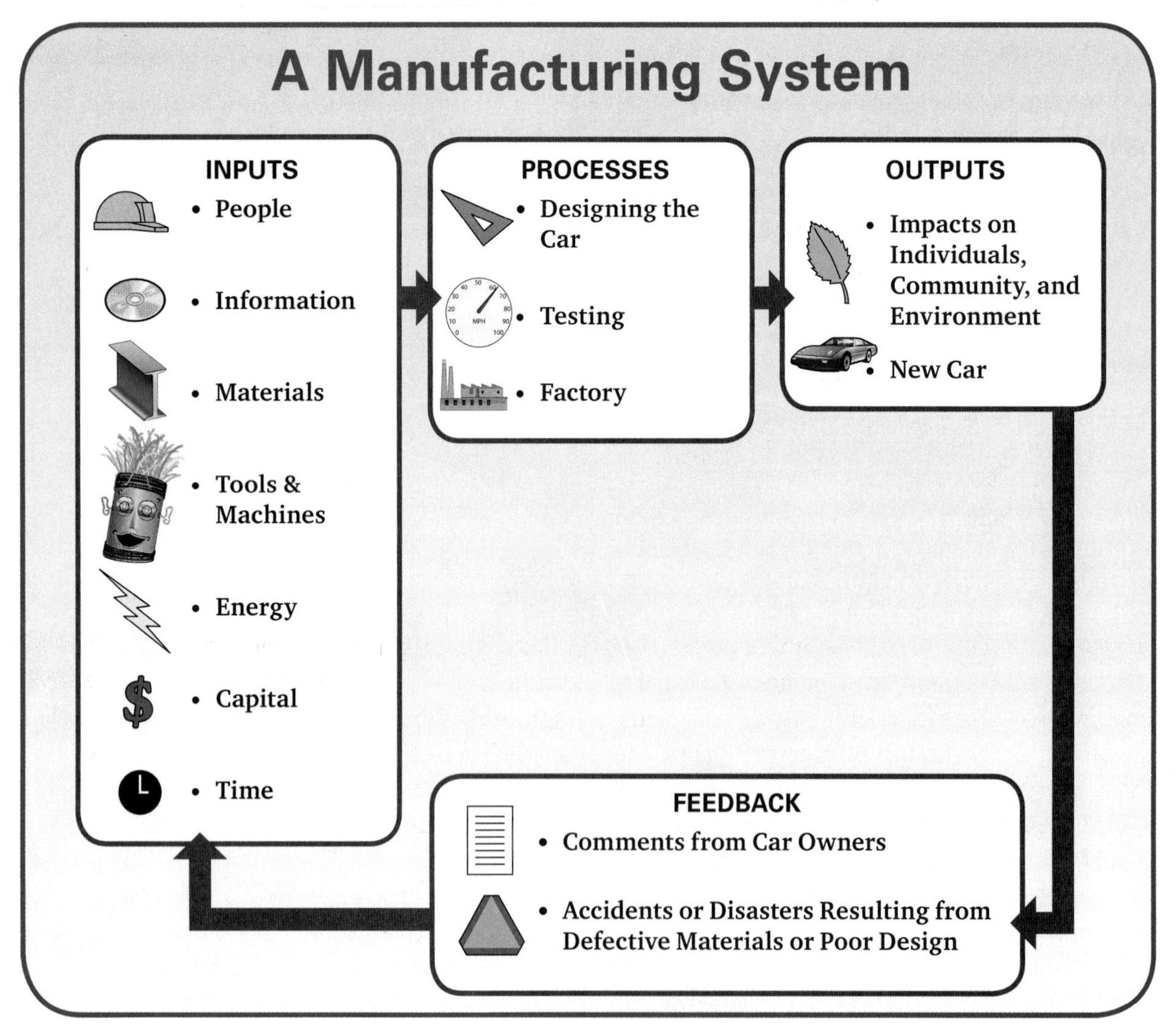

Math APPLICATION

How Earnings Become a Paycheck Suppose you just started your first job with a manufacturing company. You might be surprised at the amount on your first paycheck. There is quite a difference between your gross earnings—the amount you earn before taxes and other deductions—and your net earnings—your actual "take-home" pay.

Payroll deductions vary, depending on your income, where you live, and policies of your employer. Common deductions include income taxes (federal and sometimes state and local) and Federal Insurance Contributions Act (FICA) taxes. Your employer might also deduct payments such as union dues, medical insurance, retirement plan contributions, and savings deposits.

FICA taxes include your payment to the Social Security pension account and the Medicare account, funds that provide income and medical insurance when you retire. The FICA deduction for employees is currently 7.65 percent.

The pay stub shown here displays boxes that show the amounts withheld from a gross pay of $500. In addition to the deductions described, it includes the employee's share of a pension payment, medical insurance, and a charitable contribution that the worker chose to have held from the paycheck.

FIND OUT Answer the questions below to demonstrate your understanding of earnings.

1. Calculate the net pay for the employee whose paycheck stub is shown here.
2. If a person earns a gross pay of $400 per week, how much will the FICA deduction be?

THE TCP CORPORATION LAKE LAKOTA, MN 60041-1900
THIS IS A STATEMENT OF YOUR EARNINGS AND DEDUCTIONS.
PLEASE DETACH AND RETAIN FOR YOUR RECORDS

DEPT NO.	EMPLOYEE NO.	SOCIAL SECURITY NO.	PAY PER	PERIOD ENDING DATE	STRAIGHT TIME	1 1/2 TIME	OTHER
18	128	000-00-0000	WK	1-14-00	40 0		

HOURLY RATE	REGULAR AMOUNT	PREMIUM AMOUNT	SICK PAY	
12 50	500 00			

FEDERAL TAX	FICA	STATE TAX	COUNTY TAX	CITY TAX	
49 70	38 25	13 22			

MEDICAL INSURANCE	CHARITABLE CONTRIBUTIONS	UNION DUES	CREDIT UNION	BONDS	PENSION
5 84	1 00				14 27

TOTAL HOURS WORKED	TOTAL GROSS EARNINGS	TOTAL TAX DEDUCTIONS	TOTAL PERSONAL DEDUCTIONS	NET CURRENT PAY
40 0	500 00	101 17	21 11	

NO. 123456

ing. The company's fixed capital is invested in buildings and equipment it owns. Its working capital is the money it uses to buy materials and supplies, pay workers, buy advertising, and pay taxes. Manufacturers also need information, such as the facts about the capabilities of the machines in the factory. Time is needed to order materials, produce parts, and assemble products.

Processes include those in management and production. Managers make decisions in planning, organizing, and controlling. They make a plan of action so that things will work smoothly. Then they gather and arrange everything needed to do the job. Finally, they keep track of things. Production processes include the steps taken to make the products.

One output of manufacturing is, of course, the manufactured product. This output is expected and desirable. Other outputs, such as waste and pollution, may not be expected, and sometimes they are not wanted.

Manufacturers respond to feedback in ways that affect quality and how many products are produced. With the growing concern for the environment, for example, companies are spending money to address environmental issues.

Who Does Manufacturing?

Manufacturing is done by companies that specialize in making certain products. They exist in many different countries.

Companies and Corporations

A company is an organization formed by a group of people for the purpose of doing business. A corporation is a company that is owned by people who have bought shares in it. You probably recognize many manufacturing companies and corporations by name, like Apple Computer, Disney, and Toyota. See **Fig. 11-3**.

Some very small companies have only a few workers. Much industrial manufacturing is performed by small, locally owned companies employing a few hundred workers. These companies usually make parts that go into other larger products, such as seat covers and door panels for cars. The largest companies may manufacture many different products. The Procter & Gamble Company makes many different types of products, such as soaps, shampoos, cake mixes, and peanut butter.

Sometimes large companies own smaller companies. The smaller companies are called subsidiaries. They are separate but controlled by the same parent company. Sometimes the subsidiaries supply parts for the main product the company manufactures. For example, a company that makes telephones may also own a company that creates plastic molds, another company that makes computer telephone chips, and still another company that manufactures wire.

At other times, a parent company is diversified. That means the subsidiaries don't necessarily make the same kinds of parts or products. For example, United Technologies, a parent company, has many subsidiaries. Otis makes elevators and escalators. Carrier makes air conditioners. Hamilton Sundstrand makes mechanical systems for aircraft. Pratt & Whitney makes jet engines. Sikorsky makes helicopters. All of these companies, however, are owned by the parent corporation. All combined in one network, United Technologies is currently the 20th largest manufacturer in the U.S. and the 126th largest manufacturer in the world. Sixty-seven percent of its employees are based in countries outside the U.S.

» **Fig. 11-3** You may recognize the the manufacturer of this car by the logo on its grill. ***What is the most common way we are exposed to the names and logos of big companies?***

Sometimes several corporations, often with different specialities and located in different areas, work together as though they were one company. This arrangement, called a virtual corporation, has no employees; they all work for one of the partner corporations.

Another type of virtual corporation is one that does have employees, but it outsources all of its major functions. Suppose you and a friend come up with an idea for a new product and you establish a corporation. You hire an engineering firm to design your product. You contract with other firms to make, advertise, and sell the product. You and your friend are the heads of a virtual corporation.

The Global Market

Many companies have manufacturing plants in different parts of the world. Sometimes these locations are chosen to take advantage of lower labor costs and/or lower cost of materials. Other times, the locations are chosen so that products can be manufactured in the countries in which they are to be sold. This helps eliminate the high cost of shipping finished products.

Companies from different countries may often enter into manufacturing agreements. In some cases, the factories are managed and operated in a cooperative manner. Recently, a Canadian biopharmaceutical company established a joint venture to manufacture and market its drug and health products in Asia. It has built a new plant in China.

Rapid transportation and communications via satellite have helped products become popular throughout the world rather than in just one nation. See Fig. 11-4. Automobiles, for example, are manufactured and sold worldwide. Many products, such as clothes, appliances, and medicines, are in demand around the world. This has created a **global market**. The demand for these products and services is what drives the global economy.

Fig. 11-4 Nanjing Road in Shanghai, China is a large pedestrian mall. Stores there carry name-brand merchandise from all over the world. If you go there, you would easily recognize the same brands you find in your local mall. This is true for almost any country in the world.

This growing need for products is being helped by a gradual and steady reduction of trade barriers. For example, the European Union permits member nations to sell goods among themselves without import taxes or restrictions. A similar trade agreement, **NAFTA**, the North American Free Trade Agreement, also permits free trade among Canada, the United States, and Mexico. This decrease in trade barriers helps manufacturers find larger, worldwide markets for their products.

International Competition

One definite effect of the development of the global economy is the increase in competition for a share of the world marketplace. With goods available from all over the world, consumers have more products from which to choose.

All consumers want a good buy for their money. Worldwide competition has made quality an important issue. Often, the major differences between products produced by different manufacturers are the safety, reliability, and potential life of the product. However, to stay in business, a manufacturer must also make a profit. **Profit** is the money a business makes after all expenses have been paid. Because the selling price of a product must be competitive with similar products, the amount of profit may be small.

One method for staying competitive is to increase productivity. **Productivity** is the comparison of the amount of goods produced (output) to the amount of resources (input) that produced them. If you increase your output without increasing input, you increase your productivity. High productivity can help keep costs down.

Types of Production

The demand for products determines what will be manufactured. If fewer consumers need or want CD players, fewer will be manufactured. If fewer people take bus transportation, the demand for buses will decrease and fewer buses will be manufactured. On the other hand, if more consumers want DVD players or vehicles powered by fuel cells, more will be manufactured to meet the demand.

Consumer products that are widely used, such as candy bars and compact discs, are produced for stock. The company makes large amounts of the product and keeps them on hand. When orders come in, the products are shipped to customers.

When a product is not widely used, the manufacturer may not start producing it until the customers have placed orders. In this case, the product is produced on demand. For example, aircraft manufacturing companies make aircraft only when they have a firm order. The order may be for a single plane or for 100 planes, but the company doesn't start production until the order is placed.

A manufacturer generally chooses from three basic types of modern production systems to meet the demand for products. These are custom, intermittent, and continuous production. Each has its own advantages.

Custom Production

In **custom production**, products are made one at a time according to the customer's specifications. Each one is different. This type of production is usually the most expensive per number of products made. Custom-made products may be large and complex, such as a cruise ship, or small and simple, such as a piece of jewelry. See Fig. 11-5.

At one time, manufacturers believed that making products individually by hand would become a thing of the past. Instead, handcrafting industries have survived and grown. There are many small businesses producing specialized and/or custom-made, handcrafted prod-

Fig. 11-5 Custom-made products, such as this necklace, are often expensive and usually one of a kind.

ucts. While furniture manufacturing, for example, is becoming more automated, the most valued pieces of furniture are handcrafted.

Intermittent Production

In **intermittent production**, a limited quantity of a product is made. Then any necessary retooling or changeovers are made so a different part or product can begin to be produced. Many seasonal items, such as lawn mowers and snow blowers, are manufactured this way. Per part, this type of production, also called job lot production, is less expensive than custom production. The cost can be spread over more products.

Continuous Production

Continuous production is the system used for mass-producing products. It is also called line production or mass production. This means thousands, or even millions, of the same product are made in one steady process using an assembly line. In an **assembly line**, the product moves from one workstation to the next while parts are added.

Because changing a production line setup is expensive, continuous production is the most economical type of manufacturing system. Cars and electronic products, such as radios and computers, are made this way.

Many manufacturers today are combining the ideas of continuous and intermittent production. They are making large quantities of the same product with slight variations. Automobiles are an example. Two models may seem the same at first glance, but on close inspection, they may have many differences. The engine, seats or other interior features, and trim may vary. However, by producing many basic models at the same time, production is more efficient. Very close coordination of parts and assembly procedures is required to make this happen.

Fig. 11-6 This dimensional lumber comes in standard sizes.

Manufacturing Materials

Manufacturing often requires special materials. Several basic processes are used to turn those materials into products.

Materials may be natural, synthetic (human made), or a combination of both. Most raw materials need some refining or processing to convert them into industrial materials. Industrial materials are in a form that can be used to make products. For example, metal ore must be heated to a very high temperature to remove its impurities before it can be used to make a product.

Industrial materials are usually turned into standard stock. This means that the material is formed or packaged in a widely used (standard) size, shape, or amount that is easy to ship and to use. Standard stock includes sheets of plywood, steel, and aluminum. Bolts of cloth and barrels of liquid chemicals are also standard stock. See **Fig. 11-6**.

Science Application

Chemistry of Plastics Although a few occur in nature, most of the plastics we use today are manufactured from chemicals. Plastics are used to make clothing, packaging, building materials, and thousands of other products.

The raw, unfinished material from which a plastic is made is called a resin. Other ingredients, called additives, are often mixed with the resin. Additives give the final material its specific properties. For example, in fiberglass, glass fibers are added to a plastic resin. The fibers improve rigidity, strength, and resistance to cracking. Other chemicals, called plasticizers, make plastic materials softer and more flexible.

Plastics often include one or more stabilizers to protect the product from chemical reactions. Depending on which stabilizer is used, it may protect against attack by heat, ultraviolet light, or oxygen.

Other additives include colorants and flame retardants. Wood and metal are colored by a surface layer of paint. Plastics can be given a color that extends throughout the material by adding colorants to the resin. Flame retardants are added to carpet fibers to reduce dangers of toxic fumes and rapid spread of flames.

FIND OUT Choose one of the following plastics: polyethylene, polyurethane, or polyvinyl chloride (PVC). Research how it is used and what properties make it useful.

Scientists continue to develop new materials capable of providing specific properties not available in conventional materials. An alloy is made by combining two or more metals or a metal and a nonmetal. The new material has improved properties.

A composite is a material made by combining two or more materials. A composite is usually much stronger and more durable than the individual materials from which it was made. One widely used composite material is fiberglass. Another example is graphite-reinforced plastic, sometimes called carbon fiber. Similar to fiberglass, it is a clothlike woven mesh of carbon fibers instead of glass fibers, which are then coated with plastic. The result is a molded or shaped part that is smooth, stiff, and strong. These products are the result of chemical technologies that allow us to alter or modify materials.

Ceramics are materials made from nonmetallic minerals that have been heated to very high temperatures. Ceramic materials are strong, hard, and resistant to corrosion and heat.

SAFETY FIRST

Handling Materials. Materials must be stored, used, and disposed of properly. A Material Safety Data Sheet (MSDS) lists details for using a specific material safely. Ignoring safety guidelines can result in personal injury, damaged equipment, and damage to the environment.

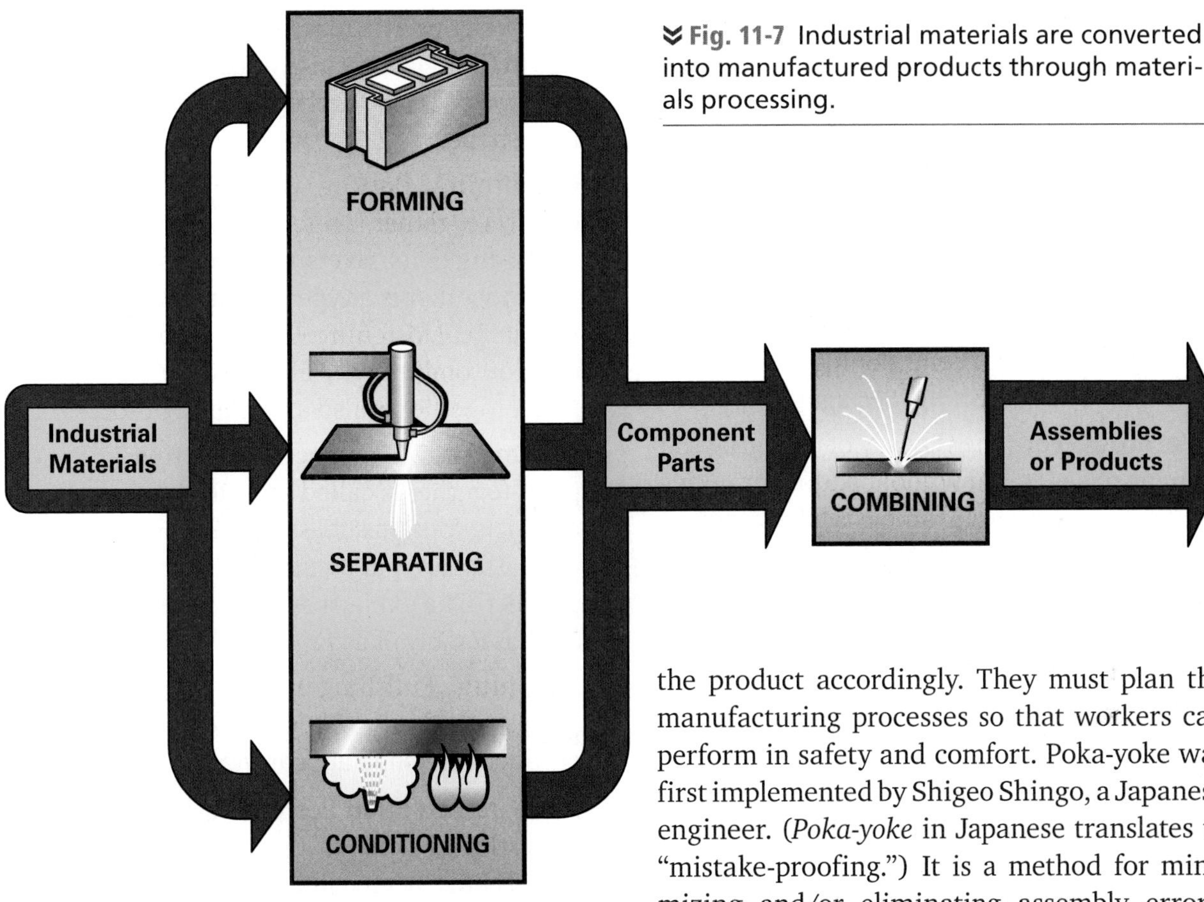

Fig. 11-7 Industrial materials are converted into manufactured products through materials processing.

Manufacturing Processes

What is the next step in manufacturing once materials are gathered? Processes are used to make products and check for quality.

Total Quality Management

Total quality management means that employees are expected to meet a performance standard for their jobs. Their job description describes how they will be measured to see if they're doing a good job. Empowerment means allowing employees to make decisions without asking the boss every time.

Designers and engineers must think about how people will interact with the product during manufacturing and must also shape the product accordingly. They must plan the manufacturing processes so that workers can perform in safety and comfort. Poka-yoke was first implemented by Shigeo Shingo, a Japanese engineer. (*Poka-yoke* in Japanese translates to "mistake-proofing.") It is a method for minimizing and/or eliminating assembly errors. For example, using color-coded parts makes it easier for workers to remember that the red wire plugs into the red socket. Instead of having separate left-hand and right-hand parts, only one part is used that can be flipped over so it fits either way. When parts are designed to clip or snap together instead of requiring screws and bolts, workers are less likely to make mistakes attaching them.

Production Processes

All the processes used to actually produce the product are production processes. These processes can be classified as preprocessing, materials processing, and postprocessing. See **Fig. 11-7.**

"Pre" means before. Preprocessing happens before any work is done on the material. When the raw material or standard stock is received at the factory, it must be unloaded, stored, and protected until used. These activities are examples of preprocessing. Note that preprocessing activities don't actually change the material.

Materials processing means changing the size or shape of a material in order to increase its usability or value. Industrial materials are turned into components or parts. Materials can be changed using several basic processes: forming, separating, conditioning, combining, assembling, and finishing. See again **Fig. 11-7**.

- **Forming.** Forming is a way of changing the shape of a material without adding or taking anything away. An example is hammering a round piece of steel into a flat shape. You still have the same amount of steel, but it is flat instead of round. Other forming methods include forging, extruding, casting, and molding.
- **Separating.** Changing the shape of a material by taking some away is referred to as separating. If you saw a two-by-four in half, you have changed its shape by separating. Cutting a piece of fabric from a large roll to make a car seat cover is another example. The amount of material has been changed; there is less than when you started. Common methods of separating include sawing, drilling, shearing, and punching.

  In one special type of separating, called noncontact cutting, there is no contact between the tool and the material. An example is cutting by laser beam. The laser melts a small amount of material along the cutting path. Other examples are water jet cutting and electrical discharge machining.
- **Conditioning.** Changing the structural properties of a material is called conditioning. For example, heat treating is done to the teeth on a gear to make them extra hard so they won't wear out as fast. Adding a catalyst to fiberglass resin produces a chemical reaction that causes the resin to harden and take shape.
- **Combining.** Putting two or more materials or parts together is referred to as combining. Using bolts, rivets, and screws to hold parts together is a type of combining called mechanical fastening. Gluing is another form of combining. Welding two pieces of metal together is also a combining process.
- **Assembling.** The process of putting parts together is called assembling. Some products contain subassemblies, or smaller units, that are put together to form the main product. The exhaust system, for example, is a subassembly of a car.
- **Finishing.** Finishing processes include such things as painting. They are done to give the product a finished appearance.

"Post" means after. Postprocessing activities include things done to the product after the materials have been changed in form or shape. Installing, maintaining, repairing, and altering are all examples of postprocessing. Handling, protecting, and storing products are also examples of postprocessing activities.

Postprocessing may also include recycling activities. Collecting glass or plastic bottles, sorting them, and grinding them into chips is an example of recycling. The chips can then be used as raw material or standard stock for manufacturing other products.

Postprocessing may also include servicing a product to keep it in good operating condition during a warranty period. Automobile manufacturers are an example.

Young Innovators

JETS TEAMS

You have a strong desire to understand how things work, you like to build things and take them apart, and you look for ways to make things better or improve them. Where can you apply your skills and develop your passion for how things work?

There are plenty of problems in the world that need real solutions from people with the type of passion you possess. If engineering the solutions to real-life problems is something that you enjoy, then perhaps JETS TEAMS may be the opportunity for you. The JETS TEAMS competition is an annual, one-day event where groups of high school students work together to synthesize key concepts in math, science, and physics and apply them to solve real-world engineering challenges.

The competition questions are based on everyday issues found in various industries and fields of study. An example might be to determine the most efficient way to lay out a factory to produce ice cream. Perhaps you will be tasked with creating a viable solution to the problem of cheap, convenient, and safe mass transportation. Then again, you might simply need to find a new use for an old technology. The questions posed challenge your knowledge of math, chemistry, physics, biology, and computer applications. You apply what you have learned in the classroom to create viable solutions to existing problems.

The TEAMS competition is given in two parts and each part is open book, open note, and open discussion. The first part consists of a series of objective multiple-choice questions. In the second part of the contest, students work to solve open-ended problems that present scenarios based on the questions answered in the multiple-choice section. Students and school teams receive local, state, and national recognition and awards from participating in the event. The JETS TEAMS competition is an excellent way for students to improve their problem-solving and teamwork skills while having fun and exploring engineering at the same time.

Impacts and Effects of Manufacturing Technology

Manufacturing is important to our society and our economy. Many people work in manufacturing to help produce products. They also buy products with the money they earn. The more products people buy, the more products are manufactured. This enables more people to work. Also, materials are worth more after they have been changed into useful products. That's value added. Their value is increased by the manufacturing process.

Cars, for example, are an output of a manufacturing system. They are so numerous and widely used that they have had an impact on nearly every part of our lives. Cars affect our:

- **Economy.** Many people's jobs depend on the automobile industry. Cars enable people to work far from their homes and to travel more for pleasure.
- **Society.** People sometimes see cars as a symbol of status, wealth, or even personality. See **Fig. 11-8**. Cars can be a way to express who we are or who we would like to be.
- **Politics.** Who should pay for road building? Should the government protect the American car industry from foreign competition? Should the speed limit be raised or lowered? All these political issues result from our use of automobiles.
- **Environment.** To build and drive cars, we use nonrenewable resources, such as metals and oil. Car exhaust pollutes the air. Auto junkyards are an eyesore, and they can pollute the land and water.

Sometimes, manufacturing can play a part in overcoming negative impacts. For example, one way to prevent environmental damage is to design products to be demanufactured (taken apart) and recycled. Manufacturers have collected information about materials, the amount of time it takes to remove parts, and what parts have the most value.

Another approach is to focus on reducing waste materials left over from the manufacturing process. For example, as part of the National Waste Minimum Partnership Program of the U.S. Environmental Protection Agency (EPA), Toyota reduces the amount of solvent used to clean paint from the robots used to paint fuel tanks by 25 percent.

Other companies have found that garbage can be a valuable industrial material. Some park benches and picnic tables are manufactured from recycled plastic. Ford uses plastic soda bottles to make grill reinforcements, door padding, luggage rack side rails, and other parts.

Fig. 11-8 This Lamborghini Gallardo SE attracts a lot of attention. ***What characteristics does this car have that might convey a symbol of status?***

CAREERS IN MANUFACTURING

There are many fields with jobs in manufacturing such as in the aerospace, chemical, computer, food, machinery, motor vehicle, pharmaceutical and medicine, printing, steel, and textile industries. Jobs are available for artistic people (hand food decorator), high-tech types (aerospace engineer), hands-on workers (assembler and fabricator), and science types (chemical plant and system operator). The U.S. Department of Labor estimates that there will be seven million new jobs in the manufacturing industry by the year 2012.

Aerospace Engineer • First-line Supervisor/Manager • Chemist Mixing and Blending Operator • Computer Hardware Engineer Electrical and Electronic Engineering Technician • Hand Food Decorator CNC Operator • Industrial Engineer • Sewing Machine Operator Assembler and Fabricator • Sales Representative • Biological Technician Welder • Baker • Inspector • Job Printer • Printing Machine Operator Customer Service Representative • Electrician • Production Worker Maintenance and Repair Worker • Crane and Tow Operator • Packager

INVESTIGATE CAREERS What industry do each of the above jobs belong to? Do some of the jobs belong to multiple industries? Create a poster for each type of manufacturing industry that displays the different jobs available.

CHAPTER 11 REVIEW

Main Ideas

- The purpose of manufacturing is to make things that people want and need.
- Like other systems, manufacturing has inputs, processes, outputs, and feedback.
- Manufacturers generally choose from three types of production systems to meet the demand for products.
- Manufacturing impacts our economy, our society, our politics, and our environment.

Understanding Concepts

1. Why is manufacturing so important to our daily lives?
2. What are the three basic types of production?
3. Describe how companies compete in a global market.
4. Name the six production processes.
5. What impacts do cars have on our economy?

Thinking Critically

1. How did an improvement in leisure time and education occur during the Industrial Revolution?
2. Do you think that the Industrial Revolution occurred in all countries around the world at about the same time? Why or why not?
3. Why would a manufacturer allow negative outputs in a product?
4. How have manufactured automobiles affected you socially?
5. Why is communication essential in a global market?

Applying Concepts & Solving Problems

1. **Assess** Evaluate the quality of a manufactured product. By assessing the quality of the product, you can also assess the process used to make it.
2. **Economic Impacts** Write a report discussing the economic effects of producing a practical pollution-free car. How would this car affect your personal life? How would it affect your local area? What would the impact be on the oil industry?
3. **Design** Create a product that can be easily produced using an assembly line.

Activity: Create Composite Materials

Design Brief

Composites are commonly used as materials in many products, such as the hull of a fishing boat or the body of a car. A composite material consists of a matrix (woven mesh) that is reinforced with a liquid that hardens and turns the matrix into a thin sheet. The sheet either remains flat or is formed into a desired shape. Examples of matrix materials include glass fibers and carbon fibers. The reinforcement material is usually a form of plastic resin.

Create several sample pieces of composite materials and test their properties.

Refer to the

HANDBOOK

Criteria

- You must create at least three examples of composite materials using three different matrices. You should make at least three samples of each type.
- You must design at least three different tests to determine the materials' strength, flexibility, and breaking point.
- You must keep a daily log of your work.
- You must document and chart your test results.

Constraints

- Your samples must not exceed four square inches in size.
- Your daily log must contain all information that can be recorded during the project.

Engineering Design Process

1. **Define the problem.** Write a statement that describes the problem you are going to solve. For example, what do you think determines a material's flexibility?
2. **Brainstorm, research, and generate ideas.** With your team, discuss possible solutions. Hint: You may want to study existing composites and matrix properties available to you. Good choices may include mesh materials like gauze, cotton cloth, and paper towels.
3. **Identify criteria and specify constraints.** These are listed on page 244.
4. **Develop and propose designs and choose among alternative solutions.** Choose the best designs that will solve your problem. For example, you may be able to determine before testing that some matrices will not produce a strong product.
5. **Implement the proposed solution.** Decide on the process you will use for making the composites and testing them. Gather any needed tools or materials. You may want to use a nonstick baking sheet as a work surface.
6. **Make a model or prototype.** Be sure to wear plastic gloves. Thin the glue with water to make it spread better.
7. **Evaluate the solution and its consequences.** After your samples have hardened, analyze each one to determine its strength, flexibility, and breaking point. What impacts and effects will result from this solution?
8. **Refine the design.** Based on your evaluation, change your composites if needed.
9. **Create the final design.** Once you have an approved design, create your final composites and test them. Be sure to document the results.
10. **Communicate the processes and results.** Present your finished composites and test results to the class. Be prepared to answer questions. Turn in your assignment to your teacher. Be sure to include the name of the activity, your definition of the problem, a description of how you solved the problem, your log sheets, and test results.

SAFETY FIRST

Before You Begin. Make sure you understand how to use any tools and materials safely. Have your teacher demonstrate their proper use. Follow all safety rules.

CHAPTER 12 PRODUCT DEVELOPMENT

⮝ Precise measurements are extremely important in making most manufactured products. In this chapter's activity, "Reverse-Engineer a Product," you will investigate how a product works and what parts and materials were used to make it.

Objectives

- Explain where product ideas come from
- Name two important factors that must be considered in the engineering of a product.
- Describe the three types of engineering analysis.
- Summarize the steps taken to market and distribute a product.
- Explain the stages in a product's life cycle

Vocabulary

- research and development (R&D)
- design for manufacturability
- working drawings
- functional analysis
- failure analysis
- value analysis
- target market
- supply chain

Reading Focus

1. As you read Chapter 12, create an outline in your notebook using the colored headings.
2. Write a question under each heading that you can use to guide your reading.
3. Answer the question under each heading as you read the chapter. Record your answers.
4. Ask your teacher to help with answers you could not find in this chapter.

The Birth of a New Product

Hardly a day goes by without a new or "improved" product appearing in stores or on TV. Where do manufacturers get all their ideas for products? How are the ideas turned into the items we buy?

In most manufacturing companies, developing new products is the responsibility of the **research and development (R&D)** department. Research is done to gather information. Development is the process of using that research information to bring the product into existence. The research and development department may be a part of the manufacturing company or an independent organization hired to create workable ideas. See **Fig. 12-1**.

Managers from R&D and other departments are often combined into management teams. These management teams work together to oversee the work done on the product. Each department has its own concerns. For example, because the goal of manufacturing is to produce and sell products that make a profit, the finance department must examine the financial risk. They determine if it makes economic sense to proceed with a new product idea.

In this chapter, we'll discuss the different parts of the product development process, from getting the idea to selling the product to the customer.

« **Fig. 12-1** These two R&D team members are working on a new package for a product.

Where Do Product Ideas Come From?

Sometimes the idea for a new product comes from an outside individual, who may then sell the idea to a manufacturer. In that case, the personality of the inventor or designer may influence the design. For example, clothing designer Ralph Lauren likes western wear, and his casual collections often feature that look.

Most product ideas, however, come from the research and development group. Scientific researchers, designers, and engineers all work within this department. They often use brainstorming and problem-solving techniques to help generate ideas.

Sometimes the R&D department conceives of a new product to solve a particular problem. Refrigerators were invented to solve the problem of food spoilage. The problem created the demand for the product. Other times, the department comes up with an idea for a product that they think will be useful or desirable. Through advertising, they plan to create a demand for it. See **Fig. 12-2**.

To protect their chosen idea, companies obtain a patent. A patent is the right, granted by the government, to produce and sell a particular product. No one else may copy it. Companies also usually check existing patents to be sure no one else has produced a product they're interested in working on.

Unfortunately, not every idea becomes a successful product. From about 60 new ideas, only one finally makes it as a new product. That's another reason manufacturers are always looking for new ideas.

Fig. 12-2 Cell phones that take pictures were introduced into the United States market in 2000. They were not a success. Later, they were reintroduced using an extensive advertising campaign. This time they caught on.

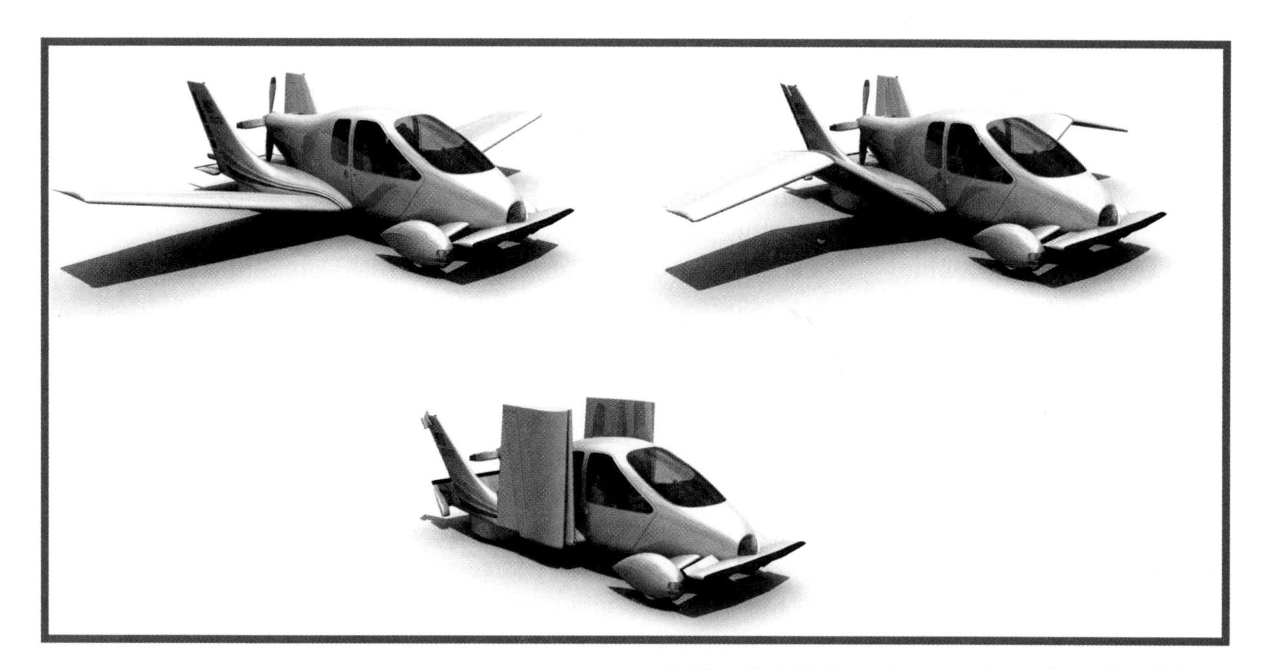

Fig. 12-3 This scale model mock-up is of the Transition Personal Air Vehicle, a design for a real flying car that you may one day see on the roads—and in the air.

What Will the Product Look Like?

Deciding what the product should look like is one of the first steps in the development process. Suppose the new product is really unusual, such as a flying car. Appearance may affect the materials and power source chosen to make the car. It may also help determine whether or not a customer wants to buy the product. As a rule, the design should be eye-catching and attractive.

Most designs are created using CAD software and computer-aided engineering (CAE) programs. Design details are stored in the computer and can be used later by engineers and other workers as the product is developed, tested, and analyzed.

Drawings

Product designers often start with many thumbnail sketches, which don't include much detail. The purpose of making thumbnails is to capture many different ideas.

Then the product designer picks the best parts of each thumbnail sketch and combines them into a larger, more detailed drawing called a rough sketch. As each new drawing is made, the product's final look becomes more and more apparent. During the design process, engineers, managers, and workers from different departments in the company may contribute suggestions.

A realistic rendering is made last. A rendering shows the designer's final ideas about the appearance of the product. It is usually shown to company managers as part of the approval process. It may also be shown to potential consumers to see if they like the design.

Mock-Ups

Designers also make mock-ups of products. A mock-up is a three-dimensional model of the proposed product. See Fig. 12-3. It looks real but has no working parts. Its purpose is only to show people what the final product will look like.

Most of the time a mock-up is built full size. However, sometimes a scale model is made. Mock-ups may be made of wood, plaster, plastic, or even cardboard. Many virtual 3D models are created using CAD software.

What Kinds of Research Are Done?

As the design concept is refined, engineers begin to tackle issues involved in making the product. First, research is done to gather information they can use. What kind of material will be light yet strong enough to make a flying car? What scientific principles can be used to keep it in the air? These are just a couple of the kinds of questions that might be answered by research.

Basic research is done to help people understand how things work. It is usually scientific in nature. Often, new information can be put to use immediately. Sometimes, however, additional knowledge must be gained first. Scientists discovered the transistor as they investigated electronics. It wasn't until later that they found uses for it, such as making radios and TVs that are more portable. See Fig. 12-4.

Applied research is done to solve a specific problem. Most research done in manufacturing is applied research. For example, finding ways to lower the fuel emissions in a vehicle is done to decrease pollution.

The same methods are used for both basic and applied research.

- Retrieving is finding information that is already known so that a manufacturer doesn't "reinvent the wheel."
- Describing is telling about present conditions. For example, if you wanted to manufacture that flying car, you'd need to know how much weight it would need to carry.
- Experimenting occurs when researchers try things out to see what happens and if a product idea will work.

Fig. 12-4 Transistors and printed circuit boards replaced old vacuum tubes. Further research led to microchips, making way for very small recorders and playback devices.

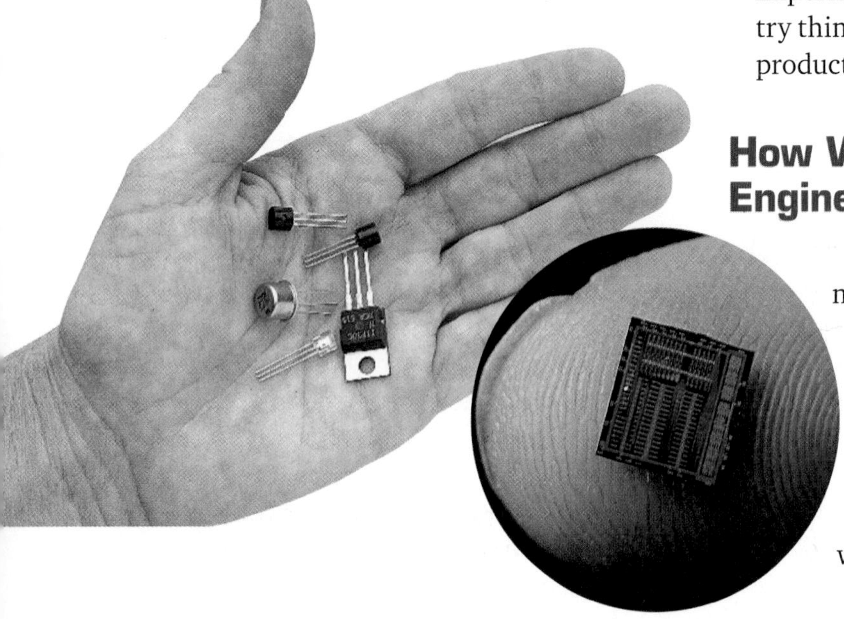

How Will the Product Be Engineered?

Product engineering involves planning and designing to make sure the product will work properly, will hold up well under use, and can be manufactured with few problems. In a process called concurrent engineering, groups of specialists from all areas of design and production work as teams.

Many factors, including all criteria and constraints, must be considered when engineering a product. Especially important are designing for a product's function and manufacturability. Then the design is finalized and analyzed. Engineers use CAE programs for these tasks because their work involves mathematics, and a computer can solve math problems much faster than a human can.

Designing for the Product's Function

Functional design ensures that a new product will work properly. Proper function often depends upon the materials used, structural and mechanical elements, and any power requirements. See **Fig. 12-5**.

Materials Engineers must specify the exact materials to be used in the product. Will the materials used to make a flying car have to be strong? What about weight? The product engineer has to know exactly what is required.

The composition of a material has an effect on its characteristics, such as strength. Certain plastics, for example, are twelve times stronger than steel. Could the wing on a flying car be made of plastic? A wing and a CD case might both be plastic products, but they have different strength requirements. Thus, different plastics would need to be specified.

Science Application

Mechanical Properties Engineers often test materials to see which ones to use for a product. One way they evaluate materials is by testing their mechanical properties.

Ductility is a material's ability to be formed and reformed.

Elasticity is the ability of a material to regain its original shape after being pulled or pushed.

Hardness is the ability to resist denting and scratching.

Plasticity is the ability to be pressed, molded, or stretched into a new shape that can be maintained.

Strength is the ability to resist forces such as compression and tension.

Depending on the product's function, different materials are used. Concrete, for example, is often used in construction. It has both strength and hardness. However, concrete would not be used to give elasticity to a product.

FIND OUT Locate products that demonstrate each kind of mechanical property. You can locate these items by testing for each property. Create a chart showing your findings. Do any of your products feature more than one property?

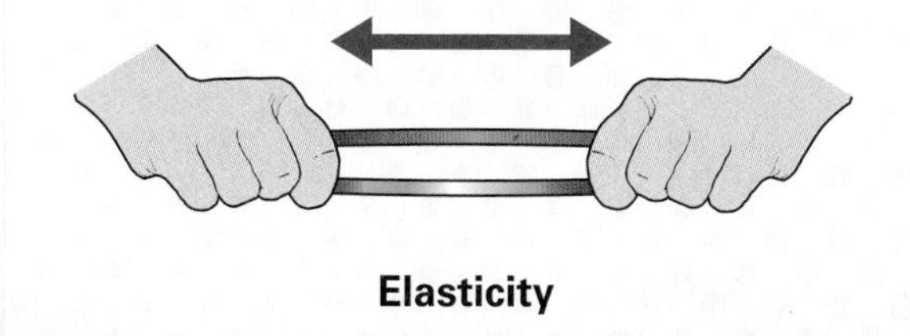

Elasticity

« **Fig. 12-5** Engineers must determine how to make a product work well. ***What product function is being discussed in this photo?***

Structural and Mechanical Elements Structural elements are frameworks and supporting members. Mechanical elements are the working parts. Such things as fasteners, the shape of a piece of aluminum, and the meshing of gears and levers can all affect product performance. Why does the cap on a pen snap on and stay there? Why does the door open when you turn a doorknob? It's because the structural and mechanical elements of these products were engineered properly. The engineer is responsible for how each part will fit and work with the other parts. See **Fig. 12-6**.

Some products, such as automobiles, must be routinely serviced. Design for servicing considers how workers will gain access to different systems and maintain them properly. The easier it is, the less costly for service centers and customers.

Power Sources If power is needed to make the product work, engineers must select the most efficient, accessible, and economical power source that can be used. What would you choose to power a flying car? Should the power source be replaceable or renewable?

Many products use electricity. Some, such as a tractor, are powered by hydraulic power. Still others may be mechanically powered, such as by one or more springs. Even compressed air is a power source. The size of each piece of electrical wire, the horsepower of each motor, and the stroke of each cylinder are all determined by the engineer.

Designing for Manufacturability

What effect will the design of the product have on the ability to actually manufacture it? How will the design affect the production of parts? Engineers answer these questions by trying to **design for manufacturability**, or design for ease in manufacturing. For example, a part may be made with rounded corners instead of square ones if it is easier to make it that way.

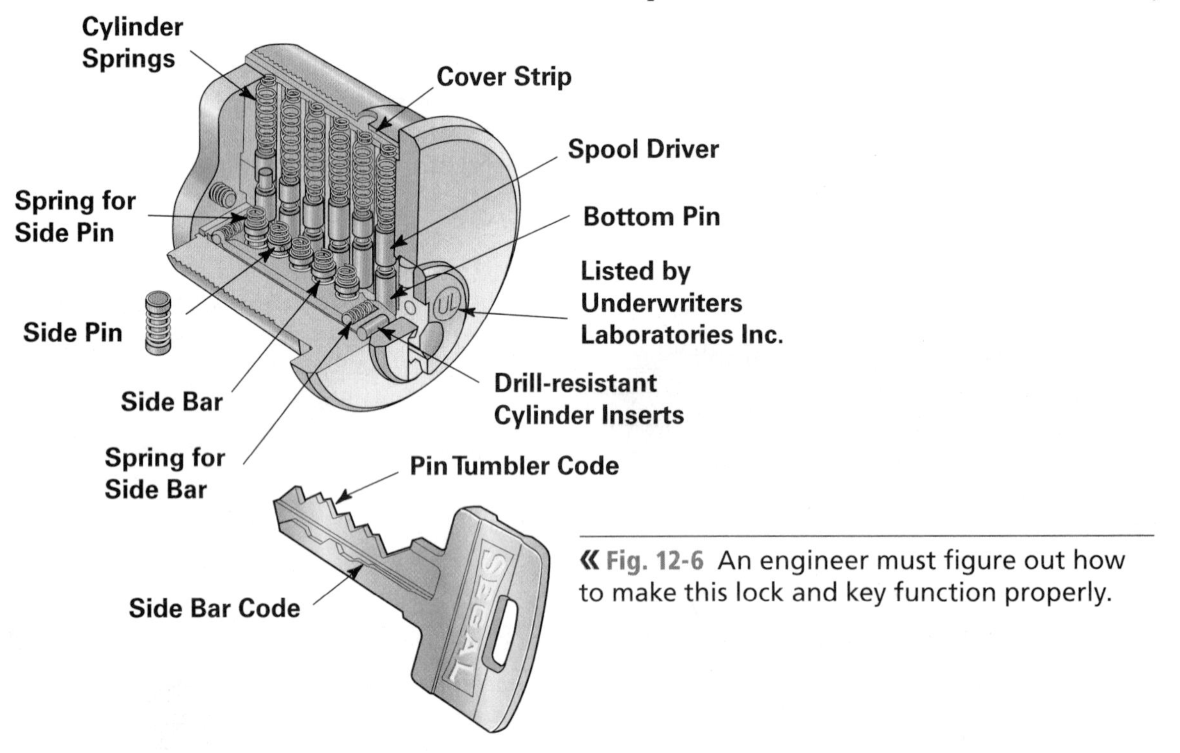

« **Fig. 12-6** An engineer must figure out how to make this lock and key function properly.

» **Fig. 12-7** Many parts inside computers can be held together with clips instead of nuts and bolts.

Engineers must also design products so that parts can be assembled more efficiently. For example, if parts can be held in place with clips that are molded into the product instead of with nuts and bolts, the nuts and bolts can be eliminated. Assembly can also be done much faster, since using clips is quicker than threading nuts onto bolts with wrenches. Many products, such as computer printers, automobile dashboards, and refrigerators, are held together with clips instead of screws or nuts and bolts. See **Fig. 12-7**. Would you be surprised to learn that many airplane parts are held together with glue?

Other factors that engineers must consider include interchangeability of parts, standardization, simplification, and modular design.

Interchangeable Parts The process of mass production involves producing large quantities of a particular product. In order to do this efficiently, interchangeable parts are required. These parts are identical. Any one of them will fit the product. Assembly is thus speeded up.

However, because of human error, differences in machines, and tool wear, there are always tiny differences. To allow for this, engineers plan for a certain amount of variation. Tolerance is the amount that a part can be larger or smaller than the specified design size and still be used.

Standardization Manufacturers who make similar products often require the same kinds of parts. Makers of bicycle tires, for example, can sell them to more than one bike manufacturer. In order to do this, however, the sizes must be the same. Agreement on common sizes of parts is called standardization.

Whenever possible, engineers specify standardized parts for a product. See **Fig. 12-8** on page 254. Instead of every manufacturing company making their own sizes of bolts, they can buy standard-size bolts. It is cheaper and faster. It also reduces the amount of engineering work that must be done on a new product.

Fig. 12-8 Engineers are aware of the wide selection of standard sizes from which they can choose. For example, these wood fasteners (nails) come in many sizes.

Simplification Reducing the number of different size parts in a product also makes the product easier to manufacture. Suppose that the design for the seats in the flying car specified 16 screws of different sizes. Then 16 different screws would have to be ordered and assembled correctly. However, if the same size screw fits in each place, just one size would be required. Assembly would be simpler, too. Any one of the screws could be placed in any one of the 16 locations.

Modular Design A module is a basic unit. In manufacturing, using preassembled units is called modular design. For example, in automobiles, units such as engines, wheels, and radiators are created as modules. It's possible to produce a variety of car models by starting with basic parts and adding different modules. The same basic car could be made with a four-cylinder engine, a six-cylinder engine, or even an eight-cylinder engine.

Communicating the Final Design

As a new product evolves, drawings and models are made. Any changes are recorded, and the final results are analyzed. Finally, finished drawings and models are presented to management for approval. Eventually, the final drawings and models are given to the production department where the product is manufactured.

A computer-generated virtual model, usually called a solid model, may be constructed using special CAD 3D modeling software. The size, shape, and other details of the parts are specified and stored in the computer. See Fig. 12-9. Other manufacturing details, such as the need for assembly lines, are recorded using manufacturing planning software. Workers who need that information to make the product can easily obtain it and make hard copies if needed.

Working Drawings As the details for the product are determined, they must be recorded in a set of **working drawings**. These drawings show exact sizes, shapes, and other details. Products are manufactured using them, so the drawings must be accurate and complete. There are three main types of drawings in a set of working drawings: detail drawings, assembly drawings, and schematic drawings.

A detail drawing specifies the details of a particular part. The detail drawing shows shape, dimensions (sizes), and locations of features like holes and bends. Tolerances are indicated. Special information about materials and surface finish may also be provided.

Assembly drawings show parts in their proper places and how they fit together. See Fig. 12-10. There are several different types of assembly drawings, but they all serve the same purpose.

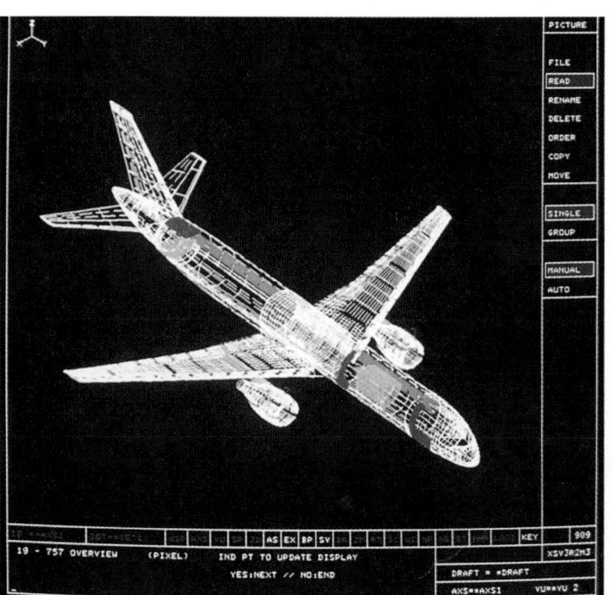

≈ Fig. 12-9 This computer model was used to "build" and analyze this jet engine before the real product was manufactured. The colors in the model represent stress levels for the different materials.

≈ Fig. 12-10 Exploded assembly drawings, such as this one, are needed for repair and service manuals. They also help workers assemble products.

KEY NO.	PART NO.	DESCRIPTION
1	34750C AAC	Frame
2	34781	Grips
3	34771C	Handlebar Assembly
4	302108	Binder Bolt
5	302107	Wedge
4-6	32752	Handlebar Stem
7	303384	Head Bearing Set
8	34784C	Fork
9	303005 CCD	Frontwheel - Less Tire (Tire Size 20x1.5)
10	303155	Front Spoke & Nipple Set (6 each)
11	303520	Axle Bearing Set
12	32375Z	Frontwheel Retainer
13	27679Z	Kickstand Assembly
14	14590C	Sprocket
15	12300C	Crank
16	32809	Pedals
17	34779C	Chain Guard
18	12834	Chain & Link
19	303333	Crank Hanger Bearing Set
20	303013 CCD	Rearwheel - Less Tire (Tire Size 20x1.75)
21	303155	Rear Spoke & Nipple Set (6 each)
22	98X250	Saddle
23	32942Z	Seat Post
24	303404Z	Seat Post Clamp Assembly
25	34350	Reflector Package
26	34792Z	Front Reflector Bracket
27	34800Z	Rear Reflector Bracket
28	34696	Handlebar Pad
29	32618	Tie Straps
30	34764	Number Plate
31	34783PA	Caliper Brake
*	64X286	Front Plate Decal
*	F-4661	Owner's Manual

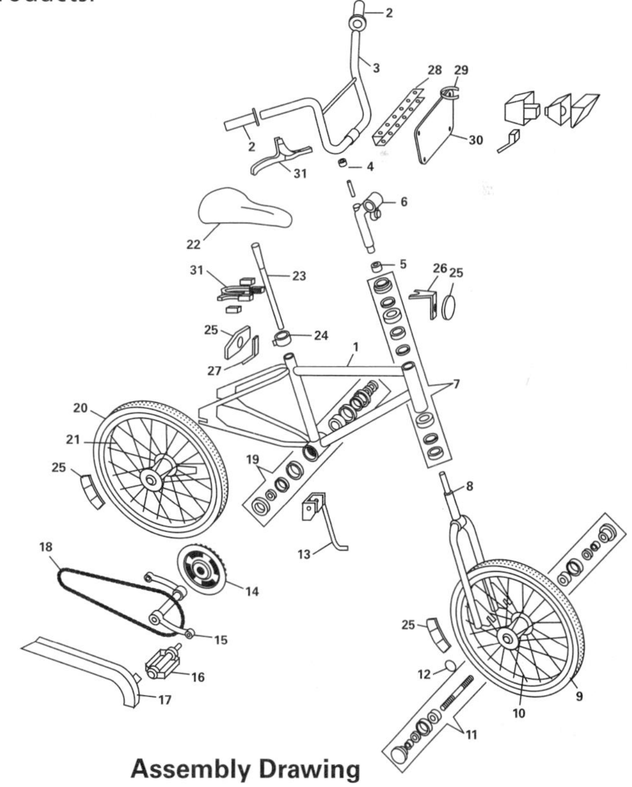

Sometimes a special drawing called a schematic is used to show the position of parts in a system—that is, the "scheme" of things. Parts are not shown as they actually look. Instead, symbols are used to represent the parts. Electrical and hydraulic systems are usually shown in schematic form.

Prototypes A prototype is a working model of the actual product. It is the first of its kind, and prototypes have been traditionally built by hand, part by part. Only one may be built or maybe a few hundred. The number depends on the size and complexity of the product. Hand-built prototypes are usually very expensive. A prototype for a new automobile, for example, can cost a million dollars or more! For this reason, more and more prototypes are being created using CAD programs and rapid prototyping software. See Chapter 7, "Graphic Communications," for more on rapid prototyping.

Prototypes can be used to check manufacturing procedures. As the prototype is being built, manufacturing engineers note any difficulties. This provides feedback for planning the processes that will be used to mass-produce the product.

Analyzing the Design

A product engineer's job includes making sure that everything about the new product is right. One very important aspect of product engineering is reliability. It's one thing for a product to work properly, but will it continue to work? The reliability of a product and its individual parts can seriously affect its success. Multiple reliability problems can quickly give a manufacturer a bad reputation. Consumers will then choose a competitor's product instead, and sales will go down. On the other hand, when a manufacturer's products have a high reliability rating, consumers are more likely to buy them. This increases demand, and manufacturers can then ask a higher price for their product.

Math Application

Manufacturing with Fractions In the manufacturing industry, working drawings are usually dimensioned in decimals. However, in your lab, drawings might be dimensioned in fractions.

For example, imagine you want to increase the diameter of the top of the wastebasket in this drawing by $\frac{5}{8}$". To find the new diameter, you add the existing diameter to the amount of increase: $12\frac{1}{4}" + \frac{5}{8}"$. The two fractions have different denominators, so you find a common denominator. In this case, change $\frac{1}{4}$ to $\frac{2}{8}$. Once the denominators are the same, you can add the numbers:

$$12\frac{2}{8}" + \frac{5}{8}" = 12\frac{7}{8}"$$

Suppose you want to cut the bottom of the wastebasket from a piece of wood measuring 11" × 22". Will you have enough wood left for another project for which you need a piece that measures 11" × 12"? To find out, you need to subtract $10\frac{1}{8}"$ from 22". First, borrow from the whole number to have another fraction to work with and then subtract:

$$21\frac{8}{8}" - 10\frac{5}{8}" = 11\frac{7}{8}"$$

It looks like you will need to find another piece of wood for the second project.

FIND OUT The outside diameter of the top of the wastebasket is $12\frac{1}{4}$". If the walls are $\frac{3}{8}$" thick, what is the inside diameter?

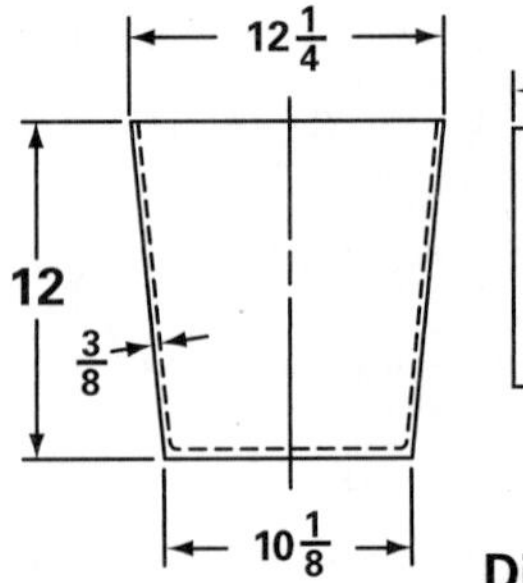

Wastebasket

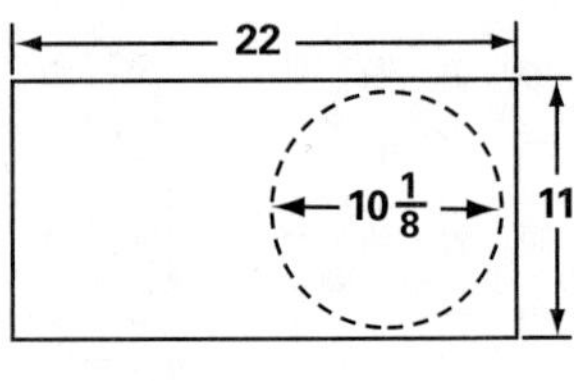

Dimensions Are in Inches

Engineers use the results of testing to make predictions about the working life of the product. For example, based on their design data and on test results, engineers can calculate and predict the life of a car battery, a motorcycle tire, or a computer keyboard. Some products are designed to last only a certain period of time or under certain conditions. This is called planned obsolescence. For example, a paper facial tissue is designed for a single use. It does not last as long as a washable cotton handkerchief.

Various methods are used to ensure that the product and its individual parts will serve the intended purpose. Prototypes are subjected to a variety of tests. The tests are designed to try out the product under extreme and difficult conditions. Some forms of testing can be done on computer models.

Fig. 12-11 Crash testing is often necessary to ensure safety in final production automobiles.

Functional Analysis During the early stages of product engineering, engineers predict (mathematically) how the product will work. After the prototype is built, it can be tested using **functional analysis** to see if the predictions were correct. For example, a prototype of a flying car might be subjected to a road test to make sure it handles well and gets good mileage. The prototype would also be tested for takeoffs and landings.

One important aspect of functional analysis is to make sure that the product is ergonomically correct. It must be comfortable and easy for humans to operate.

Failure Analysis If the prototype fails during testing, then engineers use **failure analysis**. They try to discover why it did not work.

Most prototypes are tested under certain conditions until they break down or fail. Some tests actually destroy the prototype. This is called destructive testing. Crashing a prototype car into another car in order to test its "crashworthiness" is an example. See **Fig. 12-11**. Then engineers tear the cars apart and analyze the failure very carefully. A flying car prototype, for example, might be deliberately crashed to simulate a car accident. The engineers want to see how well it can stand the impact. Depending on test results, the product may be redesigned to make it better.

Value Analysis Engineers do **value analysis** in order to reduce the cost of materials and purchased parts without sacrificing appearance or function. Every part of the new product is studied carefully to see if the right material has been selected. A certain part, for example, might be made of steel in the original design. A value analysis might reveal that the part could be made of plastic. Plastic would cost less, and the plastic part would still work as well as the steel part. Then the plastic part would have to undergo the same kind of product testing as the steel part. The engineers specify the most functional, yet lowest cost, material for every part.

Who Will Buy the Product?

While designers and engineers are busy developing the product itself, the marketing department is thinking about potential customers. They try to identify who is most likely to buy the product and in what quantities, how it should be advertised, and how it should be sold.

Identifying the Market

Manufacturers often try to make products for a certain market. A market is a specific group of people who might buy their product. A watch manufacturer, for example, might make colorful plastic watches aimed at the teenage market. The two main types of markets are the business-to-business and consumer markets.

Business-to-Business Market Multiple businesses make up the business-to-business market. For example, car dealerships, hospitals, magazine publishers, and even governments all buy office paper used in their daily operations. One business (the paper manufacturer) sells the paper to another (the car dealership). Companies also buy products to use as parts in the products they themselves make. Automobile manufacturers may buy tires from one company, windshield wipers from another, and so on.

Consumer Market Consumers are ordinary people who buy products for their own personal use—things like toothpaste, DVDs, and sneakers. The consumer market consists of many millions of people. See Fig. 12-12. Markets change as populations, incomes, and lifestyles change. For example, in more and more households, both parents have jobs outside the home. This has increased the demand for prepared foods. People in charge of marketing must be aware of these changes.

The Marketing Plan

After the market is identified, the marketing department creates a marketing plan. This plan includes a sales forecast, which is a prediction of how many products the company expects to

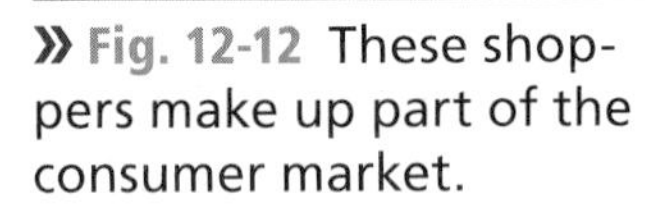

» Fig. 12-12 These shoppers make up part of the consumer market.

sell. They base the plan on market research. The plan also includes ideas for advertising and sales.

Market Research Market research includes all a company's efforts to find out what customers will buy and how much they will pay. The results indicate how well the company can expect the product to sell. Will enough people want to buy a flying car? What features do customers want? Who else makes a similar product? In what ways will our flying car be better than theirs? Reliable answers are needed to protect the company's investment.

One way of researching the market is to interview people. Potential consumers are asked their opinions about a product. If it's a food product, people may be asked whether they like its appearance or taste. The most important question asked is, "Would you buy this product?"

Another way of doing market research is test marketing. A company produces a small number of trial products and sells them in a very limited area, such as in one city or region of the country. See Fig. 12-13. If sales are good in the test market, then the company can usually expect that the product will also sell elsewhere.

Fig. 12-13 Have you ever tried a sample of a new food product? Some products are introduced in certain areas to test consumer reactions.

Advertising

Advertising includes the methods the company uses to inform the target market about a product and persuade them to buy. The **target market** is the group of consumers most likely to use and want the product. Market research is used to determine the best type of advertising to reach them. Advertising campaigns that introduce a new product are often designed during the last stage of product development. A campaign may include TV, radio, newspaper, magazine, and billboard advertising. The newest form of advertising is posting ads on the Internet.

Another effective method of advertising is to give away free samples. If you like the free sample, you are more likely to buy the product.

Sales

If a company can't sell the product, all is lost. Sales personnel are introduced to a new product or service at sales meetings. Designers and engineers give presentations about the product, explaining why the company decided to manufacture it and why customers will want to buy it. The more excited salespeople are about the product, the easier it is for them to sell it.

Many people in sales work on commission. That is, they receive a certain percentage of the amount paid for the product. The more they sell, the more money they make. This encourages them to sell more products.

The three main ways of selling products include direct sales, wholesale sales, and retail sales.

Direct Sales Sometimes a company sells its product directly to the customer. This method of sales is called direct sales. Manufacturers commonly sell direct to purchasers in the business-to-business market. For example, an aluminum producer may sell aluminum tubing directly to an air conditioner manufacturer. Direct sales are also made to ordinary consumers, but this is less common.

A type of direct selling that has grown very popular is the factory outlet store. An outlet store carries products made by one manufacturer. These products are usually last season's merchandise that did not sell well in ordinary stores. The items are usually available at reduced prices.

Wholesale Sales Wholesalers buy large quantities of products from manufacturers. Then they sell the products to commercial, professional, retail, or other types of institutions that also purchase in quantity. Wholesalers may also be involved in such activities as financing, storing, and transporting products.

Retail Sales Retailers buy products from manufacturers or wholesalers. Then they sell the products to the consumers who will actually use the product. Retail stores include department stores, chain stores, and discount stores, among others. The newest type of retail selling is done by means of the Internet. The Internet "store" may consist of only a series of computers that take orders from customers and then place those orders with manufacturers or wholesalers. See **Fig. 12-14**.

How Will the Product Reach the Customer?

The path that goods take in moving from the manufacturer to the consumer is called the **supply chain**. The chain can be short or long. It may involve only the manufacturer and the consumer, as in direct sales, or include both a wholesaler and a retailer as well. Will the product have to be stored? If so, will it require a lot of space or only a little? Can it be shipped by truck or will it require air freight? The planning and control of this flow of goods and materials is called logistics. Getting the product to the people who will buy it is called distribution.

Managing the supply chain is important because the product must be available to customers at the right time and in the right place. People, trucks, and storage facilities must be ready when needed. Interest in a new product might falter if customers can't buy it right away.

Fig. 12-14 You may have purchased products over the Internet. ***How is it different from shopping in a store?***

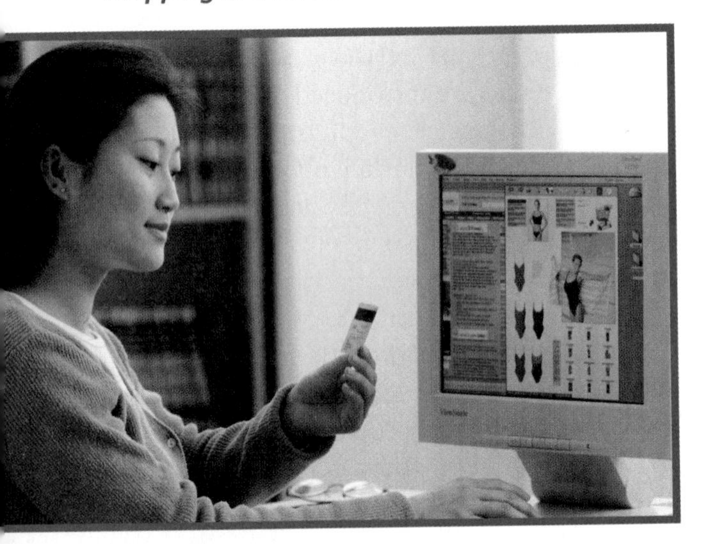

Warehousing

A warehouse is a building where products are temporarily stored until the next part of the supply chain is ready for them. Having products already made and stored ahead of time ensures that there will be enough on hand when they are needed. See **Fig. 12-15**.

The number and location of these storage facilities are important. A large manufacturer of snack foods, for example, may have warehouses located in 20 states. This makes distributing the product easier.

Some products, however, are manufactured on demand. This means they are not made until a customer places an order. They require little or no warehousing.

Fig. 12-15 This warehouse stores a large quantity of items that are ready to ship when ordered.

Transportation

Regardless of the supply chain used, some form of transportation is involved. All products need to be shipped from the factory and must eventually reach the user. Trucks and trains are commonly used, as is air freight. For products that are exported (sent out) to other countries, ships are the most common method of transportation.

How Long Is a Product's Life?

The product life cycle starts when a new product is introduced. When it is new, the product is usually expensive. It may also be unreliable. However, as more people begin to buy the product, it enters the growth period. During the growth period, improvements are made. The "bugs" are worked out. More of the products are manufactured, and the price usually comes down. For example, when DVD players were first introduced, they cost around $400. Today, many can be bought for less than $100.

When the product achieves general acceptance by the public, it has entered maturity. During this period, the price stops changing, and the product is usually very reliable.

After a while, sales start to taper off. Nearly everyone who wanted the product now has one. This period in the product life cycle is called saturation. The demand has been satisfied. More products may be available to sell than customers want to buy. Eventually, sales start declining, and if the product is not updated, it may be taken off the market.

The length of a product's life cycle varies, depending on the product. It may last a few months, a few years, or many, many years. Durable goods have a long life span. Quality furniture is one example. Nondurable goods, such as calendars, for example, have short life spans.

To be successful, manufacturing companies must continue to produce and sell products. Because of the product life cycle, each company tries to keep creating new product ideas. As one product is declining, a new product is entering the growth stage.

Main Ideas

- Many steps are involved between developing an idea for a product and distributing the product to consumers.
- Product design means planning to make sure the product will work as intended in a safe and reliable manner and that it can be produced at a reasonable cost.
- Marketing plans introduce the product in a successful way.
- The path that goods take in moving from the manufacturer to the consumer is called the supply chain.

Understanding Concepts

1. Where do most product ideas come from?
2. What two factors are most important when designing a new product?
3. List the three types of engineering analysis.
4. Briefly summarize what a marketing plan includes.
5. Explain the stages in a product's life cycle.

Thinking Critically

1. Why might a design with a known problem still be produced?
2. When in product development should quality control take place?
3. Are prototypes always full-size models? Why or why not?
4. How could identifying a target market be unethical?
5. What technologies do you use that have a short product life cycle?

Applying Concepts & Solving Problems

1. **Math** Create a working drawing of a product. Be sure to include all dimensions.
2. **Research** What level of government (local, state, or federal) grants patents? Research the topic and determine the process used to apply for a patent.
3. **Marketing** Develop a market plan for any product. Be sure to include a sales forecast and ideas for advertisements and sales.

Activity: Reverse-Engineer a Product

Design Brief

While engineers often design a product from scratch, they sometimes reverse-engineer a product. Reverse engineering involves taking a product apart to see how it works. Engineers disassemble a real object to examine all the parts, types of materials used, and how the parts fit together. All parts are measured and all the information is described and recorded. The data is combined with 3D CAD software to create a solid model.

Reverse-engineer a product to find out how it works, what kind of parts are included, and what materials were used to make it.

Refer to the

HANDBOOK

Criteria

- As you take the product apart, you must sketch each part to show its shape and special features. Identify the location of each part.
- Write down the materials used for the product.
- Create a final analysis that shows an understanding of how your product works and how it was put together.

Constraints

- The product you choose should not be overly complex. It should have no more than 20 individual components.
- Your final analysis must be easily understood by the rest of your class.

Engineering Design Process

1. **Define the problem.** Write a statement that describes the problem you are going to solve. For example, you might want to describe why you want to reverse-engineer the product you've chosen.
2. **Brainstorm, research, and generate ideas.** With your team, research the best way to disassemble your device. Hint: You may want to research reverse engineering on the Internet.
3. **Identify criteria and specify constraints.** These are listed on page 264.
4. **Develop and propose designs and choose among alternative solutions.** Develop a system for taking apart the device and recording the information.
5. **Implement the proposed solution.** Begin disassembling your product. Take note of the location of each component.
6. **Make a model or prototype.** Sketch each component and record all necessary information.
7. **Evaluate the solution and its consequences.** Do you understand how the product works and how to put it back together?
8. **Refine the design.** If you still do not understand how your product works, review the individual components some more and make sure you have all the necessary information recorded.
9. **Create the final design.** Using the information you have gathered, create your final analysis.
10. **Communicate the processes and results.** Present your final analysis to the class. Do they understand how the product works? Hand in the product, your sketches, measurements, notes, and final analysis to your teacher.

SAFETY FIRST

Before You Begin. Make sure you understand how to use the tools and materials safely. Ask your teacher to demonstrate their proper use. Follow all safety rules.

CHAPTER 13 PRODUCTION PLANNING

Objectives

- Describe planning procedures used to achieve production efficiency.
- Name the three questions that need to be asked in a make-or-buy decision.
- Explain how the layout of a manufacturing facility is developed.
- List advantages of making a pilot run.

These workers take great pride in assembling customized bicycles. In this chapter's activity, "Plan and Produce a Product," you will have a chance to create, plan, and build your own customized product.

Vocabulary

- bill of materials
- part print analysis
- process chart
- group technology
- simulation
- plant layout
- materials-handling system
- tooling-up
- pilot run

Reading Focus

1. Read the title of this chapter and describe in writing what you expect to learn from it.
2. Write each term in your notebook, leaving space for definitions.
3. As you read Chapter 13, write the definition beside each term in your notebook.
4. After reading the chapter, write a paragraph describing what you learned.

Planning the Production Process

As you read in Chapter 12, management approves the design of the product. Management also decides whether the company should proceed with production. Engineers and managers review information about the product design. They also consider future sales. If these factors look promising, then production planning begins.

Changing the form or shape of an industrial material usually requires several carefully chosen processes. Because manufacturers want to make high-quality products in the least costly way, careful and detailed planning is required. Efficiency, product requirements, and resources must all be considered, among other things.

Product Requirements

Manufacturing engineers plan production. To do so, they must know all the details about the product. They study the bill of materials and the working drawings.

A **bill of materials** is a list of the materials or parts needed to make one product. The quantity of each part or material is given. Items are listed in the order in which they are used. Figure 13-1 shows an example.

Working drawings provide valuable information. By carefully studying the drawing for a part, a process planner can begin to get ideas about how the part could be made. This is called **part print analysis**. Specified features and dimensions are identified on the drawing. Information about the finished shape and

BILL OF MATERIALS

Part Number	Part Name	Quantity
256100	Wheel	1
256110	Rim	1
256120	Hub	1
256121	Axle	1
256122	Cone	2
256123	Bearing	16
256124	Locknut	2
256125	Nut	2
256126	Washer	2
256130	Spoke	36
256131	Nipple	36

» Fig. 13-1 A bill of materials lists all the parts needed to make a product.

size of the part, the materials needed, and the tolerances is also collected during part print analysis.

Three important factors that must be considered in planning processes include volume, quality, and available equipment.

- Volume refers to the amount of parts or products that must be produced in a given time period. For example, do we need to make 100 per hour or 5 per day?
- Quality refers to the specified degree of excellence. Perhaps you've heard of something called a "grade A" product. That usually means the product matches the design specifications very closely.
- Available equipment refers to the necessary tools and equipment needed to manufacture parts and products. Sometimes this is also called capacity.

Make-or-Buy Decisions

Should the company make or buy the parts needed? Manufacturing engineers must compare the possibilities and make intelligent decisions. Three questions must be answered for every part in the product:

- **Availability.** Can the item be purchased?
- **Manufacturability.** Can our company make the part?
- **Cost.** Is it cheaper to make the part or to buy it?

Information is gathered about those parts that are available for purchase. Quality is important. Also, suppliers must be able to provide the proper quantities at the time they are needed.

Engineers also consider whether their company is able to manufacture the part. Can the tools and equipment now owned by the company be used? Do workers have the knowledge and skills needed to make the part?

Parts that cannot be purchased must be made by the company. However, make-or-buy decisions are usually based on cost. Which way will be the most cost-effective, based on all the factors?

Selecting the Processes

Manufacturing engineers make production plans for each part that will be made in the plant. They identify the processes needed, analyze and compare the advantages and disadvantages of each, select the equipment, and arrange all the processes into a logical sequence.

One common procedure for organizing this information is to make a **process chart**. See Fig. 13-2. There are several types of these charts. The amount of detail varies, but all process charts show the sequence of manufacturing steps. These steps reflect the decisions manufacturing engineers have made.

Manufacturing Resource Planning

Many production planners use a special kind of manufacturing resource planning (MRP II) software to help in the planning process. Planners enter information about the required quantities, estimated delivery time, and the plant capacity. The program then provides a plan and master schedule for production. These include other important information, such as purchasing budgets and inventory, which help control the overall manufacturing production. You will learn more about this process in Chapter 15, "High-Performance Manufacturing."

Group Technology

Another interesting procedure used in production planning is called **group technology**. This is a method for identifying and grouping similar parts that the company manufactures. Using a classification scheme, all parts are coded according to their shape and features.

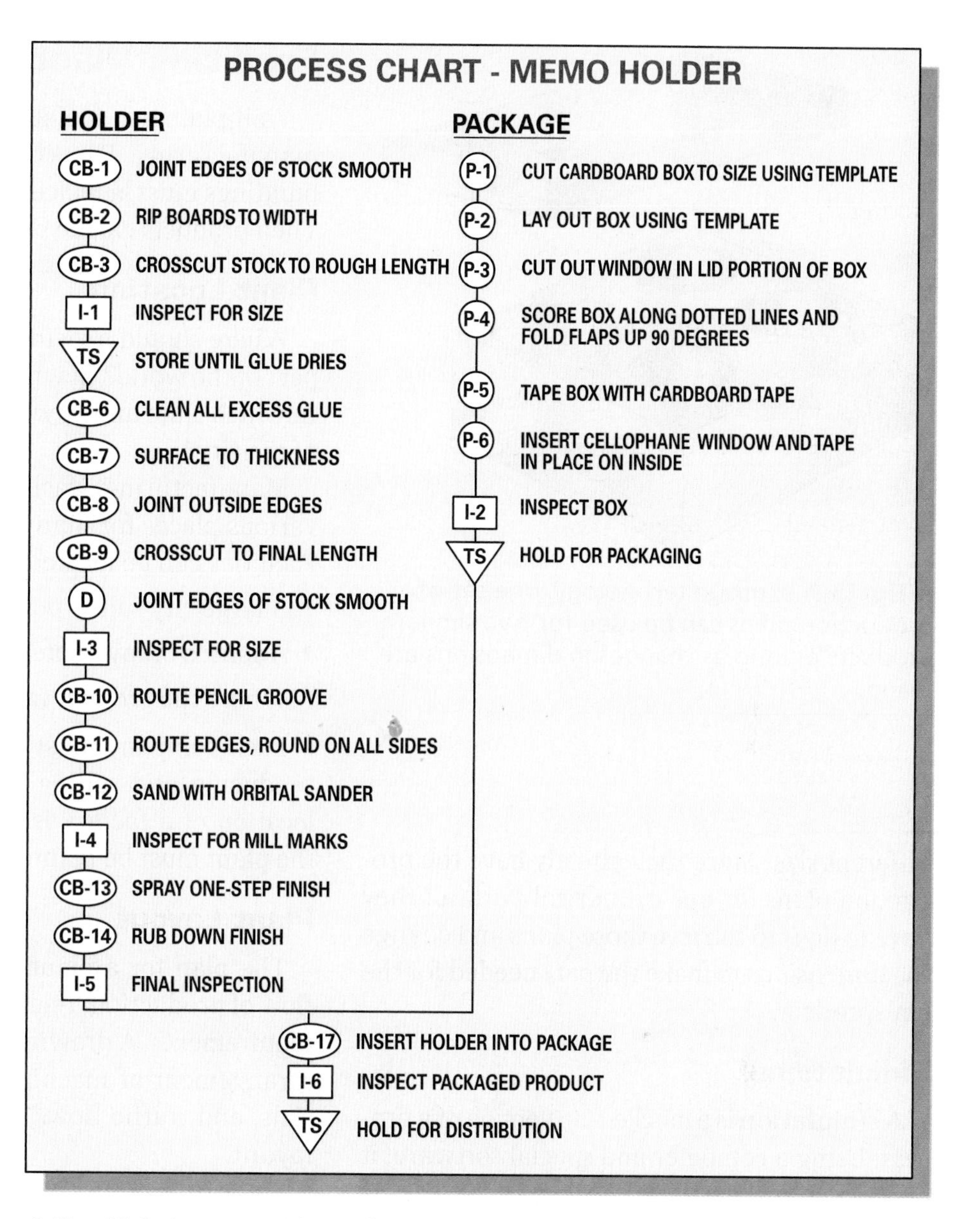

Fig. 13-2 A process chart shows the sequence of operations required to make a product.

When the company has to manufacture a new part, the engineers first check the file to see if plans for one like it already exist. Production planning can be done more quickly when the manufacturing engineers don't have to start over each time. They adapt the old plans for the new part. See **Fig. 13-3** on page 270. Computers have made this type of planning very efficient.

For example, suppose the company manufactures a certain cylindrical part with a hole through the center. For a different product, they may need to make a similar cylindrical part in a

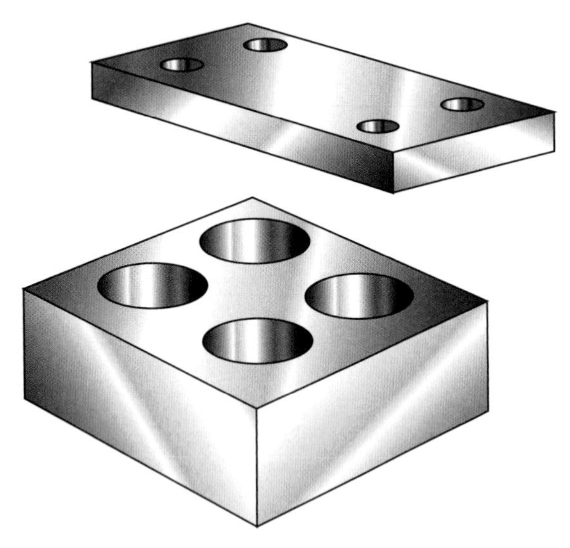

Fig. 13-3 In group technology, one set of production plans can be used for two similar products as long as changes in dimensions are made.

different size. Since they already have the production plans for one cylindrical part, all they have to do is to retrieve those plans and change the dimensions to make the part needed for the new product.

Simulations

A **simulation** is a mocked-up version of a process. Using a computer and special software, it is possible to "machine" a virtual metal part or "mold" a virtual plastic part. In the process, the engineers and planners can watch how the parts would be made. This allows them to make sure there will be no mistakes. In a similar way, a process for assembling parts can be simulated and checked for efficiency and errors.

Facilities Planning

Well-planned facilities are important in manufacturing. Everything that goes into the buildings must be placed according to the plan. Then products can be produced efficiently.

Plant Location

Where should a factory be located? In which part of the world? In which part of the country? In which state and in which city? In which part of the city?

Manufacturing facilities are located in various places for many reasons. See Fig. 13-4. Facilities can be located near:

- A large city full of potential workers
- A source of raw materials
- Where the product will be sold

Each company must decide what is best for producing and selling its products. Once the location of a facility is chosen, the interior of the plant must be planned.

Plant Layout

The plan for a plant layout is based on the flow of production, related activities, and space requirements. A drawing of this plan shows the arrangement of machinery, equipment, materials, and traffic flow. This is called the **plant layout**.

Using the process charts, a layout planner first develops a plan for production flow. It is a kind of "road map." The plan shows how materials and parts are moved into, through, and out of the plant. Sometimes lines are even painted on a factory floor to show where things go.

To make production flow more efficient, related activities are usually located near each other. For example, a drying room would be located near the painting room. There is also an emphasis on arranging machinery into workplace cells. That means placing needed tools and equipment near each other. See

» Fig. 13-4 Manufacturing plants located in big cities have the potential for a larger workforce.

Fig. 13-5. However, some activities that occur in the factory are not directly involved with production. Every factory has a shipping and receiving area. There are also offices, restrooms, and maybe even a cafeteria. These areas support the manufacturing process, but they can be kept separate.

The amount of space for each activity has to be determined. Usually this is done mathematically. Consider cafeteria space, for example. Suppose 50 people will be using the space and each person needs 12 square feet. Then 600 square feet of space will be needed altogether

» Fig. 13-5 On this automobile assembly line, all the tools needed are within easy reach of each worker.

(50 × 12 = 600). Similarly, if you know how much space is needed for each machine and you know how many machines will be used, then you can calculate how much space will be needed.

Materials-Handling Systems

Materials handling is moving and storing parts and materials. Many different and automated types of equipment are used. The equipment and its automated systems are known as the **materials-handling system**.

Conveyors are used to move materials, parts, and products from one place to another along a fixed path. See **Fig. 13-6**. There are many different types of conveyors. Each one is used for a specific purpose. Belt conveyors consist of a turning belt on which items are placed. Many supermarkets have belt conveyors at checkout counters. Roller conveyors work best for carrying heavy loads. Skate-wheel conveyors are good for transporting boxes.

A hoist has one or more pulleys and is used to lift heavy loads. A crane is actually a hoist that can be moved about within a limited area. One kind of crane travels on overhead rails. See **Fig. 13-7**. The parts move along the track suspended from the hoist.

Fig. 13-6 Roller conveyors are used to move boxes from one place to another.

Fig. 13-7 An overhead crane is used to lift large, heavy items and move them to another spot. The crane shown here can also transfer its load to another crane.

» Fig. 13-8 Industrial trucks, such as this forklift, are used to lift and move materials from one place to another.

Trucks used for materials handling are different from the trucks you see on the road. The proper name for them is industrial trucks, but most people just call them forklifts. They have forks on the front that can be raised and lowered. Forklifts are used for lifting and carrying loads from one place in the plant to another. See Fig. 13-8. Forklifts are often used to move pallets, which are special platforms on which large parts or materials are often stored. Loaded pallets can be stored on pallet racks to save floor space.

Today, computers are also used to control materials-handling systems. These computerized systems will be discussed in Chapter 14, "Production."

Putting the Plan into Action

After planning is complete, everyone involved is informed about the plan. First, product drawings and specifications, the bill of materials, the process chart, route sheets, operation sheets, schedules, and floor plans are all distributed. Further preparations are made. Workers are chosen and organized. Machines and other equipment are obtained and prepared for production. The system is tested and refined.

Organizing People

People are needed to work in production. The human resources department employs and trains workers. However, specifying the number and types of workers needed is part of production planning. See Fig. 13-9 on page 274.

Production planners decide which methods and machines to use. Then they determine how many workers will be needed and exactly what each worker will be required to do.

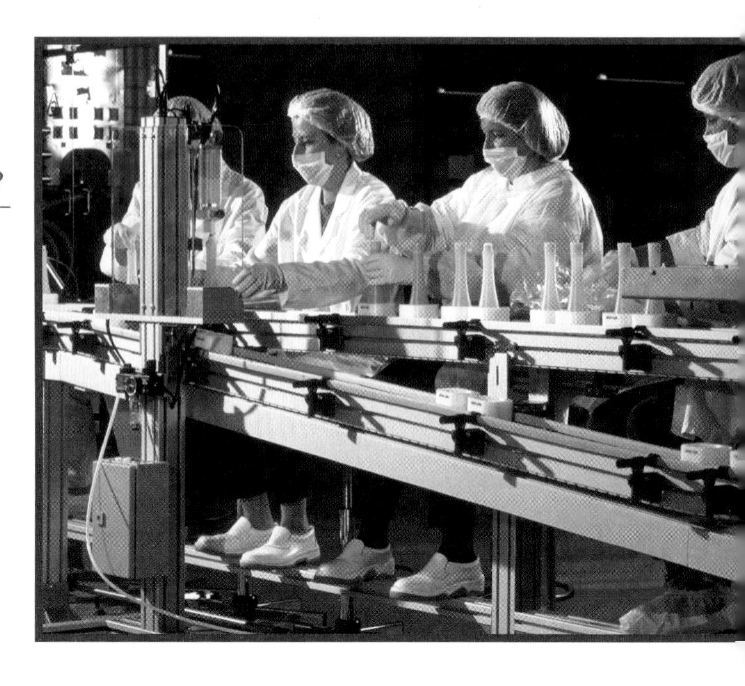

» Fig. 13-9 Production planners determine the amount and kind of workers needed. These people are working with cosmetic cream in a pharmaceutical factory. ***Why do you think they are wearing special clothes?***

Tooling-Up

Getting tools and equipment ready for production is called **tooling-up**. All the machines and tools are prepared for certain jobs. Tooling-up often includes adapting (changing) existing machines. Special tools may also be needed. These are made or purchased.

Special tools include jigs, fixtures, molds, and dies. Jigs and fixtures are used to adapt

Math Application

Costs of Workplace Illnesses People sometimes become ill while at work. When calculating the cost of doing business, manufacturing companies, like other employers, must include the expense of workplace illnesses. Not only do companies help pay for employee insurance, but they also bear the cost in lost productivity when employees are absent. When figuring out costs of workplace illnesses, calculations are performed that involve percentages and dollar amounts. Do you know what kinds of illnesses are work related?

Skin diseases caused by chemicals and other irritants account for about 13 percent of work-related illnesses. In a recent year, 66,000 cases were reported nationwide. Annual cost in lost workdays and productivity was about $1 billion.

Workplace substances, such as certain types of dust, can cause occupational asthma. The cost to U.S. companies due to increased absenteeism and reduced productivity is about $400 million annually.

On an average workday, about 15,900 workers are injured due to accidents. Approximately 5.8 million U.S. workers are injured on the job each year. The annual cost to companies is estimated to be $155 billion.

FIND OUT STE Parts, a manufacturing company with 3,400 employees, is conducting a study on work-related skin diseases. Assuming that the company's averages equal the U.S. averages, approximately how many STE employees were affected this year?

» Fig. 13-10 Most fixtures are custom-made to hold a specific type of part.

general-purpose machines to do certain operations. They are frequently used during the drilling and cutting of metal parts. A fixture clamps and holds the part in place during processing. See Fig. 13-10. A jig is like a fixture in that it holds the part, but it also guides the tool. For example, a jig might hold a piece of wood in place while it also serves as a guide for the saw being used to cut the wood.

Some machines require molds or dies to shape materials. A mold is a hollow form. Liquid material is poured into the mold cavity. As the material hardens, it takes the shape of the cavity. See Fig. 13-11.

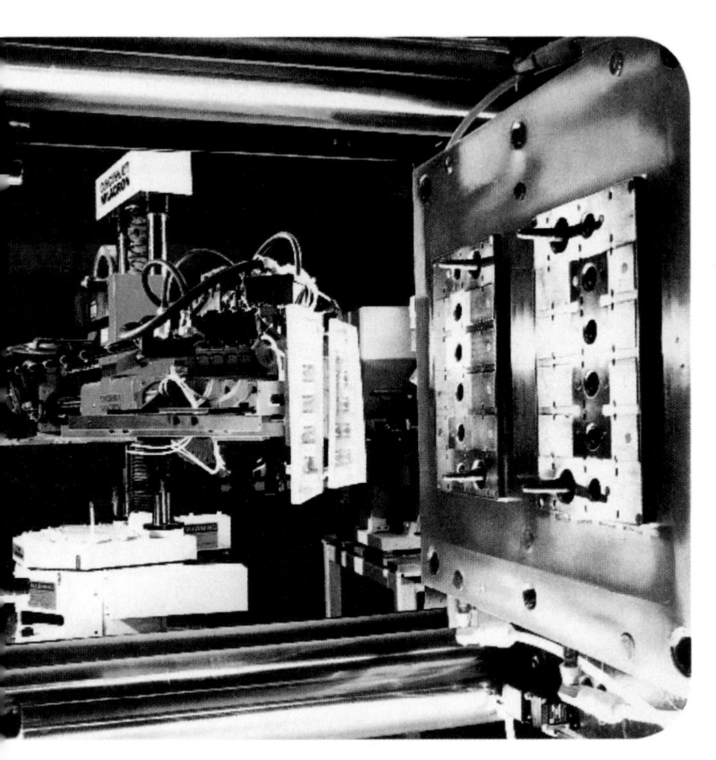

« Fig. 13-11 This robot is removing a finished part from an injection molding machine, where a heated liquid is forced into a mold and is allowed to cool and harden.

A die is a piece of metal with a cut-out or raised area having the desired finished shape. The material is then forced through or against the die to take on its shape. A punch-and-die set is used to punch or cut shapes from materials. This is called die punching. See Fig. 13-12A on page 276. Sheet metal and plastics are frequently cut to shape using a punch and die. Coins are made this way. See Fig. 13-12B.

Another kind of die is used for forming sheet material. The material is stretched over the die, or the die is pushed into the material. This is die stamping. The body of a car is shaped this way. Many times, die stamping and die punching are done at the same time with the same machine.

By simply changing the mold or die, one machine can be used to make different parts. For example, an injection-molding machine might be used to make plastic hairbrushes. By changing the mold, the same machine could be used to make toothbrushes.

» Fig. 13-12A It is important to follow all safety guidelines when working with punch and die equipment.

The Pilot Run

Will the production plan work? One way to know for sure is to conduct a pilot run. A **pilot run** is like a practice where all parts of the system are operated together before production really starts. The main purpose is to find and correct production problems. See Fig. 13-13.

Engineers watch closely and keep records. As problems are identified, corrections are suggested and tried. This process of finding and correcting problems in the operation of a system is called debugging. Because this is a pilot run, the system can be stopped while corrections are made. Actual production is not delayed.

The pilot run also allows engineers to check and adjust the timing along the production and assembly lines. They may speed up or slow down equipment to see the effects on produc-

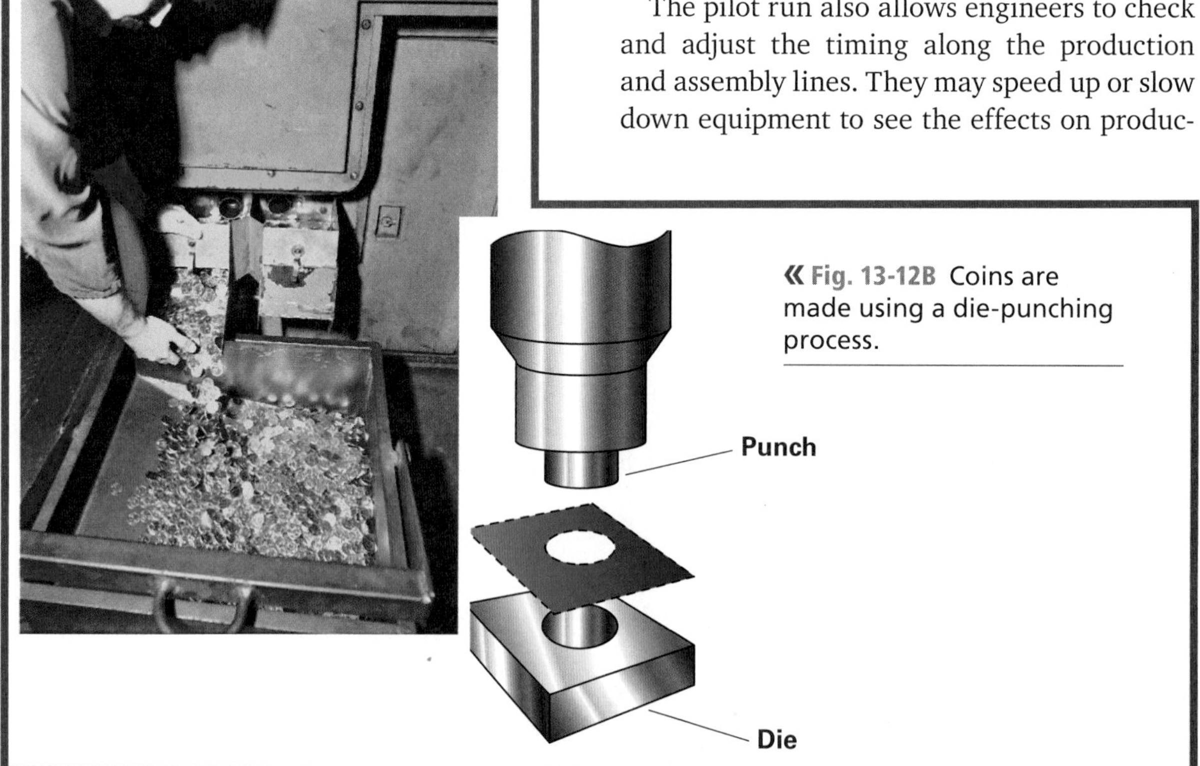

« Fig. 13-12B Coins are made using a die-punching process.

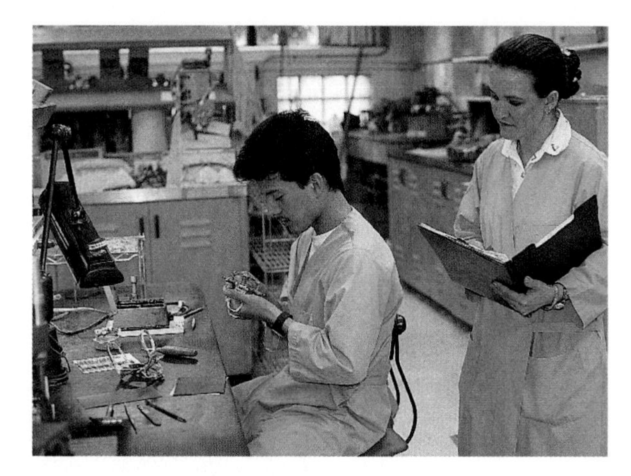

Fig. 13-13 An engineer times the assembly process during a pilot run.

tion. Speeding up may increase the production or assembly rate, but the quality of the product may be affected. Workers must have enough time to do tasks properly.

The pilot run also serves as a training time for workers. They can learn new tasks or see how their jobs have been changed. Workers need to know exactly how to make the new product. During the pilot run, workers are given operation sheets, which give step-by-step instructions for making the parts. See **Fig. 13-14.** This helps them understand how their work fits into the system.

Fig. 13-14 During the pilot run, workers learn the step-by-step procedures for making the parts for the product being manufactured.

Science Application

Matching Tools to Materials A key part of production planning is deciding which tools to use. For cutting, drilling, and other machining processes, the tool must be carefully matched to the material being processed.

For example, a drill bit designed for drilling wood can be made of fairly soft carbon steel. The wood cuts easily and does not wear the bit rapidly. Plastics need specially designed bits that reduce the friction and heat that can damage the plastic. Drilling harder metals, such as steel, requires a cobalt steel drill bit or a titanium carbide tipped bit. These materials are much harder than carbon steel.

Friction between a drill bit and a hard metal object generates a lot of heat. This heat can damage the drill bit or the part. It also creates risk for the operator. Some drill presses apply a stream of cutting oil directly to the drilling site. Oil reduces friction and it carries heat away. If your drill press does not use oil, it is a good practice to apply oil to the bit before drilling hard metals. Your job will be safer and easier, and your tools will last longer.

FIND OUT Research the tools and materials used for different manufacturing processes. Create a list of common tools and materials used. Then, based on your research, make a chart showing which tools go best with which materials.

CHAPTER 13 REVIEW

Main Ideas

- Manufacturing engineers use all the information about the product to create a production plan.
- Well-planned facilities are important in manufacturing.
- Plant location, plant layout, and the materials-handling systems must all be chosen before production begins.
- Organizing workers and tooling-up are steps in putting the plan into action.

Understanding Concepts

1. What is a bill of materials?
2. What three questions need to be answered in a make-or-buy decision?
3. Identify what would influence the location of a production facility.
4. How is a production plan used to determine plant layout?
5. Identify advantages of a pilot run.

Thinking Critically

1. What might happen in production if a supplier does not meet a manufacturer's need for parts?
2. An employee with a key production role calls in sick. How can stopping production be avoided?
3. How does a machine that can use different molds or dies affect costs?
4. Discuss several consequences of *not* performing a pilot run.
5. Imagine a product is on a tight timetable, but you are missing one component. What are your options?

Applying Concepts & Solving Problems

1. **Math** Imagine that a fire code requires that all 250 employees be able to evacuate a manufacturing plant in no more than three minutes in the event of an emergency. If a typical doorway allows about 30 people to pass through it per minute, how many doors will be needed for the factory to be in compliance with the fire code?
2. **Assess** Examine the layout of your school. How does the layout handle efficiency and safety?
3. **Compare and Contrast** Analyze all the parts from a typical classroom chair and desk. How many of the parts could be interchangeable with parts in other chairs or desks with little or no modification?

Activity: Plan and Produce a Product

Design Brief

You may have already followed a production plan without even realizing it. Have you ever built a model based on instructions or baked a batch of cookies based on a recipe? Step-by-step instructions for creating something lead to producing the finished "product."

To successfully manufacture a product, a plan must include a list of components and the processes required to make the parts, subassemblies, and final assembly.

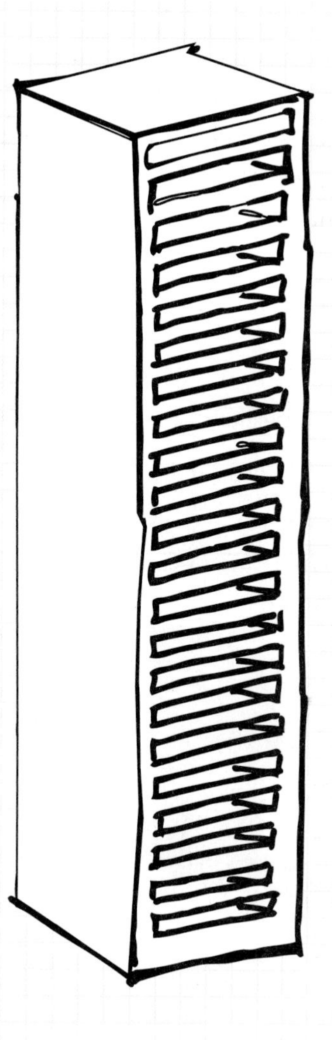

Create a production plan for manufacturing a simple device, such as a DVD holder. Then carry out production.

Refer to the

HANDBOOK

Criteria

- Your plan must include a bill of materials.
- Diagrams showing parts, subassemblies, and the final product must be provided.
- The plan must describe how many employees will be needed to make the product.

Constraints

- Your product must be something that can be created using the tools and equipment in your technology lab.
- At least four separate components need to be assembled to make your product.

Stock Cutting List

Key	Quantity	Description and Size
A		
B		
C		
D		

Bill of Materials

Part	Finished Size			Matl.	Qty.
	T	W	L		
A side panels	½″	5⅝″	13⅛″	O	2
B stiles	¾″	¾″	12⅝″	O	4
C cap	¾″	7⅜″	8¼″	O	1
D base	¾″	6⅝″	6¾″	O	1
E grille stiles	3/16″	¼″	12⅝″	O	4
F grille rails	3/16″	¼″	5⅛″	O	4
G back	⅛″	5¾″	13⅛″	H	1

Please read all instructions before cutting.

Materials Key: O–oak; H–hardboard.

Supplies: #16 × ½″ brads, stain, feet, clear finish.

Engineering Design Process

1. **Define the problem.** Write a statement that describes the problem you are going to solve. For example, what are you going to produce and how do you plan to do it?
2. **Brainstorm, research, and generate ideas.** With your team, discuss possible solutions. Hint: You may want to see what tools and materials are available in your lab.
3. **Identify criteria and specify constraints.** These are listed on page 280.
4. **Develop and propose designs and choose among alternative solutions.** Choose the best design that will solve your problem. For example, avoid designs that require materials that are hard to acquire or tasks that are too difficult for a high school student to accomplish.
5. **Implement the proposed solution.** Gather any needed tools or materials. Create your production plan.
6. **Make a model or prototype.** Make your diagrams, and confirm that all materials and tools are available.
7. **Evaluate the solution and its consequences.** Does your production plan work within the confines of your technology lab? Are all materials, tools, and processes listed?
8. **Refine the design.** Based on your evaluation, revise the plan if needed.
9. **Create the final design.** Produce your product by closely following the production plan.
10. **Communicate the processes and results.** Present your finished product and production plan to the class. Be prepared to answer questions. Turn in your assignment to your teacher. Be sure to include the name of the activity, your definition of the problem, and a description of how you solved the problem.

SAFETY FIRST

Before You Begin. Make sure you understand how to use any tools and materials safely. Have your teacher demonstrate their proper use. Follow all safety rules.

CHAP 14 PRODUCTION

Objectives

- Explain the difference between components and assemblies.
- List the purposes of packaging.
- Describe how production and product quality are controlled.
- Explain how inventory is controlled.

≈ These workers are part of an assembly line. At this stage of production, the chocolates are being packaged. In this chapter's activity, "Assemble an Assembly Line," you will have a chance to design and operate your own assembly line.

Vocabulary

- production
- component
- assembly
- inventory
- quality assurance
- International Organization for Standardization
- Gauge R&R
- acceptance sampling
- burn in

Reading Focus

1. Write down the colored headings from Chapter 14 in your notebook.
2. As you read the text under each heading, visualize what you are reading.
3. Reflect on what you read by writing a few sentences under each heading to describe it.
4. Continue this process until you have finished the chapter. Reread your notes.

Producing Products

Producing products is what manufacturing is all about. Regardless of the type of production system (custom, intermittent, or continuous), **production** is the multistep process of making parts and assembling them into products.

Components

Each individual part of a product is called a **component**. Some components are simple, such as a wire. Other components, such as a casting for an automobile engine, may be very complex. See Fig. 14-1.

As you learned in Chapter 13, "Production Planning," before production actually begins, decisions are made whether to make or buy each part. A computer manufacturing company, for example, may buy many electronic components. However, the company would probably make some of the special parts.

Assemblies

Components are assembled with other components. This means they are put together in a planned way. Assembled components are called **assemblies**. If an assembly will be used as a component in another product, it is called a subassembly. The wheels on a skateboard and

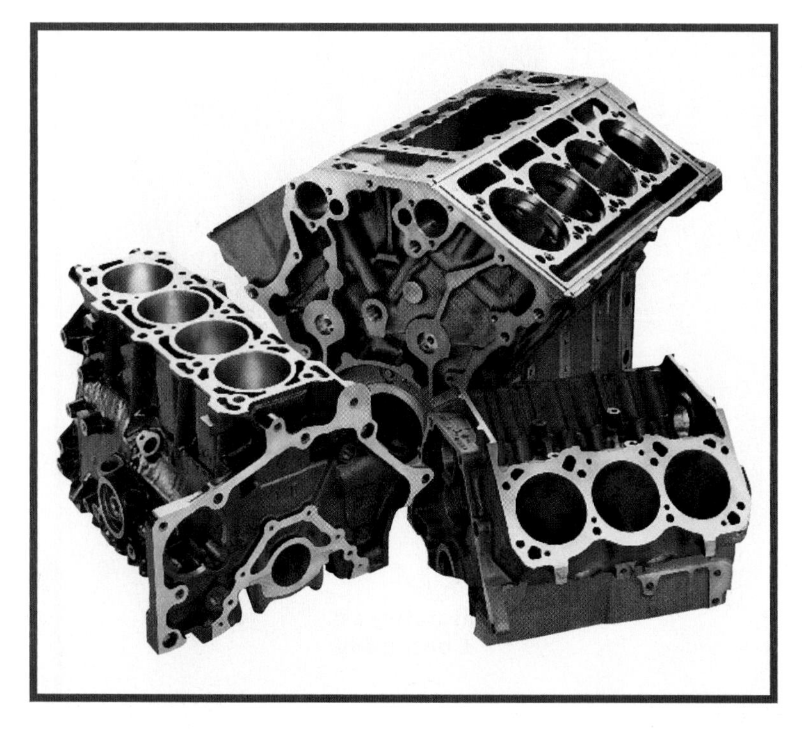

« Fig. 14-1 Castings, such as these parts for an automobile engine, can be quite complex.

≈ **Fig. 14-2** Even a small product, such as this portable electric saw, can contain many components and subassemblies.

brakes on a bicycle are examples of subassemblies. Subassemblies may then be assembled with additional components or other subassemblies. See **Fig. 14-2**.

When assembly operations begin, all the necessary parts must be available in the right quantities. Sometimes assembly work is done by hand. That is, workers pick up parts and put them together. They may glue the parts together or perform other tasks. Sometimes automatic assembly machines are used. The point at which all the parts are combined to form the product is called final assembly.

≈ **Fig. 14-3** Products are packaged for many different reasons. ***Can you identify why some of the products below are packaged a certain way?***

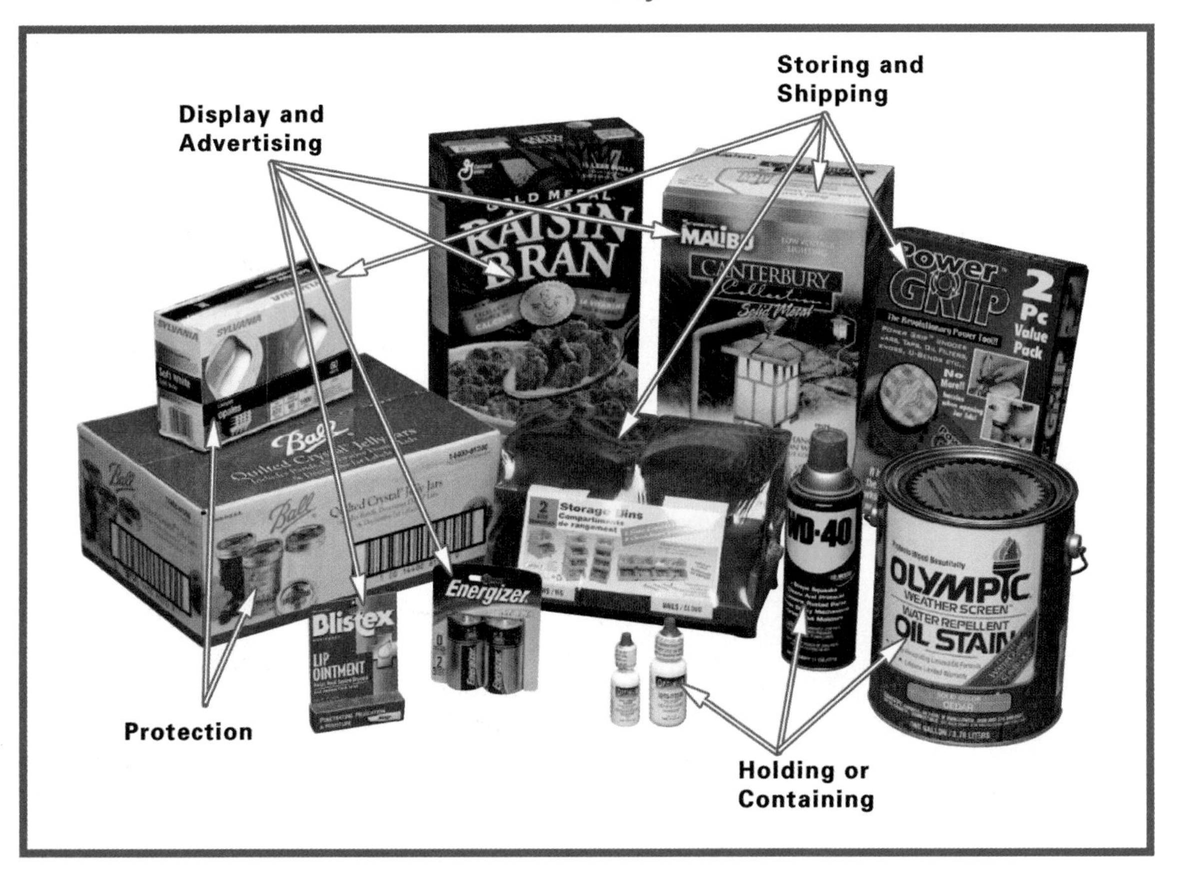

Packaging

After final assembly, many products are packaged. There are many reasons for packaging products. See Fig. 14-3. Fragile or easily broken items, such as glassware or electronics, must be packaged for protection. Items that could spoil or deteriorate, such as crackers and other foods, must be packaged to maintain freshness. Liquid items, such as beverages or paint, require tightly sealed packages to avoid leakage.

Packaging is also extremely important in marketing. The package must be designed to attract the consumer. It must be pleasing to look at. Can you think of a time when you bought one product instead of a similar one because the package made it look better? Some products require more than one package. Think of chewing gum. First, each stick of gum is wrapped in its own foil wrapper. Then several sticks are put in another package. These packages may in turn be placed into a larger package to be sold as a multipack. For shipping, the packages are placed in a large fiberboard or cardboard carton.

Fig. 14-5 A supervisor is monitoring production in a paper mill. If something goes wrong, he can interrupt the process to fix the problem.

Fig. 14-4 Both manufacturers and retailers use the information coded within each bar code.

Most manufacturing companies typically buy packages from a package manufacturer. The packages may already be printed with labels and other information. If the packages are not preprinted, they can be printed or labeled after a product is put inside.

In Chapter 5, "Information Technology," you learned about universal product codes. A UPC may be printed on the package or applied as a sticker. This code identifies the product in the package. That information can then be used for such things as product tracking and updating inventory records. See Fig. 14-4.

Production and Inventory Control

During the actual production process, someone must make sure there's enough material on hand. Someone also needs to see that the right number of parts is being made. These tasks are part of production and inventory control. See Fig. 14-5. Remember MRP II from Chapter 13, "Production Planning"? Information from the

Auto Parts, Inc.
Chicago, Illinois
MASTER PRODUCTION SCHEDULE

PRODUCT	Week Beginning					
	1/7	1/14	1/21	1/28	2/4	2/11
Serpentine Belts	100	100	100	100	100	100
Wheel Covers	–	600	1200	600	–	600
Hub Caps	–	375	–	–	375	–

Fig. 14-6 A production schedule shows what work is to be done and the dates for each specific task.

production run is entered into the system and compared to the production plan.

Production Control

Production control means controlling what is made and when. How do workers know when to start on a certain product? How many parts should be made at one time? When will products be ready to ship to the customer? A plan for controlling production provides answers to these and other questions.

The master production schedule is very important to production control. This is a chart that lists all the parts and shows how many of each one is to be made in a certain period of time. See Fig. 14-6. Usually a schedule is prepared several months in advance. It gives start and stop dates as well as the number of machines to be used. The schedule is a plan for what is supposed to happen. The actual time spent may be longer or shorter than planned. If so, the schedule is adjusted.

Controlling production also involves keeping track of what work has been done, when it was done, and who did it. After raw material is released to the production department, production control must know what is happening to it at all times. All material that is being worked on is called work in process (WIP). A system called shop floor control is often used to keep track of work that has been done. The needed information is collected on the shop floor. Workers record information about the work they've done, like start/stop times and number of parts produced. See Fig. 14-7. They may enter data directly into the computer by

Fig. 14-7 Workers enter data to keep inventory and operational records up to date. This process is part of a system called shop floor control.

scanning UPC labels at their workstation terminal or at a portable terminal. The computer records are then updated in "real-time."

Inventory Control

Inventory is the quantity of items on hand. In a manufacturing plant, inventory control means keeping track of raw materials, purchased parts, supplies, and finished goods.

- **Raw material.** For purposes of inventory control, a raw material is any material before it enters processing. For example, potatoes, corn, and cheese may be raw materials at a snack food factory.
- **Purchased parts.** These ready-made parts that the company buys are called purchased parts. A lawn mower factory might buy engines and wheels.
- **Supplies.** Different kinds of items needed to keep the plant running smoothly are supplies. These items do not become part of the product, but they are needed to support the production process. Supplies include staples, computer paper, oil, and lightbulbs.
- **Finished goods.** These are products that are completed but not yet sold. In a furniture factory, tables and chairs might be the finished goods. See Fig. 14-8.

Keeping good records of all inventories is important. When the inventory for materials gets low, more must be ordered. Without records, the company would not know how much was on hand. Inventory records are usually kept in computer files.

The purchasing department helps make sure proper inventory levels are maintained. Buyers, or purchasing agents, buy the materials and other things a company needs. They make sure the items ordered are of the proper quality and are reasonably priced. They also make sure the items are delivered at the right time. Timing is very important.

Fig. 14-8 This worker is taking inventory of finished goods.

Quality Assurance

The quality of a product is how well it is made. Manufacturing companies want to produce high-quality products. They want each product to fulfill its purpose in the best way possible. They also want to satisfy consumers. **Quality assurance** is the process of making sure the product is produced according to plans and meets all specifications. Sometimes this is also called quality control.

The level of quality must be set in advance. This level is called the quality standard. However, it is impossible to do something perfectly over and over again. There will always be slight changes, or variations, from one part or piece to another. See Fig. 14-9 on page 288. Variation occurs because of differences in workers, materials, machines, and processes. Controlling this variation to keep the best quality possible is the goal of quality assurance.

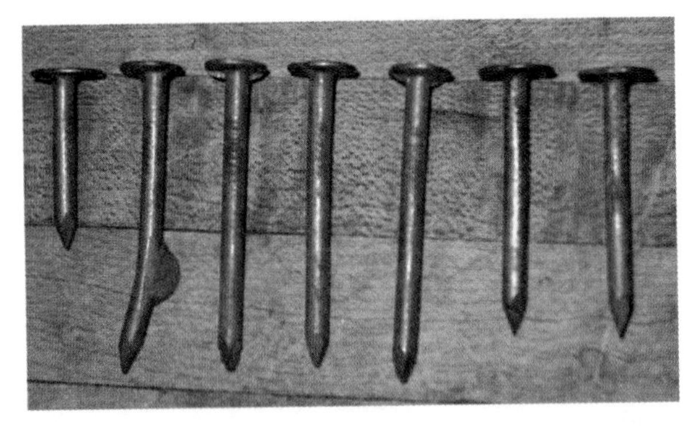

Fig. 14-9 These nails came out of the same box. The differences are called variations. The best quality parts have few variations from one to the next.

There are two basic ways to approach quality assurance: prevention and detection. Prevention involves doing everything possible to prevent variation in materials or processes before parts are made. Detection means inspecting parts or products after they have been produced to find any variations. However, preventing mistakes is better than finding mistakes. It costs less to correct errors before or during production instead of afterward.

Science Application

Storage and Product Quality Even after a product has shipped out to warehouses or storerooms, product quality can change. Foods and other manufactured items can lose quality when they are stored for any length of time, costing manufacturers millions of dollars every year. Four enemies of stored items are microorganisms, enzymes, pests, and oxidation.

Microorganisms are tiny life forms such as bacteria, molds, and yeasts. They primarily affect food, although some also work on wood and petroleum products. Microorganisms can cause decay in the product, alter the taste or safeness of a food, or destroy a product's packaging.

Enzymes are protein molecules that exist in all living things and speed up chemical reactions. Enzymes can make fruits overripe and change the way they taste.

Pests include rodents and insects. Rodents can eat or contaminate foods and chew other items. Termites can attack wood and wood by-products, such as cardboard. Moths lay their eggs on clothing, and the larvae eat holes in it.

Oxidation occurs when oxygen in the air combines with other substances. When food oxidizes, the fat or oil changes flavor, often becoming rancid tasting. Oxidation can also cause metals to rust and paint to dull.

FIND OUT How can manufacturers help control the quality of products once they have been shipped out for sale?

Standardization

One important step toward quality assurance in today's global market is standardization. Because the world's economy is so interrelated, there is a need for international standardization. The **International Organization for Standardization** (commonly referred to as ISO) promotes and coordinates worldwide standards for many things. One important ISO standard is known as ISO 9001. This set of standards is used by manufacturing companies to establish and maintain documentation of their manufacturing practices. A company can have its procedures certified or approved and then can say that it meets ISO 9001 standards. This implies that it produces quality products because it consistently follows procedures. See **Fig. 14-10**.

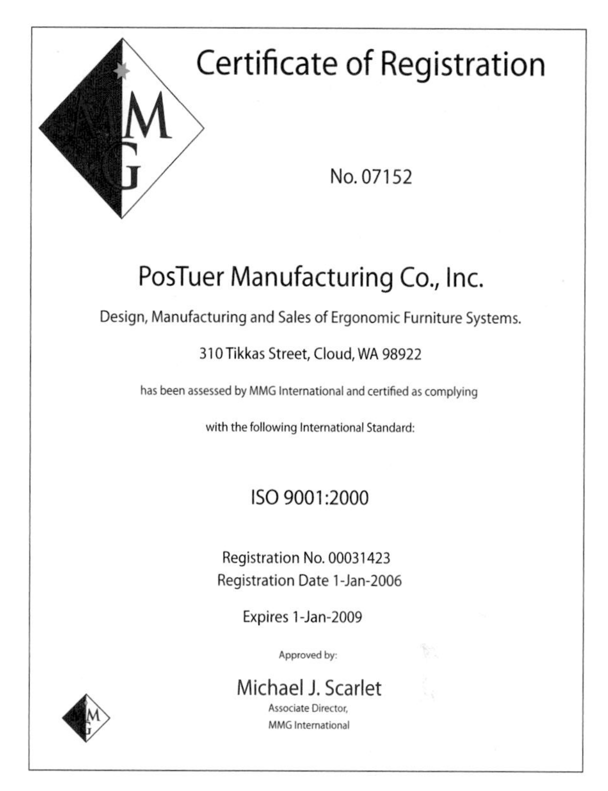
Certificate of Registration

No. 07152

PosTuer Manufacturing Co., Inc.

Design, Manufacturing and Sales of Ergonomic Furniture Systems.

310 Tikkas Street, Cloud, WA 98922

has been assessed by MMG International and certified as complying

with the following International Standard:

ISO 9001:2000

Registration No. 00031423
Registration Date 1-Jan-2006

Expires 1-Jan-2009

Approved by:

Michael J. Scarlet
Associate Director,
MMG International

Fig. 14-10 This certificate shows that the company has met the ISO standards.

A standardization effort that has been led by automobile manufacturers is referred to as TS-16949:2000. This standard requires that their suppliers be certified as producers of quality parts. The goal is to help companies develop and sustain good quality control practices at every step in their operation. If the standards are followed, the company's products and services should be of good quality.

Some countries have their own standards organizations. In the United States, for example, there is an organization called the American National Standards Institute (ANSI). ANSI does not make the standards but coordinates and organizes qualified groups who then agree on the standards.

Process Improvement

Process improvement involves continually working to improve the processes by which things are made. This can involve changes in machines, for example, or in the ways workers do their jobs. Some variation is normal in any process. With careful checking and analysis, quality assurance workers can discover the type and amount of variation that is normal for a process. Then they can monitor the process and determine when it is either in control or out of control.

Statistical process control (SPC) is one technique used in process improvement. It uses a special type of mathematics called statistics, which involves collecting and arranging facts in the form of numbers to show certain information.

Computers and other recording devices keep a record of what a particular machine is doing. This information is recorded on a control chart. See **Fig. 14-11** on page 290. The chart shows a mean and the upper and lower acceptable lim-

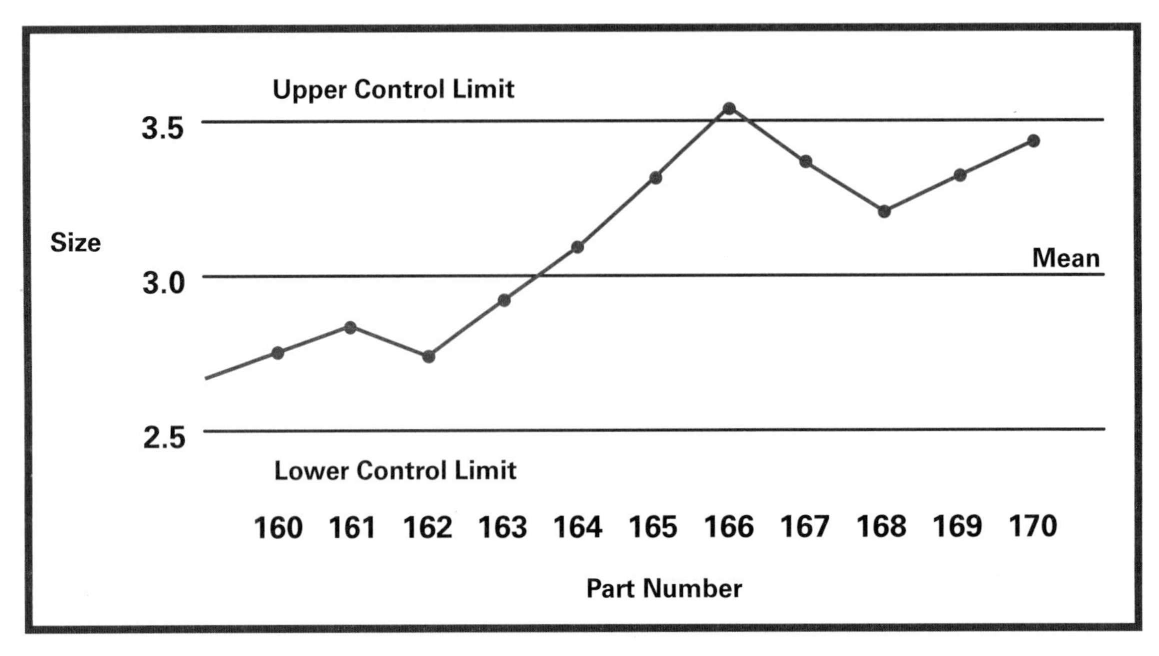

Fig. 14-11 Control charts are used to track processes. ***At what point did this process go out of control?***

its of variation from that mean. (As you learned in Chapter 12, "Product Development," this is called tolerance.) Quality assurance workers use the control charts to analyze the process. Let's look at an example.

Suppose a certain machine automatically fills empty cereal boxes with cornflakes. The label on the box says that there are 16 ounces of cereal in the box, but the boxes are not all filled exactly the same. There is a tolerance of plus one-half ounce. That means the box can actually have any amount between 16 and 16.5 ounces. By checking the weight of every 100th box and plotting that information on a control chart, it's possible to monitor the machine's correctness. As long as the machine is running within the upper and lower limits, the product is fine. If the control chart shows that the boxes are being over or underfilled, workers can stop the machine and make the necessary adjustments.

Inspection

To inspect something means to look at it and compare it to some standard. Inspectors examine a part, process, or product to see if it meets the specifications. Specifications are the detailed descriptions of the design standards for a part or product, and they may include drawings and word descriptions. These standards include rules about a product's size, shape, function, performance, and the type and amount of materials used.

Inspections are made when materials are delivered, while the work is in process, and when the goods are finished. If delivered materials don't meet the standards, they are rejected and returned to the supplier. Work in process is inspected to make sure it is being done properly and that the parts are correct. After the product is made, it is given a final inspection. Everything is checked to make sure it works and looks right.

Inspection Tools Sometimes inspectors visually inspect the part. See Fig. 14-12. Most often they use some kind of measuring device. A variety of inspection tools and devices are used to check materials, parts, and products. Some are used for measuring and others for comparing.

Various gauges are used to compare or measure sizes of parts and depths of holes. One simple gauge is a go/no-go gauge. By slipping a part into this gauge, the inspector can tell at a glance whether or not the part is the right size. See Fig. 14-13.

It is important to periodically check the accuracy of inspection tools. If a tool is broken or no longer works properly, the measurements may be inaccurate. **Gauge R&R** (repeatability and reproducibility) is used on a regular basis to check and recalibrate the tool. The process is used for checking both the accuracy of the gauge and the operator's interpretation of the data.

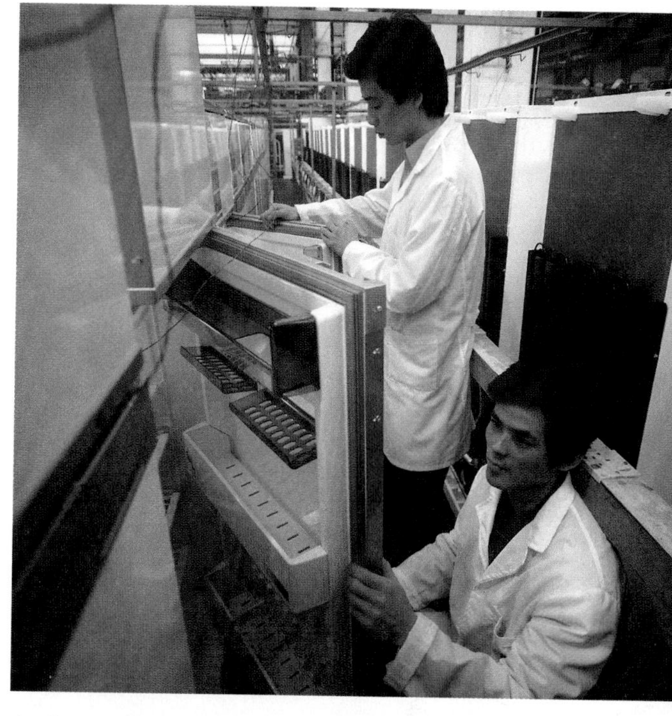

Fig. 14-12 Final inspections are done on products before they leave the factory to make sure that everything works properly. This inspector is checking a refrigerator.

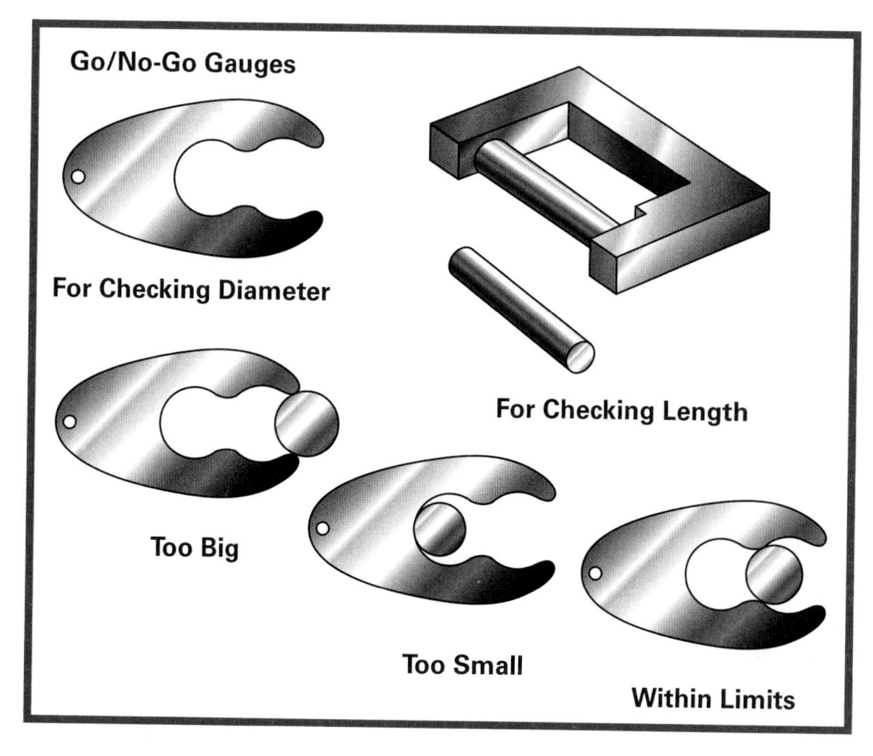

Fig. 14-13 A go/no-go gauge can be used to quickly check if the part is within tolerance limits.

Not all inspection tools are as simple as gauges. Computer-controlled devices, including coordinate measuring machines and laser scanners, can make very precise measurements. A coordinate measuring machine (CMM) has a probe that is used to touch the actual part being measured. It accurately records the part's dimensions along the X, Y, and Z axes. A laser scanner scans the surface of the part and obtains its size dimensions from the reflected light. Sizes and locations of part features are checked against the dimensions on the drawing for the part that is stored in the computer. See **Fig. 14-14**.

Fig. 14-14 This laser scanner is one inspection tool that allows noncontact precision measurements.

Math Application

Repeatability and Reproducibility Gauge R&R is a way to test a measurement device for repeatability and reproducibility. When a gauge is used to measure, there is often variance between readings.

The gauge's repeatability is defined as the ability of one user to get consistent readings. For example, suppose a person needs to measure the thickness of a part using dial calipers. The first reading may be 0.500 inches. After zeroing the calipers and taking a second reading, he or she gets 0.501 inches.

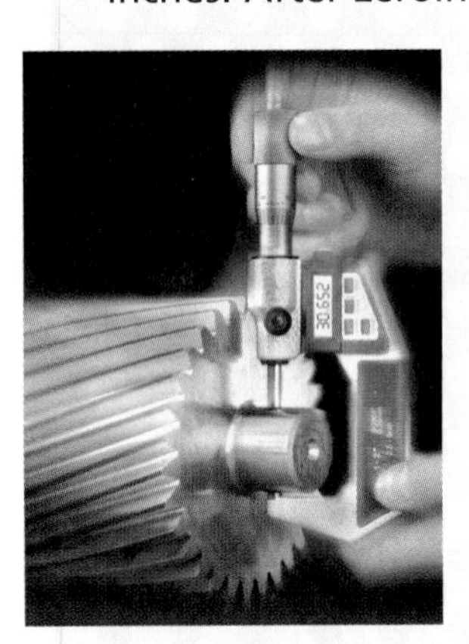

The gauge's reproducibility is defined as the ability to get consistent readings from user to user. If another user gets a reading of .500 or .501 inches, the readings were reproducible.

In order to evaluate the gauge, the users would input their data into statistical software. The software will determine the variance in the gauge, and from this the users will know if the way in which the gauge is being used is acceptable. If the measurement technique is not acceptable, then a different technique or different device should be used.

FIND OUT Are the everyday measurements you use repeatable or reproducible?

1. Measure a classroom desk twice using a tape measure. Do you get the same results? Have a classmate measure the same desk. Do your classmate's readings match yours?
2. Draw an angle and measure it twice with a protractor. Do you get the same readings both times? Have a classmate measure the same angle. Does your classmate get the same readings?

Acceptance Sampling Many products are made in large quantities, so it is not possible or practical to inspect each individual item. An inspection procedure called acceptance sampling is used in these cases. **Acceptance sampling** involves randomly selecting a few typical products from a production run, or lot, and inspecting them to see whether they meet the standards. If they do, then the whole lot is approved. If a certain percentage of the samples are rejected, the other products from that lot are also rejected.

The size of the sample depends on the lot size. Generally speaking, the smaller the lot size, the greater the percentage that should be inspected. The table in **Fig. 14-15** shows one company's sampling plan. You can see that for a lot size of two, all products are tested. However, for a lot size of 10,000 the sample size is only 125. The acceptance level is how many of the sample must pass if the lot is to be accepted. In a lot size of 280, the sample size is 20. A minimum of 16 must be acceptable. So of the 20 pieces inspected, you would reject the entire lot if more than four of the sample pieces did not meet acceptable standards.

Burn In

One special quality assurance measure is called burn in. **Burn in** means using the components of a system before putting them into actual service. Burn in is done to electronic products like computers. Electronic products that fail tend to do so in the first few hours of operation. Because of this, a computer manufacturer turns on and actually runs every computer for a few hours. Those that fail are repaired, if possible. If the computer passes the burn-in test, then it will probably last a long time.

Fig. 14-15 A sampling plan establishes how many items must pass if the lot is to be accepted.

XYZ Company Sampling Plan				
Lot Size	**Sample Size**	**Sample Percentage**	**Acceptance Level**	**Acceptance Percentage**
2-8	2	100%-25%	1 out of 2	50%
9-15	2	22%-13%	1 out of 2	50%
16-25	3	19%-12%	2 out of 3	67%
26-50	5	19%-10%	3 out of 5	60%
51-90	8	16%-9%	6 out of 8	75%
91-150	13	14%-9%	10 out of 13	77%
151-280	20	13%-7%	16 out of 20	80%
281-500	32	11%-6%	27 out of 32	84%
501-1,200	50	10%-4%	43 out of 50	86%
1,201-3,200	80	7%-3%	71 out of 80	89%
3,201-10,000	125	4%-1%	114 out of 125	50%

CHAPTER 14 REVIEW

Main Ideas

- Production is the multistep process of making parts and assembling them into products.
- Production control is the process of controlling what is made and when.
- Inventory is the quantity of materials on hand.
- Quality assurance is the process of making sure the product is produced according to plans and meets all specifications.

Understanding Concepts

1. What is the difference between a component and an assembly?
2. Identify three reasons why packaging is needed.
3. How is a master production schedule important to production control?
4. What is tracked during inventory control?
5. What happens in a burn-in test?

Thinking Critically

1. Explain how a pen is an assembly.
2. Why do many manufacturers buy packaging from other companies instead of making their own?
3. Can you remember the packaging of a recently purchased product? Did it influence the decision to buy it? Could it be improved?
4. What are some consequences if an inspection is not performed throughout production?
5. How do you think quality was checked during the production of this textbook?

Applying Concepts & Solving Problems

1. **History of Technology** Henry Ford is often credited with being the "father of mass production." Why is this so? How did his techniques affect sales of the Model T?
2. **Language Arts** Production is a multistep process that should run smoothly. Write a sequence chain to analyze a production process, such as painting a chair.
3. **Compare and Contrast** How is making a prototype different from manufacturing the same product?

Activity: Assemble an Assembly Line

Design Brief

Many identical products can be mass-produced in a short time using an assembly line. The assembly starts with the basic component and is passed along to other assembly stations where additional parts and subassemblies are added. At the end of the assembly line, you have a finished product. Automobiles, television sets, and even food are assembled this way.

With several of your classmates, set up an assembly line consisting of five stations. You will be assembling bags of trail mix snacks.

Criteria

- Four types of ingredients must be used.
- You should be able to fill 20 bags of trail mix.
- Set up five work stations—one for each ingredient and one for final assembly.

Constraints

- Include a label that lists all ingredients and the date of manufacture.
- Each bag should be assembled in a time limit specified by your teacher.

Engineering Design Process

1. **Define the problem.** Write a statement that describes the problem you are going to solve. For example, how much time do you have to manufacture the 20 bags of trail mix?
2. **Brainstorm, research, and generate ideas.** With your team, discuss possible solutions. Hint: You may want to learn more about assembly lines.
3. **Identify criteria and specify constraints.** These are listed on page 296.
4. **Develop and propose designs and choose among alternative solutions.** Choose the best design that will solve your problem. Assign tasks for each point of production. Determine when quality will be checked.
5. **Implement the proposed solution.** Gather any needed tools or materials. Arrange your assembly line.
6. **Make a model or prototype.** Assemble one bag of trail mix using the assembly line.
7. **Evaluate the solution and its consequences.** Did you make your first bag within an acceptable time? Are all components included?
8. **Refine the design.** Based on your evaluation, revise the plan if needed.
9. **Create the final design.** With your team, assemble the remaining 19 bags of trail mix.
10. **Communicate the processes and results.** Present your finished product and description of your assembly line to the class. Each team member should discuss his or her own role in the assembly line and an evaluation of the line's efficiency. Hand in your problem statement, ingredient list, and assembly line plan to your teacher. Your teacher can decide whether to distribute the trail mix to the class.

SAFETY FIRST

Food Safety. Be sure that all team members wear gloves when handling food. Your assembly line should be kept as clean as possible. Food manufacturers must keep the area sanitized.

CHAPTER 15 HIGH-PERFORMANCE MANUFACTURING

⮝ Many of today's manufacturing processes require an extra clean work environment. In this chapter's activity, "Clean Up Your Workspace Using 5S," you'll have a chance to select an area in your home and, using the 5S method, make it a more clean and efficient work area.

Objectives

- Explain what "high performance" means to manufacturing.
- Give examples of smart planning, smart production, and smart control methods.
- Identify advanced systems.
- Compare and contrast automatic factories to traditional ones.

Vocabulary

- high-performance manufacturing
- design for x (DFx)
- product configuration
- lean manufacturing
- just-in-time (JIT) manufacturing
- manufacturing cell
- supply chain

Reading Focus

1. As you read Chapter 15, create an outline in your notebook using the colored headings.
2. Write a question under each heading that you can use to guide your reading.
3. Answer the question under each heading as you read the chapter. Record your answers.
4. Ask your teacher to help with answers you could not find in this chapter.

What Is High-Performance Manufacturing?

High-performance manufacturing combines highly skilled and empowered workers, advanced technology, and new work methods to achieve very high levels of quality, efficiency, and customer satisfaction. See **Fig. 15-1**. It is sometimes called "smart" manufacturing or computer-integrated manufacturing (CIM) because computers, special software programs, and networks are used throughout the manufacturing organization, including the business side.

For CIM to work, all the computers must be connected so that they can communicate. This is done using computer networks. A network is a way of hooking together several computers. Information is transmitted from one device to another over the network. Inside a plant there is a local area network (LAN), while among plant locations there is a wide area network (WAN). Many companies also use the Internet to communicate with employees, vendors, and customers throughout the world. See **Fig. 15-2** on page 300.

Fig. 15-1 This plant manufactures motorcycles using high-performance methods.

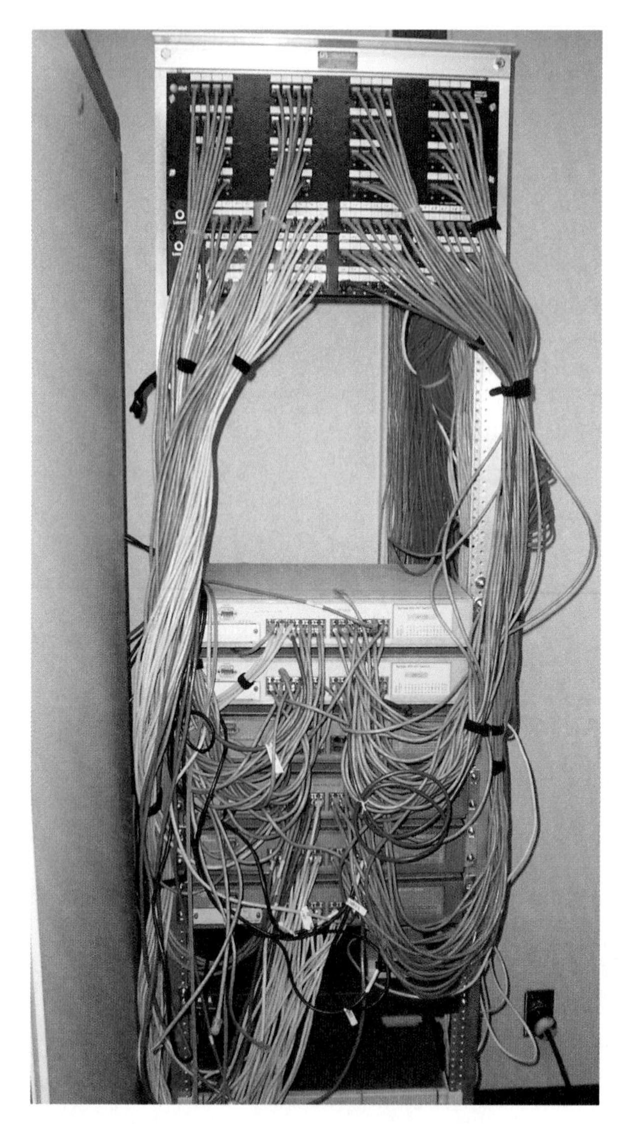

Fig. 15-2 This LAN system provides Internet access to approximately 40 employees.

Different terms may be used to indicate the different levels of computer control during production. When a computer is used to control an individual machine, such as one that drills holes, the process is called computer numerical control (CNC). When linked computers operate machines, plan production processes, and control those processes, it is commonly referred to as computer-aided manufacturing (CAM).

Some processes used in high-performance manufacturing have already been discussed in this unit because they are found in many manufacturing facilities. However, they are mentioned again here because they are key to high-performance operations.

Smart Planning

Smart planning means considering not only how the final product will work but also how to manufacture it. Computers are used for design and engineering, as well as for production planning.

Product Design and Engineering

Engineering a new product requires many mathematical calculations. Computers are used not only to perform calculations quickly but also to produce working drawings and to analyze parts. Product development is made easier with the use of computer-aided design (CAD) and computer-aided engineering (CAE) programs.

A program may be used to add geometric elements to a model. It may also be used to manage the design database, which contains specific information about the product, such as material, strength, and dimensions.

CAD can also be used to produce detail, assembly, and schematic drawings. After the drawing is created on the computer, it can be turned into a hard copy using a printer. Information from drawings can be sent directly to CAM machines that make the actual parts.

Computer simulations can show how moving parts will work. Interactions with humans, robots, and other machines can be tested. See Fig. 15-3. Simulations allow the designers to analyze and make improvements in the product before it is actually produced.

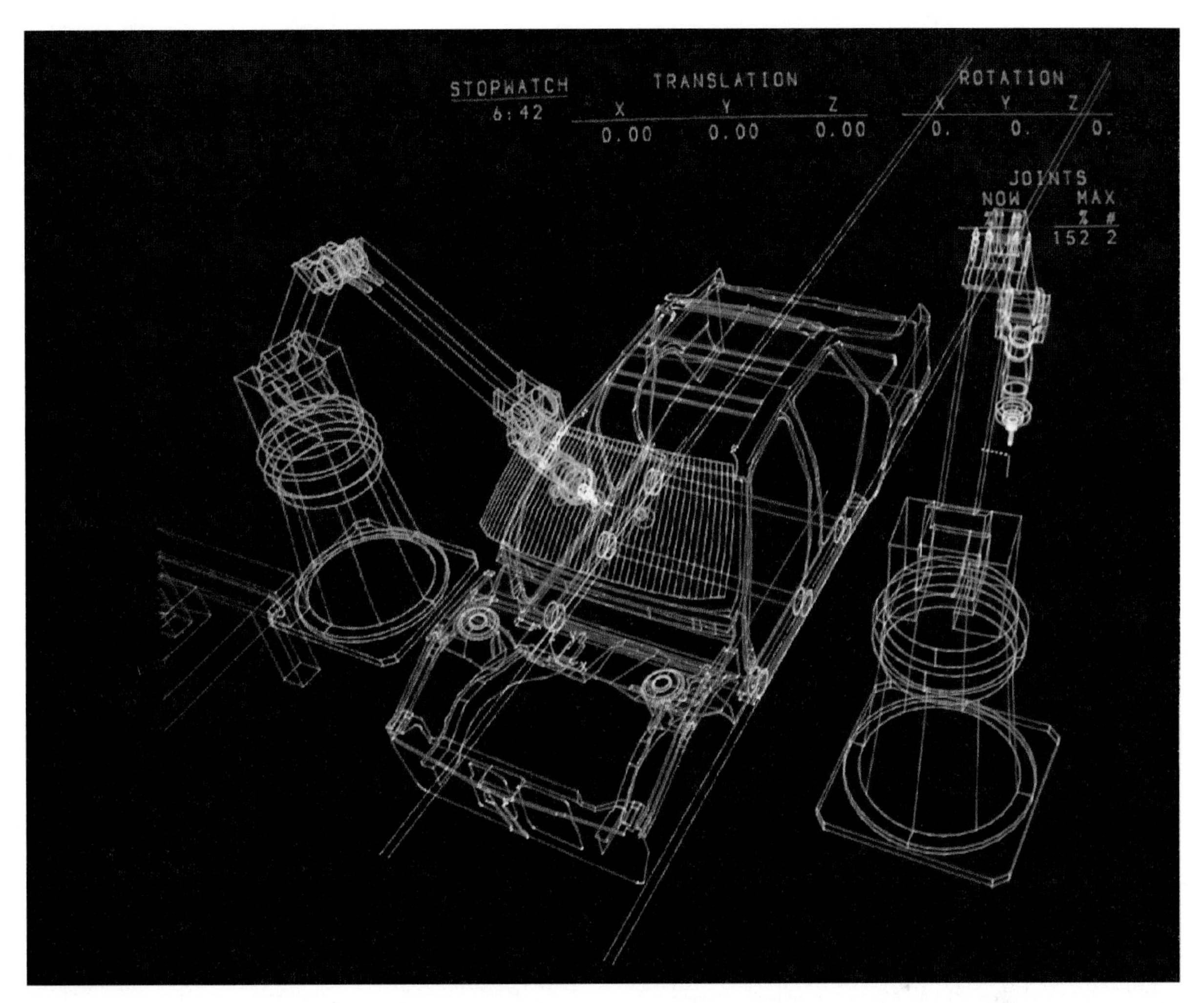

Fig. 15-3 Computer simulations help planners to anticipate the amount of time required to complete each process. Shown here is simulation of robots placing and gluing a car's windshield.

Finite Element Analysis (FEA) is a type of simulation that predicts how a component or assembly will react to environmental factors such as force, heat, or vibration. It predicts what will happen when the product is used. The wheel assembly on a car can be "road tested" using FEA. How will the car work when it crosses railroad tracks, hits bumps, and encounters water and other road hazards? Will parts rub against one another, or will vibrations loosen a component? Any needed changes can be made in the design before making real parts.

Digital Prototyping Another simulation process, digital prototyping, is used at an early stage in product development. This process examines 3D images of all the separate parts of the product at the same time in assembled form. Parts that interfere with one another can cause problems, but animation tools help engineers detect these errors by allowing an up-close and detailed view. Engineers quickly locate problems while watching the product function. Digital prototyping software can also show crosssections or cutaway views of the product, display critical measurements, and allow for changing the location or size of parts. See **Fig. 15-4**.

Dimensional Variation Analysis Not every part comes out exactly the same during mass production. Engineers also use digital prototypes to predict these variations. Dimensional variation analysis (DVA) simulates product assembly thousands of times in just minutes and calculates all the possible changes in sizes of parts. Using animation, it also shows these variations as they occur, so the engineer can quickly determine the cause.

Fig. 15-4 Using digital prototyping, these engineers can detect errors and fix them before production begins.

Rapid Prototyping As you learned in Chapter 7, "Graphic Communications," 3D models can also be created using rapid prototyping. Using the dimensions from the CAD model, the prototyping machine makes the item. The difference between a conventional prototype and a rapid prototype is that the latter usually isn't a working model. It is often only a detailed physical model that shows the designers what a product, such as a telephone, will look and feel like.

Design for x One planning concept that engineers use is called **design for x (DFx)**, where x stands for a process that can improve the design, manufacture, and use of a product. Design for manufacturability (DFM) is one of these processes. As you learned in Chapter 12, "Product Development," it means analyzing a product design and determining ways to make it easier to manufacture. See **Fig. 15-5**.

Design for assembly (DFA) focuses on reducing the time and effort needed to assemble parts. Many parts today are designed to snap together, or adhesives are used to glue parts together, eliminating screws, bolts, and nuts. The result is fewer parts, reduced labor, and faster assembly.

Design for disassembly (DFD) is important for products that must be taken apart at some time in the future. For example, batteries may have to be replaced, or some service may be required.

Design for recycling (DFR) means designing product materials and parts that can be easily salvaged and reused. Ink cartridges for ink jet printers are good examples.

Production Planning and Work Flow

Companies can save time and money if changes and adjustments to products and processes can be made *before* production begins. One process used is **product configuration**, which allows customers to specify certain components and other variations in a product.

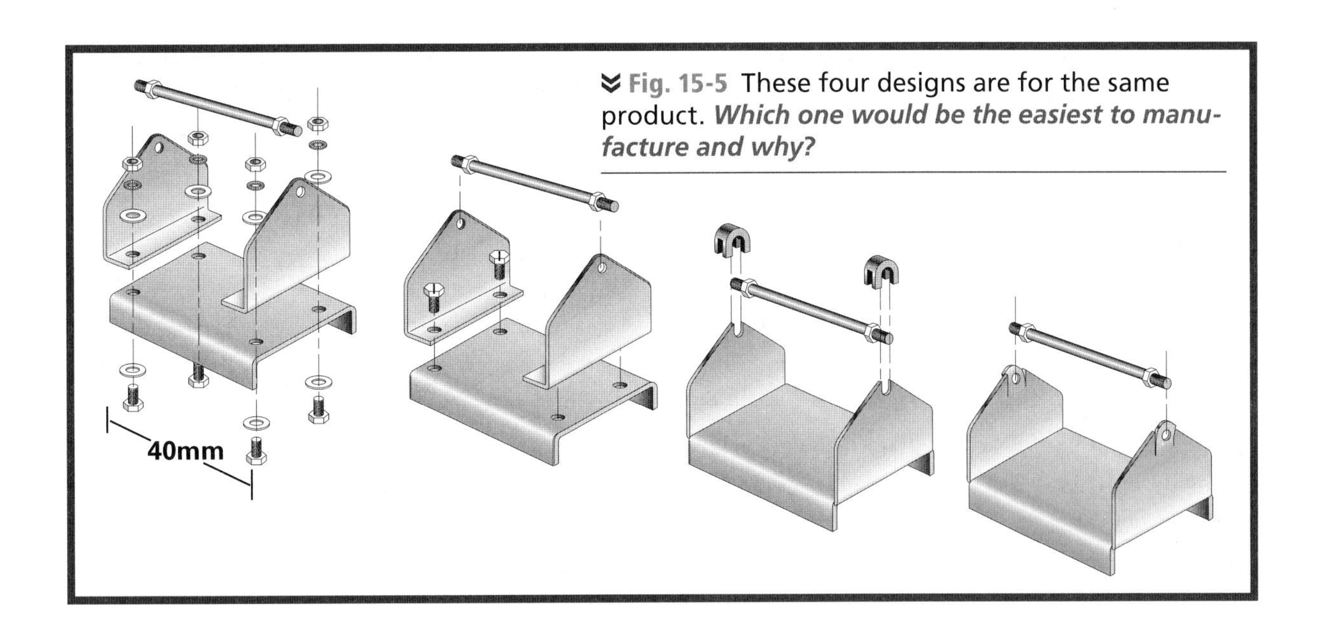

Fig. 15-5 These four designs are for the same product. ***Which one would be the easiest to manufacture and why?***

The manufacturer makes a basic model and the buyer customizes the product for individual needs. A production plan and a production schedule are automatically generated by the computer. The schedule serves as a road map for the company and the buyer as the product is built.

Product configuration offers many advantages. Customers can control the price of the product by choosing components that fit their budget. Companies can reduce the amount of time it takes to produce an order. The accuracy of orders increases as well. Most computer manufacturers use product configuration in manufacturing, selling, and servicing their computers. See Fig. 15-6. Each computer can be built containing the features that the customer has chosen.

Fig. 15-6 These computers are being fitted with components customers have requested. ***How does product configuration benefit both the consumer and the manufacturer?***

Processes and Programs for Production Planning and Work Flow	
Group Technology	A method of classifying similar parts according to shape and features. The design for a part can be altered slightly and used in a new product.
Enterprise Resource Planning (ERP)	Program that manages data used throughout the entire company.
Computer-Aided Production Planning (CAPP)	Program that allows planners to quickly determine the best ways to obtain materials and schedule production.
Materials Requirements Planning (MRP)	Program that analyzes information from the bill of materials, master production schedule, and inventory records. It then provides recommendations about resources and costs.
Manufacturing Resource Planning II (MRP II)	Program that calculates which materials, machines, and workers are needed during production, where they are needed, and when. It also analyzes costs, skills needed, and details about the job itself.

Fig. 15-7 The above processes and programs are used to make high-performance manufacturing as efficient and cost-effective as possible.

Fig. 15-8 This automatic guided vehicle is part of a system that moves materials and products around a factory.

High-performance companies rely on computers to help obtain materials and schedule production. Product configuration is but one way computers are used to help manufacturers save time and money before actual production. Additional processes, as well as computer programs, are described in **Fig. 15-7**.

Moving Materials

What happens if a product is ready to be produced, but the materials are nowhere to be found? Computers can make sure that the right material is in the right place at the right time within the factory.

Automatic Guided Vehicle Systems (AGVS) feature specially built driverless carts (AGVs) that follow a wire path installed in the floor. See Fig. 15-8. Movement is controlled by a central computer that keeps track of each vehicle's location. It directs starting, stopping, and speed, and causes the vehicle to switch from one path to another. Some AGVs can carry heavy loads such as car engines. Others are made to carry lighter loads, such as computer chips and circuit boards.

Automated Storage and Retrieval Systems (AS/RS) are special sets of tall racks with a computer-controlled crane that travels between them. See Fig. 15-9. Loads are usually on pallets. The computer instructs the crane to pick up a load. Then it selects an empty slot in one of the racks and directs the crane to travel to that spot and store the load. The crane can also be directed to retrieve or pick up loads from storage. One type of AS/RS, called a miniload, is designed to handle items small enough to fit in drawers or tote pans.

Fig. 15-9 Computer-controlled stacker cranes store and retrieve parts in an efficient manner.

Smart Production

Parts and products can be made more efficiently using computerized production methods. Some machines are still controlled by human operators. People turn handles and press levers to make the machine work. However, today, a computer program often controls machine operation.

Hydroforming

Materials are formed using casting, molding, and stamping methods. One type of high-performance forming is called hydroforming. A fluid is used to push a material into a mold to give it the desired shape. Forming the frame for a motorcycle is one example. Metal tubing is placed inside a negative mold and fluid is pumped into the tubing at a high pressure, causing it to expand outward into the mold.

Cutting Technologies

Much cutting is still done in traditional ways using blades, but computers have made several

Fig. 15-10 Waterjet cutting describes the process of forcing water through a tiny hole under high pressure.

new processes possible. Waterjet cutting is the use of a highly pressurized jet of water to cut a material. See Fig. 15-10. The water pressure in your home is about 50 pounds per square inch (psi). At that pressure, water won't cut anything. Squirting water through a very tiny hole at 50,000 psi, however, turns the water into a knife blade. The water jet follows a path guided by a computer program and cuts sheet materials, such as cloth, plywood, rubber, and plastic.

Laser cutting is another computerized way to machine parts. Lasers strengthen and direct light to produce a narrow, high-energy beam. Whatever the beam strikes becomes so hot that the material vaporizes. This concentrated, high-energy beam of light can be used to cut materials. Another use of laser cutting is to engrave the molds used to make compact discs.

Flexible Machining Centers

A flexible machining center (FMC) is a computer-controlled machine tool. It's capable of drilling, turning, milling, and other processing operations.

One place a flexible machining center might be used is in an engine factory. The FMC might work first on an eight-cylinder engine. Next, it might work on a four-cylinder engine and then on a six-cylinder engine. Ordinarily, each engine would require a change in tool setup, but an FMC handles the differences easily.

SMED

A die is a tool used for cutting or molding operations. SMED stands for Single Minute Exchange of Dies. Typically, it takes hours to change dies on a machine from one die to another. With the SMED process, dies are changed in one minute. This is done by having the new die and all the tools ready before the actual change begins. SMED reduces downtime of the machine, thus increasing productivity.

Robotics

Robots are special machines programmed to automatically do tasks that people usually do, such as moving objects from one place to another, assembling parts, welding, or spray painting.

Several types of robots are used in manufacturing, but they have common features. Most have a hand, usually called an end effector. The end effector may be a gripper for holding things or it may be a built-in tool, such as a drill or a

Science Application

Machining Tiny Products Many different high-performance manufacturing processes use computers to control tools for material machining and shaping. The science of cutting materials involves using various forms of energy and being as precise as possible.

As you know, cutting a material requires energy. With a handsaw, the energy comes from your muscles. With a power saw, energy comes from an electric motor. Precision cutters often use lasers, which are beams of light, or electron beams, which are beams of charged particles. These devices focus a great deal of energy into a small point. As the beam moves across the material, atoms and molecules are turned into vapor form, creating a narrow cut.

A human operator's vision and hand control are much less accurate than the precision of the beam. Computers and robots can provide the precise motions that are needed to produce tiny gears and other precision parts.

FIND OUT Measure the length of a thin piece of wood. With your teacher's permission, cut it in half with a crosscut saw. Now measure both cut lengths and compare the total to the original length. Repeat the process using a backsaw (miter) or dovetail saw instead of a crosscut saw. How do the lengths compare?

welder. See Fig. 15-11. Robots also have joints, usually a wrist, an elbow, a shoulder, and a waist. These joints allow robots to stretch, turn, and raise or lower themselves within a limited work area. This work area is called the work envelope. A robot can be programmed to manipulate its end effector anywhere inside the work envelope.

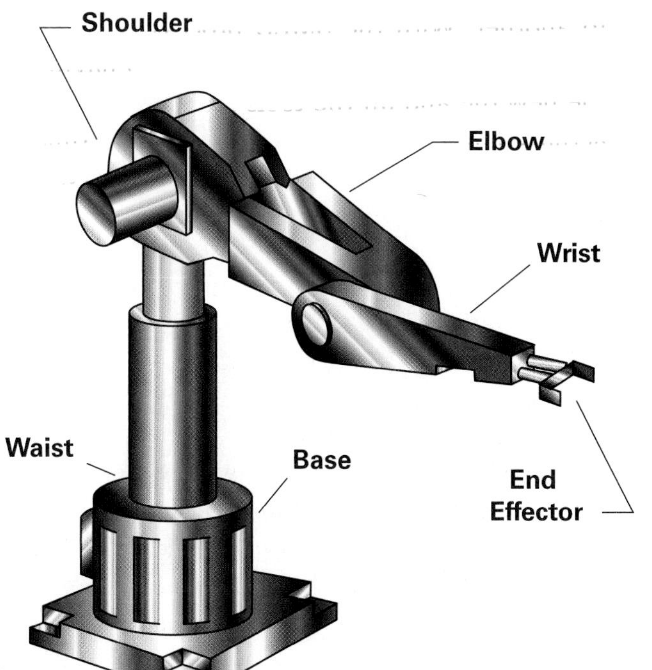

« Fig. 15-11 Robots have movable joints, much like humans do. This flexibility allows them to reach any point inside their work envelope.

Young Innovators

Digital Dueling With Bots IQ

A small plume of dust rises behind your vehicle as it races across the arid landscape in the desert heat. The finish line is just ahead, and the competition is closing fast. At the finish line, excitement overtakes you as your vehicle comes into view and you begin cheering loudly. Wait a minute, if you are at the finish line cheering, who is driving your vehicle? On any other day, the absence of a driver in a moving vehicle would be cause for concern, but not today. You are participating in the autonomous vehicle competition, part of the Bots IQ educational program.

Bots IQ is a robotics program and competition open to middle school, high school, and college students. There are actually three different competitions that you can enter. The Grand Challenge IQ (GCIQ) is the autonomous vehicle competition. In the TableTop IQ (TTIQ) task-oriented robotics competition, you build a robot to perform a specific task. You may have also heard of the BattleBots® IQ (BBIQ) battling robots competition, where two robots come to fisticuffs and duel until the last circuit board is fried.

Perhaps you will someday engineer robots that perform dangerous tasks on the manufacturing floor or security robots that help keep people safe. No matter what your ambition, Bots IQ is an excellent opportunity for you to become involved in the exciting field of robotics and gain the knowledge and skills that will help you in the future. Best of all, you get to do fun and exciting things like racing robotic vehicles and building battling robots in the process!

Even if robotics is not your calling, do not worry; there is something for everyone in Bots IQ. If you like computers, you might enjoy developing software. Perhaps managing publicity is more your style. There are also positions for those who enjoy creating artwork, team logos, and T-shirts, or any one of a number of countless jobs. Competing in Bots IQ is a true team effort.

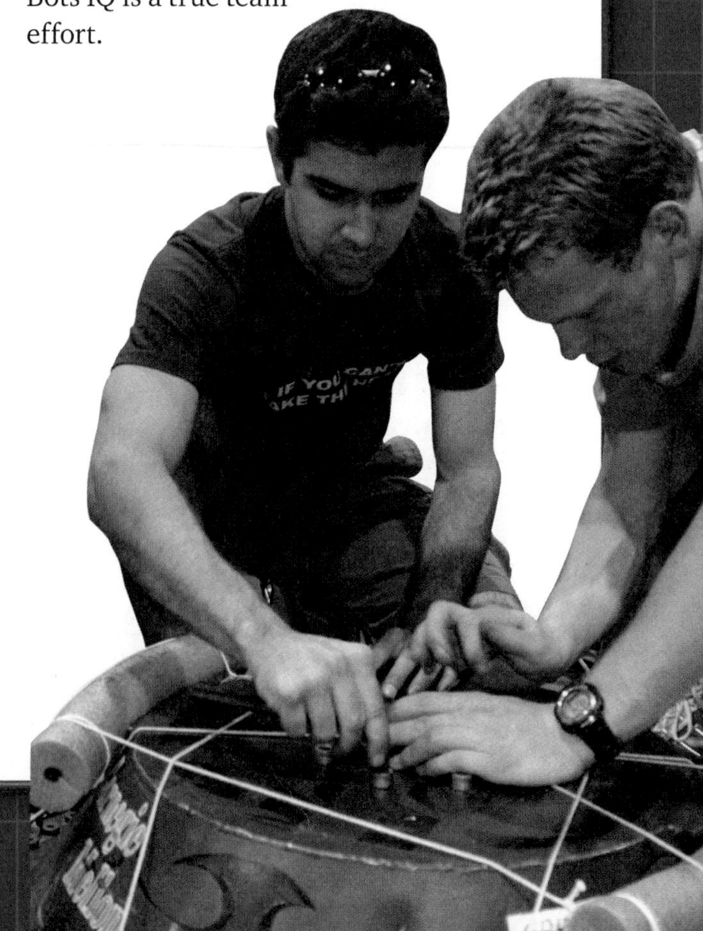

Math Application

The Geometry of Robots How do robots move? Like humans, robots have degrees of freedom. In robotics, this refers to a robot's ability to move in a particular direction. Each degree of freedom requires a separate joint.

Your hand and wrist have 22 joints, or degrees of freedom. A robotic hand usually has only up to six degrees of freedom. However, while robots have fewer joints, their wrists also have a greater range of motion. For example, your wrist can only bend about 165 degrees, but a robot's wrist can spin 360 degrees.

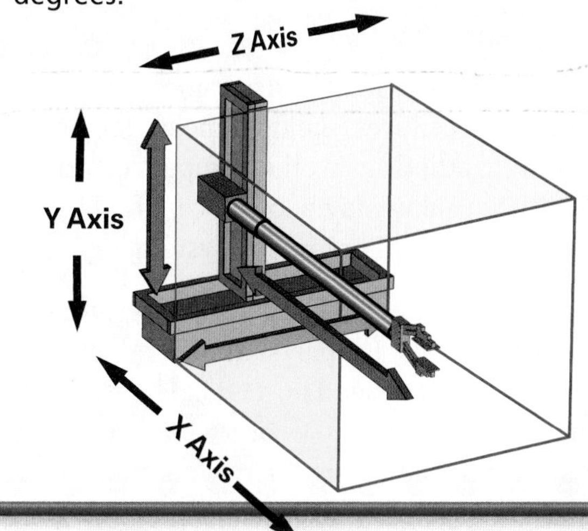

These degrees of freedom are often programmed into the robot using the axes of the Cartesian coordinate system. A robotic arm with three degrees of freedom, for example, can move along three axes: back-and-forth movement along the X axis, up-and-down movement along the Y axis, and in-and out-movement along the Z axis.

The degrees of freedom play a part in determining a robot's work envelope. A robot with a rectangular work envelope moves in a straight line and from side-to-side. Robots with cylindrical work envelopes rotate on their bases. They can extend themselves vertically and their horizontal arms move in straight lines. A spherical work envelope results from a robot with a rotating base, a main body that tilts, and a horizontal arm that moves in and out.

FIND OUT How would you find out the work envelope of your own arm? Hint: You may want to look at formulas that show volumes of geometric shapes.

One main advantage of robots is that they can be programmed to repeat a task over and over. They don't need a coffee break or rest period. Also, they are very accurate. Robots are good for doing work that is hazardous to humans, such as in paint shops filled with fumes, in high or low temperature conditions, or where very heavy loads must be lifted repeatedly. These advantages help manufacturers increase production, improve product quality, and ensure worker safety.

Automated Assembly

Most assembly machines are specially built to assemble a certain product. They are often automated. For example, automated assembly is common in the electronics industry. Small components, such as resistors or diodes, are packaged on a roll of tape and loaded into an automatic insertion machine. As circuit boards travel past the insertion machines, the parts are rapidly and accurately inserted into the right places. Computers, video recorders, and stereos are examples of products assembled in this manner.

Smart Control

A very important part of efficient manufacturing is control. Computerized production not only uses computers to automate the machines but also uses computers to monitor and control the flow and quality of work. For example, if one workstation develops a production problem, the computer adjusts schedules at other workstations. Many methods and devices are used.

A programmable controller is a small, self-contained computer used to run machines and equipment. It is housed in a heavy-duty case. The case protects it from the factory environment. The fact that it is programmable means that workers can change the way it functions. This makes it more useful than controllers that are built to do only one thing. See **Fig. 15-12**.

Fig. 15-12 This programmable controller is used to run a machine that cuts sheet metal.

One system called manufacturing execution systems (MES) operates the MRP and MRP II systems discussed earlier. It helps companies make sure all the materials, workers, and other resources are available for use during the production stage. MES also helps companies to plan so resources are paid for within the budget.

Controlling Quality

In Chapter 14, "Production," you learned about quality assurance. Manufacturers spend lots of time, energy, and money to make sure that the parts and products they produce meet design standards.

ISO 9000 is a set of international standards that establish quality assurance procedures. These procedures include measurement and testing methods as well as proper record keeping. A company may become certified by proving that it has an organized system in place.

Six Sigma is a method for finding and eliminating defects in parts and products using statistical data. See **Fig. 15-13**. The focus is on improving processes. In order for a company to reach the six sigma level, it must produce no more than 3.4 defective parts in one million!

Through special training, a person can become a six sigma expert at one of several levels: Champion, Master Black Belt, Black Belt, or Green Belt. Each level has specific requirements.

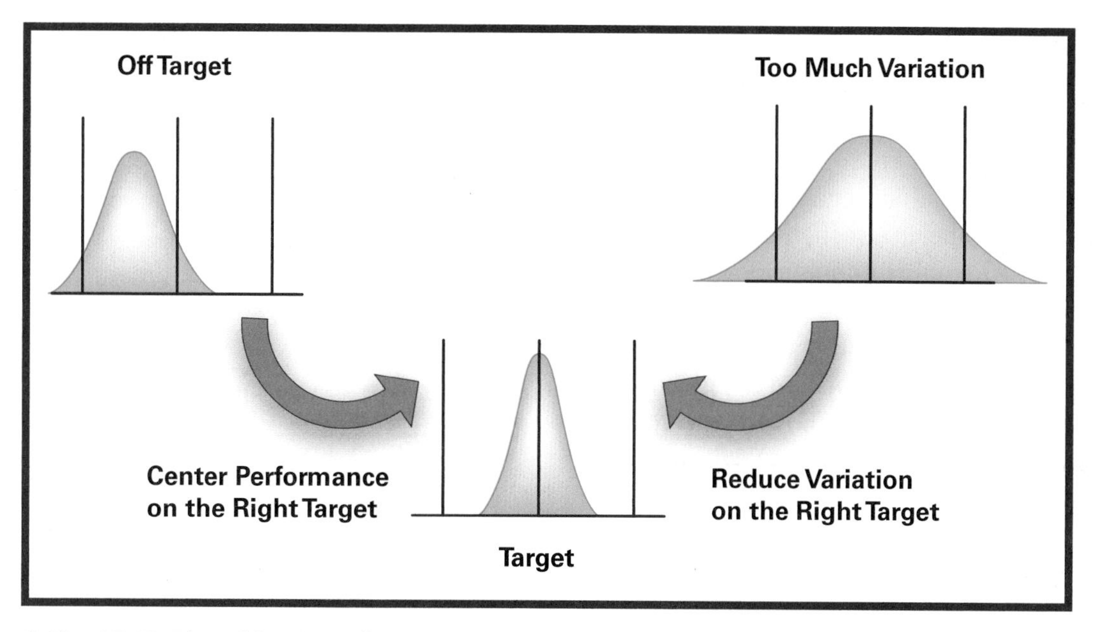

Fig. 15-13 The objective of six sigma is to reduce variations, errors, and defects in parts and products. This improvement in manufacturing leads to an increase in customer satisfaction.

Computerized Inspection Quality is often measured by computerized machines. A coordinate measuring machine (CMM) is a very accurate computer-controlled measuring device. Its main advantage is a high degree of accuracy and consistency. A CMM is usually used to measure hard-to-measure parts, such as rounded or spherical items. It measures the part and compares the measurements with design specifications that have been stored in its memory.

Other instruments include optical comparators and special scanning microscopes that magnify small parts. See **Fig. 15-14**. X-ray machines are used to see inside welded metal parts. Other devices emit sound waves to check product characteristics.

Noncontact measuring systems measure parts using cameras and laser beams. There is no need to actually touch the parts being measured.

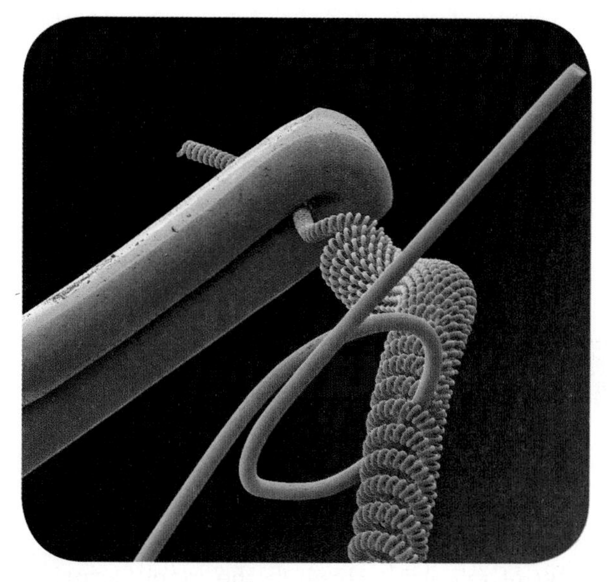

Fig. 15-14 A lightbulb filament is shown here in a colored scanning electron micrograph (SEM).

Automatic Identification

Controlling any activity requires current and accurate information. One way that current information is entered into a computer is by a process called automatic identification. A special tag or label is attached to a part or product. The tag contains an identification code. A machine can read the code to identify the part.

The best example is the universal product code (UPC), which is a form of the global trading identification number (GTIN). You've probably seen these rows of black and white lines on products you buy in the store. They are sometimes called bar codes. The computer reads the code and identifies the product. In manufacturing, the information can be used to keep track of inventory, to direct a part to the right workstation, and to otherwise control what happens to it.

You learned about radio frequency identification (RFID) in Chapter 6, "Electronic Communication." Because it can track moving objects, this technology is often used during production to check production activities. For example, the manufacturer can pinpoint the date and time a production problem occurred, having to recall only a small number of products made at the same time instead of the entire run.

Kanban

Kanban is Japanese for "card" or "signal." A kanban is sent from one workstation to another indicating material is needed. Using a kanban, a worker at workstation B, for example, sends a signal to the preceding workstation A. That card is a signal for worker A to pass one part to worker B, and so on. This establishes an orderly flow of materials through the manufacturing process. This is a version of a "pull" system. The parts are being pulled through the manufacturing process rather than being pushed.

Advanced Systems

Several advanced systems are used to control costs, quality, and production flexibility. Some companies rely on only one; others may use a combination.

5S

Organization is essential to any manufacturing system. A form of "housekeeping," 5S is based on five simple methods used to improve any process. See **Fig. 15-15**.

- **Sort.** Clean out the work area and get rid of the things that aren't needed.
- **Straighten.** Arrange things needed so they are easy to find and use. Time is not wasted searching for them.
- **Shine.** Keep tools and equipment clean and in good working order.

Fig. 15-15 Notice how clean and neat this manufacturing line is. This manufacturing plant follows 5S procedures. ***What elements of 5S can you see in this photo?***

- **Standardize.** Arrange work areas in similar ways. This helps workers to know where things are and how to carry out a task.
- **Sustain.** Establish a set of rules that are easy for people to remember and that can become good habits.

Lean Manufacturing

"Lean" is a word usually used to describe a piece of meat with very little fat in it. **Lean manufacturing** means reducing the amount of waste in the production of parts and products. It represents the ultimate goal of smart manufacturing.

Lean manufacturing activities focus on the entire process from customer order to delivery. It attempts to reduce seven areas of waste:

- Waiting for the next process step
- Correcting defects or errors
- Making more parts or products than needed
- Overprocessing, or doing too much
- Wasted motion by workers
- Unnecessary transportation of materials
- Keeping too much inventory on hand

Kaizen

The Japanese term *kaizen* literally means "change for the better" and focuses on activities for eliminating wasted movement, time, and materials, thereby improving productivity. Everyone in the organization is involved. To do kaizen, manufacturing engineers analyze a specific process, such as assembling a bicycle. They look at all the steps in the process, as well as the final result. Then they try to think about it from a larger point of view using a systematic way of thinking. Finally, they try not to blame anybody for mistakes but, instead, focus on fixing the problems. Kaizen is a version of total quality management, which is discussed in more detail in Chapter 11, "Manufacturing Fundamentals."

Fig. 15-16 These work cells are involved in manufacturing airplanes.

JIT Manufacturing

In high-performance manufacturing, computers are used to keep track of inventory, to order materials, and to schedule deliveries. Computer control helps make **just-in-time (JIT) manufacturing** practical. In JIT, materials are delivered as they are needed—just in time. This reduces the need for warehouse space. It also means fewer workers are needed to organize and keep track of the materials. However, the system must be set up carefully. If the right materials in the right amount do not arrive at the right time, production is halted.

Manufacturing Cells

Manufacturing cells, or work cells, are small groups of machines and people working together as a unit to produce a product from start to finish. See Fig. 15-16. Each member of the team is trained to do all of the jobs. All of the machines are controlled by computers. Machines can be moved in and out to make customized products.

Flexible Manufacturing Systems

Flexible manufacturing systems (FMS) are groups of manufacturing cells and flexible machining centers. The cells and FMS are tied together by an automated materials-handling system and by computer control. Most of these machines can be adjusted and changed as product needs change.

One-Piece Flow

One-piece flow describes the continuous flow of parts and materials within a manufacturing cell. In a more traditional manufacturing process, parts are made in batches of about 50 at a time. When they are complete, they are moved to the next workstation. In one-piece flow, each part is moved on as soon as it is finished. This process is especially effective in a cellular manufacturing environment.

Supply Chain Management

Perhaps one of the most important aspects of high-performance manufacturing is supply chain management (SCM). A **supply chain** is the sequence of suppliers and processes necessary to deliver finished products to customers. The major groups involved are manufacturers, service providers, distributors, and retail stores. All of their activities must be planned and controlled. Computer communications, manufacturing, logistics, and transportation are all included. Many types of computer software, such as CAD and ERP, are used to manage the supply chain. See Fig. 15-17.

The Automatic Factory

A factory in which almost everything is done automatically by machines is an automatic factory. There are no people making products. All parts are made by automatic machines. All assembly work is done by automatic assembly machines. All materials are moved by using automatic materials-handling equipment. Quality control checking is also all

Fig. 15-17 Many types of computer software are needed to efficiently manage the supply chain. All of its activities must be planned and controlled.

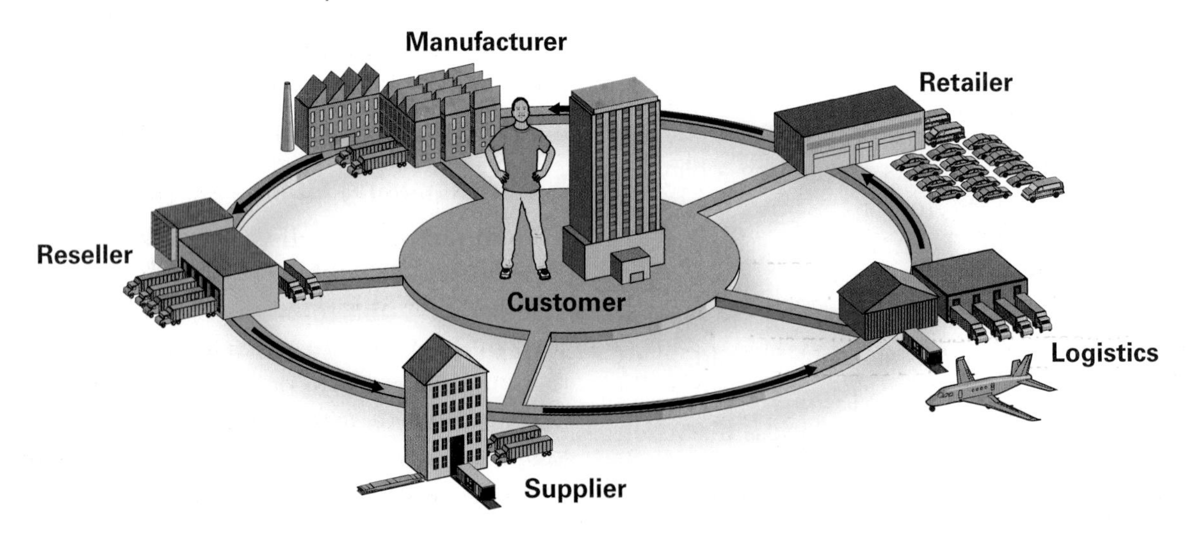

» Fig. 15-18 "Moto-man" robots weld the bodies of these automobiles in a Japanese plant.

done automatically. The factory may be dark inside, because most machines do not need lights to operate.

In an automatic factory, all the various systems work together. The information collected from all the subsystems is kept in a main computer. The main computer directs the other computers. All records are kept in computer database. A copy of a report or drawing can easily be printed at any time from any of the company's locations worldwide.

One manufacturing plant that is completely automatic already exists. See Fig. 15-18. It is a Japanese manufacturer of heavy machinery used throughout the world. The plant operates in the dark, except when humans are needed to make repairs or changes in manufacturing procedures.

In the future, more automatic factories may be used for certain types of products. The cost of such plants is very high, and they can be used only to manufacture products that will require few design and tooling changes. The products must be in demand for many years to make the investment in an automatic plant worthwhile.

Will people be needed at all in such a factory? Yes; they will be needed to control, service, and repair the computers and other equipment. However, far fewer workers will be needed in automatic factories than in today's factories.

Factories in Space

In space, natural conditions such as gravity are different. Objects are almost weightless because the forces of gravity acting upon them are very small. This condition is called microgravity. In experiments, researchers have studied how some products might be improved if they are made under microgravity. For example, microgravity permits droplets of metal to be formed into nearly perfect spheres, making superior ball bearings. Reduced gravity also makes it possible to make mixtures from materials of different densities, which is not possible on Earth because the materials separate. In space, however, new metal alloys can be made because heavier materials do not settle to the bottom. Other materials and products that can be produced better in space than on Earth include crystals for electronic applications, ceramics, glass, and new medicines.

Experts predict that manufacturing in space will involve the use of both manned and unmanned automatic factories. Although it is now possible to manufacture some products in space factories, it will be a long time before doing so is economically feasible.

CHAPTER 15 REVIEW

Main Ideas

- High-performance manufacturing combines highly skilled workers, advanced technology, and new work methods.
- Smart planning means considering not only how the final product will work but also how to manufacture it.
- Computerized production uses computers both to control the machines and to monitor and control the work.
- Several advanced systems are used to control costs, quality, and production flexibility.

Understanding Concepts

1. How is high-performance manufacturing, or CIM as it is sometimes known, different from traditional manufacturing?
2. List advantages of using product configuration.
3. How do robots help improve safety?
4. What is ISO 9000?
5. How is an automatic factory unique?

Thinking Critically

1. How might digital prototyping have been used in designing a car?
2. Some technology courses are taught as modules. What manufacturing structure is a modular course similar to?
3. What would the advantage be for a manufacturer to be ISO certified?
4. What kinds of jobs would be available for someone at an automatic factory?
5. Can you think of any other "housekeeping" steps that would fit in 5S?

Applying Concepts & Solving Problems

1. **Math** If a robotic arm can move in a rectangle 14 inches on the Y axis, 10 inches on the Z axis, and 12 inches on the X axis, what is the size of its work envelope?
2. **Manufacturing Activity** Steelco Parts has headquarters in Sacramento, as well as one factory each in Atlanta, Denver, and Hong Kong. Create a diagram that shows how the networks of the company would be linked.
3. **Science** List at least five environments where robots could be used in place of humans for scientific work.

Activity: Clean Up Your Workspace Using 5S

Design Brief

As you read in this chapter, 5S is based on five easy ways to improve productivity by means of simple housekeeping techniques. Any activity can be streamlined by applying 5S concepts.

Do you remember each part of 5S? Sort means to clean out the work area and get rid of things that aren't needed. Straighten means to arrange things so they are easy to find and use. Shine means keeping things clean and in good working order. Standardize refers to arranging the work areas in similar ways to help workers know where things are. Sustain means to establish rules that are easy for people to remember so they can develop good work habits.

Select an area in your home where you spend time, such as your bedroom, the family room, bathroom, garage, etc. Study the area and use the 5S method to make it a more efficient work area.

Refer to the

Criteria

- Choose an area that you use frequently, preferably on a daily basis.
- Use all five 5S steps in order to improve the area.
- You must document your work with a written report and "before" and "after" drawings or photographs of the work area layout. You may or may not decide to include these comparison shots in a slide presentation format.
- Before you make changes, you must time at least one activity that takes place in the area. Then you must time it again after changes are made.

Constraints

- Limit your project to only one room.
- You must obtain permission from a parent or other person in authority in your home before you begin the project and before you make changes.

Engineering Design Process

1. **Define the problem.** Write a statement that describes the problem you are going to solve. For example, which 5S steps are most needed in the area?
2. **Brainstorm, research, and generate ideas.** Think about possible solutions. Hint: You might want to ask a family member to help you by performing an activity in the area you'll be improving. Then observe the person and note potential changes you could make. For example, are the tools in a garage kept close to where they are used? Time the activity for the "before" results.
3. **Identify criteria and specify constraints.** These are listed on page 318.
4. **Develop and propose designs and choose among alternative solutions.** Choose the best design that will solve your problem. You might want to make several sketches of layouts and lists of other improvements for the area.
5. **Implement the proposed solution.** Decide on the design you will use. Gather any needed tools or materials.
6. **Make a model or prototype.** Create a final layout and drawings that show your solution.
7. **Evaluate the solution and its consequences.** Discuss your layout with those who use the area. Will your solution solve the problem? What impacts and effects will result from this solution? Make notes.
8. **Refine the design.** Based on the evaluation, change the design if needed.
9. **Create the final design.** Apply the 5S steps to the work area. Take any photographs or make finished drawings. Time the work for the "after" results.
10. **Communicate the processes and results.** Write your report and present your project to the class. Turn in your drawings, photos, notes, and other documentation to your teacher. Be sure to include the name of the activity and your definition of the problem.

SAFETY FIRST

Before You Begin. Be sure not to lift heavy items by yourself. Make sure you understand how to use any tools and materials safely. Have your teacher demonstrate their proper use. Follow all safety rules.

UNIT 5

CONSTRUCTION
ENGINEERING & DESIGN FRONTIERS

In 1908 Sears, Roebuck and Co. introduced a “kit home” in a mail-order catalog. These ready-to-build homes could be ordered for as little as $500. Between 1908 and 1937, the company sold more than 100,000 kits. Today these houses are almost thought of as collectors’ items. In a small town in Illinois there are 152 of these Sears, Roebuck homes in a twelve-block area.

New Designs and Methods

Douglas Gauthier and Jeremy Edmiston are trying a different approach to the kit home idea. Their kit houses are not the traditional styles we are used to. Their first model in a new series of kit houses is called BURST 003. It was designed by using computer 3D modeling. The house is a three-bedroom, two-bath, 1,000 square foot bungalow. It is made with 1,100 pieces of plywood that have been cut very precisely with a laser. In the BURST 003 design, there are no beams, posts, or nails, and each piece of plywood is different. The plywood is held together by a series of stainless-steel fasteners.

The 3D model is sliced into each structural piece. The pieces are then numbered. With another computer program, the pieces are arranged on plywood sheets in such a way as to waste as little plywood as possible. This practice is called "nesting." The real pieces are then laid out and cut by a laser beam.

CHAPTER 16

CONSTRUCTION FUNDAMENTALS

Objectives

- Define and give examples of the four major types of construction.
- Name and describe six types of structures.
- Name and describe the three types of tunnels.
- Explain the reasons for building a canal.
- List the seven types of bridges.

Vocabulary

- construction technology
- residential construction
- stick construction
- industrial construction
- commercial construction
- public works construction

This cloverleaf intersection will allow you to go in five different directions on three different freeways. **Do you think you could design a freeway like this? In this chapter's activity, "Design a Highway," you will have a chance to try.**

Reading Focus

1. **Read the title of this chapter and describe in writing what you expect to learn from it.**
2. **Write each term in your notebook, leaving space for definitions.**
3. **As you read Chapter 16, write the definition beside each term in your notebook.**
4. **After reading the chapter, write a paragraph describing what you learned.**

What Is Construction Technology?

Construction technology is the design and building of structures. Some structures have been so beautifully designed and built that they are considered great works of art. Other structures have become important historical or cultural symbols. See **Fig. 16-1**.

Structures are built for many purposes. Some, such as houses and apartment buildings, provide us a place to live. Other structures are places to work, shop, or learn. Think of all the different buildings in your community. What are their purposes?

Structures do not have to be buildings. Highways and bridges are also structures. These support transportation. What other structures can you name that are not buildings? What is their function?

Impacts and Effects of Construction Technology

Like other technologies, construction affects our lives in many ways, both positive and negative. For example, we would have far less variety at the grocery store without the highways that link farms, food processors, and cities. Yet highways can contribute to the decline of cities (people move out to the suburbs), and the increased traffic adds to air pollution.

The construction industry is one of the top 10 areas of job growth in the United States. It is estimated that the industry will employ 7.8 million people by 2012. What do you think that means for our country's economy? What could happen if construction slows down?

Fig. 16-1 For over 200 years, the White House has stood as a symbol of the home of the President of the United States. Construction began in 1792. President Washington oversaw the construction of the house, but he never lived in it.

EVOLUTION OF Construction Technology

The construction technology we have today evolved from simpler tools, techniques, and materials.

« Construction tools have changed greatly. The first hammer was probably a rock. Attaching a stick to the rock made it a better hammer.

» Building techniques are often determined by the tools and materials that are available. In the past, shelters were often simply a few tree limbs leaning against each other and covered with animal skin.

2550 BCE

Initially, construction projects were just planned out in one's head. As the projects became more complicated, it was necessary to illustrate the plans on clay tablets or paper. Think of all the detailed plans and figures that must have gone into the building of the pyramids!

226

Early construction projects were limited to what nature provided. Tree limbs, animal skins, and rocks are examples. The Romans developed concrete, a mixture of cement, sand, gravel, and water. Many of the bridges, roads, and aqueducts they built centuries ago are still used today.

In addition to wood and concrete, the construction industry today uses engineered woods, composites, and metal alloys. Human-made materials make new types of designs possible.

Today's self-powered pneumatic nailers make it possible to work faster and with less effort. Do you think a nailer would have been helpful to the Stone Age builder?

1500

1960

1995

2000

1980

Can you imagine making all the drawings and calculations for a major project by hand? With today's computer software, the job is faster and more accurate.

If you look at some of today's buildings, you can see similarities to early techniques. However, the tree limbs have been replaced by steel and the skin by a new engineered material. What other techniques are carryovers from the early days of construction?

The Construction System

Like other technological systems, a construction system has inputs, processes, outputs, and feedback. These all work together to achieve the goal of constructing a finished structure. See **Fig. 16-2**.

The inputs to a construction system include the seven technological resources that will be used in designing and building the structure: people, information, materials, tools and machines, energy, capital, and time.

The processes of construction include everything from planning the project and choosing the site to turning over a finished building to its new owner. Perhaps you have watched the gradual progress of a construction project, such as a new home or a shopping mall, in your own community. If so, you have seen a number of construction processes going on.

The main output of a construction system is, of course, the finished structure. Depending on the purpose of the particular system, this can

Fig. 16-2 The parts of a construction system are inputs, processes, outputs, and feedback. When they all work properly, the result is a finished construction project.

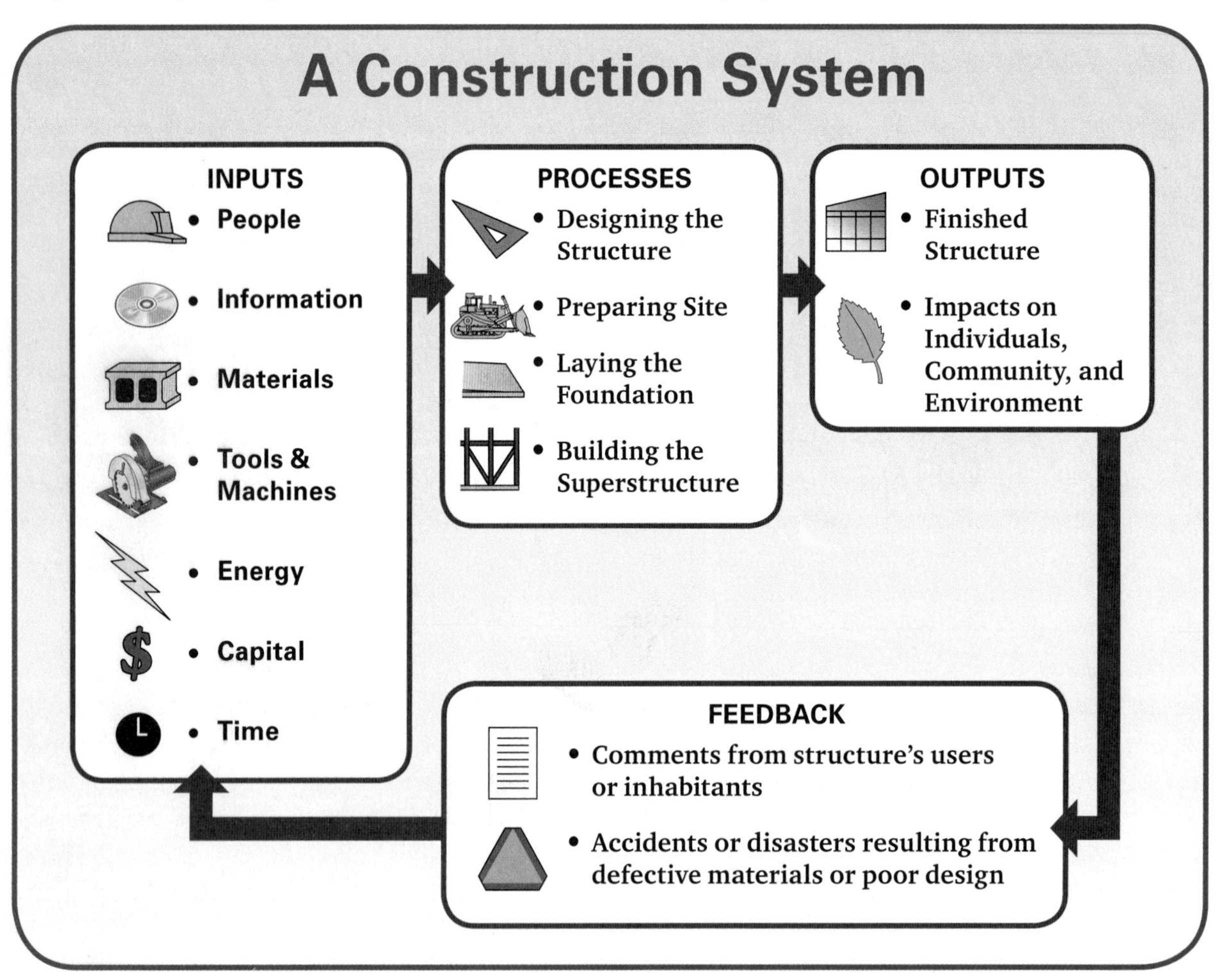

range anywhere from a backyard tree house, to a new dam or bridge, to a whole new development of residential or commercial buildings.

Feedback on construction systems is important. Even though people try to plan the best possible structures, they sometimes overlook important factors. People using a building, for example, might realize there is not enough ventilation, and the system will need to be improved.

Types of Construction Projects

Each construction project falls into one of the four major types of construction:

- Residential
- Industrial
- Commercial
- Public works

Fig. 16-3 Stick construction is still the most common way to build a house. However, the "sticks" being used today are often metal rather than wood.

Residential Construction

Residential construction refers to the building of structures in which people live. Most residential structures are single-family or private homes. However, residential construction also includes the building of small multifamily units, which are residences that have two or more apartments or dwelling areas. Most residential construction is done by fairly small construction companies.

In the early years of the United States, people on the frontier built log houses. Most early settlements were built near heavily wooded areas, so logs were plentiful. In the mid-1800s a new way of building developed in which smaller and lighter pieces of wood were used for home construction. The wood was assembled into a framework, which was then covered with other building materials to make the walls, floors, and roof. This method of building was called **stick construction**. See **Fig. 16-3**. Today stick construction is still the most common method for home construction and in many cases for commercial construction too, even for interiors of skyscrapers.

Industrial Construction

Industrial construction includes the building of manufacturing plants and other industrial structures. This type of construction is usually planned by specialized engineering firms. Industrial structures are usually built by large construction firms that have many employees.

Fig. 16-4 New York City has hundreds of skyscrapers. ***What are the two materials that are most often used in constructing skyscrapers today?***

Commercial Construction

Commercial construction involves building structures that are used for business. Supermarkets, shopping malls, restaurants, and office buildings are examples of commercial construction. Commercial projects are usually large-scale construction projects that involve millions of dollars and many workers.

A skyscraper is an example of commercial construction. See **Fig. 16-4**. One reason developers prefer to build skyscrapers is because they do not require a lot of land. In a large city, commercial property in a desirable location can be very expensive. Therefore structures are built up rather than out. In the suburbs building sites may be less expensive, so a larger "footprint" for the building is possible. These buildings are usually longer and wider but have fewer floors than the skyscrapers.

The building materials and techniques used in commercial and industrial construction are somewhat different from those used in residential construction. These structures often have steel frames and many concrete parts.

Public Works Construction

Public works construction involves building structures intended for public use or benefit. This type of construction includes large projects such as dams, highways, bridges, tunnels, sewer systems, airports, hospitals, schools, and parks.

Types of Structures

As you've seen, construction technology includes many types of structures. Here are brief descriptions of the major types. The other chapters in this unit will describe how these structures are planned and built.

Buildings

A building is usually defined as a structure with a roof and walls. It may be temporary or permanent. Schools, stores, houses, and theaters look very different from each other, but they are all buildings.

Roads

Road construction includes the building of highways, streets, and other types of roads. Major roads through and around cities are highways. They can vary in width from two lanes to eight. An expressway is a limited-access highway. Interstate highways pass through more than one state. See **Fig. 16-5**.

Fig. 16-5 The Big Dig, located in Boston, is considered one of the largest and most expensive road construction projects ever in the United States.

Math APPLICATION

Calculating Distance When reading an ordinary road map, you often need to calculate the distance from one location to another. Road maps usually have a small box with a map scale and a brief explanation of how to read the mile markers. The easiest method is simply to measure the distance with the scale that is provided and then calculate the miles.

Use the scale by placing a piece of paper along the scale and marking the distance for 10 miles and 20 miles. Then lay the paper scale over the map, measuring 10 or 20 mile increments until you have covered the distance. Your total of increments determines the approximate mileage.

Another way is to measure with a ruler and multiply that by the number of miles per inch. In this illustration one inch equals 23.6 miles.

FIND OUT Using that scale, if the distance is 2.5 inches on the map, how far is that in miles?

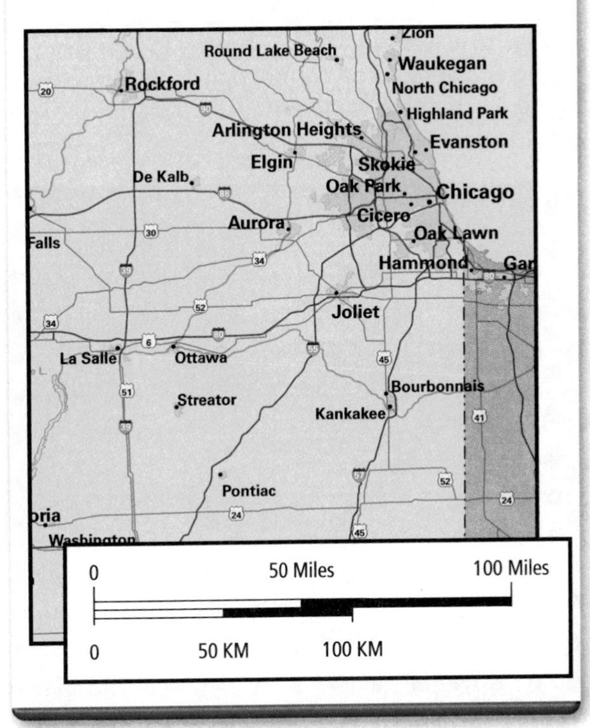

Tunnels

A tunnel is an underground passageway. Tunnels are built to allow people, vehicles, or materials to pass through or under an obstruction. See **Fig. 16-6**. For example, tunnels may be built under busy city streets for subways. They may be built under rivers or through mountains for railroads or highways. A tunnel may also be built to carry water around a dam.

There are three common types of tunnels: earth, immersed, and rock. Earth tunnels are constructed in soil or sand. The earth tunnel is the most dangerous to build. Immersed tunnels are premanufactured sections that are floated to the tunnel site. Here, they are sunk into trenches that have been scooped out of the bottom of the waterway. The sections are then connected to form the tunnel. A rock tunnel is blasted or drilled through rock with a large boring machine.

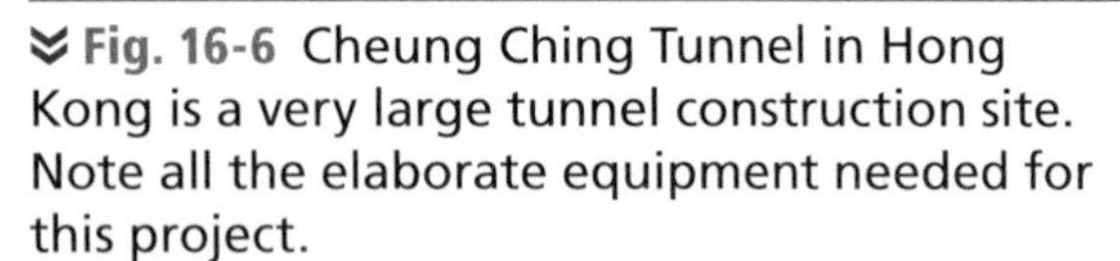

Fig. 16-6 Cheung Ching Tunnel in Hong Kong is a very large tunnel construction site. Note all the elaborate equipment needed for this project.

Fig. 16-7 The Three Gorges Dam in China is the largest dam in the world.

Dams

A dam is a structure placed across a body of water to control or block its flow. The water that collects behind the dam creates a reservoir. A reservoir is a lake in which water is stored for use. This is one of the main reasons for building a dam: to provide a dependable water supply for nearby communities. See **Fig. 16-7**. Dams are built from earth, concrete, steel, masonry, or wood. Usually a combination of materials is used.

Canals

Canals are artificial waterways built for irrigation or navigation. Irrigation canals carry water from a place where it is plentiful to another place where water is needed. Such canals supply water to land that otherwise could not be used to grow crops. Navigation canals connect two bodies of water. See **Fig. 16-8**. Navigation canals may also be constructed when a river has a portion that either bends too much or is too shallow to navigate. The canals allow ships to bypass those parts of the river.

Bridges

Bridges are structures built to allow people and vehicles to pass over something else. When we think of bridges, we most often think of those built over water. However bridges may also extend over valleys, highways, or railroad tracks. The different types of bridges are beam, arch, truss, cantilever, suspension, cable-stayed, and movable. See **Fig. 16-9** on page 332.

» **Fig. 16-8** The Panama Canal connects the Atlantic and Pacific Oceans. It is 50 miles long and will save a boat 7,872 miles by not going around Cape Horn.

Science Application

Experimenting with Cable Sag Try this experiment and you will be able to feel how the span and sag on a suspension bridge are affected by the distance between the towers.

Attach a weight to the middle of a small rope. Now hold your hands fairly close together, letting the weight make the rope sag. Notice how heavy the weight feels. Now move your hands out wider and notice the amount of thrust it takes and how heavy the weight feels. The weight is the same, but the sag and span have changed. This is why the cables on long spans and small sags must be bigger. On shorter spans with larger sags, the span can be made with thinner cables.

FIND OUT Why do suspension bridges have an extra set of cables at each of the ends?

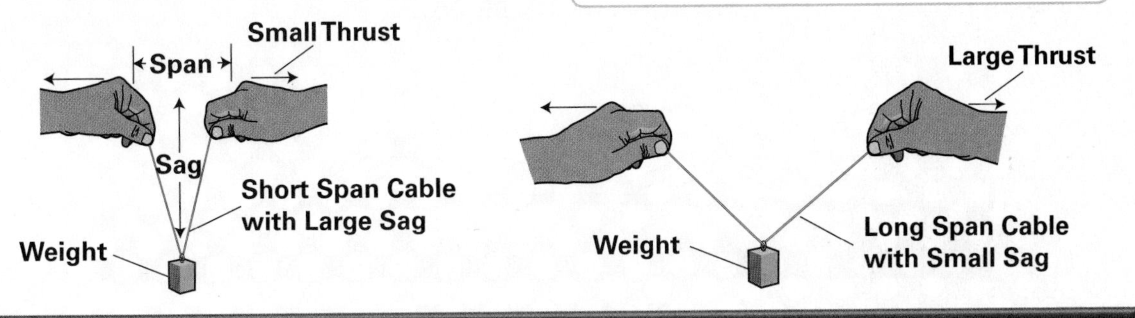

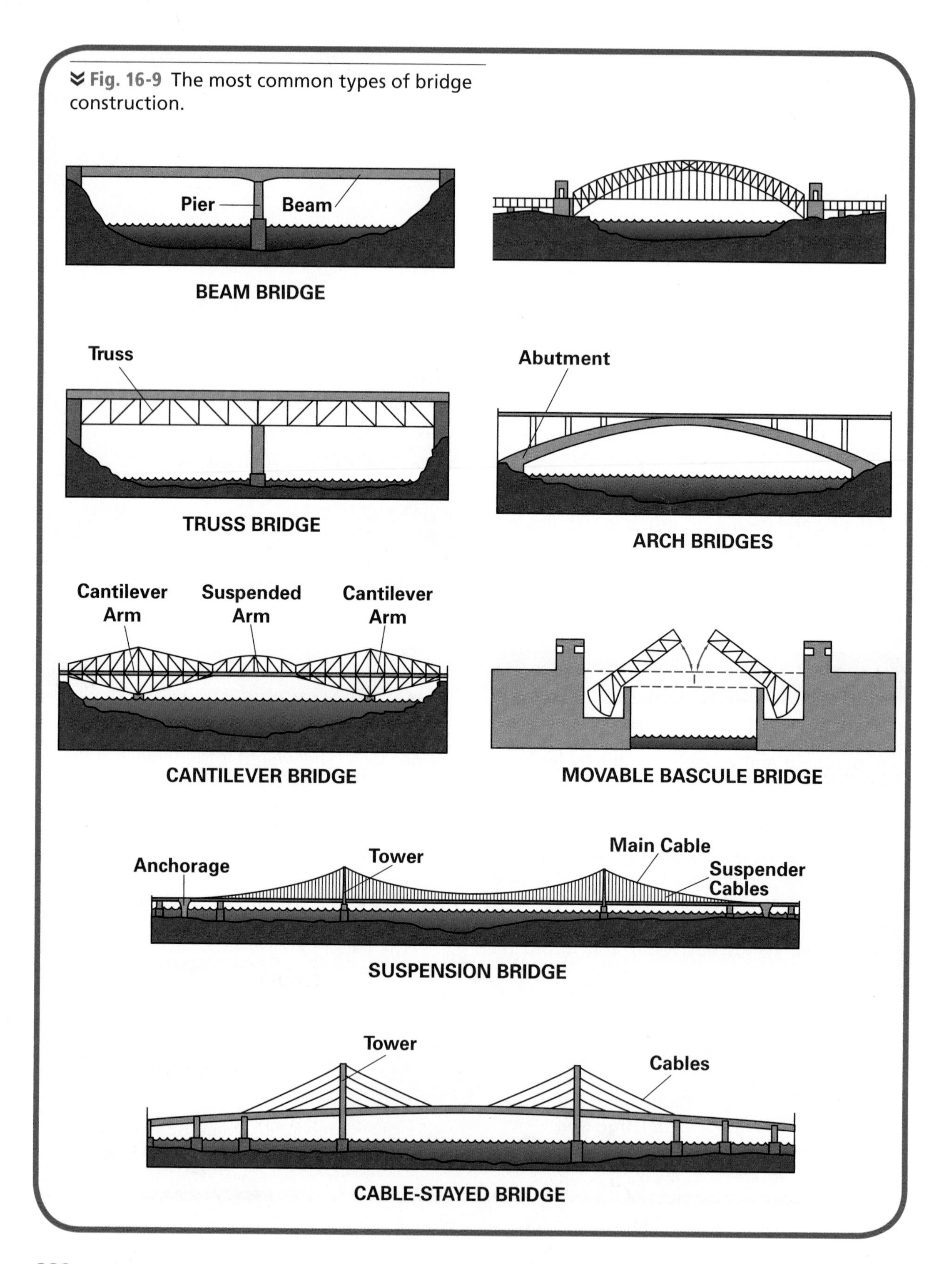

Fig. 16-9 The most common types of bridge construction.

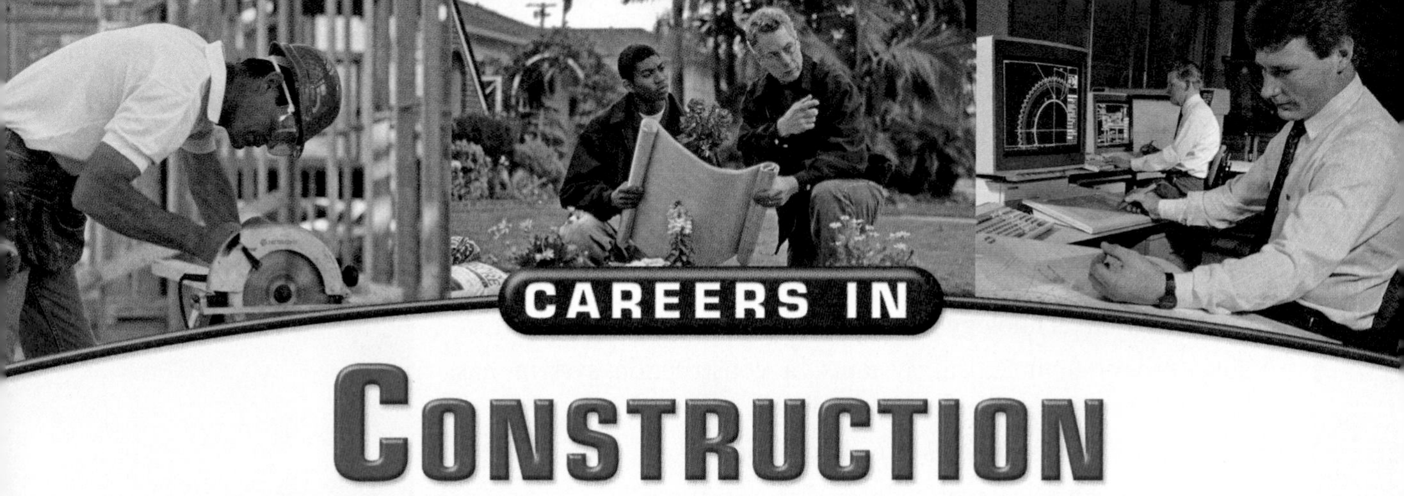

CAREERS IN CONSTRUCTION

There are about 70 different job titles in the construction industry. There are jobs for the "hands-on" type of person (bricklayer, for example), high-tech types (CAD technician), nature lovers (landscape contractor), and number crunchers (cost engineers). The U.S. Department of Labor estimates that there will be one million new jobs in the construction industry by the year 2012.

Carpenter • Architect • Bricklayer • Contractor • Landscape Architect

Structural Engineer • Millwright • Plumber • Electrician • Surveyor

Project Manager • Electrical Engineer • Construction Manager

Civil Engineer • Roofer • Painter • Mechanical Engineer • CAD Operator

Developer • Equipment Operator • Estimator • Ironworker • Plasterer

Safety Official • Scheduler • Sheet Metal Worker • Steel Erector

INVESTIGATE CAREERS Pick three construction careers. Write a report focusing on each career's description and the amount of education needed.

CHAPTER 16 REVIEW

Main Ideas

- Construction technology is the design and building of structures.
- Like other technological systems, a construction system has inputs, processes, outputs, and feedback.
- There are four major types of construction.
- Structures can be categorized into six major types.
- The construction industry includes a wide variety of jobs.

Understanding Concepts

1. What are the four major types of construction? Give an example of each.
2. Name and describe the six types of structures.
3. Name and describe the three types of tunnels.
4. Give two reasons for building a canal.
5. List the seven types of bridges.

Thinking Critically

1. What impacts might occur if a highway that goes directly through the center of a town were to be replaced with one that would bypass the town?
2. What needs to be considered when deciding what type of bridge to build?
3. What factors would you consider if you had to decide whether or not to build a dam in a specific location?
4. Discuss negative and positive impacts of construction (highways, dams, etc.) on farming.
5. What construction careers might be good for someone who likes to work outdoors?

Applying Concepts & Solving Problems

1. **Math** Nick wants to install a 13' × 24' driveway. If he intends to pour the concrete 4" thick, how many cubic yards of concrete will he need?
2. **Social Studies** Which cities are home to these famous structures: the Chrysler Building, the Space Needle, Petronas Towers, the Golden Gate Bridge, the Parthenon? Report one significant fact about each structure.
3. **Compare and Contrast** Both asphalt and concrete are used to surface roads. Research the reasons for using one or the other. Create a display or poster explaining their advantages and disadvantages and the maintenance procedures used.

Activity: Design a Highway

Design Brief

Highway construction occurs in response to local or regional transportation needs. Roads may have become more heavily traveled, accidents may have increased, or traffic tie-ups may be a daily commuter problem. New roads or changes in the existing roads could solve these problems.

Road construction will have impacts on the community. In many cases, buildings have to be torn down, utility poles moved, and private property bought. Traffic noise and parking problems must also be considered. Each community must consider many issues when designing a highway.

CHALLENGE

Select a road in your community that needs improvement. Make a map of it, redesign it, and create a new map showing your improvements.

Refer to the

Criteria

- You must create two maps, one of the current road and one showing your improvements.
- The maps must use the appropriate civil engineering symbols and show traffic patterns, road design features, and environmentally sensitive areas next to or within five miles of the runway.
- The maps may be drawn with technical drawing instruments or on a computer using CAD software.
- The maps must show the key features along the road.

Constraints

- If the maps are drawn manually, they must be at a scale that will fit onto a 17" × 22" sheet of paper.
- If the maps are drawn on a CAD system, they must be at a scale that will fit onto an 8½" × 14" sheet of paper.
- Use color or shading to show key features and environmentally sensitive areas.
- Document information about any environmentally sensitive areas along the road.

Engineering Design Process

1. **Define the problem.** For the road you have selected, write a statement that describes the problem you are going to solve. For example, are accidents or traffic congestion the problem?
2. **Brainstorm, research, and generate ideas.** With your team, discuss possible solutions. Hint: You may want to study similar local roads that have already been improved.
3. **Identify criteria and specify constraints.** These are listed on page 336.
4. **Develop and propose designs and choose among alternative solutions.** Choose the best design that will solve your problem. For example, wider shoulders and guardrails might be used in a design for decreasing traffic accidents.
5. **Implement the proposed solution.** Decide on the process you will use for making the maps. Gather any needed tools or materials. Then make a map of the original road.
6. **Make a model or prototype.** Create a separate map that shows your proposed changes.
7. **Evaluate the solution and its consequences.** Compare the map of the original road to your design. Will your solution solve the problem? What impacts and effects will result from this solution?
8. **Refine the design.** Based on your evaluation, change the design if needed.
9. **Create the final design.** Once you have an approved design, create your final map.
10. **Communicate the processes and results.** Present your finished design to the class. Show the old and new maps and explain the advantages of your design. Be prepared to answer questions about possible impacts. Turn in your assignment to your teacher. Be sure to include the name of the activity, your definition of the problem, a description of how you solved the problem, and your drawings.

SAFETY FIRST

Before You Begin. Make sure you understand how to use the tools and materials safely. Have your teacher demonstrate their proper use. Follow all safety rules.

CHAPTER 17 PLANNING CONSTRUCTION

⮝ This unique library was built using many of the techniques of green construction. In this chapter's activity, "Design a Green Building," you will have a chance to plan, design, and build a scale model of a green building.

Objectives

- Explain how construction projects are planned.
- Discuss regulations that apply to construction.
- Identify factors to consider when selecting a site.
- Describe architectural working drawings.

Vocabulary

- private sector
- public sector
- infrastructure
- city planner
- smart growth
- zoning law
- building code
- structural material

Reading Focus

1. Write down the colored headings from Chapter 17 in your notebook.
2. As you read the text under each heading, visualize what you are reading.
3. Reflect on what you read by writing a few sentences under each heading to describe it.
4. Continue this process until you have finished the chapter. Reread your notes.

Initiating Construction

Every stage of construction projects requires the skills, hard work, and cooperation of many people. See Fig 17-1. Throughout this chapter you will learn about the jobs of people involved in planning construction projects.

Most buildings and other structures are built for ordinary people. These people make up the **private sector** (part) of our economy. As you learned in Chapter 16, there are four types of construction—residential, industrial, commercial, and public works. The private sector initiates (begins) three of those types. For example, a family may need a home (residential). A person in business may need a store or a warehouse (commercial). A manufacturing company may need a factory or other facility (industrial). Private funds are used to pay for the design and construction of these projects.

The public sector of our economy is responsible for public works construction. This **public sector** includes municipal (city), county, state, and federal governments. People are appointed to or hired by the government to serve on boards or in agencies, bureaus, departments, or commissions. These people are responsible for initiating construction projects such as highways, post offices, and fire stations. Tax money is used to pay design and construction costs.

Fig. 17-1 The more complex the structure, the more workers are required, and the more important planning becomes. The project shown here probably needed to have a lot of pre planning before construction began.

Constructing new buildings is very expensive. For most families, buying a home is the most expensive purchase they will ever make. Most nonresidential construction projects are even more expensive. Careful construction planning can make the difference between a business making a profit or going bankrupt. See **Fig. 17-2**. Poorly conceived public works projects can waste millions of dollars collected from taxpayers.

All construction, both in private and in public sectors, makes up the physical infrastructure of our communities. An **infrastructure** is the underlying base or basic framework of a system. Constructing an adequate infrastructure is essential for well-planned growth.

Community Planning

A community must be carefully planned to best meet the needs of the people who live there. The planning process is usually overseen by **city planners** who have studied all aspects of community development. Many larger communities have planners on their permanent staffs. Smaller communities usually hire planners on a temporary basis. Planners work closely with city, county, and state officials and various governmental agencies. In the course of their work, they study the following:

- Size and character of the population
- Local economy
- Nature and quantity of natural resources such as oil, gas, water, timber, and farmland
- Transportation facilities
- Educational facilities
- History and culture of the area

After learning all they can about a community, planners identify areas of potential growth. They also identify potential problems that might limit future growth and development and work to find solutions. Then planners make recommendations for future community development.

Citizen representatives and elected officials sit on planning commissions and boards. These people study all recommendations or plans carefully. Before deciding whether to initiate any construction, they consider the potential impacts on the following:

- Lives and property of people living in the area
- Health, safety, and general welfare of the people in the area
- Local economy
- Level of employment
- Property values
- Taxes

Some communities make plans based upon a concept called **smart growth**. The goal of smart growth is to create more livable, attractive, and economically strong communities. People want

Fig. 17-2 This abandoned construction site was never completed, and millions of taxpayer dollars were wasted.

to reduce urban sprawl, protect open spaces and farmland, and offer mass transportation systems that will reduce the pollution caused by automobiles. Smart growth is part of a movement known as "green construction." You will learn more about green construction methods in Chapter 19, "Constructing Buildings."

Alteration and Renovation

Some construction takes place in order to maintain, alter, or renovate structures. This may improve the structure or change its use. For example, a family may want to add a sunroom to their house or replace an aging roof. A community may decide to install a swimming pool in one of its parks.

Fig. 17-3 Many structures in New Orleans are being rebuilt due to devastation caused by Hurricane Katrina. ***What storm-resistant features do you think might be included in the construction of this house?***

Controlling Construction

No matter what type of structure is desired or who desires it, it must meet the requirements and standards set up by the community in which it is to be built.

Communities are divided into residential, commercial, and industrial zones. Appointed officials set up special **zoning laws** that tell what kinds of structures can be built in certain areas. These laws may specify such things as the following:

- Maximum property size
- Maximum height of a building
- Number of families that can occupy a house
- Number of parking spaces a commercial building must provide
- Distance structures must be from the property's boundary lines

Zoning laws are designed to protect homeowners from traffic, noise, and other environmental problems. Zoning regulations also attempt to limit noise and traffic congestion near hospitals.

In addition to zoning specifications, all structures must meet certain building codes. **Building codes** are local and state laws that specify the methods and materials that can or must be used for every aspect of construction. Building codes may vary from location to location based on climate and geological conditions. Buildings in geologically active areas, such as portions of California for example, need to be reinforced to prevent them from collapsing during earthquakes. All buildings in areas where hurricanes may strike must be built to withstand extreme winds. See Fig. 17-3. To make sure each structure is constructed according to these codes, the structure must be inspected throughout the construction process and when construction is completed.

Site Selection and Acquisition

Two basic decisions must be made before construction can begin. One is choosing the best site. The other is choosing the best design. These two decisions are not usually made independently. The design may be influenced by the nature of the site. The site choice must be suitable for the design. Once the best site has been chosen, it must be acquired.

Selecting a Site

In addition to being suitable for the structure's basic design, the site must also meet a number of other important criteria.

- **Location.** Is the site in the city or the country? Is it near roads or highways? For stores and restaurants, is it an area where there will be a lot of potential customers?
- **Size.** Is the site large enough? If not, can an adjoining site be acquired?
- **Shape.** What is its shape? Will the planned structure fit well on the site?
- **Topography.** Topography refers to the site's surface features. Is there a lake or stream? Are there hills, gullies, large rocks, or trees? See Fig. 17-4. What is the nature of the soil? Is it dry sand or wet clay?
- **Utilities.** Utilities are services such as water, electricity, natural gas, waste disposal, and telephone service. Are these available at the site? If not, how much will it cost to have them installed?
- **Zoning.** Will zoning laws permit the type of planned structure to be built there?
- **Cost.** Is the price of the site reasonable and affordable?

The Survey

During the site selection process, prospective buyers may sometimes consult a survey or may have a survey made of a piece of land being strongly considered. A survey is a drawing that shows the exact size and shape of the piece of property, its position in relation to other properties and to roads and streets, the

« Fig. 17-4 Houses built on steep hillsides require special foundations. ***What features give these homes extra support?***

» **Fig. 17-5** The spot where the laser beam strikes the measuring tool indicates the elevation.

height (elevation) of the property, and any special land features (rivers, streams, hills, gullies, trees, etc.). The survey also includes a written description of the property.

Laser and GPS equipment are rapidly replacing traditional instruments to measure elevations and distances on construction sites. A laser beam is precisely located, and the surveyor observes where the laser beam strikes a level rod to determine the elevation at that spot. See **Fig. 17-5**. Surveyors use the laser instrument to measure distances by aiming it at a mirror device called a prism. The laser's light bounces right back to another instrument called an electronic distance meter or EDM. The EDM translates the time difference into a precise measurement, whether the prism is 100 feet away or one mile away.

GPS technology can also precisely measure elevation and distances. The surveyor uses a precision receiver with a sensitive antenna to pick up GPS radio signals.

After all the measurements are taken, a stake is driven into the ground at each corner of the property to mark the boundaries. These stakes will later be used as points of reference when workers identify and mark the exact location of the structure.

Acquiring the Site

Once the site has been selected, the land or property must be acquired. The simplest way to acquire land is to make a direct purchase from the owner. Doing this may not be as easy as it seems. For example, the land may be excellent farmland that the owners don't want to see used for buildings. The land may be located in or near a major city, where land is scarce and very expensive. Quite often, the owner may simply not want to sell or may ask for more money than the buyer is able or willing to pay.

The right to own property is precious. However, sometimes land may be needed for public purposes. Perhaps public housing or a new road, school, or dam is needed, but a landowner may refuse to sell. If there are no acceptable alternatives, governments can take legal steps to force the owner to sell by exercising its power of eminent domain. The power of eminent domain is a law that states the government has the right to

Science Application

How a Laser Works Lasers show up in many technologies, including those used in construction. Laser beams are used in levels, cutting tools, and measuring instruments. How is this useful light produced?

As you learned in your science classes, all things are made of atoms. At the center of an atom is its nucleus. Electrons whirl around the nucleus in different orbits. Sometimes an electron jumps from one orbit to another. When this happens, the electron loses energy in the form of light.

When light energy travels, it does so as waves radiating in all directions. As the light waves move through the air, they bump into one another, which scatters them. This is why light gets dimmer the farther it moves from the source. When something prevents the light waves from scattering and forces them to march in harmony in only one direction, laser light is formed. Most lasers use red or infrared light. However, engineers have also designed blue lasers that produce shorter wavelengths; they can work faster and in tighter places.

Laser light can be produced using a crystal, a semiconductor, gas, chemicals, or dyes. A ruby laser consists of a crystal inside a tube similar to a fluorescent lightbulb. A photographic flash is beamed inside the tube. As the light strikes the crystal, it excites the crystal atoms. Electrons in the atoms leap to other orbits and give off additional light. The ends of the crystal act as mirrors to reflect this light back and forth until the light waves are all moving "in step." Finally, a shutter is opened and the light waves leave the tube as a powerful laser beam.

FIND OUT Research to learn how a gas or semiconductor laser works. Report your findings to the class.

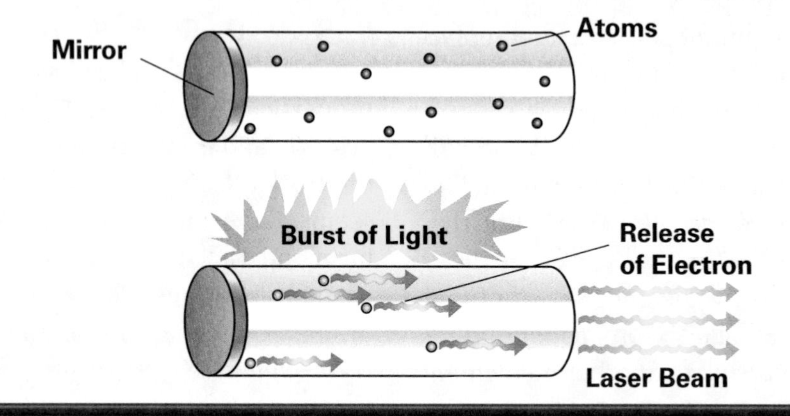

buy private property for public use. (The rights or needs of all come before the rights or needs of one.) See **Fig. 17-6**.

The legal process for taking over land that an owner has refused to sell is called condemnation. However, individual rights must be protected. The government must prove that the property is needed for legitimate public use. If it does, then the owner has to sell and must be paid a fair price for the land.

» **Fig. 17-6** This house must be removed to allow construction of a new city housing project. The owner did not want to sell but was forced to move under eminent domain.

Construction Design

You have already learned about the engineering design process. Similar steps are followed for all types of construction.

Identifying Specific Needs

Suppose a family needs a new home. They may decide to have one custom built. This means the house will be designed and built especially for them. To begin the construction process, they hire an architect to design the structure and develop the plans for building it. The family chooses someone who is experienced in designing custom-built houses.

A meeting is held between the architect and the family. The family members may already have an idea of what they want. At the meeting, the architect asks questions to identify specific needs or problems:

- How large is the family? What are the age and sex of each member? This may determine how many bedrooms and bathrooms the house should contain.
- Does someone in the family work at home? If so, an office or den may be needed. Perhaps soundproofing or special wiring (such as for telephones, computers, or other equipment) may need to be installed.
- What are the family's special interests or hobbies? For example, if a family member enjoys woodworking, there may be a need for a workshop.
- How much money can the family afford to spend? Money is the single most important factor and influences nearly all other decisions.
- Where is the family going to live? Suppose they plan to live in town. Land there may be scarce and lot sizes rather small. The architect may recommend a two- or three-story house. If, on the other hand, the family plans to live in the country, a sprawling, one-story house may be preferred.
- Is the family concerned with energy use or the environment? Many families want to construct a home that uses as little energy as possible or uses only easily renewable nontoxic materials.

A similar procedure is followed in other types of construction. For example, suppose a company needs a new office building. Company representatives meet with the architect. Questions are asked to identify company needs. For example, how many people will be working in the new building? What types of work will be done? Is the company planning to expand in the near future?

Developing Ideas

After the architect has learned about needs, he or she begins jotting down ideas. At this stage, the architect thinks about what building materials will be used in the structure. Some make sketches of possible designs and may use architectural CAD software. See **Fig. 17-7**. Some early ideas may be discarded almost immediately. Others may be saved and reviewed later. As more facts are gathered, certain ideas show greater promise. Frequently, various sketches are combined. The most promising ideas and sketches move into the next stage of design.

Refining Ideas and Analyzing the Plan

As architects refine their ideas, they may prepare renderings or models. This depends on the type of project and the people making the decisions.

For individual residential projects, the architect will usually again meet with the clients to show them the preliminary ideas and together agree on which ideas seem to best meet the buyer's needs.

In many construction projects, one or more engineers may be asked to analyze the preliminary plans. Engineers make sure the design is structurally sound. They may also help determine how the structure will be built and what materials should be used. Sometimes, ideas must be eliminated because the structure simply cannot be built strong enough with existing structural materials. Basically, engineers deal with matters related to the function and strength of structures. Questions they need to consider might include the following:

- What is the soil type, and how much weight can it carry? Will special supports be needed?
- What are climate and geological conditions? See **Fig. 17-8**.
- Could different materials or construction techniques be used to reduce costs and yet maintain quality?

Soon an improved finished (though still preliminary) design is developed. Most architects complete this stage on computers with CAD software.

« Fig. 17-7 These rough sketches made by an architect show several possible ideas for a new building and how it might be placed on the building site.

» Fig. 17-8 In areas that get heavy snow, the roof must be sturdy and set at an angle so the snow will slide off.

Renderings of the refined ideas help clients visualize what the structure will look like. Sometimes models are made. However, this is usually done only when a presentation must be made before a group.

Virtual reality technology now allows architects to further refine the design process. Powerful computers and software provide clients with the ability to "walk through" 3D versions of planned buildings. The software can even simulate the location of furniture and the light provided by the sun at various times of the day and year. See Fig. 17-9. This allows architects and their clients to make changes that might better meet the clients' needs before construction begins.

Preparing Final Plans and Specifications

The architect's refined plans are reviewed by the client and the engineers. When plans are approved, final drawings and specifications are prepared.

« Fig. 17-9 This virtual system allows clients to experience what it might be like to walk through a new area before construction even begins.

Working Drawings These drawings contain information needed to construct a project. They are drawn to scale. In scale drawings, one measurement is used to represent another larger or smaller measurement. For example, one-fourth inch (¼") may represent one foot (1'). Measurements given on the drawings are for the actual size, however.

There are five main types of architectural working drawings.

- A site plan shows where the structure will be located on the lot. Boundaries, roads, and utilities are included.
- A floor plan shows the locations of rooms, walls, windows, and other features. See **Fig. 17-10.**
- Elevations show the finished appearance of the outside of the structure. A separate elevation is made for each side of the structure.
- Detail drawings show any features that cannot be shown clearly on other drawings or that require more information.
- Section drawings show how something looks when sliced by a cutting plane, such as a fastening detail.

Fig. 17-10 This floor plan shows where rooms, windows, doors, stairs, and other features will be located in the finished house.

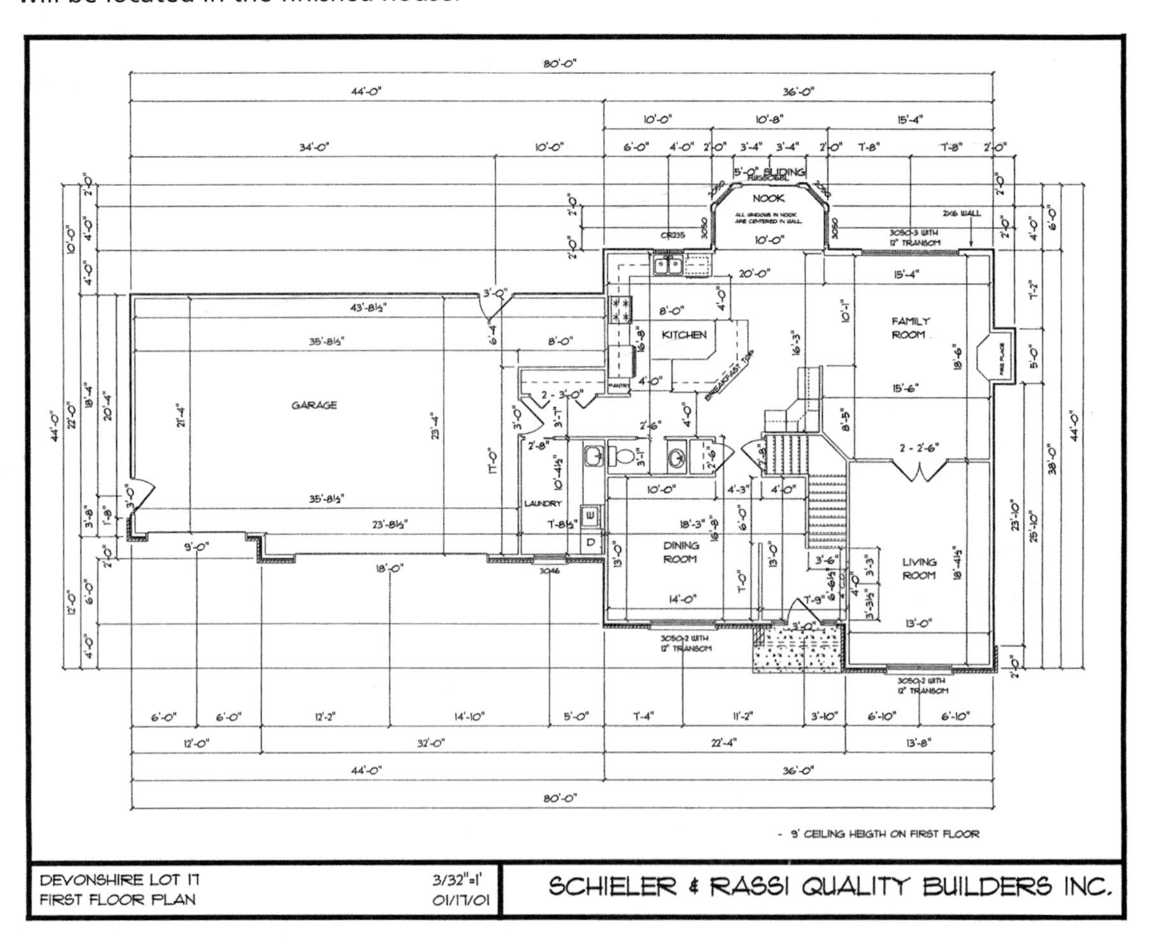

Specifications In addition to all the working drawings, a set of specifications must be prepared. Specifications are written details about what materials are to be used, as well as the standards and government regulations that must be followed. They describe or list the size, number, type, and (if appropriate) model number and color of every item to be included in the finished building. Construction details and materials that could not be shown on the drawings are given in the specifications. Contracting firms use specifications when calculating costs and when building the structure.

Selecting Building Materials

Many different kinds of building materials are available. The size and nature of the project under construction determine what is needed.

Structural materials are those used to support heavy loads or to hold the structure rigid. They are chosen for strength and stiffness. Wood, steel, aluminum, concrete, masonry, plastics, and adhesives are often used as structural materials.

Wood

Wood has many practical advantages as a structural material. Wood is fairly durable, and it can be a renewable resource. It can readily be cut, shaped, and fastened together with nails, staples, screws, bolts, or adhesives. Wood can be used as conventional solid-sawn lumber, or it can be used as the main ingredient in modern engineered wood materials.

Math Application

Scale Drawings Architects, engineers, and other construction planners often use scale drawings. The dimensions on the drawing are proportional to the dimensions on the full-size object. The drawing includes a note that shows the scale used to make it. For example, in the scale drawing of a house, the scale might be shown as 1″ = 2′. These numbers represent the scale factor of the drawing. A length of one inch on the drawing represents a length of two feet on the house.

To convert a measurement on the drawing to the full-scale measurement, let *x* equal the full-scale length and use the following formula:

$$x = \text{drawing length} \times \frac{\text{full-scale factor}}{\text{drawing factor}}$$

For example, if the scale factor of the drawing is 1″ = 2′-4″ and you measure 5 inches on the drawing, you can calculate the length of an object.

First, convert your numerator to inches. Then multiply the numerator and denominator by 5. Convert your answer back into feet and inches to achieve the answer of 11′-8″.

First, convert to inches:

2″ -4″ = 28″

Next, apply the formula:

$$x = 5 \times \frac{28}{1}$$
$$= 140$$
$$= 11''\text{-}8''$$

FIND OUT Calculate the full-scale lengths.

1. 2 inches on a drawing with a scale 1″ = 2′-6″.
2. 3 centimeters on a drawing with a scale 1 cm = 0.5 m.

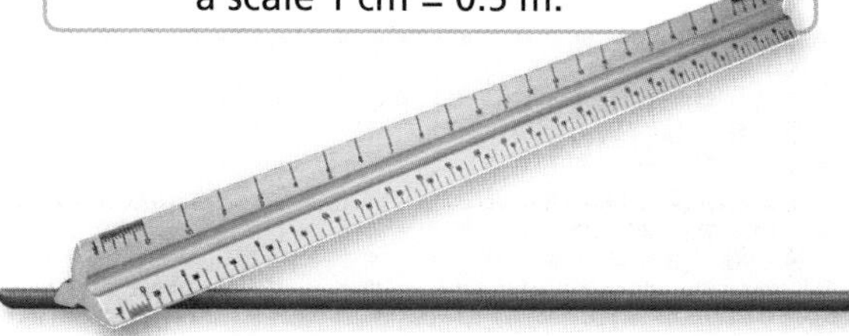

Conventional solid-sawn lumber is wood that is sawed and planed to standard sizes. Dimension lumber measures between 2 and 5 inches thick. See **Fig. 17-11**. It is classified by its dimensions as it is cut from the log before it is planed and dried. Board lumber measures less than ½ inch thick and 4 or more inches wide. Both types of lumber are commonly available in even-numbered lengths, such as 6, 8, 10, or 12 feet.

Engineered wood materials have been developed to stretch our wood supply. Plywood, particleboard, waferboard, and oriented-strand board (OSB) are engineered wood panels made by mixing woods from small, crooked trees that would otherwise be unprofitable to harvest. Glulams are beams made from layers of solid-sawn lumber glued together. Laminated veneer lumber is engineered dimension lumber made of smaller pieces of wood glued together with waterproof adhesive and bonded under heat and pressure.

These engineered materials are up to 30 percent stronger than conventional lumber of similar sizes. In addition, they are a more efficient way of using our wood resources. About 80 percent of a tree can be used when making these wood products. Only about 50 percent of a tree is used when it is harvested for conventional lumber.

Wood has some disadvantages as a construction material. Depending upon the conditions to which it is exposed, it may burn, warp, split, or rot. Untreated, it can be attacked by termites and other insects.

» **Fig. 17-11** The frame of this house is made from wood. Standard sizes of lumber are used.

Metals

Steel is an outstanding structural metal material made by combining iron with small amounts of carbon. Its major advantages are that it's very strong and rigid (stiff). Steel can be formed into various structural shapes, such as tees (shaped like the letter T) and I beams (shaped like the letter I). See **Fig. 17-12**. Steel can also be formed into wire, which can be woven into extremely strong rope or cable. Steel units can be welded together or fastened with bolts or rivets.

» **Fig. 17-12** Steel beams, girders, and other forms are being used to frame this tall building. Some skyscrapers are more than one-quarter-mile high. They must be built to withstand many stresses, such as wind.

Steel also has disadvantages. Ordinary steels may rust or corrode. They must be protected by either being painted or enclosed in concrete. When steel is exposed to temperatures above 700 degrees Fahrenheit (371 degrees Celsius), it will rapidly lose its strength. Therefore, it is usually covered with a fireproof material, especially in high-rise buildings.

Aluminum is used when light weight and resistance to corrosion are important. Like steel, aluminum can be formed into a variety of shapes. Aluminum parts can be welded or fastened together with rivets or bolts.

A variety of new, lightweight, ultrastrong steel and aluminum materials is giving architects and design engineers greater freedom to experiment with new design and construction techniques. Examples can be seen in steel and glass towers and clear-span structures.

Concrete

Concrete is a mixture of sand, gravel, water, and Portland cement. When properly cured, it is a very strong material. Curing is not simply "drying." It is a chemical reaction called hydration that makes the concrete hard and strong. Concrete can be used to build such structures as tunnels, highways, large buildings, and bridges.

Concrete has many advantages as a structural material. Besides being strong, it can be poured into molds to form almost any kind of shape. Huge panels or entire walls can be made of concrete. Separate concrete units can be joined to form structures, such as high-rise buildings.

Concrete also has some disadvantages. It is heavy. It must be properly prepared, poured, finished, and cured. If even one of these processes is not done correctly, the concrete may crack or crumble.

Masonry

Masonry is a broad term that includes both natural materials (such as stone) and manufactured products (such as bricks and concrete blocks). See **Fig. 17-13** on page 352. The material selected depends on the type of project. Most masonry structures are held together by some kind of cement mortar or other bonding material.

Masonry has the advantage of being fireproof and is also a sturdy structural material. It is usually more costly than wood, however, and the bricks or stones must be laid properly. This means much skilled labor is involved. In addition, the mortar that holds the brick or stone together tends to crumble over time and must be replaced periodically.

Plastics

Plastics are used in many areas of construction. New, strong, lightweight plastics and plastic composites are beginning to be used in place of other heavier structural materials. Because they are also waterproof, economical, resistant to corrosion and rust, and can be formed into any desired shape, new plastics might someday replace many more of the materials with which we now build.

Plastic pipe is used in most plumbing systems. Some bathtubs and sinks are made from fiberglass, a composite made from plastic resin and spun glass fibers. Plastics are also being used for roofing materials, insulation, liquid storage tanks, exterior siding, protective coatings, and fasteners. A new type of "lumber" made from recycled plastics is being used to make outdoor items such as picnic tables and decks.

Geotextiles, also called engineering fabrics, are large pieces of plastic cloth. They are spread on the ground as an underlayment, or bottom layer, for roadbeds. They are also used on slopes along highways to help keep soil in place and prevent erosion. Buildings can be made from these materials. The Denver International Airport terminal building has a textile roof. See **Fig. 17-14**.

Adhesives

Adhesives are materials that are used to bond together, or adhere, two objects. Glue is a common type of adhesive. New, stronger adhesives that can bond almost any combination of materials are being used to make engineered lumber and to bond many of the new plastic construction materials.

« **Fig. 17-13** This mason is laying blocks of concrete to make a foundational wall.

Fig. 17-14 Geotextiles were used to create the roof of the Denver International Airport terminal.

Strong adhesives are replacing nails to help save time on the job. For example, plywood or drywall can be fastened to studs (boards used to frame walls) more quickly with adhesives than with nails.

Other Materials

Other materials used in construction include roofing materials, vinyl siding, insulation, drywall or gypsum, electrical wiring and lighting, plumbing supplies and fixtures, and heating and cooling systems.

Structures may also include prefabricated parts. Floors, walls, and even entire rooms can be built in factories and shipped to the site.

CHAPTER

17 REVIEW

Main Ideas

- Construction projects must be carefully planned to meet the needs of the people in a community.
- Various standards, zoning laws, and building codes regulate construction.
- Before construction begins, both the site and the design must be chosen.
- When plans are approved, final drawings and specifications are prepared.

Understanding Concepts

1. Describe the role of smart growth in planning a community.
2. Explain how construction is regulated by zoning laws.
3. List seven important factors in selecting a site.
4. What do working drawings contain?
5. What are structural materials?

Thinking Critically

1. What building codes might be needed in the southeastern U.S.?
2. Why might a family wish to renovate or alter a residence?
3. If a town begins to rapidly grow, how might the structure of the local high school need to be changed?
4. Discuss why you think laser survey tools are replacing levels, transits, and other traditional tools.
5. Suppose you are an architect in charge of designing a new school. Make a list of questions that would need to be answered.

Applying Concepts & Solving Problems

1. **Design** Create working drawings of a room in your house. Make sure the dimensions are to scale.
2. **Science** Some people think that concrete must dry out to gain strength and rigidity but concrete can actually harden under water. How is that possible?
3. **Research** Find out more about Habitat for Humanity. How does this organization plan construction? Write a report detailing your findings.

Activity: Design a Green Building

Design Brief

Green architecture, also called earth-friendly architecture or sustainable architecture, aims to save energy, protect the environment, and make healthier, more pleasant places to live and work. Hartley Nature Center in Duluth, Minnesota, is an example of a green building. Its features include the following:

- Solar-heated fresh air intake
- Roof-mounted photovoltaic cells
- Passive solar heating
- Natural lighting
- An air-to-air heat exchanger
- Motor-operated windows that vent hot air in summer
- Ventilation that moves hot air around in winter
- A geothermal heat pump

Plan, design, and build a model of a green building.

Refer to the

Criteria

- Create a floor plan of your model that shows all major features.
- Your model should include mock-ups of at least four green features, such as a soil roof.
- Include labels or signs on your green features that tell their purposes and advantages.
- Your floor plan may be drawn with technical drawing instruments or on a computer using CAD software.

Constraints

- If you are creating a large building, such as an office building, model only one floor or one section.
- Build your model to a scale that will fit within the space constraints given by your teacher.
- If your floor plan is drawn on a CAD system, it must be at a scale that will fit onto an 8½" × 14" sheet of paper. If drawn by hand, it must be at a scale that will fit onto a 17" × 22" sheet of paper.

Engineering Design Process

1. **Define the problem.** Write a statement that describes the problem you are going to solve. For example, what green features will be most advantageous in the type of building you want to design?
2. **Brainstorm, research, and generate ideas.** With your team, discuss possible solutions. Hint: You may want to take a look at Web sites for green buildings. If possible, visit a green building in your area.
3. **Identify criteria and specify constraints.** These are listed on page 356.
4. **Develop and propose designs and choose among alternative solutions.** Choose the best design that will solve your problem. For example, if energy conservation is one of your goals, you might want to choose a design with solar heating.
5. **Implement the proposed solution.** Decide on the best design. Gather any needed tools or materials. Then make a floor plan.
6. **Make a model or prototype.** Begin to build your model, but allow for possible changes.
7. **Evaluate the solution and its consequences.** Will your solution solve the problem described in Step 1? What impacts and effects will result from this solution?
8. **Refine the design.** Based on your evaluation, change the model and floor plan if needed.
9. **Create the final design.** Once you have an approved design, create your final floor plan and model.
10. **Communicate the processes and results.** Present your finished model to the class. Explain the advantages of your design. Be prepared to answer questions about possible impacts. Turn in your assignment to your teacher. Be sure to include the name of the activity, your definition of the problem, a description of how you solved the problem, and your floor plan and model.

SAFETY FIRST

Before You Begin. Make sure you understand how to use any tools and materials safely. Have your teacher demonstrate their proper use. Follow all safety rules.

CHAPTER 18 MANAGING CONSTRUCTION

Objectives

- Identify the responsibilities of a contractor.
- Explain how the amount of a construction bid is figured.
- Describe the three types of schedules.
- List the three main concerns of inspectors when monitoring construction.

Vocabulary

- contractor
- bid
- subcontractor
- schedule
- purchasing agent
- Occupational Safety and Health Administration (OSHA)

You can buy paint in almost any color imaginable. There are unlimited numbers of colors, shades, and finishes. In this chapter's activity, "Estimate and Bid a Construction Project," you will have a chance to estimate the cost of materials and labor needed to paint several rooms in a house.

Reading Focus

1. As you read Chapter 18, create an outline in your notebook using the colored headings.
2. Write a question under each heading that you can use to guide your reading.
3. Answer the question under each heading as you read the chapter. Record your answers.
4. Ask your teacher to help with answers you could not find in this chapter.

Who Manages Construction Projects?

Management procedures may vary according to the size and type of construction project. Large projects are usually overseen by an architect or engineer. See Fig. 18-1. Smaller projects may use existing designs that have already been made and checked by architects and engineers. These projects are then overseen by a **contractor** who owns and operates a construction company. Contractors are responsible for the actual building of projects. They must follow the plans and specifications developed by the architect(s) and/or engineer(s). Whoever manages the construction—the architect, the engineer, or the contractor—is responsible for seeing that the owner's wishes are carried out.

The contractor is sought after planning is complete. The person or group planning the project announces or advertises that the job is available. Qualified contractors submit bids for the job. A **bid** is a price quote for how much a contractor will charge. The amount equals the contractor's best estimate of what it will cost to build the project according to the owner's plans and specifications, plus the amount of profit the contractor hopes to make. Generally, the contractor who submits the lowest bid is awarded the contract. A signed contract gives

» Fig. 18-1 Engineers and architects often oversee very large construction projects.

the chosen contractor the right to build the project. However, a contractor with a history of poor-quality construction may not be awarded the contract even if he or she submits the lowest bid. The contractor must control costs in order to stay at or below the bid price. At the same time, the contractor must use quality materials and make sure that high-quality work is done in order to develop (and keep) a good reputation. Since a contractor's reputation often influences whether or not he or she gets the job, it is very important for a contractor to carefully manage construction projects.

Fig. 18-2 A subcontractor was hired to construct this unique type of foundation. It is formed using polystyrene foam that is left in place once concrete is poured. The plastic foam acts as insulation.

Contractors may also hire subcontractors. **Subcontractors** specialize in certain types of construction work. See **Fig. 18-2**. For example, the electrical systems in a building are usually installed by subcontractors.

Scheduling Construction

Time is another critical resource in construction technology. Construction jobs are often dependent on one another, so if one job is not done on time, it will affect the progress of many other jobs. For example, if the ground is not prepared, workers cannot build the foundation. If the foundation is not done, the frame cannot be constructed.

Because workers, materials, and equipment must come together at the right times, the contractor must prepare schedules. A **schedule** is a plan of action that lists what must be done, in what order it must be done, when it must be done, and, often, who must do it. See **Fig. 18-3**. Of course, schedules must have some degree of flexibility and allow time for unexpected events like bad weather or design changes.

The three main schedules that control a construction project are the work schedule, materials delivery schedule, and financial schedule.

Work Schedule

Before a useful work schedule can be prepared, the contractor must first analyze the job. That is, he or she must know and understand every phase of the total project. A contractor must know what must be done and where, how it should be done, and when each phase must be finished. Also, the contractor must decide who should do each job. By studying both the drawings and the specifications, a contractor can determine the following:

- Number of workers needed
- Skills or crafts needed
- Equipment needed

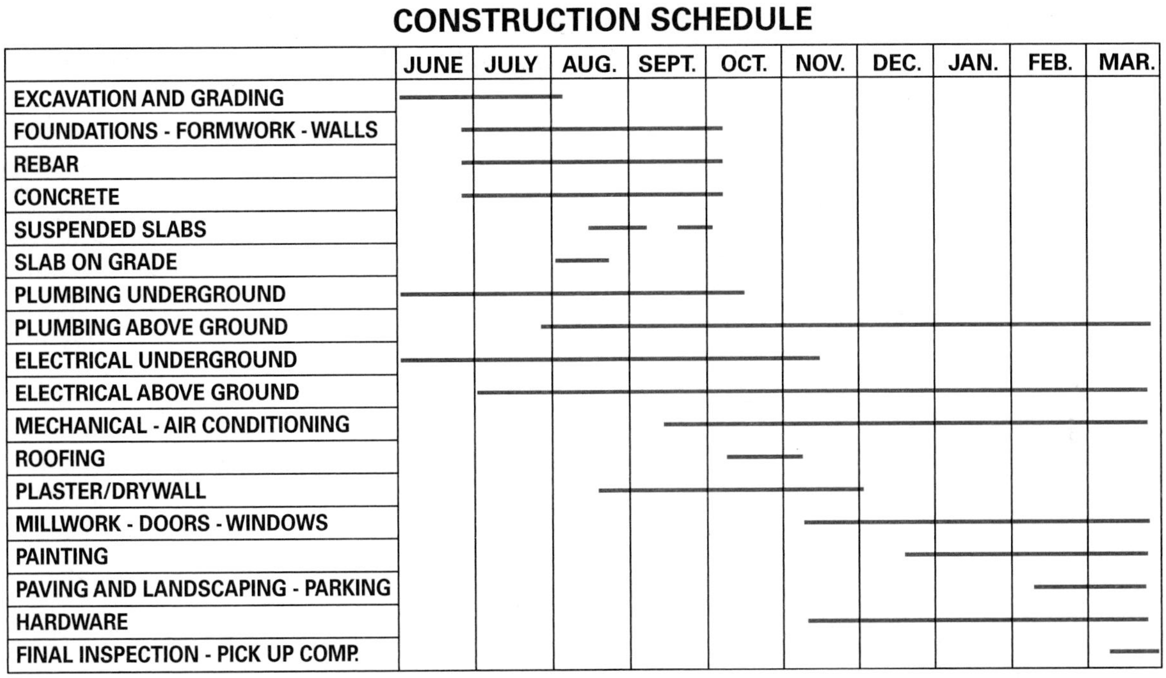

CONSTRUCTION SCHEDULE

	JUNE	JULY	AUG.	SEPT.	OCT.	NOV.	DEC.	JAN.	FEB.	MAR.
EXCAVATION AND GRADING										
FOUNDATIONS - FORMWORK - WALLS										
REBAR										
CONCRETE										
SUSPENDED SLABS										
SLAB ON GRADE										
PLUMBING UNDERGROUND										
PLUMBING ABOVE GROUND										
ELECTRICAL UNDERGROUND										
ELECTRICAL ABOVE GROUND										
MECHANICAL - AIR CONDITIONING										
ROOFING										
PLASTER/DRYWALL										
MILLWORK - DOORS - WINDOWS										
PAINTING										
PAVING AND LANDSCAPING - PARKING										
HARDWARE										
FINAL INSPECTION - PICK UP COMP.										

Fig. 18-3 Bars have been drawn across this schedule to indicate the time allowed for each task. Some jobs can overlap; others must be finished before the next task can begin.

- Time required for each process
- Sequence of jobs

Using this information, the contractor can then prepare a written work schedule. Computer software is often used to help plan work schedules as well as materials delivery and financial schedules.

Materials Delivery Schedule

Suppose a contractor was hired to build a brick house. Then, on the day the bricks were to be laid, only the bricklayers show up. The bricks were not delivered.

To avoid such problems, the contractor prepares a materials delivery schedule. This is a list of every kind of material needed to complete the project. It includes the quantity, style, color, and price of each material. It also shows where and when the materials must be delivered.

Financial Schedule

A contractor pays wages to workers on a regular basis. Materials and supplies must also be paid for. This means a regular and dependable source of money is needed.

How and when a contractor will be paid is worked out in advance with the people who initiated the project. Generally, contractors are paid certain amounts as the work progresses. The amount is usually a percentage of the total price of the job. However, what if money that was expected does not arrive in time? Suppose no arrangements were made? Then the contractor could be in trouble.

To avoid these problems, the architect or engineer prepares a financial schedule. It lists amounts and dates for payment. The financial schedule is reviewed and approved by the owner and, possibly, the bank. Construction

Math Application

Logistics The management of moving materials at a facility or construction site is called logistics. This includes scheduling shipments to and from the site. Logistics is an important part of any production because scheduling problems can be very costly. If materials arrive too early, there is a cost for handling and storage. If they arrive too late, production may stop entirely.

For example, assume the average cost in wages and benefits for employees at a site is $18 per hour. What is the cost for a delay in receiving materials that causes 10 employees to be unable to do their jobs for 5 hours?

$$10 \times \$18/\text{hr} \times 5 \text{ hours} = \$900$$

This accounts only for the amount lost in wages. There might be additional costs due to other factors such as lost sales, equipment idle time, and time needed to restart construction.

In order to schedule shipments for efficiency, you need to know the schedule for construction. In addition, you need to have communication between different people handling the materials. That means that sending information is also part of logistics.

FIND OUT A building project needs a shipment of steel beams at noon on Tuesday. If it arrives late, six workers will be idle at a cost of $40 per worker per hour. If it arrives early, the trucking company charges $150 per hour for the truck and driver. Which is more costly—arriving three hours early or three hours late?

contracts often include financial penalties if construction is not complete by a certain date. However, contractors may also receive bonuses if work is completed before a certain date.

Monitoring Construction

Careful scheduling alone is not enough to ensure a successful operation. In order to make sure that all the terms of the contract are properly met, the project must be carefully monitored by the contractor or those hired to do the monitoring. To monitor a construction project means to watch over and inspect it to ensure safety and quality.

Monitoring Materials

A **purchasing agent** (buyer) is responsible for obtaining the right materials at the right price. The materials must be of the proper kind and in the correct quantity. They must be reasonably priced. The purchasing agent works closely with the various materials suppliers.

The purchasing agent also prepares and monitors the materials delivery schedule. He or she checks all materials as they are received. Materials must be not only those that were ordered, but they must also be delivered on time to the correct site. They must be in good condition, of acceptable quality, and in the quantity ordered. See **Fig. 18-4**.

Monitoring Job Progress

Keeping a construction project on schedule requires careful supervision and close monitoring of the work schedule. Suppose bad weather interrupts work. The contractor must find ways to make up for time lost. If needed materials are not on hand, the supplier must be notified at once.

Monitoring Quality

Quality is vital in all aspects of construction systems. Contractors are often legally responsible for quality. For example, if a bridge should later collapse, the contractor may be sued. See Fig. 18-5. Contractors monitor the quality of both the materials and the work to make sure that everything is being constructed according to the plans and meets all specifications.

Fig. 18-5 This Boston tunnel collapsed in 2006, causing a fatal accident. After the accident, inspectors found many problems with the construction of the tunnel.

Fig. 18-4 Purchasing agents must keep track of materials. All materials must be checked to be sure they are of the correct quality and quantity.

Suppose the quality of the work does not meet the standards described in the original contract. Then the owner may not be required to pay for it. The unacceptable work has to be done over and wasted materials replaced. Making extensive repairs is very costly and can put the entire project way behind schedule. The contractor may have to bear the cost of the repairs.

Quality cannot be added or attached to a structure later. It must be part of it from the beginning. Everyone involved in the project must be committed to quality. Most contractors encourage all their employees to take pride in their work and help monitor every stage of construction projects as part of quality control. Committed workers who are proud of the structures they build are essential to quality construction. This continuous monitoring by and commitment from workers has enabled some construction companies to become known for their dedication to quality.

Inspecting Construction

All work done must meet the requirements of the contract and the standards set by building codes. To ensure this, the structure is inspected carefully by someone who knows what the correct results should be and what conditions should be met.

Inspections are usually done on a regular basis during the building process. This way, problems can be spotted early and corrected without much loss of time or money.

It is the contractor's responsibility to notify the inspector at certain points. Construction cannot legally proceed to the next step until the inspector checks the work already done.

« **Fig. 18-6** Plumbing and electrical systems must closely follow building code requirements. The walls may not be enclosed before these inspections are completed.

Who Inspects?

The types of inspectors depend upon the nature of the project. Inspections may be conducted by quality-control specialists or local building inspectors. Sometimes insurance agents or bank representatives inspect projects. All projects are inspected by representatives of the government.

All levels of government have specific safety regulations. The **Occupational Safety and Health Administration (OSHA)** has been established by the federal government. It is part of the U.S. Department of Labor. OSHA sets safety standards for the workplace. Its representatives visit sites to see that those standards are met.

What Do Inspectors Look For?

It is usually not necessary to inspect all parts of a project. Trying to do so would be difficult, expensive, and time-consuming. Inspections are usually limited to the more critical parts of the project. For example, inspectors routinely check structural, electrical, and mechanical elements. See **Fig. 18-6**.

Inspectors check the quality and appropriateness of the materials being used. They make sure work is being done properly. In short, inspectors are concerned with the three main aspects of any construction project: materials, methods, and quality. Quality must be present in the design of the structure, in the engineering, in the materials used, and in the way the work was done.

Science Application

Second Law of Thermodynamics When building a house, it's important to install an energy-efficient furnace. Do you know how houses are heated? During cold weather, the furnace in a structure must turn on and off repeatedly to keep the interior at a certain temperature. The warmth does not last. This behavior of heat is described by an important scientific law called the second law of thermodynamics. (Thermodynamics is the study of the transfer of heat.)

The law states that heat (or energy) seeks to disperse (scatter) itself throughout a closed system (such as a house). This dispersal always increases; it never decreases.

To understand this law, it helps to know about entropy. Entropy is a measure of disorder (randomness) in a system. An example is smoke. Suppose a piece of bread in the toaster jams, and it burns. Smoke pours out. At first, the smoke is concentrated in the section of the room where the toaster is located. Gradually, it spreads out. Thanks to entropy, the entire house soon smells like burnt toast.

Entropy is also a measure of how much energy is contained in something. You know that energy is the ability to do work. If your flashlight battery is new and fully charged, it is able to do a lot of work (produce light). Its entropy value is low. If you forget to shut the flashlight off and the battery runs down, it is no longer able to do much work; its entropy has increased.

FIND OUT Research the other laws of thermodynamics. Create a poster or display that defines each and illustrates how all the laws work.

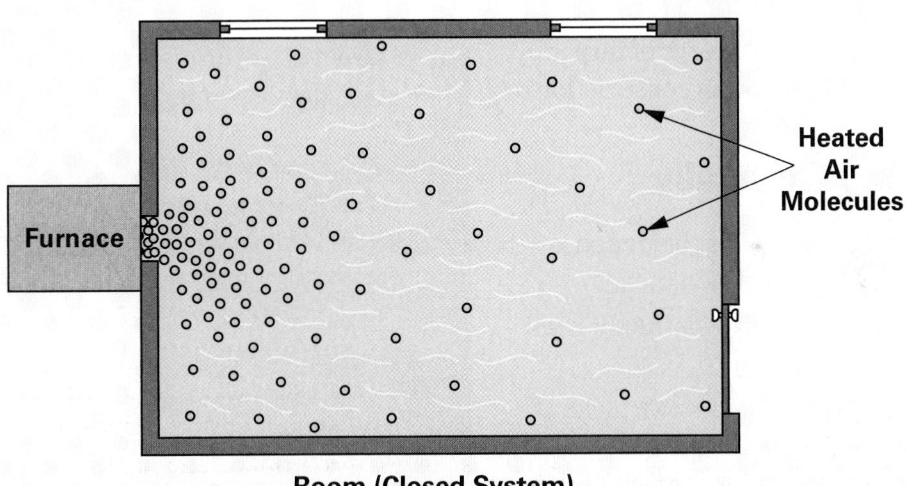

Room (Closed System)

CHAPTER 18 REVIEW

Main Ideas

- Contractors are responsible for the actual building of projects.
- The three schedules used for a construction project are the work schedule, materials delivery schedule, and financial schedule.
- To monitor a construction project means to watch over and inspect it to ensure safety and quality.
- Inspections are done on a regular basis during construction.

Understanding Concepts

1. Identify two factors that typically determine the size of a contractor's bid.
2. A work schedule should contain what kind of information?
3. What should be included on the materials delivery schedule?
4. What does OSHA stand for?
5. List three main concerns of inspectors when monitoring construction.

Thinking Critically

1. Why might the lowest bidder on a project not be awarded the job?
2. What traits would a contractor need to do a good job?
3. What are the risks to a contractor if his or her estimates are way off?
4. Why would replacing an inspector halfway through construction lead to problems?
5. A contractor is involved with more than just the construction of the physical building. What other aspects of a house does the contractor help manage?

Applying Concepts & Solving Problems

1. **Math** If a contractor wishes to make eight percent profit on a job costing $125,000, how much should be bid?
2. **Research** Investigate the roles of a contractor, architect, and engineer in a joint construction project. Create a presentation showing the responsibilities of each person.
3. **Activity** Create a work schedule for this week's school assignments. Your work schedule should include all appropriate elements discussed in this chapter.

Activity: Estimate and Bid a Construction Project

Design Brief

In order to know how much a construction job will cost, an estimate is prepared for all the materials and labor required, plus profit. The estimate is based on information found in the architectural drawings and specifications. The cost of painting a given area, for example, will vary depending upon its size, because the larger the area, the more time and materials that will be needed.

For this activity, you will estimate the cost of materials and labor needed to paint several rooms in a house. You will then submit a bid for the project, which will compete with other bids submitted by your classmates.

Refer to the

HANDBOOK

Criteria

- Use the information shown in the floor plan and Tables A, B, and C to prepare your estimate.
- Calculate the total cost of painting each room indicated in Table C—the cost of the paint (for walls, trim, doors, and window frames) plus the labor costs (assume that the painter earns $18.50 per hour).
- Your costs should reflect that each area will need two coats of paint. (Note that estimate amounts are given for one coat.)
- Add 10 percent to your totals to allow for profit.

Constraints

- You must complete a bid estimate form similar to Table D. Your teacher may require you to include additional information.

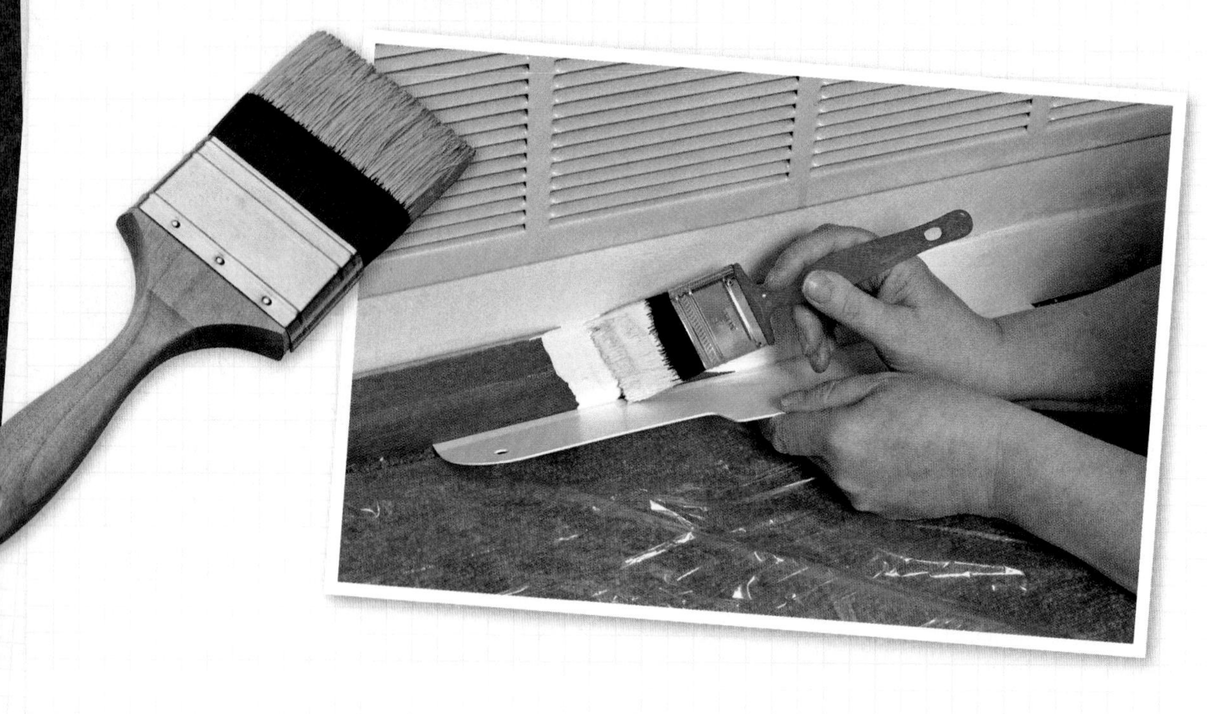

Table A. Approximate Paint Requirements

Room Perimeter	For Walls with 8′ Ceiling Heights	For Ceiling
30 ft.	$\frac{5}{8}$ gallon	1 pint
35 ft.	$\frac{3}{4}$ gallon	1 quart
40 ft.	$\frac{7}{8}$ gallon	1 quart
45 ft.	$\frac{7}{8}$ gallon	3 pints
50 ft.	1 gallon	3 pints
55 ft.	$1\frac{1}{8}$ gallons	2 quarts
60 ft.	$1\frac{1}{4}$ gallons	2 quarts
70 ft.	$1\frac{3}{8}$ gallons	3 quarts
80 ft.	$1\frac{1}{2}$ gallons	1 gallon
Baseboard trim	$\frac{1}{6}$th the amount required for the room	
Window frames (average size)	$\frac{1}{4}$ pint each	
36″-wide doors	$\frac{1}{2}$ pint each	
*For one coat		

Table B. Labor Required for Interior Painting

Baseboard trim	2.5 ft. per minute
Walls and ceiling (with roller)	5 sq. ft. per minute
Window frame (average size)	$\frac{3}{4}$ hour per coat
Door	$\frac{1}{2}$ hour per coat

*Does not include time for preparing surfaces or setting up equipment.
*Labor for trim is given in linear feet while labor for walls and ceilings is given in square feet.

Table C. Job Requirements

Room	Wall Color	Trim Color	Ceiling Color	Door Color	Window Frame Color
Family room	yellow	white	white	white	white
Living room	blue	white	white	white	white
Dining room	green	white	white	white	white
Bedroom A	blue	white	white	white	white
Bedroom B	pink	white	white	white	white

Table D. Construction Bid Estimate Form

Paint Color	Quantity	Unit Cost	Total Cost	Extended Cost
				Total: $
Room	**Hours of Labor**	**Unit Cost**	**Total Cost**	**Extended Cost**
				Total: $
		Total bid for the job: $		

Engineering Design Process

1. **Define the problem.** Write a statement that describes the problem you are going to solve. Be sure to include all its parts.
2. **Brainstorm, research, and generate ideas.** Consider possible solutions. Study the tables provided to be sure you understand what they say.
3. **Identify criteria and specify constraints.** These are listed on page 368.
4. **Develop and propose designs and choose among alternative solutions.** The bid estimate form you design will help direct your work. Choose the best design that will help you solve your problem.
5. **Implement the proposed solution.** Decide on the process you will use for making your estimates and preparing your bid.
6. **Make a model or prototype.** Create a bid estimate form you think will work best. Figure at least one trial estimate to be sure you understand the process.
7. **Evaluate the solution and its consequences.** Check your trial estimate to be sure you have left nothing out.
8. **Refine the design.** Based on your evaluation, change your methods if needed.
9. **Create the final design.** Complete your estimates and your bid form.
10. **Communicate the processes and results.** Submit your bid and estimates to your teacher. Be sure to include the name of the activity, your definition of the problem, and a description of how you solved the problem. A student committee will evaluate all the bids, and the lowest, most accurate bid(s) will be the winner(s).

SAFETY FIRST

Estimating Safety Costs. When a contractor creates an estimate and bid, he or she must factor in costs for ensuring the safety of the construction crew. For example, when creating an estimate for painting a house, safety goggles should be used by the painters.

CHAPTER 19

CONSTRUCTING BUILDINGS

Objectives

- Compare and contrast the ways to clear a site for construction.
- Explain how foundations and superstructures are constructed.
- Describe how interiors are finished.
- Identify post-construction tasks.

Vocabulary

- foundation
- superstructure
- stud
- joist
- rafter
- roof truss
- sheathing
- utility
- insulation
- prefabrication

⮝ Carpenters use many different types of tools to build a house. In this chapter's activity, "Use Construction Tools," you will have a chance to research, use, and make a presentation about a specific construction tool.

Reading Focus

1. **Read the title of this chapter and describe in writing what you expect to learn from it.**
2. **Write each term in your notebook, leaving space for definitions.**
3. **As you read Chapter 19, write the definition beside each term in your notebook.**
4. **After reading the chapter, write a paragraph describing what you learned.**

Preparing the Construction Site

Construction of a structure begins with preparing the site. If the site has not already been surveyed to establish and mark the property lines, surveying will need to be done as the first step in site preparation. Next, the site is cleared of anything that might interfere with construction. Then, the building's position on the site is laid out.

Clearing the Site

The site must be cleared of anything in the way of new construction. This might include trees, old structures, rocks, and/or excess soil.

Trees the owner wants to keep are marked. Then, bulldozers can be used to remove unwanted trees. See **Fig. 19-1**. Owners and companies who are environmentally aware try to save as many trees as possible.

Fig. 19-1 Bulldozers remove any unwanted trees and shrubs and help level the site. Special training is required to operate and maintain heavy equipment. Safety checks are made daily.

The technique used for clearing an area depends on the site. Two common ways of clearing a site are demolition and earthmoving.

Demolition is used to rid the site of old buildings. It involves destroying a structure with machines or explosives. A crane may be used to swing a large, steel wrecking ball against large structures. A bulldozer may be used to demolish small structures. Tall buildings, smokestacks, and similar structures may be demolished quickly by blasting. See **Fig. 19-2**. The remains are then hauled away in dump trucks. Some of the rubble may be bulldozed to low areas of the site and used as fill.

Not everything that is removed from a construction site is destroyed. For example, signs, trees, and even whole houses can sometimes be saved and relocated. Doors, windows, lighting fixtures, and cabinets are often saved before a building is demolished.

Earthmoving In the process known as earthmoving, excess soil and rocks are cleared away and the remaining soil is leveled and smoothed with large and powerful heavy equipment. Cranes, bulldozers, backhoes, and graders are examples.

Laser and GPS equipment have greatly improved the accuracy and speed of large earthmoving machinery. This allows the operator to work without the assistance of extra workers who used to stand outside and help the operator move the machinery into position using hand signals.

Occasionally, swampy areas or ponds must be drained or pumped dry. Then the area is filled in or covered with a deep layer of earth or sand. Also, small creeks or streams may need to be blocked or redirected.

Laying Out the Site

Laying out the site is the process of identifying and marking the exact location of the structure on the property. The site plan indicates how far the building is to be from the edges of the property. Using the site plan as a guide, workers take measurements from the stakes surveyors placed at each corner of the property to the corners and edges of the proposed building. They mark these new locations with additional stakes.

« **Fig. 19-2** Special demolition experts are responsible for blasting. They are careful to control the explosion and not damage any nearby structures.

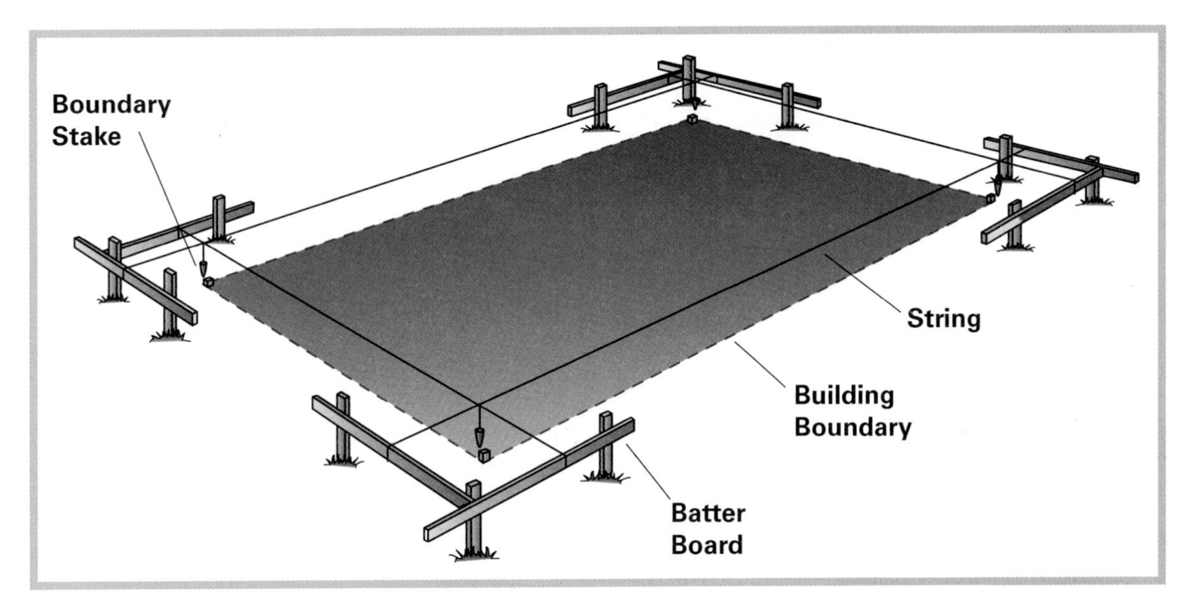

Fig. 19-3 Batter boards and string identify the area to be excavated for the foundation.

The boundaries of the proposed building can be marked using batter boards. Batter boards are boards held horizontally by stakes driven into the ground four to five feet outside the building's boundaries. Strings attached to the batter boards cross directly over the boundary stakes at the building's corners. See Fig. 19-3. The corner stakes can then be removed, and the batter boards and strings guide excavation.

The Foundation

Structures have two major parts: the foundation and the superstructure. The **foundation** is the part of the structure that rests upon the earth and supports the superstructure. A foundation may be above or below ground level, depending upon the type of structure and the local climate. Foundations are usually made from concrete. The **superstructure** rests on the foundation and usually consists of everything above ground.

Excavating for the Foundation

Excavating, or digging, for the foundation is done by heavy equipment such as backhoes, front-end loaders, and trenchers. The size, shape, and depth of the excavation depend upon the design of the building. For example, a ranch-style house may be designed to rest on a simple concrete slab. It will require a wide but shallow area below ground level. A two-story house with a full basement will require a wide but deep opening. The excavation for a skyscraper extends deep into the ground.

The soil may be checked by an engineer to make sure it will be able to support the structure. If the soil is soft or very loose, it may have to be compacted (packed down to make it firm). Soil can also be made firm sometimes by adding certain chemicals to it.

Parts of the Foundation

The two main parts of the foundation are the footings and the walls. Footings lie below the foundation wall and distribute the structure's weight to the ground. See **Fig. 19-4**. A footing is usually twice as wide as the foundation wall so it can distribute the weight over a wider area. (See inset in **Fig. 19-4**.) Footings are made of reinforced concrete.

The foundation walls transmit the weight of the superstructure to the footing. Foundation walls may be made from concrete blocks or from poured concrete that has been reinforced with steel. In buildings with basements, the foundation walls become the basement walls.

Building the Superstructure

What is the next stage of construction after the foundation is finished? Work begins on the superstructure. The superstructure is usually a framed structure or a load-bearing wall structure.

Framed Structures

A framed building has a main "skeleton," or framework, that supports the weight of the building. The framework consists of various structural members and other supports fastened together. These give the building its particular shape. The members may be made of concrete, steel, or wood.

Most houses have a wood-frame structure. As you know, this method of building is called stick construction. In a wood frame, the members may be nailed, screwed, or bolted together. The walls, floors, and roof are framed using the following framing members. See **Fig. 19-5** on page 378.

- **Studs** are parallel, evenly spaced, vertical boards that frame exterior and interior walls. Studs are nailed at the top and the bottom to horizontal boards called plates.
- **Joists** are parallel, evenly spaced, horizontal boards that support floors and ceilings.

Fig. 19-4 The foundation consists of footings and foundation walls.

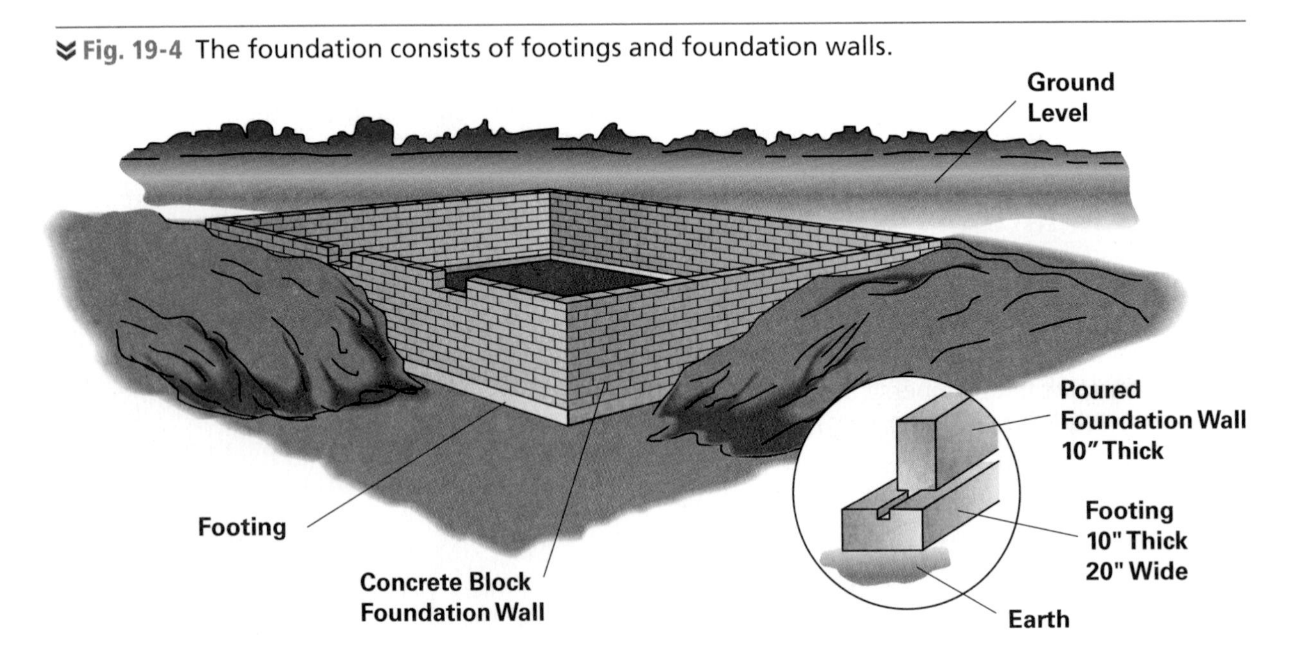

Math Application

Number of Studs in a Wall If you know the length of a wall (in feet), you can determine the number of studs that are needed to frame the wall. You also need to know the spacing between the studs.

Assume you are calculating studs for a solid wall, without windows or doors, which has a standard spacing of 16 inches from the center of one stud to the center of the next. One foot is ¾ (.75) of the distance between the studs, so you can find the number of 16-inch intervals by multiplying the number of feet by 0.75. You also need to add one stud at the end. Now you can find the number of studs.

$$\text{number of studs} = (0.75 \times \text{length}) + 1$$

If the value of (0.75 × length) has a decimal part, use the next higher number. The gap between the final stud and the end of the wall will be less than 16 inches in this case. For example, calculate the number of studs for a wall that is 11 feet long:

$$\text{number of studs} = (0.75 \times 11) + 1 = 8.25 + 1 = 9.25$$

Because the value is not a whole number, you use the next higher number. The wall will have 10 studs.

Some projects place studs at 2-foot intervals, instead of 16 inches. You can use a similar calculation:

$$\text{number of studs} = (0.5 \times \text{length}) + 1$$

Again, if there is a decimal part to the calculated value, you use the next higher number.

FIND OUT Calculate the number of studs for the following:

1. An 18-foot wall with 16-inch intervals.
2. A 14-foot wall with 2-foot intervals.

- **Rafters** are sloping roof-framing members cut from individual pieces of dimension lumber. They extend from the ridge (the horizontal beam along the roof's peak) downward past the side walls of the building. **Roof trusses** are preassembled triangular units used to frame the roof instead of rafters. The sloping sides of the truss serve as rafters. The base of each triangular frame forms a ceiling joist. See Fig. 19-6 on page 378.

A steel frame is used for large industrial or commercial structures, such as skyscrapers and office buildings. See Fig. 19-7 on page 379. The steel for these frames is prepared in a fabricating shop. Then it is delivered to the site. Ironworkers assemble and erect the steel framing members on the site according to the building plan. The steel parts are bolted, riveted, or welded together.

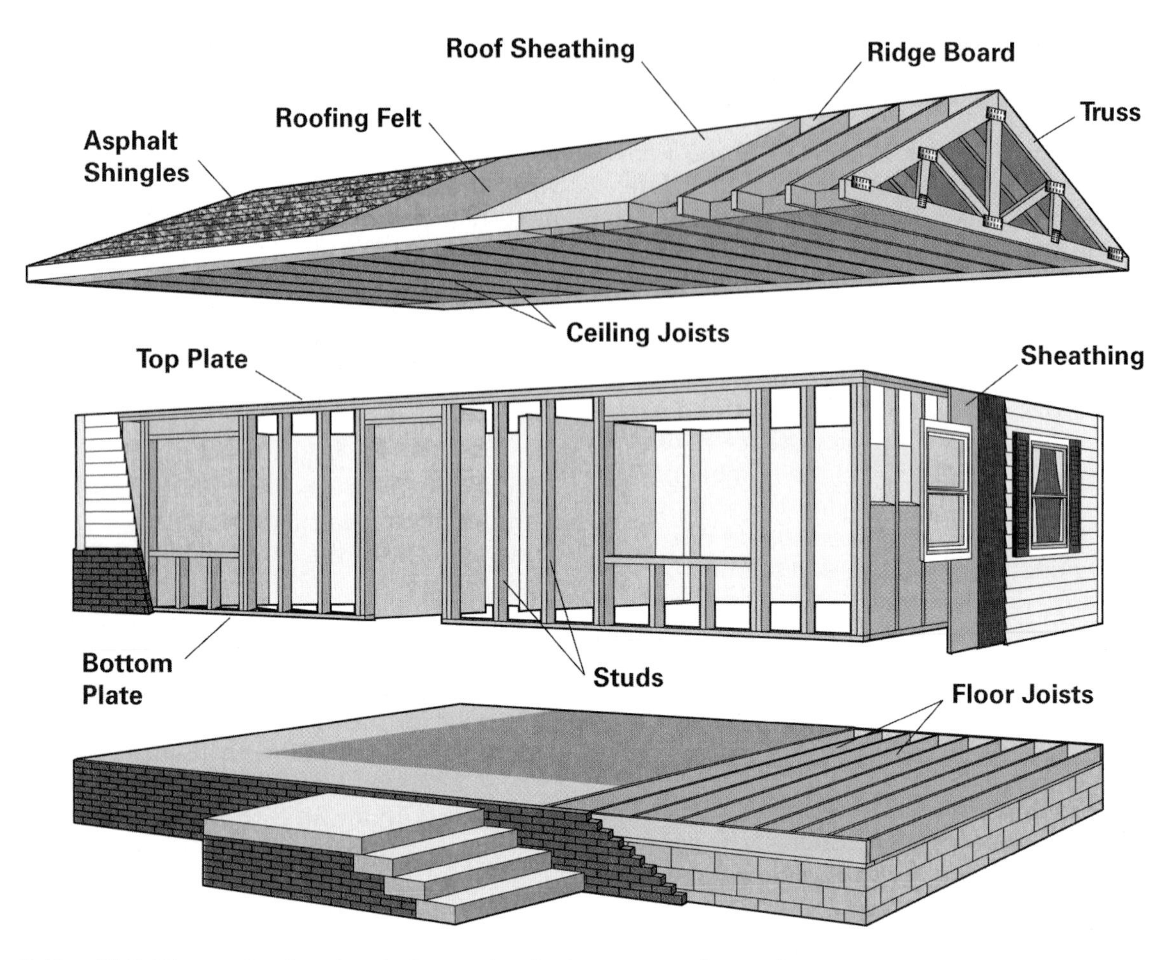

Fig. 19-5 The main structural elements of a house are shown here.

Fig. 19-6 Using prebuilt roof trusses saves time when constructing a roof. Trusses include both the rafters and ceiling joists.

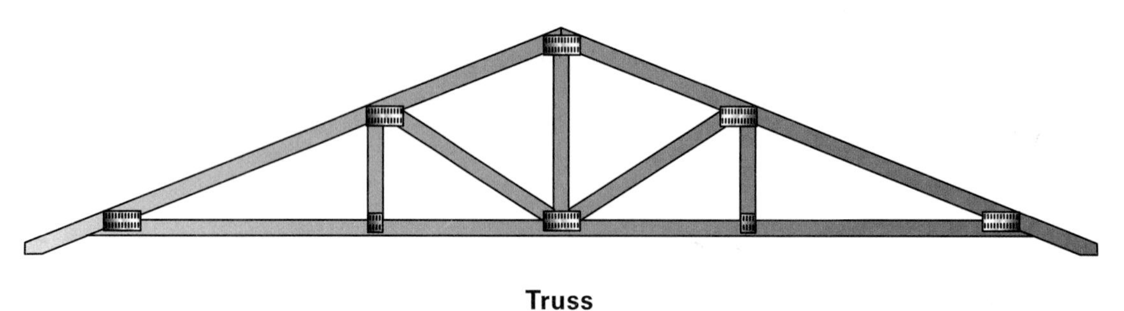

Steel is also coming into use for residential construction. Steel is not harmed by insects or moisture, and it is recyclable. Walls, trusses, and floor sections can be prebuilt at the factory or on the site. As high-quality lumber becomes scarcer, steel may grow more popular for houses.

Load-Bearing Wall Structures

In load-bearing wall structures, the heavy walls support the weight of the structure. There is no frame. Bearing walls are usually made of concrete blocks, poured concrete, or precast concrete panels. Bearing wall construction is best suited for low buildings of one or two stories. Large supermarkets and building supply stores are often built this way.

Fig. 19-7 A steel frame is used for most large buildings.

Science Application

Forces Acting on Roof Trusses In the past, house roofs were built by framing carpenters. They built the supports for the roof one board at a time. Today, most houses use preassembled roof trusses that are delivered ready to lift into place.

A roof truss is made of several parts. The top chords and the bottom chord form the basic triangle shape. The rest of the parts are braces, which distribute the forces on the truss and make it stronger.

A force is a push or a pull on an object. The parts of the truss experience two kinds of force: compression and tension. Tension is a pull that can cause parts of the truss to come apart. Compression can cause bending or warping. To see the effect of compression, try pushing down on a yardstick held vertically and observe what happens.

The braces within a truss and between adjacent trusses distribute the forces and limit distortion. If the bracing is damaged or removed, the truss is much weaker. That means there may not be enough support for the weight of the roof. Damaged trusses must be repaired or replaced.

FIND OUT Make a rectangular frame using 1″ × 2″ boards about two feet long. Use one nail at each corner. What happens if you apply forces to the rectangle? Experiment with different ways to add braces to the frame to increase its strength.

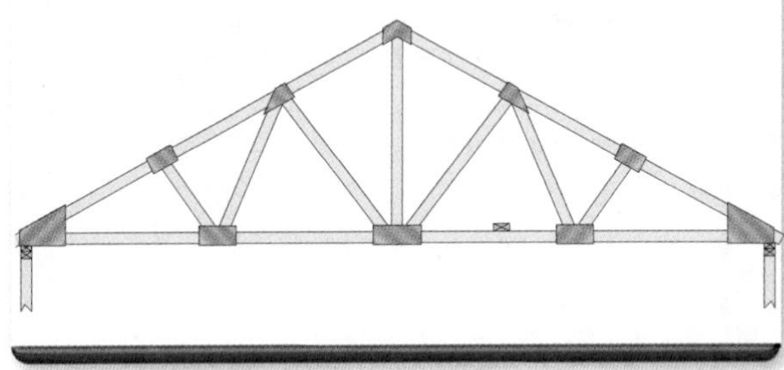

Enclosing the Superstructure

The wall and roof frames of most framed buildings are first enclosed with layers of sheathing. **Sheathing** is a layer of material, such as plywood or insulating board, that is placed between the framing and the finished exterior. After the windows and exterior doors are installed, decorative finish materials such as wood paneling, vinyl or wood siding, stone, or brick are placed over the wall sheathing. The sheathing over the roof frame is first covered with roofing felt. Then roofing materials such as asphalt, wood shingles, or tiles are applied.

Installing Floors

A subfloor, consisting of sheets of plywood or oriented-strand board, is nailed or glued to the floor joists. The subfloor serves as a base for the finish flooring. It also provides a surface for workers to walk on when completing other parts of the building. An additional layer of plywood, called underlayment, is nailed or glued to the subfloor before the finish floor is applied. See **Fig. 19-8**.

Installing Utilities

Has someone in your household ever mentioned paying for utilities? **Utilities** refer to service systems in a building. They include the following:

- Electrical systems
- Plumbing systems
- Heating, ventilating, and air-conditioning (HVAC) systems

Utilities are installed in two stages. First they are roughed in. Parts such as wires, pipes, and ductwork are placed within the walls, floors, and ceilings before the interior surfaces are enclosed. Ductwork for HVAC is usually installed first, then piping, and finally, wiring. See **Fig. 19-9**.

Fig. 19-8 Flooring consists of several layers.

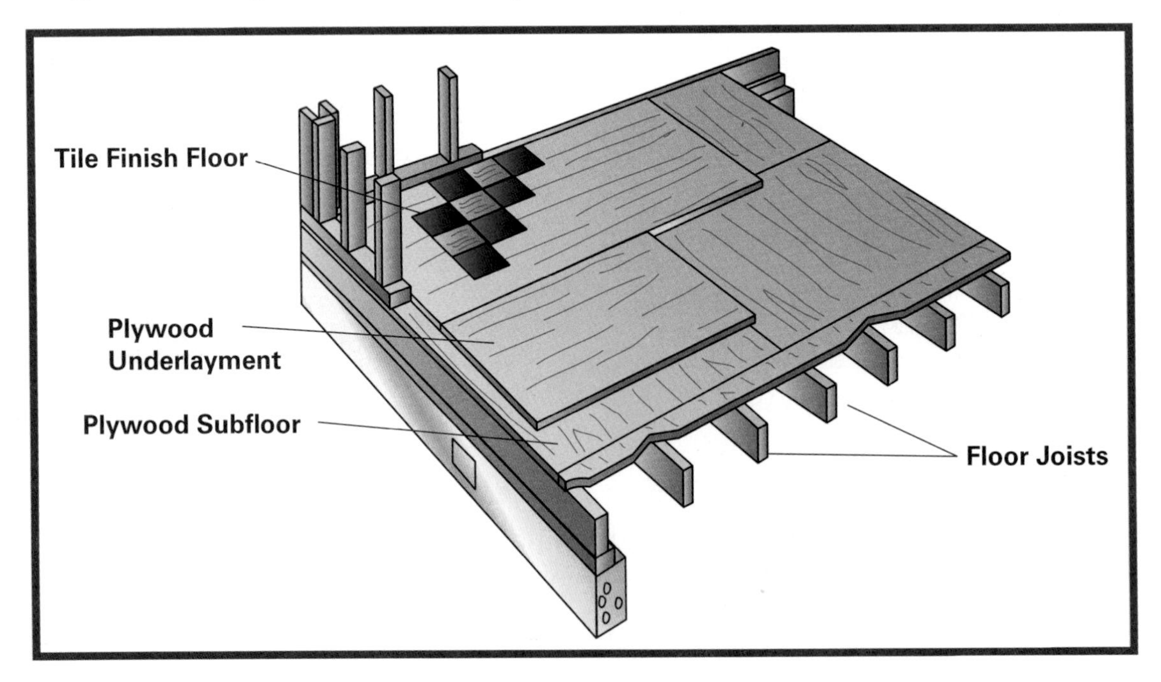

» Fig. 19-9 Wiring is installed in the wall after the plumbing and ductwork are installed. Then the wall is enclosed. ***Why is the wiring installed last?***

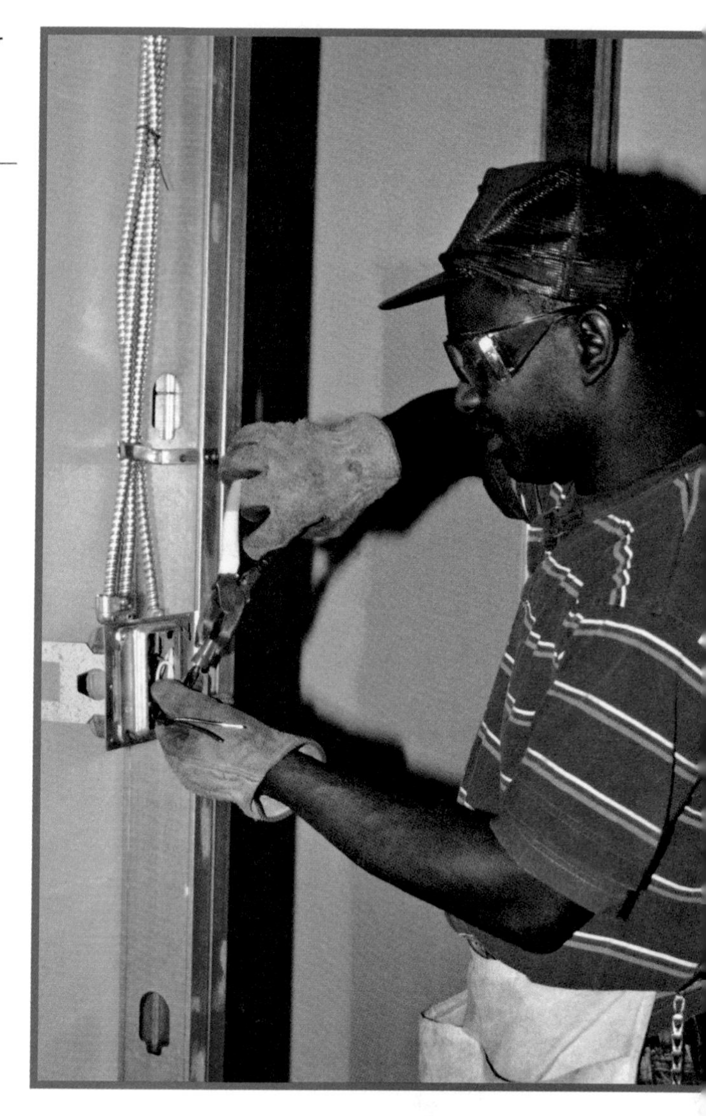

After the interior walls are enclosed, the utilities are finished. Such things as light switches, plumbing fixtures, and temperature controls are installed. This also includes the wiring for "smart" houses, those that use a centralized computer system to control such things as security, energy management, entertainment, communications, and lighting.

Installing Insulation

Insulation is material that is applied to walls and ceilings to help keep heat from entering the building in the summer and leaving it in the winter. This helps make the house more energy-efficient. The ideal amount of insulation used in a building depends on local climate. For example, a house in Seattle, Washington, where the weather is moderate year-round, requires much less insulation than one in northern Minnesota, where winters are very cold.

Insulating materials are labeled according to their R-value. A material with a higher R-value has better insulating qualities. Insulation comes in various forms.

- **Batts or blankets.** These are thick fiberglass sheets or rolls, with a paper or foil backing, designed to fit snugly between framing members.
- **Rigid panels.** These are large sheets of plastic foam or natural fibers.
- **Loose fill.** Loose fill is fibrous or granular material that is blown into place using a special hose.

Finishing the Interior

After the utilities are roughed in and the insulation is installed, the interior is ready to be finished.

Ceilings and Walls are often first enclosed with drywall. Drywall is a general term used for plasterboard or wallboard. It is a heavy, rigid sheet material. Drywall is fastened directly to the wall studs and ceiling joists with screws, nails, or adhesives. See Fig. 19-10. Holes must be cut in the drywall for outlets, switches, and fixtures.

A filler is placed into any dents as well as the spaces or seams between the drywall panels. These areas are then taped and another layer of filler is applied. When the filler is dry, the surfaces are sanded smooth. Now the walls and ceilings can be finished with wallpaper or paint. Paneling and tile are also popular wall finishes.

Floor Coverings Finish flooring is usually installed over the underlayment after the walls and ceilings have been finished. This is done to avoid damage to the flooring. Wall-to-wall carpeting, sheet vinyl, and vinyl or asphalt tiles are commonly used for floor covering. Wood, ceramic tile, and flagstone are also popular. The various materials are installed in different ways. Special fasteners or adhesives may be required.

Fig. 19-10 Drywall is often used to finish walls. Paint or wallpaper may then be added.

Trim and Other Finish Work Trim includes the woodwork, baseboards, and moldings used to cover edges and the joints where the ceilings, walls, and floors meet. Trim is also applied around doors and windows. Most trim used in homes is wood trim. Plastic and metal trims are also available.

Room doors and closet doors are put in place next. Then, hardware is installed. Hardware includes items such as doorknobs and towel bars.

Installation of accessories is the last indoor finishing task. Kitchen and bathroom cabinets are major accessories. Plumbing and electrical fixtures are also major accessories. Shelving, countertops, and other built-ins are also installed as part of finish work.

Prefabricated Structures

Components of structures and even whole structures are now being built in factories. The term used to describe these processes and products is **prefabrication**. The prefabricated parts or sections are shipped from the factory to the construction site, where they are assembled. Site preparation and the building of the foundation will already have taken place while the structure itself was being built at the factory. Prefabrication methods include panelized construction, modular construction, and manufactured housing.

Panelized Construction In panelized construction, the floors, walls, and roof all consist of prefabricated panels. See Fig. 19-11. These panels are shipped to the construction site, where they are assembled to produce the framed and sheathed shell of the structure. Prefabricated roof trusses and floor joists are usually used with the panels. Insulation, utilities, and siding may be either preinstalled or added at the site in the traditional manner.

Modular Construction In modular construction, entire units, or modules, of structures are built at factories and shipped to the site. The modules are then assembled to produce a finished structure.

This construction method has been used to build motels. Identical modules—complete with plumbing, electricity, heating, and air conditioning—are delivered to the site and assembled. In a very short time, the motel is set up and can turn on its "Vacancy" sign.

Manufactured Housing Many new single-family homes are manufactured entirely in factories and then shipped to the site. They are called manufactured housing. Usually, these homes are built in two or more sections.

Green Building

Would it surprise you to know that calling a building green refers not to its color but to its impact on the environment? In an effort to limit the impacts of construction technology on the environment, people are becoming more interested in constructing green buildings. The U.S. Green Building Council (USGBC) has developed several definitions for green building. USGBC's Leadership in Energy and Environmental Design, or LEED, standards establish how new construction can be certified as "green." See Fig. 19-12 on page 384. The standards include the following areas:

- Sustainable site planning
- Safeguarding water supplies and efficient use of water
- Energy efficiency and renewable energy
- Conservation of materials and resources

Green building can start with smart growth planning to ensure that the location of new buildings has positive impacts on the local environment and economy.

Fig. 19-11 Wall panels go in place quickly and save time when built in the factory and then taken to the construction site.

Less than 1 percent of the water on Earth is freshwater, capable of easily being purified for drinking. Green building uses less water during the construction process, prevents water pollution from runoff, and produces structures that use water very efficiently.

Buildings account for 36 percent of our total energy consumption and 65 percent of electricity consumption. Green building methods can significantly reduce the consumption of energy.

Many of the building materials used in ordinary construction projects are not being used sustainably. This means they are being used up faster than they can be replaced. Green

» Fig. 19-12 Ford Motor's River Rouge complex in Detroit, Michigan, is an example of green construction.

buildings use materials such as bamboo, which is easily regrown, for flooring and recycled steel for roofing.

Postconstruction

After construction is complete, several other tasks remain. The site must be finished and the project transferred to the client.

Finishing the Site

The site as well as the structure must be finished. The two major types of outdoor finishing that need to be done are paving and landscaping.

Several areas around homes and other buildings, such as driveways, walkways, parking lots, and patios, need to be paved. These are usually made of concrete or asphalt. Their locations, sizes, and shapes are indicated on the working drawings. Workers mark these areas on the ground. Then the soil is carefully leveled and compacted. Where concrete is used, forms that will hold the concrete while it cures may be set up. Finally, the concrete and/or asphalt is installed.

Landscaping begins after all the debris from construction has been removed. Landscaping involves changing the natural features of a site to make it more attractive. See **Fig. 19-13**. It includes shaping and smoothing the soil with earthmoving equipment, as well as planting trees, shrubs, grass, and other vegetation. Landscaping is done according to a landscaping plan. This plan shows how the finished site should look.

Transferring the Project

The last step in the construction process is formally transferring the completed project to the owner. After construction is complete, the project is given a final inspection. This is done to make certain that the terms of the original contract and specifications have been fulfilled. The quality of the work must also be acceptable.

A number of important legal matters must be attended to. All outstanding bills for the construction of the project must be paid by the contractor. There must be no claims of money owed against the property.

After all requirements have been satisfied, the final payment is made to the contractor. Then the keys to the building are given to the new owner. The new owner now takes over full responsibility for the property.

Maintenance

One of the major responsibilities that new owners assume is maintenance of the building.

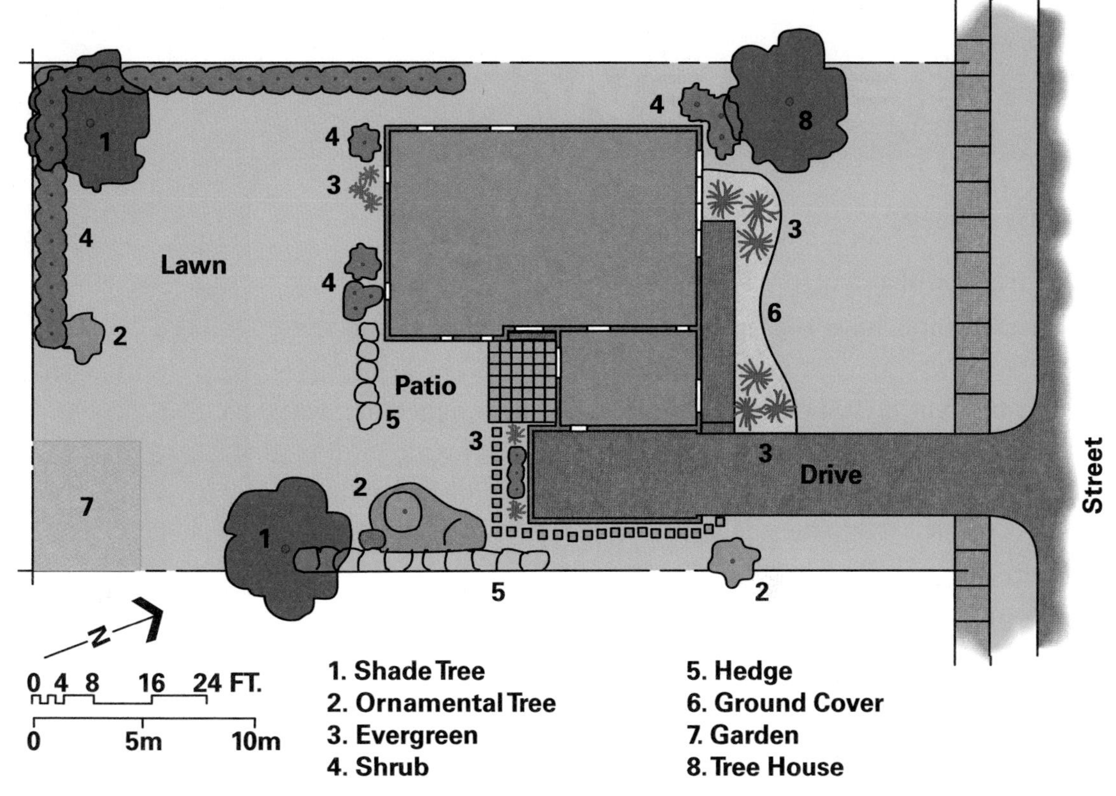

Fig. 19-13 Landscaping plans are prepared to show how the finished site will look.

Hundreds of choices are made in the design and construction that affect the durability of buildings. For example, asphalt shingles cost less than metal roofing, but shingles generally need to be replaced much sooner. See **Fig. 19-14**. Wood siding can be very beautiful, but it usually needs to be coated with stain and wood preservatives or paint on a regular basis. All these choices affect a building's maintenance needs.

No building materials last forever, so all buildings will always require at least some maintenance throughout their life span. Many individual homeowners take on these tasks themselves. Owners of large buildings generally employ workers whose job is to perform all necessary maintenance. Well-built and well-maintained buildings can sometimes last for centuries and become valued historic landmarks.

Fig. 19-14 As a roof ages, it may develop leaks that can damage the interior of a home and weaken its structure. Careful maintenance by homeowners ensures that a house will last for many years.

CHAPTER

19 REVIEW

Main Ideas

- Construction of a structure begins with preparing the site.
- Most structures have two major parts: the foundation and the superstructure.
- Green building describes a form of construction designed to have the least negative effect possible on the environment.
- After construction is completed, the site must be finished and the project transferred to the client.

Understanding Concepts

1. What are the two main ways to clear a site?
2. Describe the two major parts of a structure.
3. What step comes after roughing in the utilities and installing insulation?
4. When finishing ceilings and walls, what is done first?
5. Why is continued maintenance necessary in building construction?

Thinking Critically

1. What legal problems could arise if a site is surveyed incorrectly?
2. How can green buildings be better for the local economy?
3. How could construction crews keep an area clear for demolition?
4. Besides construction, what industries use a lot of insulation?
5. What needs to be considered when planning a landscape?

Applying Concepts & Solving Problems

1. **Math** A fifty-story office building has a base measuring 100 by 150 feet. How many square feet make up each floor? How many square feet is the entire building?
2. **Design** Create a scale model of a stick construction house.
3. **Science** Obtain information on a large piece of earthmoving equipment. Identify at least six simple machines used to create the equipment.

Activity: Use Construction Tools

Design Brief

Four basic types of tools and machines are used in construction.

- Hand tools are simple tools powered by humans. Hammers, saws, and screwdrivers are examples.
- Portable power tools are powered by electricity or air and are small enough to carry easily. Examples include power saws, drills, nailers, and staplers.
- Light equipment can be moved fairly easily. Ladders, sawhorses, and laser equipment are common examples.
- Heavy equipment includes large machines such as excavators.

Research a common hand tool, power tool, or measurement device used in construction and by homeowners for home maintenance. You will then give a presentation to the class on the tool's purpose, proper use, and maintenance, as well as on any safety precautions.

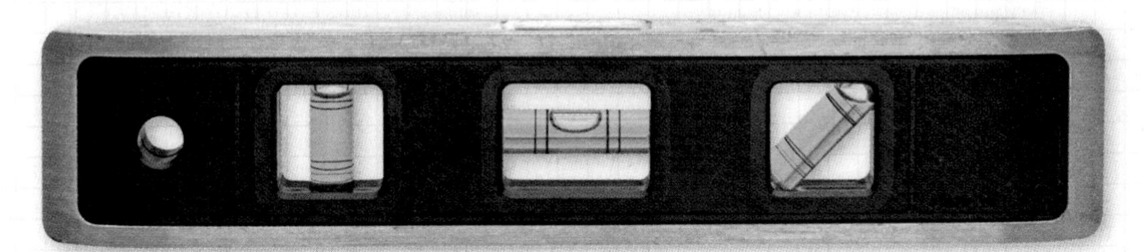

Criteria

- You must research the tool's purposes, proper use, and maintenance, plus any safety precautions for using it.
- Your presentation may be a simple talk with few props or something more elaborate with photos, drawings, signs, and other graphic aids. You may also use presentation software.
- Your presentation will be judged on the basis of its completeness and the range of sources you used.

Constraints

- Your presentation must last no longer than seven minutes.

Engineering Design Process

1. **Define the problem.** Write a statement that describes the problem you are going to solve.
2. **Brainstorm, research, and generate ideas.** Consider possible solutions. Hint: You may want to visit a home improvement store and do-it-yourself sites on the Internet. Review different props that might help you describe how the tool is used and maintained.
3. **Identify criteria and specify constraints.** These are listed on page 388.
4. **Develop and propose designs and choose among alternative solutions.** Choose the best design for your presentation that will solve your problem. For example, you may want to show your tool in use by means of animation.
5. **Implement the proposed solution.** Gather any needed tools or materials. Then write a script and make rough sketches for graphics.
6. **Make a model or prototype.** Create a prototype of your presentation.
7. **Evaluate the solution and its consequences.** Does your prototype meet the criteria and constraints? Practice giving your presentation for a friend and time it with a stopwatch. Will your solution solve the problem?
8. **Refine the design.** Based on your evaluation, change the design if needed.
9. **Create the final design.** Once you have an approved design, create your final presentation.
10. **Communicate the processes and results.** Give your presentation to the class. Be prepared to answer questions. Turn in your assignment to your teacher. Be sure to include the name of the activity, your definition of the problem, and a description of how you solved the problem.

SAFETY FIRST

Before You Begin. Make sure you understand how to use the tool safely before you handle it. If you have questions about it, ask your teacher.

CHAPTER 20

LARGE-SCALE CONSTRUCTION

≈ This astronaut is working on the exterior of the space shuttle. In this chapter's activity, "Create a Model Space Structure," you will have a chance to build a three-dimensional scale model of a space station.

Objectives

- Identify the stress forces that are exerted on structures.
- Explain how roads, dams, canals, tunnels, pipelines, and bridges are built.
- Describe the seven types of bridges.
- Explain the challenges of construction in space.

Vocabulary

- tension
- compression
- shear
- torsion
- load
- Bernoulli effect

Reading Focus

1. **Write down the colored headings from Chapter 20 in your notebook.**
2. **As you read the text under each heading, visualize what you are reading.**
3. **Reflect on what you read by writing a few sentences under each heading to describe it.**
4. **Continue this process until you have finished the chapter. Reread your notes.**

What Is Large-Scale Construction?

Construction involves many projects besides buildings. Large-scale construction includes massive projects such as highways, dams, canals, tunnels, pipelines, bridges, and construction in space. Skyscrapers may also be considered a kind of large-scale construction. Large machines like cranes are often used to build them. You learned a little about large-scale construction in Chapter 16, "Construction Fundamentals." This chapter provides more detailed information.

All structures must withstand certain forces that act upon them. The forces that act upon large structures are usually much stronger than those acting upon an ordinary house. See **Fig. 20-1**. You learned about some of these forces in Chapter 19, "Constructing Buildings." **Tension** is the force that attempts to stretch objects or pull them apart. **Compression** is the

Fig. 20-1 Structures must be built to withstand the many forces that act upon them. This bridge was destroyed by wind.

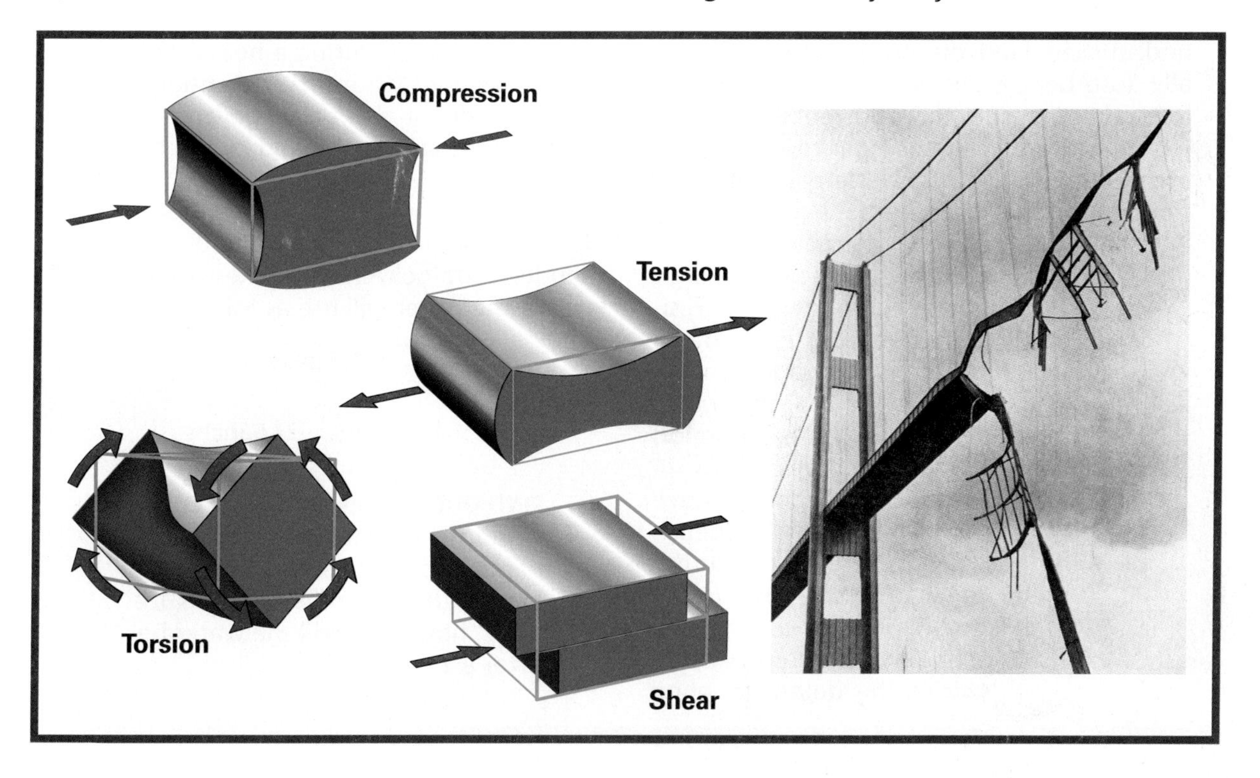

force that tries to squeeze, or crush, objects. **Shear** pushes the parts of an object in opposite directions. **Torsion** is a twisting force applied to an object, like wringing a towel.

These forces act on structures by means of the loads placed on them. A **load** can be any kind of weight or pressure. All structures must withstand two basic types of loads. Dead loads consist of all the materials used to build the structure itself, such as drywall, roofing materials, and plumbing. Dead loads usually do not change unless the structure itself is changed. Live loads change or move. People in a building or vehicles on a bridge are examples. Environmental loads are a special kind of live load. They include such things as wind, temperature changes, snow or rainfall, and earthquakes.

Engineers and architects work together to design structures and choose building materials to make sure that loads are distributed among many structural members. When the stresses are shared, no single structural member is ever stressed to its failure point.

Roads

Road construction involves the building of any type of road, including streets and highways. Before actual construction begins, the route is determined by environmental, financial, and land use issues, as well as by the topography (the land's surface features, such as hills, streams, large rocks, and soil type).

Math Application

Capturing Rainwater Water that falls on undisturbed land during a storm is generally absorbed by the soil. Buildings, parking lots, and other structures prevent water from reaching the soil. Water from these covered areas, known as impervious areas, can cause flooding during heavy rains.

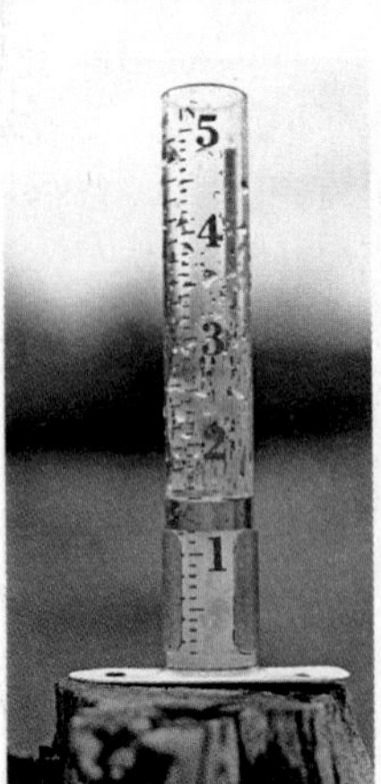

Large construction projects, such as schools, large office complexes, or residential subdivisions, often include a detention pond or basin to hold storm water and release it over a period of time, perhaps 12 to 24 hours. By slowing the flow, these ponds also allow debris and soil to settle out of the water before it flows into a stream or other channel.

Engineers determine the size of the detention pond based on the drainage area and on how much rain may fall during a heavy storm. To find the level of water in a detention pond, use the ratio of the drainage area to the pond surface area multiplied by the amount of rain.

For example, a 2-acre pond captures runoff from a 50-acre development. If a summer thunderstorm delivers 1.5 inches of rain, the level of the water will rise as follows:

$$\frac{50 \text{ acres}}{2 \text{ acres}} \times 1.5 \text{ inches} =$$

$$25 \times 1.5 \text{ inches} = 37.5 \text{ inches}$$

FIND OUT A detention basin collects water from a shopping center and parking lot that cover 120 acres of land. If the area of the basin is 1.2 acres, how deep will the water be after a 1" rainfall?

Fig. 20-2 The topography of the roadway determines how the base will be prepared.

Field survey information is used to assist in the determination of the proposed route. After the acceptable route has been determined and the project has been awarded to the construction contractor, a survey team stakes the center line and edges of the planned roadway.

The route must first be cleared of any trees, vegetation, and large rocks. Then the roadway route is graded to meet the design specifications that have been determined by the project engineers. The design is controlled by the physical features of the land and by the proposed speed limits. Earthmoving equipment is used to remove excess soil from areas that are too high and place it in areas that are too low. See **Fig. 20-2**. These fill areas must be compacted to keep the soil from settling in the future. This forms the subgrade.

Two major types of surfacing used for the finished roadway are flexible and rigid. See **Fig. 20-3** on page 394. Flexible roadbeds use materials such as asphaltic concrete (commonly known as asphalt) to create the smooth finished surface, which can flex, or give slightly, to absorb heavy loads. Usually, a thick bed of gravel or similar granular material is used as a subbase. It is placed on the subgrade before the asphalt pavement is installed. Depending upon the subgrade properties, this subbase may be up to two feet thick. The subbase transfers and spreads the loads on the highway into the soil below.

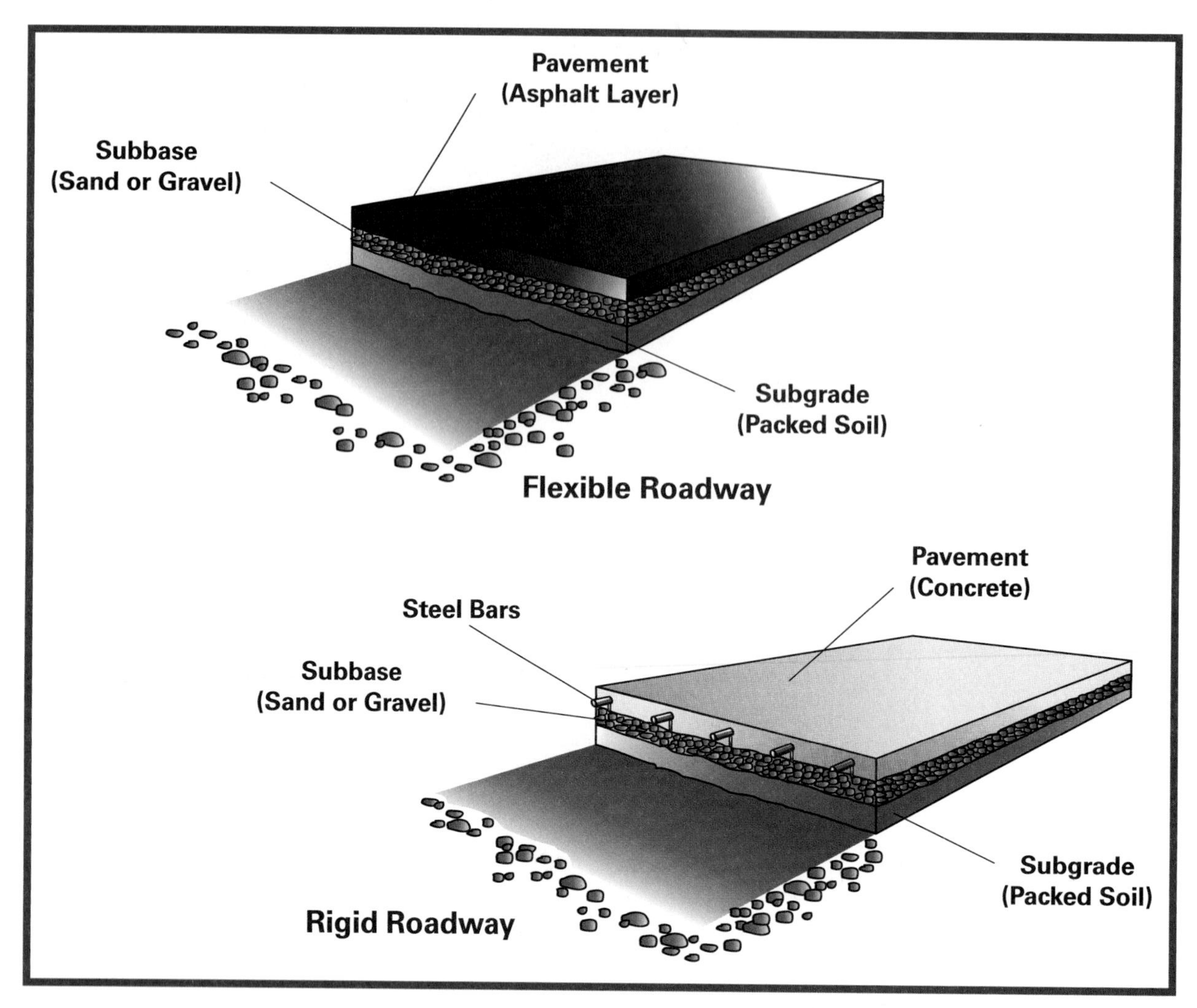

Fig. 20-3 Shown here are cross-sections of a flexible and a rigid roadway.

Portland cement concrete (PCC) is an example of rigid roadway surfacing. PCC pavement may be either reinforced or nonreinforced. Reinforced pavement is produced by placing steel reinforcing bars on metal supports so that the steel is in the middle of the poured PCC slab. Nonreinforced PCC pavement sometimes has steel bars across joints to keep those joints from moving upward or downward. Loads applied to rigid pavement are spread through the slab and transferred to a large area under the slab.

Once the finished surface has been installed, lights, pavement markings, signs, and possibly traffic control signals are added to complete the road.

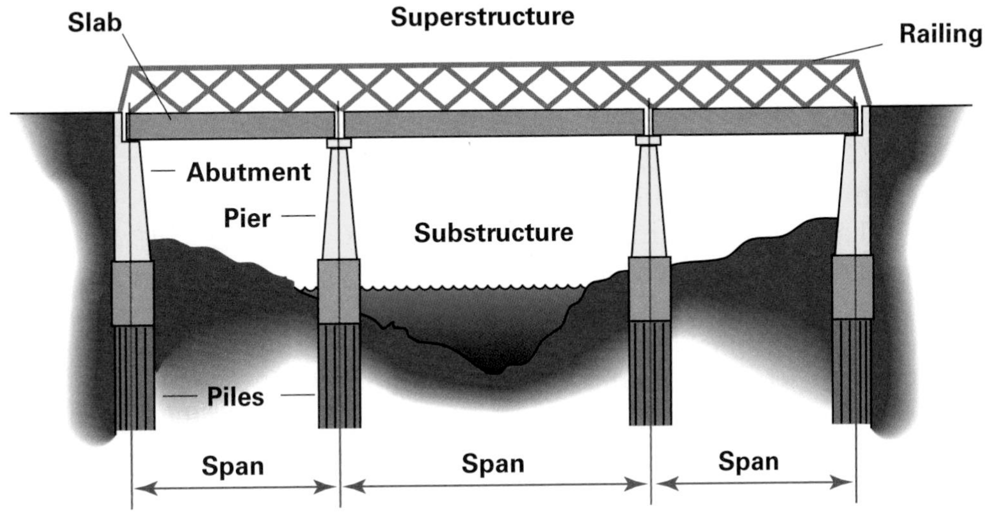

Fig. 20-4 The parts of a typical bridge structure.

Bridges

Bridges allow people and vehicles to pass over an obstacle. Because they must span long distances and resist stress forces, bridges have long been a testing ground for materials engineering. The first bridges were undoubtedly made of natural materials, such as a tree that had accidentally fallen over a stream. Later, humans made primitive bridges using plant fibers twisted into rope. Engineers of the Roman Empire used stones to craft beautiful arch bridges. The invention of iron led to bridges with long unsupported spans. Today, advanced metal alloys, composites, and adhesives are changing the way bridges are designed.

Because of their design, bridges must sometimes withstand a great deal of stress caused by wind. The faster the wind is blowing, the more force it places on everything in its path. Bridges bear a second wind-induced stress described by the **Bernoulli effect**. An increase in wind speed causes a decrease in pressure. This is a destructive force on bridges. In 1940, the Tacoma Narrows Bridge suffered wind-induced structural collapse. (See again Fig. 20-1.) The design of the bridge caused wind to travel over and under it at different speeds. These different speeds formed high and low pressure zones, which caused the bridge to shake so violently that it eventually collapsed.

Substructure of Bridges

The substructure of a bridge may consist of abutments, piers, piles, and a road deck. See Fig. 20-4.

Abutments are the supports at the ends of a bridge. They not only support the bridge, but also the soil at the ends of the bridge. Abutments are usually made of reinforced concrete.

Young Innovators

Inventing a New Bridge Design

Bridge design contests are an exciting way to apply what you have learned about engineering and design in a practical application. Designing a bridge that is not only aesthetically pleasing but highly functional as well will take all of the skills that you have acquired. There are many different bridge-building contests available. Some are very simple and involve constructing a basic bridge out of commonly available materials. Others, such as the West Point Bridge Design Contest, employ sophisticated software programs that allow you to design and test your bridge based on real-world bridge design data and applications.

At one high school, the Technology Education Department holds an annual Bridge-Building Contest. Teams of students have three weeks to build the bridge and then it will be destroyed!

This isn't as bad as it sounds. In fact, it's a huge event. Students gather for an all-school assembly and enthusiastically cheer for their favorite team as, one by one, the bridges are tested and destroyed.

Recently a new division has been added to the contest. In the past, the winner was always the bridge that could hold the most weight before breaking. That part of the contest is still in effect, but the new division will be a contest in which teams can enter the most innovative bridge. In order to compete in the "innovative" division, the bridge must still hold a minimum weight to qualify. However, it will be judged by its innovative design rather than weight tolerance. And if it survives the entry weight test, it won't be destroyed!

It will be interesting to see how many teams will enter this new contest. Do you think there will be more "civil" engineers or more "design" engineers? Which division would you enter?

Piers are the vertical structural supports placed between abutments in longer bridges. They are positioned to keep the longer bridge deck from sagging. The distance between each pier or between a pier and an abutment is called a span. (Note that the word *span* is also used to mean "extend across," such as "the bridge spans the Spoon River.")

Supports (piers and abutments) for a bridge must rest on a solid surface. When the earth under the bridge is not solid, piles are used. Piles are wood, metal, or concrete members that are driven down into the earth to a solid base. Piers and abutments are placed on top of these piles.

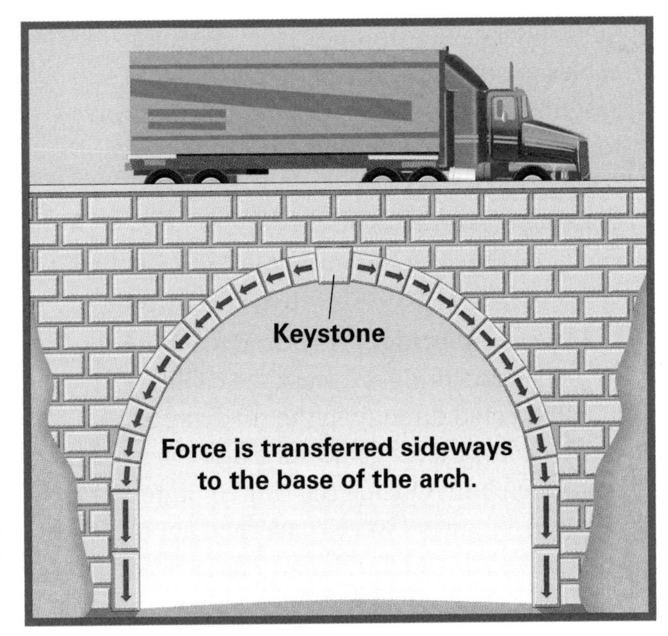

Fig. 20-5 From the keystone of the arch bridge, the weight transfers out to the sides of the arch and then to the base.

Types of Bridge Superstructures

Do you remember the seven common types of bridges? They are shown in Chapter 16, "Construction Fundamentals." The type of bridge constructed depends on how long the bridge must be and on the weight it must support.

- **Beam bridge.** Piers support beams that support spans of reinforced concrete slabs. The farther apart the piers, the more weight the beam has to support, and these spans are usually less than 250 feet. The beams may consist of reinforced concrete, rolled steel girders, or fabricated steel. This is the most frequently used type of bridge because it is normally the least expensive.
- **Arch bridge.** The arch bridge has great natural strength. The load is transferred along the arch (curved portion) to the abutments or piers at the end of the arch. Single or multiple arches may be used, depending on the distance the bridge must span. Today, arch bridges are made of steel or concrete. They can span up to 800 feet. See **Fig. 20-5**.
- **Truss bridge.** A truss is a triangular framework. Trusses may be used above or below the roadway to support the bridge. Trusses are also used with other bridge types, such as suspension bridges, to give them additional support.
- **Cantilever bridge.** This bridge is a complex version of a truss bridge. Beams called cantilevers extend from each end of the bridge. They are connected by a section called a suspended span. To visualize this kind of bridge, think of two diving boards at opposite sides of a pool being connected by placing a board across the top of them.
- **Suspension bridge.** Two tall towers support main cables that run the entire length of the bridge. The cables are secured by heavy concrete anchorages at each end.

Suspender cables dropped from the main cables are attached to the roadway. Truss systems are often used beneath the roadway to help the bridge resist bending and twisting. These bridges are used to span long distances (over 2,000 feet). See **Fig. 20-6**. Probably the most famous suspension bridge is the Golden Gate Bridge in San Francisco.

- **Cable-stayed bridge.** This bridge is similar to suspension bridges except the cables are connected directly to the roadway. Up until recently, most cable-stayed bridges have been built outside the United States because engineers were concerned about their strength and durability. The Leonard P. Zakim Bunker Hill Bridge, built as part of Boston's Central Artery/Tunnel Project, is a recent example of a cable-stayed bridge.
- **Movable bridge.** This bridge is designed so that a portion of the roadway can be moved to allow large water vessels to pass underneath. Bascule bridges open by tilting upward. Lift bridges have a section of roadway that moves up between towers. Swing bridges have a section that moves sideways.

Fig. 20-6 The Akashi-Kaikyo Bridge in Japan is the longest suspension bridge in the world. It is 12,828 feet long, and its towers are 928 feet tall. ***Can you identify the trusses in this photo?***

Dams

Egyptians built the first large dam around 2,800 B.C.E. Dams are built across a river to control or block the flow of water. Dams serve many purposes, including flood prevention, maintaining water supplies for drinking and farmland irrigation, and generation of electricity. See **Fig. 20-7**. Dams are usually built of a combination of materials, including earth, concrete, steel, masonry, and wood.

The three main parts of a dam are its embankment, outlet works, and spillway. The embankment blocks the flow of water. Outlet works are used to control the flow of water through or around the dam. When water is needed downstream, gates of the outlet works are opened to allow water to flow through. A power plant may be part of the outlet works. The spillway acts as a safety valve that allows excess water to bypass the dam when the reservoir becomes too full due to flooding. If water could not bypass the dam during flooding, the dam would break.

Constructing a large dam is a complicated process, and a dam is built in carefully planned stages. Temporary watertight walls, called cofferdams, must be built to keep the construction site dry. As work progresses and construction in the area protected by the cofferdam is completed, another cofferdam is built farther out in the river, and the first cofferdam is removed.

There are four main types of dams.

- **Arch dams.** These work well for narrow, rocky locations. The arched shape is effective in holding back the water in the reservoir and can be made with less material than any other type of dam.
- **Buttress dams.** These dams may be flat or curved. A series of supports, or buttresses, brace the dam on the downstream side. Most buttress dams are made of reinforced concrete.
- **Embankment dams.** These are massive dams made of earth and rock that rely on their heavy weight to hold back the water. They are the most common dams in the U.S.
- **Gravity dams.** These dams also use mass to resist the thrust of water based on their own weight. Most gravity dams are expensive to build because they require large amounts of concrete.

Fig. 20-7 The Three Gorges Dam in China is the largest hydroelectric dam in the world. It stands 181 meters tall and was designed to control flooding of the Yangtze River.

Canals

Canals are artificial waterways. They are usually built either to move water for irrigation or to allow ships and boats to travel between two bodies of water.

Irrigation canals help agriculture by moving water to where it is most needed by crops. Sometimes these canals allow crops to be grown in areas normally too dry to farm.

Navigation canals usually connect one navigable body of water to another, such as a river to a lake. See **Fig. 20-8**. Sometimes navigation canals are built to bypass portions of a river that either bends too sharply or is too shallow for large ships to navigate.

Fig. 20-8 Irrigation canals help make more land available for farming.

Construction of a canal requires a great deal of earthmoving. Once the excavation is complete, clay, concrete, or asphalt is usually used to line the canal. This prevents leaks and washing away of the soil.

Navigation canals may also require the construction of locks if the two waterways being connected are at different elevations. A lock is an enclosed part of the canal that is equipped with a gate. See **Fig. 20-9**. The level of water within the lock can be changed in order to raise or lower ships from one water level to another. For example, the Welland Canal, which connects Lake Erie and Lake Ontario, contains eight locks along its 28-mile route.

Tunnels

A tunnel is an underground passage built to allow people and vehicles to pass under an obstacle. Engineers have been perfecting the techniques of tunneling since ancient times. Modern tunnel construction uses massive equipment to remove huge amounts of earth and rock.

By the 19th and 20th centuries, tunnels were being constructed for railroads and motor vehicles to pass through mountains and other barriers. Earth tunnels are constructed in soil or sand. Because soil and sand are unstable, these tunnels are hazardous to build. As they are dug, concrete sections may be installed to prevent collapse.

Fig. 20-9 Locks allow ships to be raised and lowered when crossing between bodies of water that are at different levels.

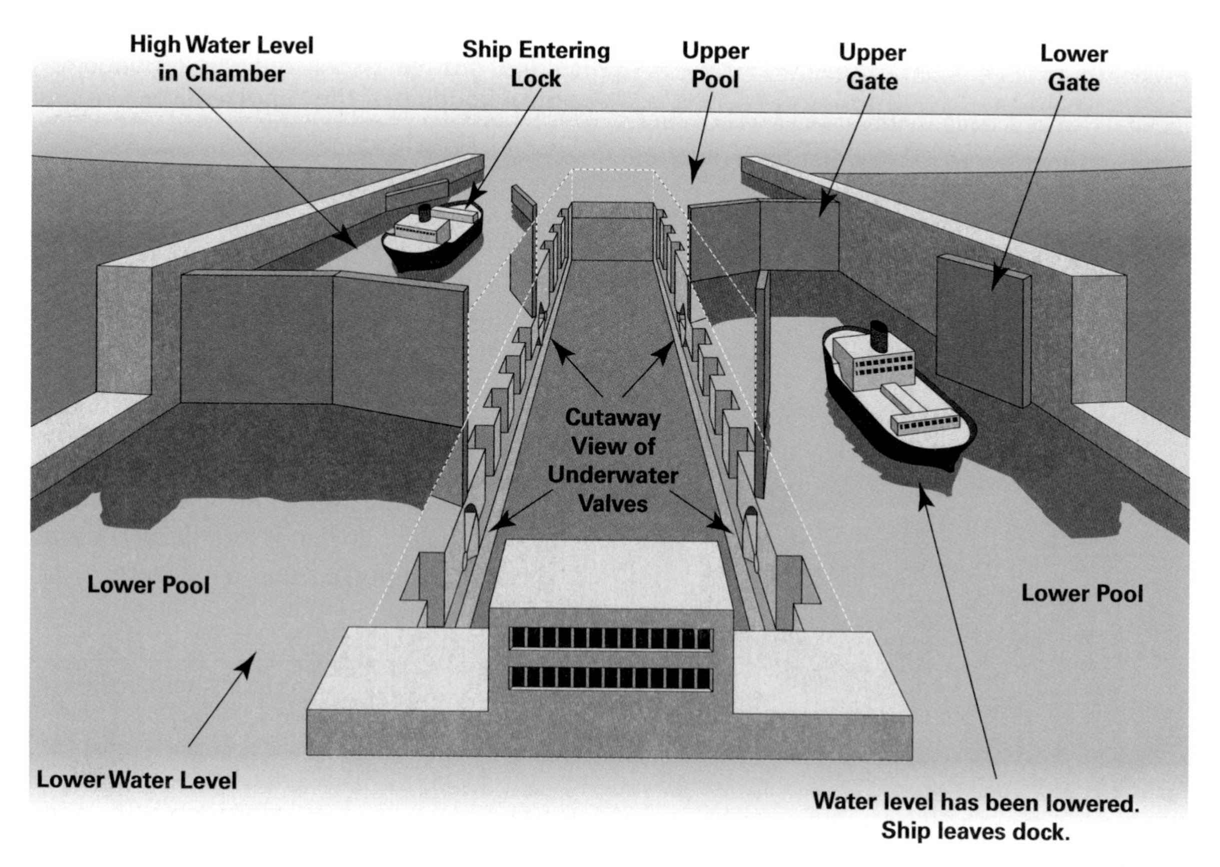

» **Fig. 20-10** Two huge boring machines such as this one were used to construct the Chunnel, which took three years to build.

For immersed tunnels, which run under waterways, premanufactured sections are floated to the site. There, they are sunk into trenches that have been scooped out at the bottom of the tunnel. The sections are then connected to form the tunnel under the waterway.

In rock tunnels, material is removed by blasting or by giant boring machines. Conveyors and boxcars carry away the rock. This method was used to cut the tunnels that make up the "Chunnel," the tunnel that runs beneath the English Channel and links England to France. See **Fig. 20-10**.

Today, as cities expand, tunnels are increasingly being used to relieve traffic congestion, to transport water and sewage, and to lay power and communication lines. A recent example of a tunnel being used to relieve traffic congestion is the Central Artery/Tunnel Project in Boston, Massachusetts. Replacing a highway system built to accommodate 75,000 cars, it includes an eight-lane tunnel three-and-a-half miles long capable of carrying 250,000 vehicles.

An innovative engineering technique called tunnel jacking was used in the construction of this project. Engineers froze the ground by piping very cold salt water under the city streets to slowly freeze it. Three feet of frozen ground could then be excavated at a time using hydraulic jacks, which pushed prepoured sections of tunnel into place. This allowed surface transportation systems to continue with very little disruption. See **Fig. 20-11** on page 402.

Pipelines

Pipelines are an efficient way to transport products such as crude oil, refined petroleum, and natural gas. Most pipelines are buried underground. Above-ground pumping stations along the pipelines are used to maintain the pressure needed to keep the product moving.

Once the route for a pipeline has been surveyed and marked, backhoes and trenchers dig the trenches that will hold the pipe. Sections of pipe one to four feet in diameter are manu-

Fig. 20-11 The "Big Dig" is a huge 11 billion dollar highway construction project which is partially beneath the Boston Harbor.

factured in factories. These are transported to the site and laid next to the trench. Then, the sections of pipe are welded together. Cranes then lift the pipe and place it in the trench. See **Fig. 20-12**.

Before being covered with soil, the joined sections must be inspected for leaks. This is usually done with X-ray or ultrasound equipment. After any necessary repairs are made, the trenches holding the pipeline are covered. The earth must be packed down all around the pipe. Otherwise, the ground may later "settle," causing the position of the pipes to shift. This could damage the welded joints, resulting in leaks. A tamper may be used to pack the earth as the trench is gradually filled. A roller or the wheels of a heavy tractor may be used to finish tamping and leveling the surface.

Construction in Space

Building a structure in space is not like building one on Earth, where thousands of years of history and experience have made design and construction problems relatively simple to solve. Many factors in space construction make any space structure a difficult project.

Huge panels of solar cells are needed to produce power. Space structures must provide all of the environmental and life-support systems needed by humans surrounded by the hostile environment of space. To perform any construction tasks in space, complex space suits and specialized tools and machines must be used. Even a tiny leak in the structure or in the suit could be fatal. In addition, all tools and materials must be blasted into orbit with expensive rockets or shuttles.

Fig. 20-12 Pipelines are assembled on site and placed into the trench with a crane.

The International Space Station

In 1998, the United States and 15 other nations began to construct the International Space Station (ISS). See **Fig. 20-13**. The purpose of the station is scientific research.

Since all components must be brought up in shuttles and other spacecraft, construction is taking place in stages. When completed, the station's habitable portion will be as large as a five-bedroom house.

The framework for the station is a backbone of trusses that have attachment points for additional modules and external equipment. Tasks must be done in orbit under microgravity. People, tools, and parts must be tethered to keep them from floating away into space. Inside, waste particles created by drilling and other operations must be carefully collected or they will contaminate the interior environment.

Fig. 20-13 The ISS orbits 220 miles above the earth at a speed of 17,500 mph, circling our planet once every 90 minutes. You can watch it pass overhead. Its flyby schedule is available on NASA's Web site.

Science Application

Newton's Laws of Motion Have you ever imagined working in a zero-gravity environment? Being weightless can make construction tasks much more difficult than doing the same tasks under normal gravity. Newton's laws of motion can be used to help explain how astronauts move in space.

Newton's first law states that an object at rest will stay at rest until a force acts on it. Once the object is in motion, it will stay in motion until an opposing force acts on it. As long as gravity is not acting on an astronaut, he or she will stay suspended in midair. To move, the astronaut must push off something else, such as a wall or other object.

Newton's second law states that when a force acts on an object, the object will start to move. The greater the force, the greater the change of movement. If an astronaut wants to get from one end of the room to another, he or she must push off with the appropriate force. To get to the other end of a room requires greater force than simply trying to get to a nearby tool.

Finally, Newton's third law states that for every action, there is an equal or opposite reaction. If an astronaut strikes a bulkhead with a tool during construction, the astronaut will be pushed back in the opposite direction. Astronauts may have to be tied down to complete a specific task.

FIND OUT Science and math often work together. What mathematical formula is often used in relation to Newton's second law of motion?

CHAPTER 20 REVIEW

Main Ideas

- Road construction involves the building of any type of road, including streets and highways.
- Bridges allow people and vehicles to pass over obstacles.
- Dams control or block the flow of water.
- Canals are artificial waterways.
- A tunnel allows people and vehicles to pass under an obstacle.
- Pipelines are an efficient way to transport products such as water, crude oil, refined petroleum, and natural gas.

Understanding Concepts

1. What are four forces that can act upon a large structure?
2. Name three ways tunnels can aid in the expansion of cities.
3. Why is it dangerous to have unequal wind speeds surrounding a bridge?
4. Describe a suspension bridge.
5. What is the purpose of the International Space Station?

Thinking Critically

1. Which force acting upon a structure most often involves friction?
2. The Bernoulli effect is a problem for bridges. What field of technology most *benefits* from this effect?
3. Of the four kinds of dams, which kind is most like what a beaver would build?
4. List consequences of not performing regular maintenance on a pipeline.
5. What might cause some humans to one day live in space?

Applying Concepts & Solving Problems

1. **Science** Design a simple device that demonstrates the Bernoulli effect.
2. **Technology and Society** Write a report describing why a certain construction technology evolved. For example, you might research the construction of dams, tunnels, or highways. In your report, discuss technological advancements and economic and social factors that influenced the technology's evolution.
3. **Maintenance** Interview a school official to discuss maintenance of your school's structure.

Activity: Create a Model Space Structure

Design Brief

Many people believe that space exploration will move more quickly if private companies get involved and make it profitable. Already some "space tourists" have traveled on shuttle missions and visited Mir and the International Space Station. Recently, SpaceShipOne completed the first privately funded human spaceflight and is ready for further tests for commercial flights. These are the first of many steps that may lead to constructing a space station orbiting the moon or Mars. What would such a space station so far from home be like? Would it be different from the International Space Station?

Solar Panel
Top Lifts Off
Greenhouse
Manufacturing
Power Plant
Passage
Passage
Open
Storage
Living Area
Satellite Repair
Hatch
Research Laboratory
Docking Area Below

In this activity, you will research the needs that must be met to make a space station livable and functional. You will then design and build a 3D scale model of a space station. Finally, you will write a paper summarizing your design process.

Refer to the

HANDBOOK

Criteria

- The work areas on the station should include a research laboratory, a manufacturing facility, and an area for satellite repair.
- The living areas should include provisions for eating, sleeping, personal hygiene, and exercise.
- Life support areas should include a greenhouse and a power plant.

Constraints

- Your design must be for a low-gravity environment. People will be able to stand, work, or sleep on the "ceiling" as well as on the "floor."
- Your model should be no larger than two square feet.
- Designs and floor plans may be done manually or using drawing software.

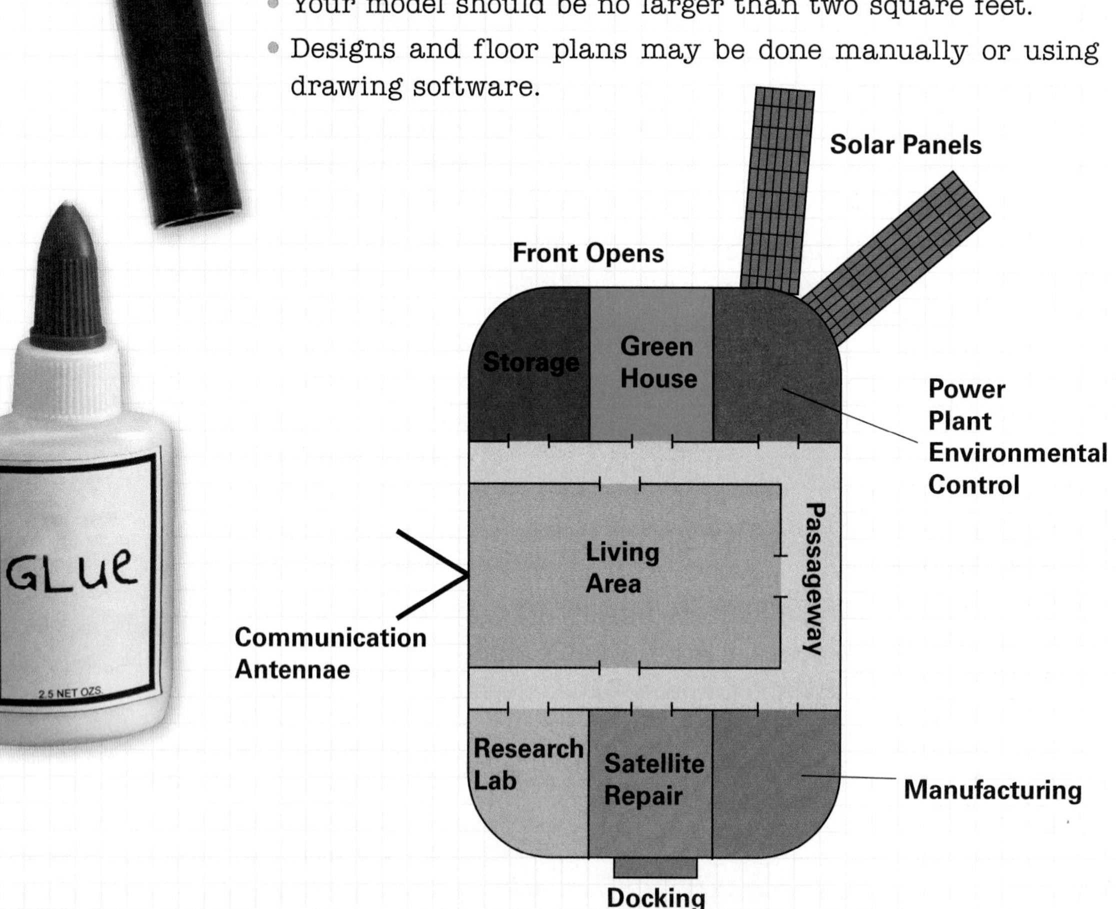

Engineering Design Process

1. **Define the problem.** Write a statement that describes the problem you are going to solve. For example, what must be included in your model?
2. **Brainstorm, research, and generate ideas.** Think about possible solutions. Hint: You may want to study the floor plan of the International Space Station.
3. **Identify criteria and specify constraints.** These are listed on page 406.
4. **Develop and propose designs and choose among alternative solutions.** Make several thumbnail sketches of possible designs. Choose the best design that will solve your problem.
5. **Implement the proposed solution.** Gather any needed tools or materials. Then make a floor plan for each level in your station.
6. **Make a model or prototype.** Create a preliminary model of your station.
7. **Evaluate the solution and its consequences.** Compare the preliminary model to your sketches and the design brief. Will your solution solve the problem? What impacts and effects will result from your design?
8. **Refine the design.** Based on your evaluation, change the design if needed.
9. **Create the final design.** Once you have an approved design, create your final model. Write a report describing your design and the process you used.
10. **Communicate the processes and results.** Present your finished design to the class. Explain the advantages of your design. Be prepared to answer questions about possible impacts. Turn in your assignment to your teacher. Be sure to include the name of the activity, your definition of the problem, your report, and your model.

SAFETY FIRST

Before You Begin. Make sure you understand how to use the tools and materials safely. Have your teacher demonstrate their proper use. Follow all safety rules.

UNIT 6

TRANSPORTATION ENGINEERING & DESIGN FRONTIERS

People the world over are concerned about the effects that fuel emissions from cars are having on Earth's atmosphere and life on our planet. Experts estimate that there are more than one-half billion cars in operation around the globe today, most of them spewing out greenhouse gases, acid-rain-producing emissions, and other harmful products.

The good news is that consumer demand for "clean" vehicles is generating intense competition among automakers. Some of them are coming up with radically new car designs.

Innovative Cars

Some researchers are focusing on hydrogen fuel cell technology. Automaker Honda is preparing to launch a sleek, roomy hydrogen-fueled sedan. One big challenge is how to provide cheap, safe hydrogen fuel. Honda is working on this problem, too. Its engineers are developing a "home energy" station that would produce hydrogen from natural gas while generating electricity for the home.

Another focus of research is battery-powered cars. These cars are clean, but they require frequent recharging and can be underpowered. Tesla, a California-based automaker, is working to minimize these limitations. The two-seater Tesla Roadster can travel 250 miles without a charge, and its speedometer can top out at 130 miles per hour.

A company called ZAP ("Zero Air Pollution") is marketing a line of small, "lean and clean" cars. Its Smart Car model is powered by a gasoline combustion engine, but the car is so small and lightweight that it gets about 60 miles to the gallon. It seats two and can top out at 85 miles per hour.

ZAP also markets an "e-bike"—a cross between a car and a bicycle. The car runs on battery power supplemented by driver pedaling. It's strictly for tooling around town at 20 miles per hour.

Could cars run on air? Yes—just ask engineers with a French company called MDI. They've developed the "MiniCat," a car that literally runs on air. The specially designed engine uses the force of compressed air instead of burning fuel to power its pistons, so there are no harmful emissions. At present, the car must be plugged in after about 90 miles so the onboard electric compressor can refill the air tanks, but someday service stations may offer highly compressed air at the pump.

CHAPTER 21 TRANSPORTATION FUNDAMENTALS

Objectives

- Describe a basic transportation system.
- Explain how GPS works.
- Describe time and place utility.
- List several positive and negative impacts of transportation systems.

Vocabulary

- transportation system
- way
- route
- on-site transportation
- global positioning system (GPS)
- bulk cargo
- break bulk cargo
- time and place utility
- gridlock

Riding public transportation is very economical for you and is also good for the environment. In this chapter's activity, "Make a Transportation Documentary," you will have a chance to create a documentary video about a single transportation system in your community.

Reading Focus

1. As you read Chapter 21, create an outline in your notebook using the colored headings.
2. Write a question under each heading that you can use to guide your reading.
3. Answer the question under each heading as you read the chapter. Record your answers.
4. Ask your teacher to help with answers you could not find in this chapter.

What Is Transportation?

To transport means to carry from one place to another. A **transportation system** is an engineered and organized way of moving goods and people using vehicles. A vehicle is any means or device used for transportation.

Transportation affects your life in many ways. Did you have milk on your cereal this morning? The milk was carried by means of pipeline and highway transportation from the cows to the dairy. Were you driven in a car or bus to school? If so, you used highway transportation. Did you receive a letter or catalog in the mail today? For part of its journey to you, it probably traveled on a plane using air transportation. See Fig. 21-1. Was any article of clothing you're wearing today made in a foreign country? If so, it came by a water vessel and reached the store where you bought it by train and/or truck.

Transportation Systems

Few systems are larger than the transportation system. Like all systems, transportation systems have inputs, processes, outputs, and feedback. Generally, they need seven basic types of inputs to operate: people, information, material, tools and machines, energy, capital, and time. See Fig. 21-2 on page 412.

« Fig. 21-1 Overnight delivery services use air transportation to carry shipments from city to city. Then drivers organize very efficient highway routes to carry the shipment to the right person.

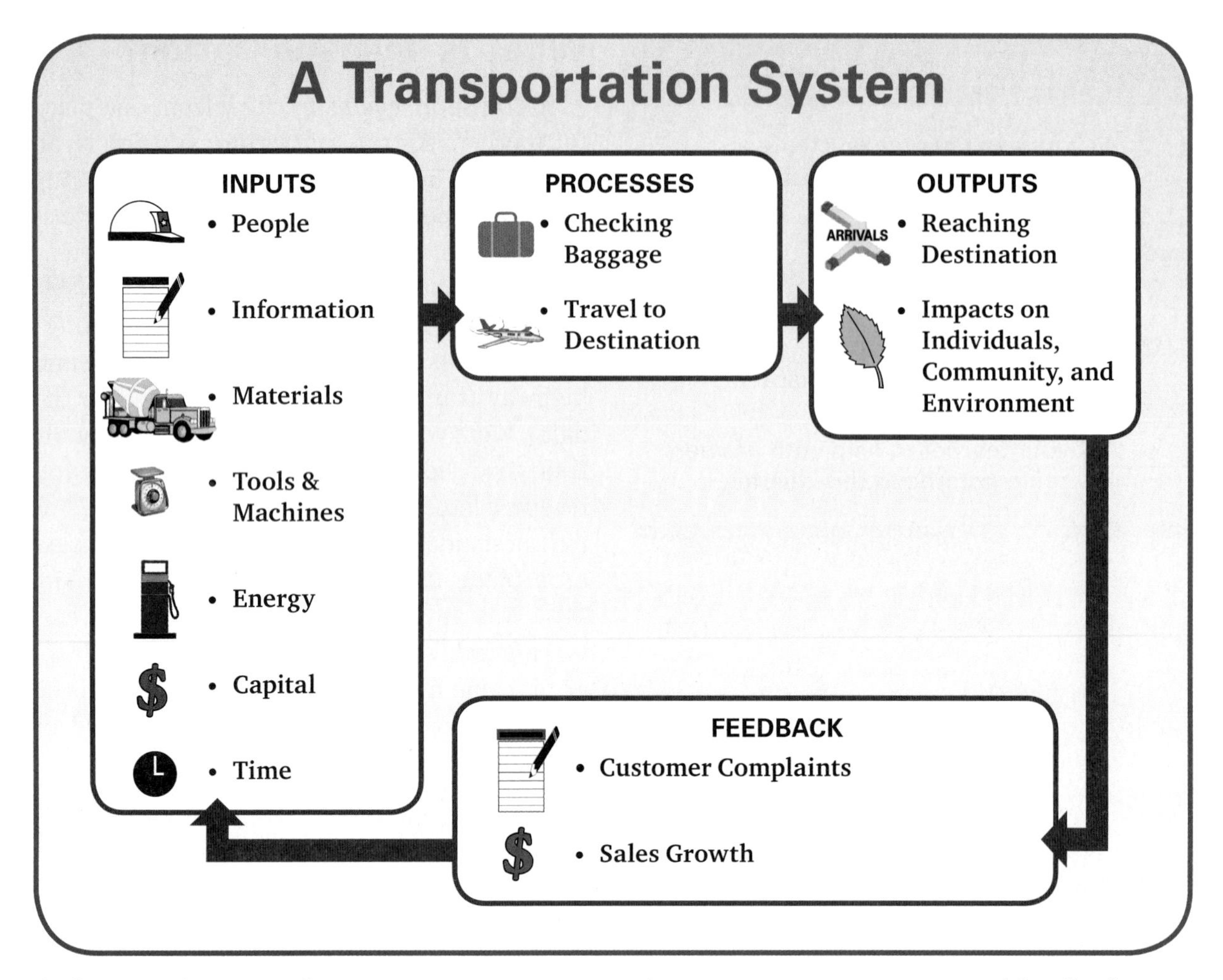

Fig. 21-2 The parts of a transportation system are inputs, processes, outputs, and feedback. When they all work properly, transportation occurs.

The people involved in transportation are either providers or users. When you ride the bus you are a user. Operational engineers, aircraft mechanics, dock workers, and train or truck drivers are all providers.

Workers need information about how to safely operate tools and vehicles or how to load cargo. Drivers need information provided by road signs and maps. Materials needed include water used to cool engines and clean vehicles and the shipping containers used to protect freight. Tools and machines are used to repair complex vehicles. Energy is used in the form of gasoline, diesel fuel, propane, and kerosene. Electricity powers some vehicles and supplies the energy to operate traffic signals, streetlights, and computers. See **Fig. 21-3**. Capital is used to buy materials and supplies, maintain vehicles, and pay employees. Time includes how long it takes to get from one place to another. Saving time often means saving money. Transportation systems that save time usually replace slower methods.

Processes include management processes and production processes. The three main management processes are planning, organizing, and controlling the transportation system. Production processes are the most visible part of a transportation system. The flight of an airplane is an example. Passengers must be loaded, and the plane must be fueled and guided to its destination.

In transportation, the basic output is that passengers and/or cargo are moved from one place to another. Other outputs include profits earned by transportation companies and pollution caused by burning fuel.

As in all systems, feedback from transportation can take many forms. For example, if a package is shipped to a specific place on a specific day, a feedback loop checks if it arrives on time. This feedback might be as simple as a phone call to the destination or as complex as a computer system that tracks the package around the world.

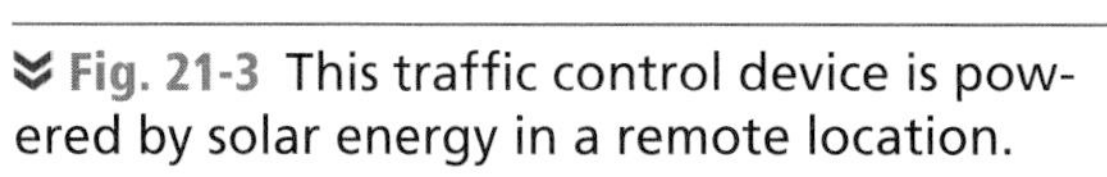

Fig. 21-3 This traffic control device is powered by solar energy in a remote location.

Fig. 21-4 Pipeline ways are used for such things as utilities.

Using Transportation

We generally think of transportation in terms of vehicles. However, other factors play a part in transportation systems.

Ways and Routes

Vehicles are usually operated on ways and routes. **Ways** are specific areas set aside for use by transportation systems. See Fig. 21-4. A way may be a strip of land or measured altitude above the earth. Pipeline, highway, and railroad right-of-ways are examples. **Routes** are ways, paths, or roads a vehicle travels to get from one place to another. The route a plane takes to fly from Chicago to Dallas is one example.

Fig. 21-5 The Segway Human Transporter is an example of on-site transportation at some locations, such as airports.

Air, water, highway, and some railroad transportation systems are operated on ways and routes that are owned or controlled and maintained by local, state, or federal governments. Most of the railroad and pipeline industries own or lease and maintain their own ways. Part of the atmosphere over a particular location is its airspace. For safety, aircraft moving through airspace follow specific routes, or airways. Different routes are set aside for commercial, general, and military aircraft. The Federal Aviation Administration (FAA) is the government agency that controls all air traffic above the United States.

Transportation within a particular city or state is called *intra*city or intrastate transportation. Transportation between one city or state and another is termed *inter*city or interstate transportation.

On-site transportation occurs within a limited area, such as a building or at a construction or assembly yard. Moving from one story of a building to another in an elevator is one example of on-site transportation. Carrying your schoolbooks from your locker to the classroom is another. See **Fig. 21-5**.

The Global Positioning System

The **global positioning system (GPS)** is a navigation system that was first developed for ships and planes. It consists of 27 satellites orbiting the earth at an altitude of about 11,000 miles. Three are on standby in case some fail. The satellites are spaced evenly so that a minimum of four are above the horizon everywhere on Earth, 24 hours a day. See **Fig. 21-6**. Signals from a vehicle allow the satellites to track it. Then, using mathematical calculations, GPS computers determine the vehicle's location.

Using GPS, travel is safer. Planes can fly closer together with less chance of accidents. Travel is also more cost-effective. Pilots do not stray from the most direct route, which saves fuel.

Math Application

Determining Cost of Travel There are many ways to travel. Space shuttles transport astronauts into outer space. Airplanes and ships take people around the world. Travelers use trains, buses, and cars when crossing the country. People walk and bike for leisure and take subways and regional rail systems to work and school.

The transportation industry is the largest industry in the world. In the United States, over 3 percent of the U.S. gross domestic product and about 7.5 percent of all jobs are related to transportation.

With so many types of transportation available, it is important to understand which modes are appropriate for particular trips. For example, taking a taxi from New York City to Los Angeles would be expensive. It is more cost-effective to take an airplane or drive your own car.

FIND OUT Compare the cost of driving a round trip between New York City and Los Angeles to flying the same distance. Assume that the distance is 2,800 miles, the cost of gasoline is $3 per gallon, and your car gets 25 miles per gallon. You can find the cost of a round-trip airline ticket online. Note the wide range of prices. Which mode costs less? Make a list of other expenses you may incur on the trip, such as for hotels and meals. Will this change your answer? As a class, compare and discuss your results.

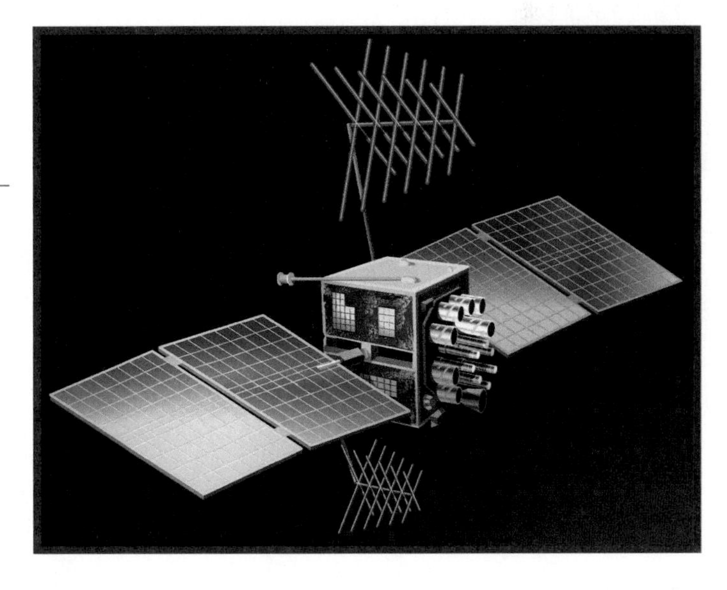

» **Fig. 21-6** GPS satellites circle the earth every 12 hours.

Travel can be faster as well. Ambulance drivers waste no time searching for the right street. In an effort to improve profits and lower costs, dispatchers can actually follow the operation of a specific truck's engine hundreds of miles away and determine the required horsepower needed to pull the truck's load on different terrain, thus saving fuel.

Today GPS can be used for almost every mode of travel. Some hikers and backpackers, equipped with special devices, have used it to find their way around in the wilderness. Farmers can use GPS when plowing fields. GPS is also being used on the International Space Station to provide more accurate speed and position data.

Cars are now being built that include dashboard computer screens on which city maps can be displayed. The car's exact position can then be seen on the screen. Some computers can even suggest the best route from one spot to another, based on GPS data.

Fig. 21-7 This load of gravel is considered bulk cargo.

Support Facilities

Transportation support facilities include a variety of terminals and warehouses, rail yards, landing strips, ports and docks, repair shops, and hangers. Your nearby fuel station is a support facility.

In every mode of transportation, the costs of operating the support facilities are greater than the costs of operating the vehicles. Most of the people employed in the transportation industry work at support facilities.

Cargo

Except for passengers, everything else transported, from furniture to zoo animals, is cargo, or freight. The transportation industry classifies cargo as bulk cargo and break bulk cargo.

Bulk cargo is loose cargo. It may be a solid material, such as sand and flour, or a liquid, such as oil and corn syrup. Bulk cargo is not packaged and it is usually not mixed with another cargo transported in the same vehicle. Single shipments of bulk cargo may completely fill a very large ship or the boxcars in a mile-long railroad train. See **Fig. 21-7**.

Break bulk cargo consists of single units or containers of freight. Books, bicycles, and computers are packaged and shipped as break bulk cargo. Almost all items in the stores where you shop were delivered as break bulk cargo.

It generally costs more to transport goods as break bulk than as bulk. This is because packaging costs are higher and more handling of each piece is required. However, cost is not the only factor that must be considered when determining how freight should be transported. How it is distributed to the final user plays a major role. For example, a school could charge less for milk if each student filled his or her own glass from a milk tank truck (bulk). Can you imagine the problems that this would cause? It is more convenient to provide milk in small containers (break bulk).

Science Application

Forecasting Weather Have you ever canceled travel plans due to poor weather conditions? Weather can affect not only plans for travel but also the shipping of products. When deciding how to transport items, companies must take weather into consideration.

How can people predict weather conditions when transporting themselves or products? They rely on forecasting. Meteorologists use science and technology to forecast the weather, often weeks in advance. They gather data about temperature, air pressure, moisture, wind direction, and wind speed. For example, if a storm front is moving 250 miles a day and is 1,000 miles away from your town, then a forecaster can predict that it is likely to be in your area in four days. Meteorologists also track weather patterns, which usually go from west to east.

Meteorologists also use radar to help anticipate changes in the weather. Doppler radar images can show the direction and speed of a snowstorm or rainstorm as well as the amount of precipitation that is falling. Doppler radar can also be used to show the formation of tornadoes. People can be alerted about tornadoes before they hit.

Knowing about weather patterns can help you plan transportation. For example, you might want to avoid traveling through the mountains during the period of the year when snow is the heaviest in that region. When you observe forecasts, you can make transportation a much smoother process.

FIND OUT Research current weather in at least two of the following states: California, Montana, Florida, Maine, New Mexico, Indiana, Louisiana, Texas, and Pennsylvania. Write a report on how that weather affects transportation.

Creating Economic Value

Transportation affects the economic or monetary value of services and products. Lumber piled at a sawmill has less value and thus a lower cost than the same lumber transported to a local retail lumberyard. The services of a person may increase in value if that person has to move to another location to complete a task. For example, an electrician may travel to an area damaged by storms to help people rebuild. In these cases, transportation satisfies the need for a person or item to be moved to another location.

Transportation must also happen at the right time for value to be increased. Oranges are not grown in Canada. Orange growers in the southern U.S. transport their fruit to Canada. There would be no added value if it took too long to transport the oranges and many of them were rotted when they arrived. If an electrician arrived in a disaster area after all rebuilding was completed, he or she would not be needed.

A change in value caused by transportation is called **time and place utility**. For the most value, items should be delivered or work completed at the time needed.

EVOLUTION OF Transportation Technology

People have developed many ingenious ways of getting from here to there.

The first vehicles were boats made from animal skins stretched over frames of wood or bone. By 3500 BCE, Egyptians were navigating the Nile River on boats equipped with sails and oars.

Steam engines were used to power ships, trains, and road carriages.

3500 BCE

2000 BCE

1700s

1800s

The first carts were probably made by adding wheels to sledges. Originally, the wheels were slices of logs. Spoked wheels, introduced around 2000 BCE, were a major innovation that led to lightweight, fast horse-drawn chariots.

The development of smaller, more efficient steam engines made powered transportation feasible. In 1769, Nicolas-Joseph Cugnot built the first automobile, a three-wheeled vehicle powered by steam and designed to move artillery pieces.

Automobiles with gasoline-powered internal combustion engines were developed. Innovations came rapidly, and by 1898 there were more than 50 automobile companies.

late 1800s

Space travel began in 1957 when the rocket-powered *Sputnik 1* became the world's first artificial satellite. By 1969, human beings had landed on the moon.

1969

1903

Orville and Wilbur Wright made the first sustained, powered, and controlled airplane flight. It lasted 12 seconds.

2000+

Economic and environmental issues have spurred research into alternative power systems for transportation. Development continues on hybrid vehicles and fuel cells.

Impacts and Effects of Transportation Technology

Like other technology systems, transportation has both negative and positive impacts. Let's consider the positive first.

Positive Impacts

Improvements in the technology of transportation have had a tremendous impact on our society. Developments in transportation have supported our country's growth. The existence of highways, waterways, and railroads enabled people to explore and settle all its regions. Today we are still exploring using transportation. We travel down into the depths of the oceans in submersible vehicles and we send spacecraft to Mars and beyond. See **Fig. 21-8**.

Transportation brings benefits to our daily lives. Producers and manufacturers expand their markets by transporting their products, which generally causes their sales to increase. As sales increase, the demand for the product or one like it increases. Increased demand also

Fig. 21-8 This submarine can reach depths of nearly 2,000 feet. ***What features might this vehicle be equipped with to withstand severe water pressure?***

encourages others to go into business. This is business growth.

Transportation may also create new businesses. The invention of the automobile, for example, brought about the manufacture of tires and the development of automobile insurance. Pollution caused by vehicle exhaust has inspired scientific and engineering efforts to correct the problems. New businesses have been built to dispose of such things as old tires.

As people increasingly move from one location to another, services and products become available to more people, allowing them to choose among various brands of products. Each provider tries to offer the best price.

Because of transportation, the countries of the world are able to trade products. American producers have sold countless shiploads of timber, corn, and other commodities to foreign nations. American businesses have in turn purchased shiploads of electronic goods, sport shoes, and toys from China, India, and most of the other producing countries of the world. None of this could have happened without the services of ships, trains, and trucks.

Improvements in vehicles, ways, and facilities have enabled people to travel longer distances in less time. Business people meet more customers. Families can travel to a wide variety of places and see more things. Students from one area in the U.S. attend colleges in another.

Modern transportation has also improved and expanded public services. Many cities have public transportation, such as subways, bus systems, and ferries. In emergency situations, specialized vehicles quickly bring help and medical treatment right to the needed location.

SAFETY FIRST

Rules of the Road. Teens are involved in more auto accidents than other groups. When driving, make safety a priority. Not only is your own well-being at risk but also the well-being of other drivers.

Negative Impacts

Just as transportation can increase the value of an item, it can also cause the value to decrease. If too much of the same product is available at a particular place, the price (value) goes down.

Because so many people in the U.S. rely upon the automobile for their transportation, most major cities experience large traffic jams each day. See **Fig. 21-9**. Some traffic jams, even on highly traveled interstate highways, can result in **gridlock**. No vehicle can move in at least one direction.

Fig. 21-9 Traffic gridlock such as this one costs many hours of lost travel time.

The costs of transportation have become higher. People must pay taxes for the construction and maintenance of public roads and transportation facilities. The cost of fuel and the price of postage have and will continue to increase. Other costs cannot be measured as easily. Homes and land must somehow be given up to make way for new highways and airports. Thousands of injuries and deaths occur each year due to transportation accidents.

Transportation also creates pollution. Discarded vehicles and tires cause land pollution and are unsightly. Vehicle exhaust pollutes the air.

One of today's biggest engineering and design challenges is to reduce negative impacts and make transportation economical, environmentally friendly, and safe. See **Fig. 21-10**.

Fig. 21-10 Jaguar's Pedestrian Deployable Bonnet System is designed to reduce damage and injuries if the car hits an object or person. At impact, the bonnet (or hood) of the vehicle pops up a few inches to create cushioning between the bonnet and the engine.

Transportation Technology

Transportation technology covers many interesting careers. There are jobs for the "hands-on" type of person (ship engineer, for example), for high-tech types (aircraft pilot), and for people who love to drive (truck drivers). The U.S. Department of Labor estimates that there will be more than four million new jobs in the transportation industry in the near future.

Automotive Technician • Aircraft Pilot • School Bus Driver • Sailor Airfield Operation Specialist • Taxi Driver • Courier • Marine Oiler Air Traffic Controller • Motorboat Operator • Railroad Conductor Freight Mover • Supervisor/Manager • Industrial Truck Operator Locomotive Engineer • Subway Operator • Waste Collector Water Vessel Captain • Truck Driver • Ship Engineer • Laborer Crane and Tower Operator • Vehicle and Equipment Cleaner Machine Feeder and Offbearer • Dragline Operator • Pump Operator

INVESTIGATE CAREERS Using the information from this chapter, classify each job by its mode of transportation. Research these jobs and write a report on what jobs might be similar despite differences in transportation modes.

CHAPTER 21 REVIEW

Main Ideas

- A vehicle is any means or device used for transportation.
- The transportation system is used to move both people and cargo.
- Transportation affects the value of services and products.
- Like other systems, transportation has both positive and negative effects.

Understanding Concepts

1. What is a transportation system?
2. How does GPS work?
3. Describe time and place utility.
4. Compare and contrast ways and routes.
5. Identify two positive and two negative impacts of transportation technology.

Thinking Critically

1. What do you think inspired the creation of traffic signals?
2. In which transportation system does the cargo move while the vehicle stands still?
3. Identify the type of commuter service that would be best for your area.
4. What might be a solution for an area where gridlock is common?
5. What are some positive impacts of bicycle transportation?

Applying Concepts & Solving Problems

1. **Math** Search the Internet to learn more about the mathematics used in GPS. Write a report on your findings.
2. **History of Technology** Why were cars originally referred to as horseless carriages?
3. **Design** Create a map that shows the various ways and routes in your area. Each way and route should be color-coded to signify a specific mode of transportation.

Activity: Make a Transportation Documentary

Design Brief

Your town is serviced by many transportation systems. Bicycles, cars, and trucks are common forms of transportation in a community. Air and rail transportation may also be used. Each transportation system has technological, social, economic, and environmental impacts on your community.

Create a documentary video that identifies at least three technological, social, economic, and/or environmental impacts of a single transportation system in your community. For example, you might choose road transportation, and one impact you might document is the effect of new road construction on businesses that must be closed along that roadway until work is finished.

Refer to the

Criteria

- Your video must identify at least three impacts of your chosen transportation system.
- You must create a script for any narrative and a storyboard that details important scenes in your documentary.
- If available, you should use computer software to edit your documentary.
- Your video should include a title and list of credits.

Constraints

- Your video must be at least 10 minutes long and no more than 15 minutes long.
- The transportation system you choose must be one that actually services your community.

Engineering Design Process

1. **Define the problem.** For the transportation system you have selected, write a statement that describes the problem your documentary is going to solve. For example, how can a video document ways the system interacts with your community?
2. **Brainstorm, research, and generate ideas.** With your team, discuss possible solutions. For example, each team member may do research on one of the impacts your documentary will cover.
3. **Identify criteria and specify constraints.** These are listed on page 426.
4. **Develop and propose designs and choose among alternative solutions.** Choose the best design that will solve your problem. For example, you may decide to make an interview with a traffic police officer part of your video rather than an animation of traffic patterns.
5. **Implement the proposed solution.** Decide on the design you will use. Gather any needed tools or materials. Then write the script.
6. **Make a model or prototype.** Create a storyboard of the main scenes in your video.
7. **Evaluate the solution and its consequences.** With your teammates, review your script and storyboard for potential problems.
8. **Refine the design.** Based on your evaluation, change the plan if needed.
9. **Create the final design.** After you have an approved storyboard, create your video.
10. **Communicate the processes and results.** Show your video to the class. Be prepared to answer questions. Turn in your video, storyboard, and script to your teacher. Be sure to include the name of the activity, your definition of the problem, and a description of how you solved the problem.

SAFETY FIRST

Before You Begin. Make sure you understand how to use the video camcorder safely. Have your teacher demonstrate its proper use. Follow all safety rules.

CHAPTER 22 THE STRUCTURE OF TRANSPORTATION

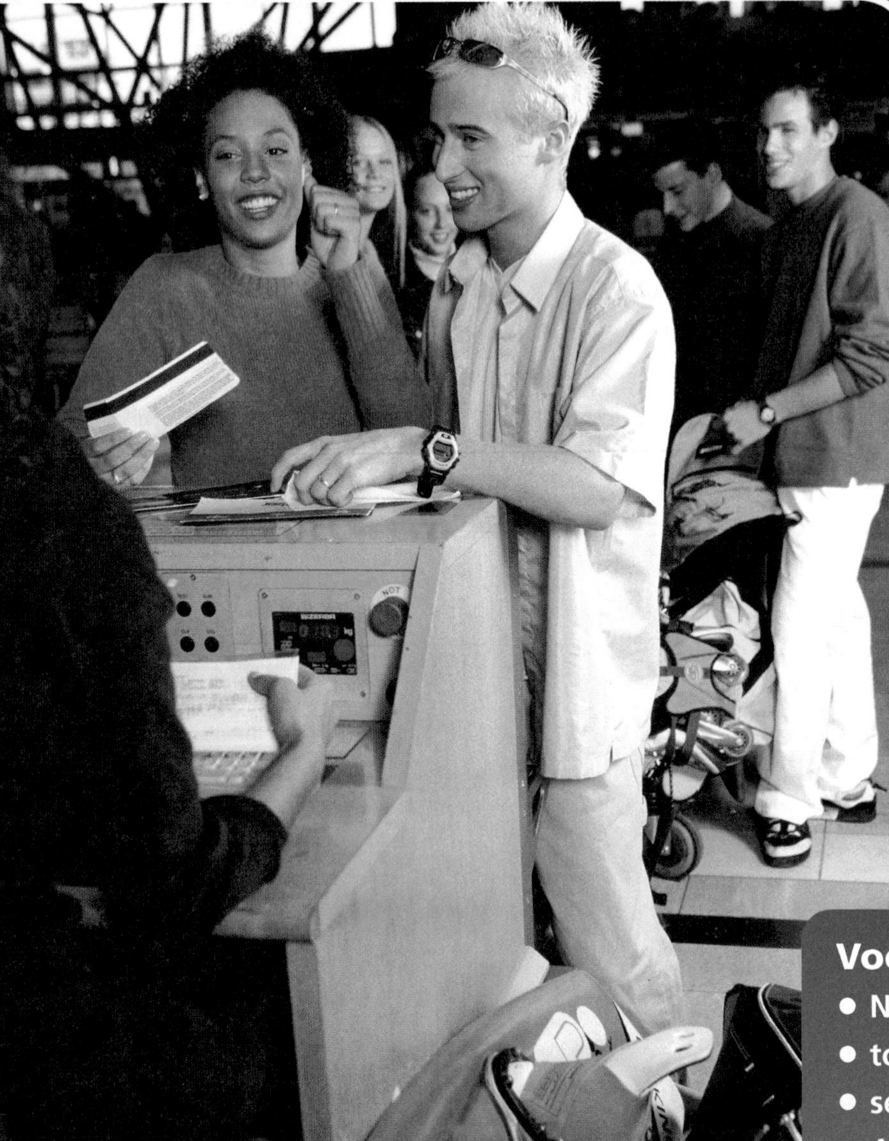

Objectives

- Name examples of vehicles in each mode of transportation.
- Identify facilities needed for each mode.
- Describe intermodal transportation.
- Explain containerization and list its advantages.

Vocabulary

- NASA
- ton mile
- sea-lane
- AMTRAK
- classification yard
- tractor-semitrailer rig
- slurry
- intermodal transportation

When traveling, it is important to choose the most appropriate mode (or modes) of transportation. In this chapter's activity, "Plan a Trip," you will compare modes of transportation and determine the best way to get to a selected destination.

Reading Focus

1. Read the title of this chapter and describe in writing what you expect to learn from it.
2. Write each term in your notebook, leaving space for definitions.
3. As you read Chapter 22, write the definition beside each term in your notebook.
4. After reading the chapter, write a paragraph describing what you learned.

Modes of Transportation

There are many kinds of transportation systems. They can be grouped into modes, or forms. Five modes are

- Air and space
- Water
- Rail
- Highway
- Pipeline

Air Transportation

In 1903, Wilbur and Orville Wright made the first sustained, powered, and controlled airplane flight. Now, more than a century later, air travel is the most popular form of long-distance transportation.

Air transportation is also the fastest mode of Earth-centered transport in operation today. Airplanes can travel across the United States from coast to coast in five hours or less. See **Fig. 22-1**.

Fig. 22-1 NASA's research scramjet, *X-43A*, recently broke world speed records, flying at nearly 7,000 miles per hour. Researchers predict that scramjet technology will revolutionize air travel, perhaps within the next 10 years.

Types of Air Transportation

Air transportation is divided into three basic types: commercial aviation, general aviation, and military aviation. Companies involved in commercial aviation provide air transportation for a profit. Most commercial planes are large. The European Airbus 380 can carry up to 800 passengers. The more people that travel in a plane, the higher the profits are.

General aviation includes all privately owned airplanes used for personal or business reasons. General aviation planes are usually smaller than commercial vehicles.

Military air transportation includes many kinds of aircraft, large and small. Helicopters, fighter planes, bombers, surveillance aircraft, and test planes are some examples.

Aircraft

Airplanes, helicopters, airships, and rocket vehicles are among the different examples of aircraft that transport various types of cargo and passengers. Not all aircraft remain aloft using the same means. How they are powered will be discussed in Chapter 23, "Powering Transportation."

Facilities and Support Services

Transportation facilities provide services that keep a transportation system working. In air transportation, many ground vehicles are needed. Some ground vehicles are equipped for cleaning the plane. Others move passengers and baggage. The planes themselves are moved by still other ground vehicles.

The airport is the hub of air transportation. See Fig. 22-2. Aircraft move on taxiways and runways. Terminals, control towers, automobile parking lots, hotels, fire stations, and airplane hangars make up different parts of an airport. Fuel storage areas are usually located nearby.

Fig. 22-2 Large urban airports, such as Hartsfield International Airport in Atlanta, Georgia, provide fuel and other ground services for planes as well as facilities for passengers.

Space Transportation

Space travel has progressed more slowly than air transportation. The first human to orbit the earth, Russian cosmonaut Yury Gagarin, made the trip in 1961. In July 1958, the National Aeronautics and Space Administration, which we know as **NASA,** was founded to plan and operate the U.S. space program. In 1969, Neil Armstrong and Edwin "Buzz" Aldrin landed on the moon in an Apollo spacecraft. The moon is still the farthest point reached by manned spacecraft.

Presently, space transportation is controlled by the government and used for exploration and research. However, this is changing as private organizations are developing spacecraft for commercial purposes. See Fig. 22-3.

Fig. 22-3 In 2004, *SpaceShipOne* competed for and won the $10 million X-Prize as the first private, manned, commercial craft able to carry passengers into suborbital space. The purpose of the X-Prize is to promote space tourism.

There are two basic types of space transportation systems—manned and unmanned. Manned systems carry humans. Unmanned systems do not have human crews or passengers.

Manned Systems

Manned spaceflights are still much less common than unmanned flights. There are two basic types: nose-cone-mounted spacecraft and space shuttles. They must provide everything necessary to keep their passengers alive in the vacuum of space.

Nose-cone-mounted spacecraft were the first type used to transport humans. These spacecraft consist of sealed capsules attached to a booster rocket. Booster rockets are rockets used to push a payload. A payload is anything transported. When the rocket reaches space, the capsule separates from the booster rocket. The capsules have no wings and no ability to fly through the atmosphere on their own. The rocket burns up as it falls back to Earth.

Space shuttles can be considered the first true spaceships because they take off, maneuver in space, and fly back through the atmosphere for landing. Shuttles have their own powerful engines, but they still need the help of a booster rocket to escape Earth's gravity. Each shuttle can be used over and over again.

Unmanned Systems

Unmanned spacecraft usually consist of hollow containers mounted on booster rockets. The containers carry the payload. Once in orbit, the containers open and the payload is maneuvered into position with tiny rocket engines. These vehicles and their booster rockets can be used only once. The most common cargo on unmanned vehicles is satellites. Most are for communication. Others are for scientific research. See Fig. 22-4 on page 432.

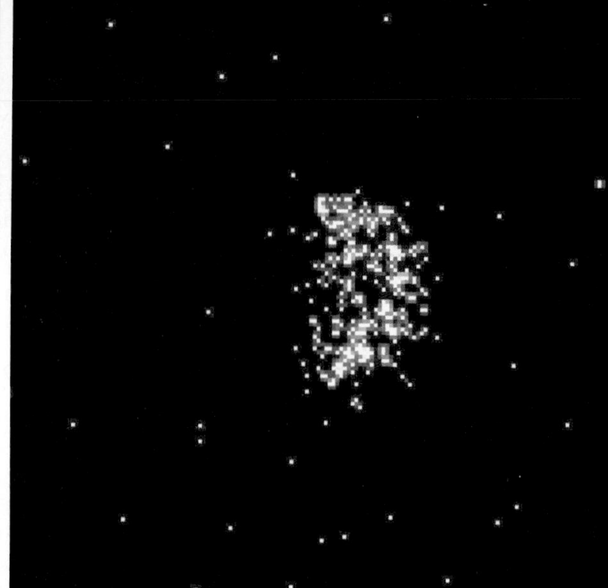

« Fig. 22-4 The *Chandra X-ray Observatory* satellite is designed to observe X-rays from high-energy regions of the universe, such as the remnants of exploded stars.

Space transportation is by far the most expensive transportation system. It costs thousands, sometimes millions, of dollars to put just one pound of payload into space. The future of space transportation will be determined by how much money governments and private organizations are willing to spend to explore space.

Water Transportation

Water transportation is restricted to areas where there are navigable waterways. A waterway that is navigable is wide enough and deep enough to allow ships or boats to pass through.

Water transportation costs less than most other forms of transport. Cost is figured by the ton mile. A **ton mile** is the cost of moving one ton a distance of one mile. Travel, however, is slow.

Types of Water Transportation

Most water transportation involves moving cargo. Cargo movement that takes place on water within the United States is called domestic inland shipping. Foreign shipping is either importing (bringing in) goods from another country or exporting (sending out) goods to another country.

Passenger lines include luxury or cruise ship lines, paddlewheel riverboats that offer short recreational trips, and public transportation. Lake and river ferries are examples of public water transportation.

Vessels

There are several different types of water transportation vehicles, or vessels. The "work-horse" of domestic water transport does not even have an engine. It is a large floating box called a barge, or a lighter. Barges are tied securely together to form a tow, which tow-boats and tugboats then push and pull through major rivers, canals, and lakes. Barges can even be loaded onto ships and transported across oceans. Barges are used to ship bulk cargo such as petroleum products, coal, and grain.

Towboats and tugboats are relatively small vessels. They operate like railroad locomotives. These boats move barge tows and larger ships.

Other types of ships include containerships, tankers and supertankers, general cargo ships, and bulk carriers. See **Fig. 22-5**. Cruise ships are like floating hotels. These passenger ships contain restaurants, recreation facilities, state-rooms (bedrooms), water treatment plants, storage warehouses, and communication stations.

Fig. 22-5 This large ocean-going tanker carries natural gas.

Science Application

How Vessels Float When you push a large ice cube down into the water, it always resurfaces. Do you know why? The answer is buoyancy, which is the upward force a fluid places on an object. According to Archimedes' principle, the buoyant force is equal to the weight of the fluid displaced by a submerged body.

Why doesn't everything float? An object's ability to float depends on its density, which is a measure of how packed together its molecules are. A stone is denser than water, so it sinks. A log, however, has a lower density and displaces enough water to create an upthrust necessary for it to float.

Vessel design makes use of the science of buoyancy. Some hulls are shaped to displace the greatest amount of water, creating enough of an upthrust that the vessel floats. Hulls that aid displacement are usually on large vessels that sit deep in the water.

The hulls of smaller boats may be designed to hydroplane. At high speeds, the hull rises partially above the water. The amount of drag and friction are reduced because less surface area is in contact with the water.

FIND OUT Test various objects' ability to float. Write a report detailing your findings.

Recreational boats take passengers on short trips for such things as scenic tours. Ferries carry people, and generally their cars, across rivers, lakes, and bays.

Waterways and Sea-Lanes

Water transportation follows either inland waterways or sea-lanes. Inland waterways are navigable bodies of water such as rivers and lakes. Channels (deep paths) must be kept deep enough for large vessels and marked with buoys, lighthouses, and other devices. In the United States, the U.S. Army Corps of Engineers is responsible for channel maintenance. Locks and dams, needed to keep rivers navigable, are also built and maintained by the Corps of Engineers. See Fig. 22-6.

Sea-lanes are shipping routes across oceans. The countries of the world agree in general on certain boundaries and locations of sea-lane routes.

Fig. 22-6 Twenty-nine locks are located along the length of the Mississippi River. These locks raise and lower boats traveling along the river.

Facilities

Major water transportation cargo facilities include ports, docks, and terminals. These places provide areas for loading and unloading cargo, needed services and supplies, cargo warehouses, and ship company offices. Passenger ships require the same facilities, but they are more comfortable than those built for cargo ships.

Rail Transportation

Some advantages of rail transportation can be seen as you look at a train. Trains are strong and tough! They have steel wheels that move on steel rails. These are strong enough to efficiently transport large, heavy loads.

Trains offer other advantages, too. A loaded railroad car is a "rolling warehouse." It protects the shipper's cargo during transport. In addition, a single train engine hauling several cars uses a lot less fuel than the several trucks it would take to haul the same amount of cargo. Also, trains do not get into traffic jams in crowded cities, which can be a big time-saver.

Passenger Service

All long-distance rail passenger service in the United States is provided by **AMTRAK**. The AMerican TRavel trAcK system is owned by the federal government, which also owns the engines and cars. However, the trains are operated by the railroad lines that own the track the trains are using.

In the United States, most trains carry cargo rather than passengers. However, high-speed passenger systems that will travel over 300 miles per hour are currently being developed. Some of these will be used to link various American cities. Similar systems are already being used in Europe, China, and Japan. See Fig. 22-7.

Fig. 22-7 The *Tokaido Shinkansen*, also called the bullet train, travels at speeds up to 185 mph between cities on Japan's main island. Similar trains may soon be built in the U.S.

Regional and city rapid transit trains now provide commuter service in some large cities. Commuter service is regular back-and-forth passenger service. For example, in New York City, many students travel to and from school on commuter trains.

Freight Service

Trains that carry freight serve over 50,000 towns and cities in North America. There are two basic types of freight trains: unit trains and regular freight trains.

A unit train carries the same type of car to the same place time after time. Some unit trains are owned by the customers, not the railroad lines. The customers pay the railroad to pull their cars.

Regular freight trains offer a variety of short and long haul services. Cars are arranged on the train according to the type of cargo and final destination. The cars may be switched to several different trains before they are finally unloaded.

Rolling Stock

Rail transportation vehicles are called rolling stock. The three basic groups of rolling stock include engines, railroad cars, and maintenance vehicles.

Engines (locomotives) pull the train. Most engines use diesel engines to turn electrical generators. The electricity produced then powers traction motors that turn the wheels. See Fig. 22-8. These combinations of engines and motors are called diesel-electric locomotives.

Railroad cars can transport many different forms of cargo. The most common car is the boxcar, an enclosed car that can carry almost anything that can be packed into it. Other cars, such as those that carry liquefied gas, are more specialized.

Maintenance vehicles are used to keep rails clean and safe. Track inspection cars, brush-cutting machines, and hoist cars are examples of maintenance vehicles.

In the United States, all tracks are built to the same gauge (56½ inches between rail centers). This allows engines and cars from one line to interchange and roll on the tracks of any other line.

Facilities

Classification yards, shops, and terminals provide service facilities for rail transportation. **Classification yards**, also called switchyards, are where trains are taken apart and put together. A train pulls into the yard and the cars are disconnected and sorted according to their destinations. The reclassified cars are rolled onto tracks with other cars going in the same direction. Modern classification yards use computers to efficiently sort and classify cars. When the trains are made up, road crews connect the engine to the train. See Fig. 22-9. The type of engine used is determined by the length and weight of the train and the terrain it must cross.

Each railroad line operates many repair and maintenance shops. Some are portable and can be moved to where they are needed.

Terminals serve as business offices and meeting places for train crews. Freight terminals are loading docks and storage places for cargo. At passenger terminals, people buy tickets, check their baggage, and wait for trains.

« Fig. 22-8 The engineer controls the speed of the locomotive and monitors engine performance.

Highway Transportation

The most common mode of transportation is highway transportation. Cargo and people are transported over millions of miles of highways, streets, and roads. The major advantage of highway transportation is the independence given to the operator. As long as there is a road going there, a person can travel almost anywhere at any time, day or night. A major disadvantage is that highways all across the United States are becoming overcrowded, leading to increased traffic accidents, traffic jams, and air pollution.

Fig. 22-9 Computers help sort rail cars, but human workers do the hard work of putting the train together.

Personal Transportation

People commonly use highway transportation for their own personal travel. This is especially true in the United States, where people use highways more than in any other country in the world. They drive cars, small trucks, vans, SUVs, recreational vehicles, motorcycles, and bikes.

Commercial Transportation

Commercial use of highways is very big business. Commercial users may travel intracity or intrastate as well as intercity or interstate.

Commercial transport can involve passengers or cargo. Bus and taxi companies typically offer commercial passenger service. However, the most common commercial use of the highways is for freight, or cargo, service.

As you travel the highway, you see many trucks with the same company name. These companies, referred to as motor freight carriers, own the vehicles and hire people as drivers. Some offer regularly scheduled pickup and delivery.

In recent years, a new type of commercial trucking has become popular. The owner of the truck is also the driver. He or she is an owner-operator. An owner-operator usually owns only one truck. This vehicle may be both office and home for the owner. Owner-operators hire out their vehicles and their driving services.

Freight-Hauling Vehicles

A variety of vehicles are used in the trucking business. The cargo and the length of the trip determine the type of vehicle used. Single-unit trucks are one-piece vehicles. They are also called straight trucks. The engine is in front, and the drive wheels are in the rear. The bed is permanently mounted over the drive wheels. Single units are generally used for carrying one type of cargo and for local hauling.

A **tractor-semitrailer rig** (also called an eighteen-wheeler) is a combination of a tractor and a semitrailer. The tractor is the base unit that pulls the semitrailer, which contains the cargo. Since the trailers have no front wheels, they are called semis. The rear wheels of the tractor support the front end of the trailer

Fig. 22-10 More and more vehicles are using our nation's highways. This tractor is hauling three trailers. ***Why do you think tractors are only allowed to haul three trailers in certain areas?***

through a "fifth wheel" hook-up arrangement. Most semitrailers can be hooked up to any tractor. These trucks can haul many different loads, depending upon the type of semitrailer.

Sometimes additional trailer units are connected to allow a single tractor to pull more freight. Some states allow as many as three trailer units behind a tractor. See Fig. 22-10. The total length could be over 140 feet.

Facilities

Commercial use of highway transportation continues to increase. The need for physical facilities is also increasing. Services must be provided before, during, and after trips. Cargo must be loaded, unloaded, and stored. Terminals and truck stops offer these services.

Good road systems are necessary for transportation to be effective. Developed in the late 1950s, the United States' interstate highway system, designed for safe travel at 65 or more miles per hour, is used by millions of automobiles, motorcycles, and trucks. See Fig. 22-11.

Highways are built and maintained by local and state governments using tax money collected on gasoline and diesel fuel. Some states help pay for their highway systems with toll roads, where drivers must pay a certain fee, or toll, to use the road. The heavier the vehicle is, the higher the fee per mile.

Pipeline Transportation

Over one million miles of pipelines cross the United States and Canada. The pipes are made of either steel or plastic. They range in size from two inches in diameter to fifteen feet in diameter. Most carry natural gas or petroleum products. See Fig. 22-12.

Fig. 22-11 The Interstate Highway System links every state in the continental United States.

Math Application

Estimating Cargo Shipping Costs Determining which type of cargo carrier to use depends upon three main factors: cost, distance covered, and speed. These factors are influenced by the type of cargo to be shipped. For example, a load of steel beams would be shipped differently than a human organ transplant. The beams would probably go by ship, depending on their destination. The human organ would be shipped by air freight. Today, many shipping companies use freight consolidators, companies that combine more than one mode in order to achieve the best time and price.

The average speed (miles per hour) for each type of transport can be calculated using the formula:

speed = distance ÷ time

FIND OUT Suppose you have to ship two tons of watermelons from New Orleans to New York. Using the chart, determine the best choice of cargo carrier. Calculate its speed using the formula. Why did you choose that carrier?

Carrier	Cost Per Ton	Miles Traveled	Travel Time to NY
Ocean freight	$90	1,800	7 days
Railroad	$210	1,350	3½ days
Truck	$324	1,000	34 hours
Air cargo	$2,500	975	2¾ hours

Gathering pipelines are used to collect the cargo from the suppliers. These lines meet at central holding tanks and pumping stations. Transmission pipelines are the main long-distance lines that transport the cargo. The batches end up at terminals where cargo can be stored temporarily. Distribution pipelines deliver the cargo from the terminals to the customers.

Pipeline transportation offers four special characteristics.

- It is the only mode in which the cargo moves while the vehicle stands still.
- Most pipelines are buried in the ground, so they are unseen and quiet. Also, they cause no traffic congestion or accidents.

Fig. 22-12 This pipeline carries oil from the drilling fields of Alaska to southern states.

- Pipelines are laid out in straight lines across the country. This decreases the transportation time.
- Theft from pipelines is difficult. Also, damage to or contamination of cargo is rare.

Using Pipelines

Certain types of cargo, such as oil products, natural gas, coal, wood chips, grain, and gravel, can be transported very economically through pipelines. However, pipeline transportation is not as flexible as other modes. Service depends upon how close a customer is to a line.

Cargo is shipped in batches. Solid materials are moved in a slurry. A **slurry** is a rough solution made by mixing liquid with solids that have been ground into small particles. The batch is put into the system. As it goes in, the amount is measured and recorded. The material is then pumped through the lines. Pipeline transportation is one-way. If products also need to flow in the opposite direction, two lines are installed.

Facilities

The facilities that service the pipelines are above ground. Pumping stations use pumps to move the cargo down a pipeline. Natural gas may be compressed up to 2000 pounds per square inch. It does not travel more than 15 miles per hour.

Liquids such as crude oil travel 2 to 5 miles per hour. To keep cargo moving, pumping stations pump the batches every 30 to 150 miles along the pipeline.

Control, measuring, and exchange stations are also located along pipelines. These facilities make sure the customers safely receive the correct size batches of their cargo.

Pipelines can become clogged. To prevent this, a scraper tool called a pig is regularly pushed through the pipeline. See **Fig. 22-13.** Some pigs have brushes on the outside. Others carry instruments that take readings on conditions inside the pipeline.

Fig. 22-13 Pipeline pigs are actually scraping tools that are pushed through the pipes to clean them.

Intermodal Transportation

In the past, transportation companies were very protective of their business. Every transportation company competed with every other kind of transportation company. A truck company competed not only with other truck companies, but also with railroads, ship companies, and airlines. Today there is still competition, but other things are quite different.

The service offered by each of the different modes is becoming more specialized. Companies cooperate and often own other types of companies. Railroads also own and operate trucks, ships, and barges. Airlines own trucking lines and automobile rental agencies.

The process of combining transportation modes is called **intermodal transportation**.

Efficiency is increased because much less time and labor are spent in loading and unloading passengers and cargo. Costs are also reduced.

Intermodal Cargo Transportation

Intermodal cargo transportation includes trailer on flatcar, containerized shipping, and other specialized methods.

Trailer on Flatcar (TOFC) involves highway and rail modes. Truck trailers full of cargo are carried on railroad flatcars. This is also called piggyback service.

A trucking company uses tractors to move semitrailers to the loading dock of the company wanting to ship the cargo. At the loading dock, the semitrailers are carefully loaded and sealed by the shipper.

The trucking company moves the semitrailers to a railroad yard. Here, they are loaded onto specially built flatcars. The flatcars are connected to trains and are moved by the railroad to another yard close to the final destination. There, the semitrailers are unloaded and connected to tractors. These units travel on the highways to the customers' loading docks. The semitrailers are unsealed and off-loaded.

Containerization (also called containerized shipping) involves loading cargo into large boxes or other containers before it is transported. See **Fig. 22-14**. Containerization is used by all modes but pipeline.

The standard container is basically a large metal box that is eight feet high, eight feet wide, and forty feet long. A smaller container is used in air transportation. The frame of the box is strong and the corners are reinforced. Holes in the corners provide a means of grasping the box during loading, securing, and unloading operations.

Containers can be made into semitrailers by placing them on a frame that has ordinary wheels. Containers can also be loaded directly onto railroad flatcars. When they are fastened directly to the flatcars, the method of transportation is called COFC for container on flatcar. Some oceangoing ships, called containerships, are specially designed and built to carry containers. The airlines use containers that are shaped to fit into different models of aircraft.

Fig. 22-14 A standard container can be carried on a highway truck frame, a containership, or a special rail flatcar.

There are many advantages to containerized shipping. The most important is less handling of cargo. Items are packed only once into the container, and the container is locked and sealed. Then the container may be attached to a truck tractor or a railroad flatcar for transport. It might also be stored on board a ship. The items never leave the container until they arrive at their destination. Using a container saves a great deal of labor and money. Also, there is less chance of theft. The chances for damage to cargo from dropping, bumping, or exposure to bad weather are also reduced.

Other Cargo Transportation It is not practical to build pipelines for moving liquids to every city and town. By using rail or highway modes, pipeline companies are able to expand their range of service to almost anywhere. Some tank truck cargo companies actually call themselves "rolling pipelines." Oil is often shipped in this way.

Milking machines collect milk from the cows at a dairy farm. The milk travels through pipelines to holding tanks. From there, it is piped into tank trucks, which transport it from the farm to the dairy. At the dairy, it is pumped through hoses and pipelines. See **Fig. 22-15**.

Another form of intermodal transportation moves coal and gravel. As these materials are gathered, they are loaded onto very long conveyors. A conveyor is a continuous chain or belt that moves materials over a fixed path. Finally, the conveyor dumps its load into trucks or railroad cars. The conveyor loads the cars so fast that the train can keep moving.

Fig. 22-15 For health reasons, milk is never exposed to air from the time it leaves the cow until it reaches the dairy for processing. This truck is unloading milk cargo at a cheese factory in Ursem, the Netherlands.

The hopper car unit trains travel to the customer's location, where they are unloaded. Hopper doors on the bottom of each car may be opened and the load dropped onto a conveyor. Another unloading method involves rolling each car onto a section of mechanical track. The track and car rotate together until the car is upside down. The load dumps out all at once. The procedure is repeated until all cars are unloaded. This is a very fast method of unloading cargo.

Other forms of intermodal cargo transportation include the vacuum-operated transfer of flour or grain among ships, railcars, and trucks. This system is much like cleaning with a vacuum cleaner.

The U.S. military has an even broader system of intermodal transportation. A container of supplies may be shipped by highway, rail, water, and air. Finally, it may be dropped by parachute from an airplane into a camp that is hard to reach.

Intermodal Passenger Transportation

Passenger and cargo intermodal transportation are quite different. The transfer of cargo between modes is the task of the transportation company. Passengers, however, must be responsible for themselves and search out the connecting links between modes. Someone may take a city bus from home to a city transit train station. At the airport, he or she might take a plane to another city and then catch a taxi to a final destination.

More and more cities are proposing direct people-mover links between airports and city centers, tourist areas, and other high-traffic locations. As a result, future passengers will have to do less of the work necessary to arrange transportation. Until then, intermodal passenger transportation will continue to be supplied by taxicabs, bus lines, and private cars.

Young Innovators

Diving Deep with MATE ROV

The water is cold, the tension is high, and the pressure is mounting as your remotely operated vehicle (ROV) dives deeper and deeper beneath the surface. You watch the monitor carefully as the dim shape ahead grows to fill the screen. It is the instrument package that has been collecting data critical to research, but it has failed to release from its hold. You guide the vehicle carefully and find the manual release loop and trigger it. The instrument package bobs to the surface before being picked up by the support vehicle—mission accomplished!

Sound like fun? Take the plunge and dive into the deep in the MATE International ROV Competition! If you have ever been interested in underwater observation and exploration, then this is your opportunity to get involved.

The Marine Advanced Technology Education (MATE) Center and the Marine Technology Society's (MTS) ROV Committee have partnered to bring about a competition for students from middle schools, high schools, home schools, community colleges, and universities. You and your team are challenged to design and build an underwater vehicle capable of performing real-world

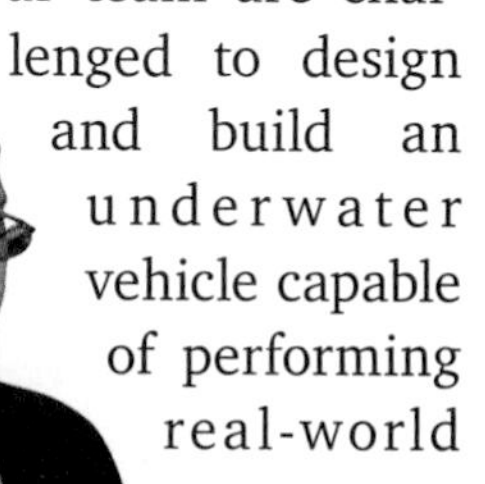

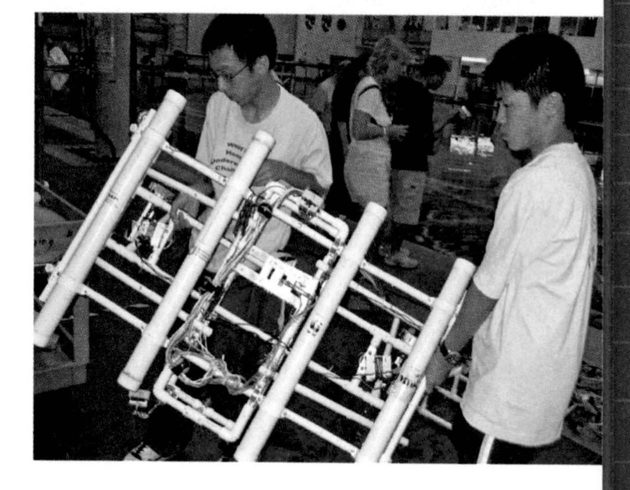

ROV tasks, such as installing a complex network of instruments that make up an underwater ocean observatory.

Your team must also prepare a technical report that documents the process you went through to design and build your vehicle and a poster display that highlights your ROV and your team. In addition, your team must make an engineering presentation on how your vehicle works and the challenges that you faced in building it to a panel of industry professionals.

Each year, the MATE Center organizes several regional contests and an international competition. These exciting competitions are open to students with an interest in engineering, technology, and undersea science and exploration. The MATE ROV competition is an excellent way to learn teamwork skills and technological literacy that can be applied to your future education and career. Best of all, you get to have fun in the process.

CHAPTER 22 REVIEW

Main Ideas

- Five basic modes of transportation are air and space, water, rail, highway, and pipeline.
- Air transportation is the fastest form of Earth-centered transportation, and land transportation is the most common.
- Each mode of transportation has facilities designed to meet the needs of the system.
- Pipeline transportation is the only system in which the vehicle is stationary while the cargo moves.

Understanding Concepts

1. What is the most common mode of transportation?
2. How is the cost of water transportation figured?
3. What kinds of facilities are needed for rail transportation?
4. What is intermodal transportation?
5. Identify at least three advantages of containerization.

Thinking Critically

1. Identify facilities for space vehicles.
2. Why are so many spaceflights still unmanned?
3. What do you consider the most severe limitation of water transportation systems?
4. Is it practical to develop airplanes large enough to carry standard size containers? Explain.
5. How many axles do you think an eighteen-wheeler has?

Applying Concepts & Solving Problems

1. **Math** If a pipeline is capable of moving 1,000 gallons a minute, how much material can be moved each day?
2. **Assess** Cars are often rated according to fuel efficiency. Research how cars are rated and then assess the fuel efficiency of a household vehicle.
3. **Hypothesize** Why do you think car manufacturers are interested in making cars that weigh less?

Activity: Plan a Trip

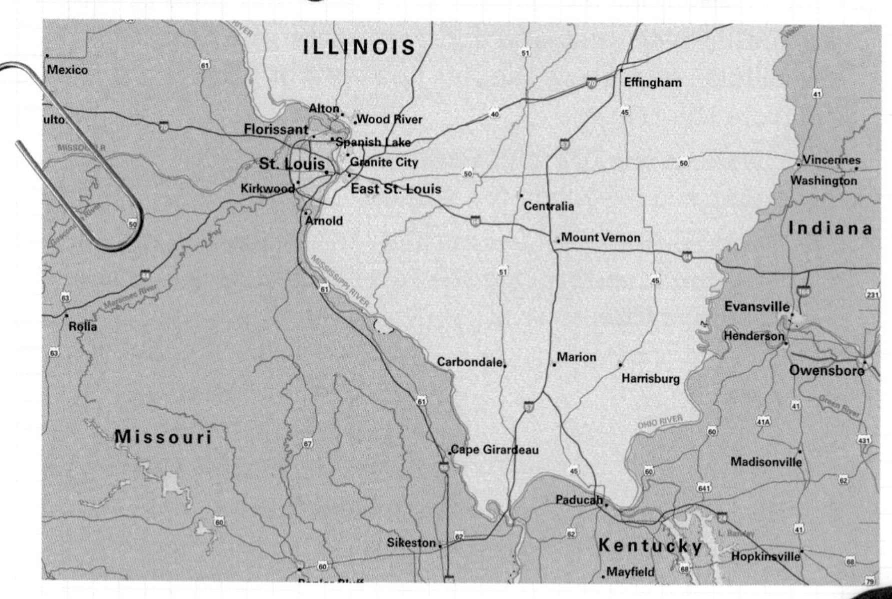

Design Brief

Selecting a mode of travel may depend on many different things. Cost may be important, as well as time spent. The modes available are also a factor. You cannot take a train if no trains service your community. If you decide to take a car, you must factor in cost for gas, as well as any wear and tear on the vehicle (oil, tires, etc.). Each traveler must weigh all the factors and decide which are most important. Then a carefully considered choice can be made.

Plan a trip to a destination about 500 miles from your home. Research the modes of travel available; travel costs for each mode, including any related costs such as parking and meals; and the time required to make the trip. Create a chart comparing the modes and write a brief report indicating your choice of mode(s) and why you made that selection.

Refer to the

HANDBOOK

Criteria

- Your chart must provide all information related to direct (normal travel) cost and indirect costs (such as parking and meals), as well as the total amount of time required to complete the trip.
- Your written report must include your choice of travel mode and your reasons for making that choice.
- Cite your sources of information. In the case of Internet sources, you must provide the Internet address (URL) and the Web page title.

Constraints

- Your route must be shown on a photocopy of a map.
- The destination you have chosen must be reachable by at least two modes of transportation.

Mode	Direct Cost	Indirect Cost	Time
Car	fuel:	motel:	
Train		meals:	
Airplane	ticket:		

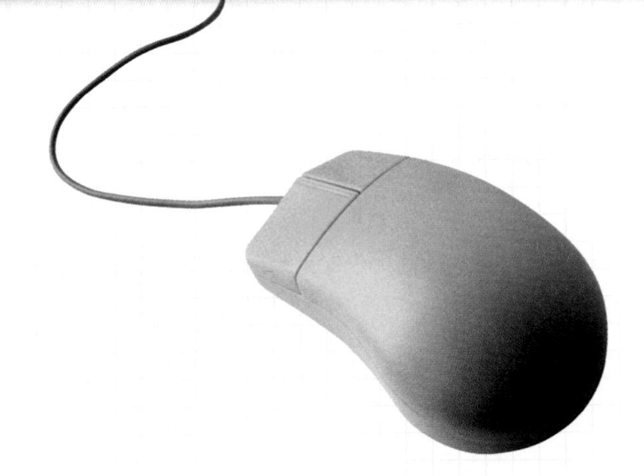

Engineering Design Process

1. **Define the problem.** Write a statement that describes the problem you face (choosing the best way to take a trip), and any obstacles you may face, such as cost, time, etc.
2. **Brainstorm, research, and generate ideas.** Make a list of possible modes of transportation. Find out which modes are available in your area.
3. **Identify criteria and specify constraints.** These are listed on page 446.
4. **Develop and propose designs and choose among alternative solutions.** Choose the best design that will solve your problem. For example, you may decide to use more than one mode for your trip. Design a chart that will let you record the appropriate information.
5. **Implement the proposed solution.** Decide on the design you will use. Gather any needed tools or materials. Then create your chart.
6. **Make a model or prototype.** Fill out all the information needed in your chart.
7. **Evaluate the solution and its consequences.** Review the information from the chart. Do you have all the necessary information? Can you tell which mode(s) of transportation will work best for your trip?
8. **Refine the design.** Based on your evaluation, change the chart if needed. If you can't make a decision on which mode to take, more information may be needed.
9. **Create the final design.** Write your report, detailing your choice of transportation and your reasons for selecting it.
10. **Communicate the processes and results.** Show your chart to the class. Read your report to the class, detailing why you chose the mode of transportation that you did. Turn in your chart, a map showing the route, and your report to your teacher.

SAFETY FIRST

Before You Begin. You may find much of the information you need on the Internet. However, do not give out personal information without being sure that the request is legitimate. Also, do not feel pressured to pay for information. The information you need should be obtained for free.

CHAPTER 23 POWERING TRANSPORTATION

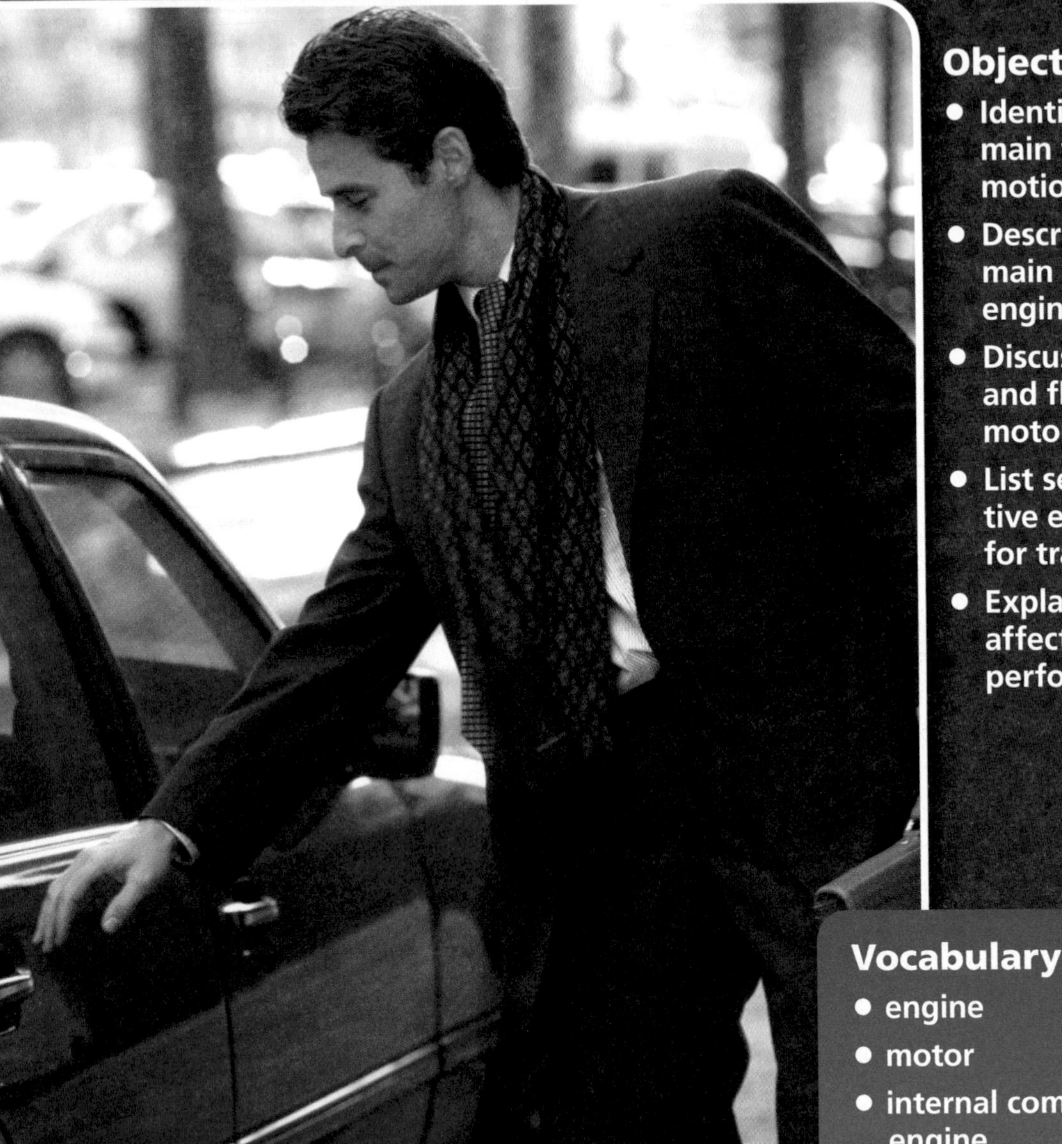

Objectives

- Identify the three main types of motion.
- Describe the two main types of engines.
- Discuss electric and fluid-powered motors.
- List several alternative energy sources for transportation.
- Explain how design affects vehicle performance.

Vocabulary

- engine
- motor
- internal combustion engine
- diesel engine
- external combustion engine
- maglev system
- aerodynamics

All forms of transportation require a power source to get you from one place to another. In this chapter's activity, "Design Transport for the Future," you will develop an alternative transportation system, including its power needs, for use within your community.

Reading Focus

1. Write down the colored headings from Chapter 23 in your notebook.
2. As you read the text under each heading, visualize what you are reading.
3. Reflect on what you read by writing a few sentences under each heading to describe it.
4. Continue this process until you have finished the chapter. Reread your notes.

Prime Movers

Prime movers supply the power (the use of energy to create movement) in transportation. A prime mover is a basic engine or motor.

The terms engine and motor are often used to mean the same thing. This is not wrong, but it is possible to be more exact. In this book, we will use the following definitions:

An **engine** is a machine that produces its own energy from fuel. Many different types of engines are used in transportation. Most engines burn fuel to create heat. (This burning is called combustion.) The primary fuels used are fossil fuels—gasoline, diesel fuel, oil, natural gas, propane, and coal. The heat creates pressure. The engine uses the pressure to produce mechanical force, or motion. There are two basic types of engines: internal combustion engines and external combustion engines.

A **motor** is a machine that uses energy supplied by another source. Motors used in transportation are either electric or fluid powered.

Types of Motion

Three kinds of motion are produced by engines and motors. They include reciprocating motion, rotary motion, and linear motion. See **Fig. 23-1**.

Reciprocating Motion

Most transportation vehicles are powered by reciprocating internal combustion engines. These include most highway and rail vehicles, propeller-driven aircraft, and smaller water vehicles.

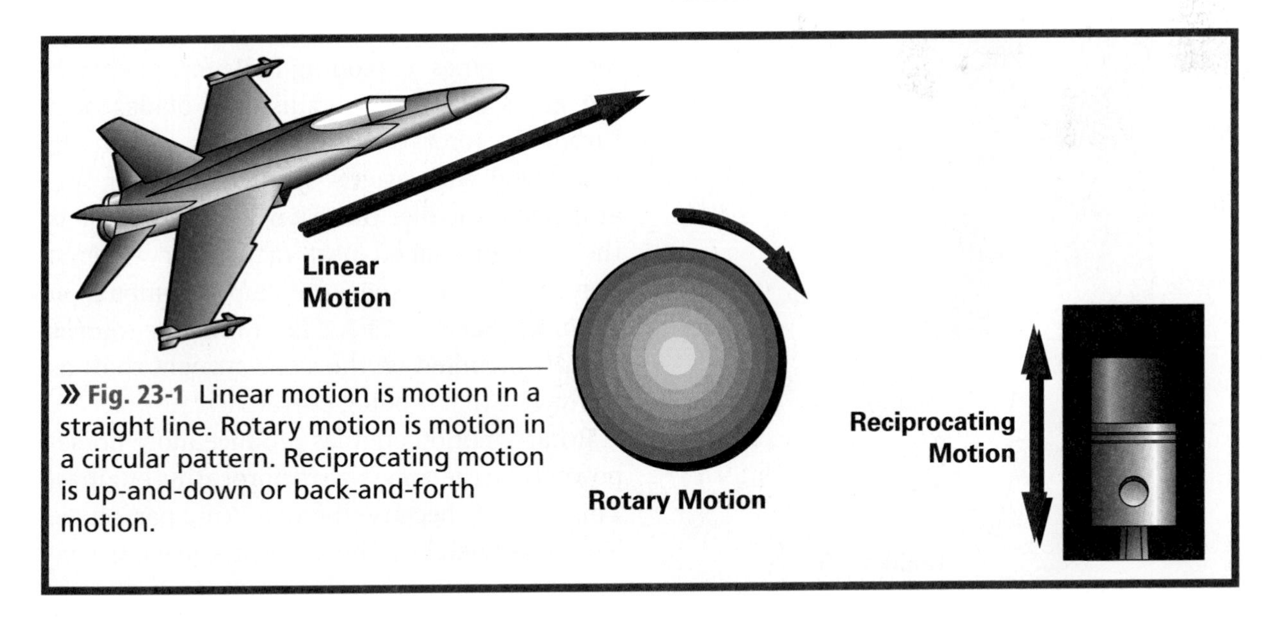

» **Fig. 23-1** Linear motion is motion in a straight line. Rotary motion is motion in a circular pattern. Reciprocating motion is up-and-down or back-and-forth motion.

During operation, the reciprocating motion (up-and-down or back-and-forth motion) of a piston is changed into the circular or rotating motion of a crankshaft. See **Fig. 23-2**. Not all of the energy transmitted by the piston through the connecting rod becomes crankshaft power. Some power is lost due to friction. The parts rub together and create resistance to motion.

Engines are classified by the number of piston cylinders—four, six, or eight—and their arrangement. The speed of the engine is determined by the size of the parts. The larger the piston (weight and diameter), connecting rods, and crankshaft, the slower the speed of movement. Very large reciprocating internal combustion engines used in locomotives or on ships may turn at only 100 to 400 revolutions of the crankshaft per minute (RPM). Very small reciprocating internal combustion engines, like those used in model airplanes, may turn at well over 20,000 RPM.

Fig. 23-2 In a reciprocating engine, the piston moves up and down, turning the crankshaft in a circular motion.

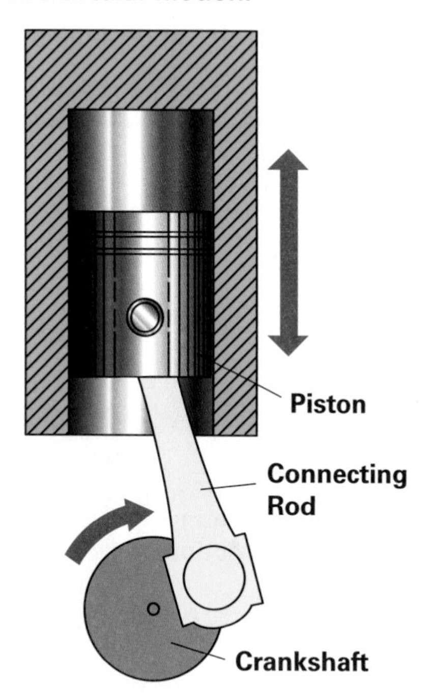

Reciprocating engines have several major advantages. The RPM of the engine can be quickly and easily controlled. This makes it well-suited for highway stop-and-go driving. The cost of manufacturing a reliable engine is low. Also, the cost of rebuilding or repairing it is low compared to the cost of repair for other types of internal combustion engines.

There are also disadvantages to this engine. Only a limited amount of energy can be economically produced by a reciprocating engine. The amount of horsepower that can be provided by each piston is limited. High-horsepower engines usually have many pistons. They have more moving parts than smaller engines, and they weigh more. These factors make large engines more expensive to produce. The faster the RPM of the engine, the shorter its usable life span. Reciprocating-motion internal combustion engines use quick-burning fuels. These fuels are more expensive than slow-burning fuels used in some other engines.

Rotary Motion

Rotary motion is circular motion. In rotary-motion engines, exploding fuels form expanding gases that push against the blades of a turbine or rotor, causing it to turn. A turbine is a wheel with evenly spaced blades or fins attached to it like those on a fan. A rotor is the triangle-shaped part of a rotary engine that revolves in a specially shaped combustion chamber. See **Fig. 23-3**. The turbine or rotor is usually mounted on the same straight shaft as the device that is being powered.

Rotary-motion engines produce more horsepower than the same size reciprocating engines. This is partly because their working parts have less mass (weight). They also produce less fric-

» Fig. 23-3 These drawings show the operation of a rotary engine.

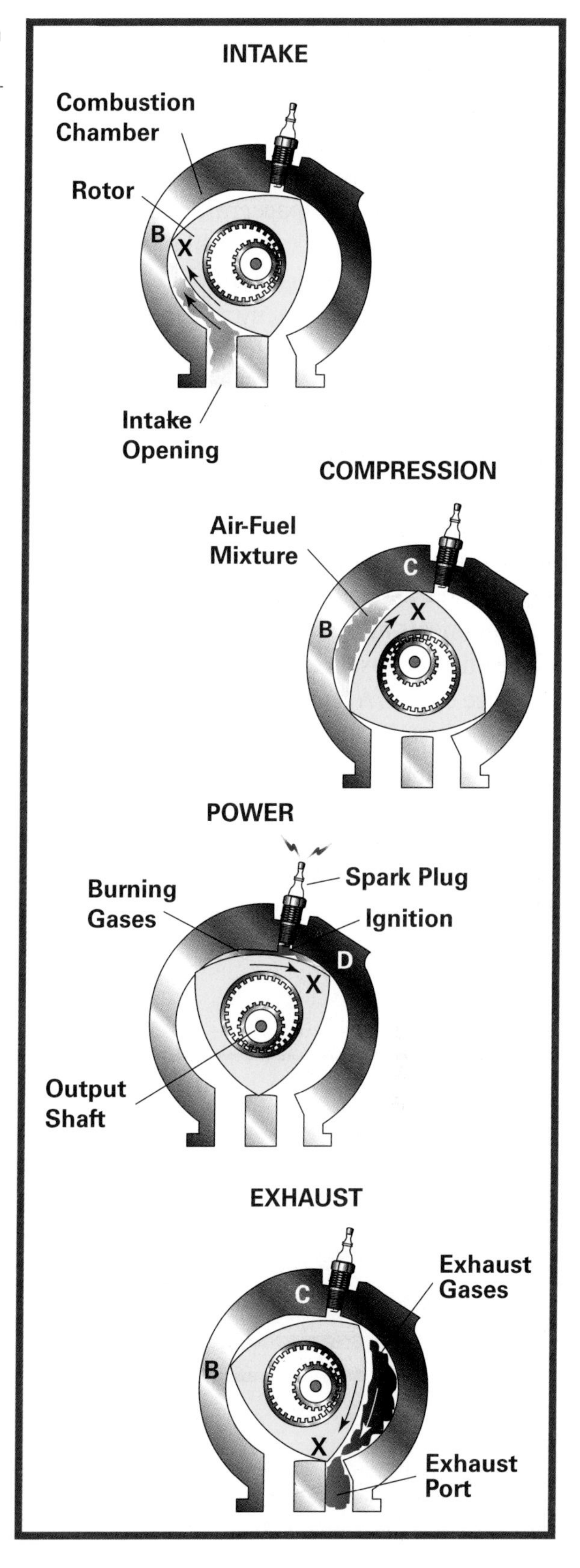

tion. Rotary-motion engines work best when the same RPM is held for extended periods of time.

Some locomotive engines use gas turbine engines. The turbines operate the electric generators that power the trains. Gas turbines used on ships turn the propeller shaft that moves a ship through the water.

Rotary motion engines have some advantages. They have fewer moving parts to maintain or cause friction than in a reciprocating engine. Horsepower can be increased by enlarging the rotor or turbine. No additional parts are required. A wide variety of fuels can be burned to produce the expanding gases. A constant RPM can be maintained for very long periods of time without damage to the engine parts. They also have some disadvantages. Quick changes in RPM cannot be made efficiently. The cost of manufacturing and rebuilding rotary engines is greater than for reciprocating engines. Because of the high RPM attained, the rotor or turbine must be kept well lubricated and in perfect balance.

Linear Motion

Linear motion is motion in a straight line. Linear-motion engines are generally referred to as jet or rocket engines. They are the most powerful types of engines.

In a jet engine, power is created by the exploding fuel and expanding gases. The gas is expelled out the back of the vehicle. The force of the gases escaping rearward pushes the vehicle forward. This resulting forward force is called thrust. The more gas that is produced, the faster the vehicle will go. Linear-motion internal combustion engines are used on large aircraft, rockets, and spacecraft. See **Fig. 23-4**.

Rocket engines work much like jet engines. Since nothing can burn without oxygen, all engines require oxygen mixed with fuel for combustion. The main difference between jet and rocket engines is that jet engines take in air from the atmosphere. Since rocket engines are usually designed for use in spacecraft outside Earth's atmosphere, a rocket engine must carry all of its own oxygen and fuel supplies. The fuel supply may be either solid fuel or liquids that are carried as two separate chemicals. When the chemicals are combined, combustion takes place and hot gases expand through a nozzle to produce thrust. A liquid fuel rocket engine may be turned on and off. Once a solid fuel rocket is started, combustion will continue until all the chemicals are used.

The linear-motion engine has several advantages. Very high horsepower or thrust can be developed. Several different types of solid or liquid fuels can be burned to make the expanding gases. There are no moving parts needed to transfer the power to the vehicle.

Linear-motion engines also have disadvantages. Manufacturing and rebuilding costs are high, and rocket engines designed for outer space can be used only once. No directional change of thrust or motion is possible. The force is generated in one direction only.

Internal Combustion Engines

Internal combustion engines are the most common engines used in transportation. They power most of the cars, trucks, buses, and motorcycles on the road today. In an **internal combustion engine**, fuel is burned within the engine itself.

Engine Systems

To understand how internal combustion engines function, let's examine a small, single-cylinder, air-cooled engine. Other types of internal combustion engines are more complex but operate in basically the same way.

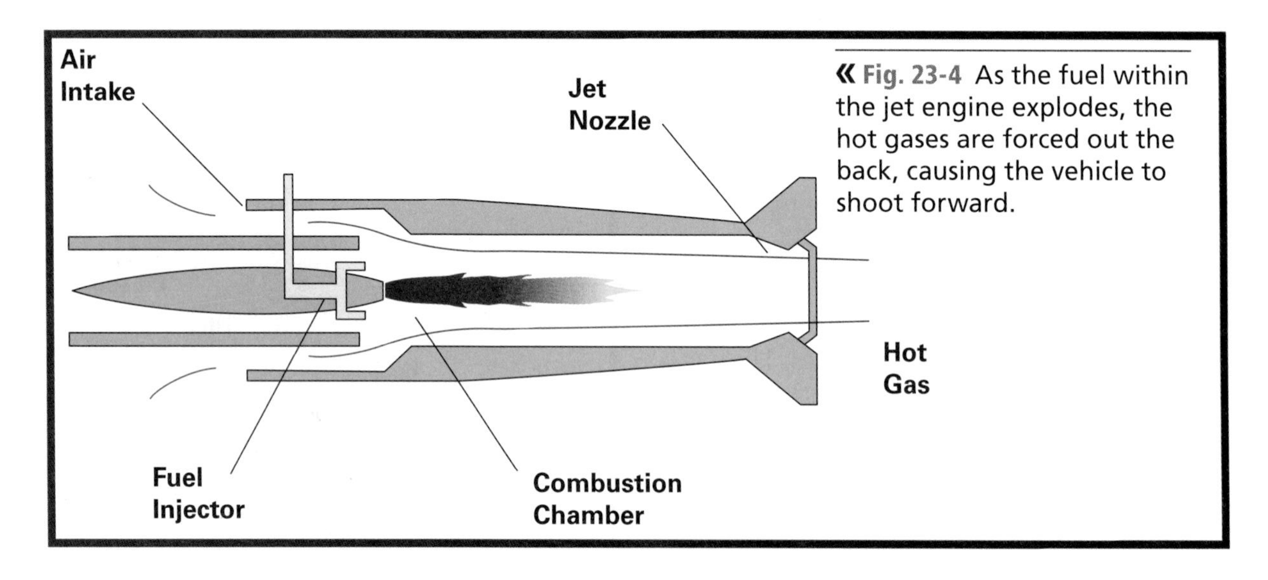

« **Fig. 23-4** As the fuel within the jet engine explodes, the hot gases are forced out the back, causing the vehicle to shoot forward.

Most engines have nine systems that include the following:

- Mechanical system
- Lubrication system
- Air induction system
- Fuel system
- Ignition system
- Charging system
- Starting system
- Cooling system
- Exhaust and emissions system

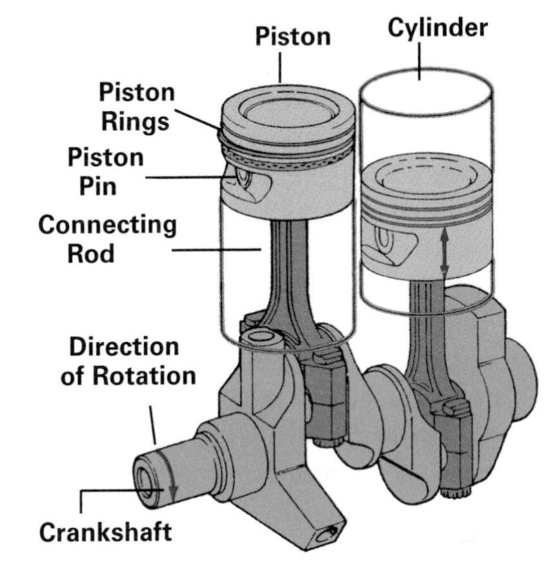

Fig. 23-5 This drawing shows the basic parts of an internal combustion engine.

The mechanical system contains the piston, connecting rod, and crankshaft. The reciprocating motion of the piston produces rotary motion in the crankshaft. See **Fig. 23-5**.

The function of the lubrication system is to decrease friction. Lubricating (oiling) parts makes them slippery. On some small engines, a small oil pump pushes oil through channels to the moving parts. Other engines rely on simple splash-and-dipper devices to oil the parts. Still other engines are lubricated by oil that is mixed into the fuel and supplied by the fuel system.

The air induction system directs air into the engine. An air filter prevents dirt from entering the engine.

The fuel system mixes the fuel with air and supplies it to the cylinder for combustion. See **Fig. 23-6**. A carburetor or fuel injectors mix the air and fuel accurately to ensure that

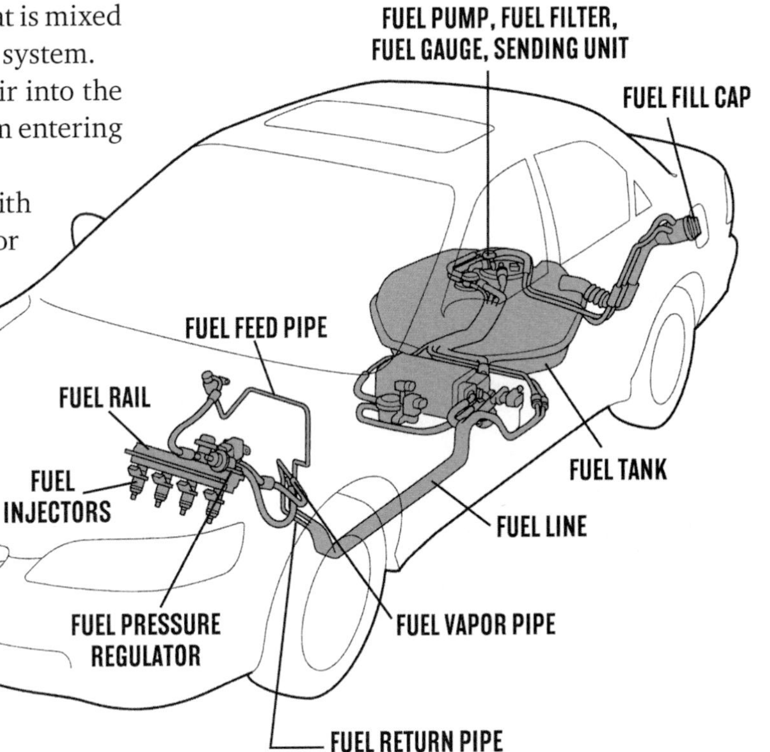

Fig. 23-6 The fuel system delivers fuel from the fuel tank to a metering system. There, fuel is mixed with air and delivered to the cylinders.

the engine operation is efficient. Intake valves open and close to control when the air-fuel mixture flows into the cylinder. Exhaust valves and pipes allow spent gases to escape from the engine, and a muffler quiets the engine noise.

The ignition system's job is to ignite the air-fuel mixture at the correct moment. Except for diesel engines, the ignition system includes spark plugs that supply a spark to begin combustion. A coil increases the spark plugs' spark. Breaker points and a condenser or other electronic device switches the electricity flow, and wires carry the electricity. A generator or magneto produces electricity from motion of the engine.

The charging system includes the battery, generator, and other components needed to supply the electrical current. If the mechanical, lubrication, fuel, and ignition systems are all working correctly, the engine can be started.

Fig. 23-7 With a rope starter, the operator turns the crankshaft by pulling the rope. This starts the engine, which then keeps the crankshaft turning.

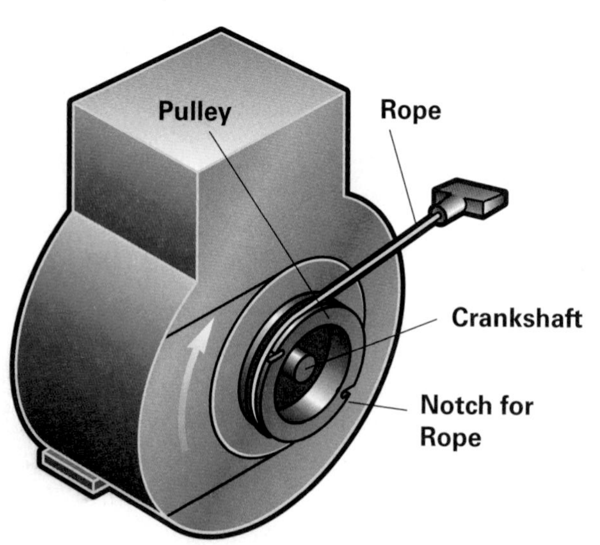

The starting system may be as simple as the one on a lawnmower. A rope on a pulley is connected to the flywheel (the part in a small engine that is set into motion by the action of pulling a starting rope). It may also be as complex as a motor that works from a storage battery. Most small engines have recoil (rewind) rope starters. See Fig. 23-7. The starting system must stop operating once the engine is running.

While the engine is running, heat from combustion and friction must be kept under control. This is the job of the cooling system. Small engines are usually air-cooled. An air-cooling system includes a fan (usually part of the flywheel) and metal shrouds (covers). The shrouds contain and direct air movement.

Larger engines are usually water-cooled. This type of system is more complex. It contains a water pump, radiator, hoses, and a water jacket. The water jacket is usually built into the crankcase and cylinder head.

The exhaust system vents engine emissions through the tailpipe. Emission pollutants are reduced by the emission control system before being vented.

Engine Operation

Internal combustion engines can be either two-stroke-cycle or four-stroke-cycle designs. Almost all automobiles and large motorcycles use four-stroke engines.

Four-Stroke-Cycle Engines are engines that complete four strokes of the piston for one combustion cycle. See Fig. 23-8.

- **Intake stroke.** The air-fuel mixture is drawn into the cylinder.
- **Compression stroke.** The mixture is compressed by moving the piston up in the cylinder.
- **Power stroke.** The spark plug fires, and the burned gases expand and force the piston down.

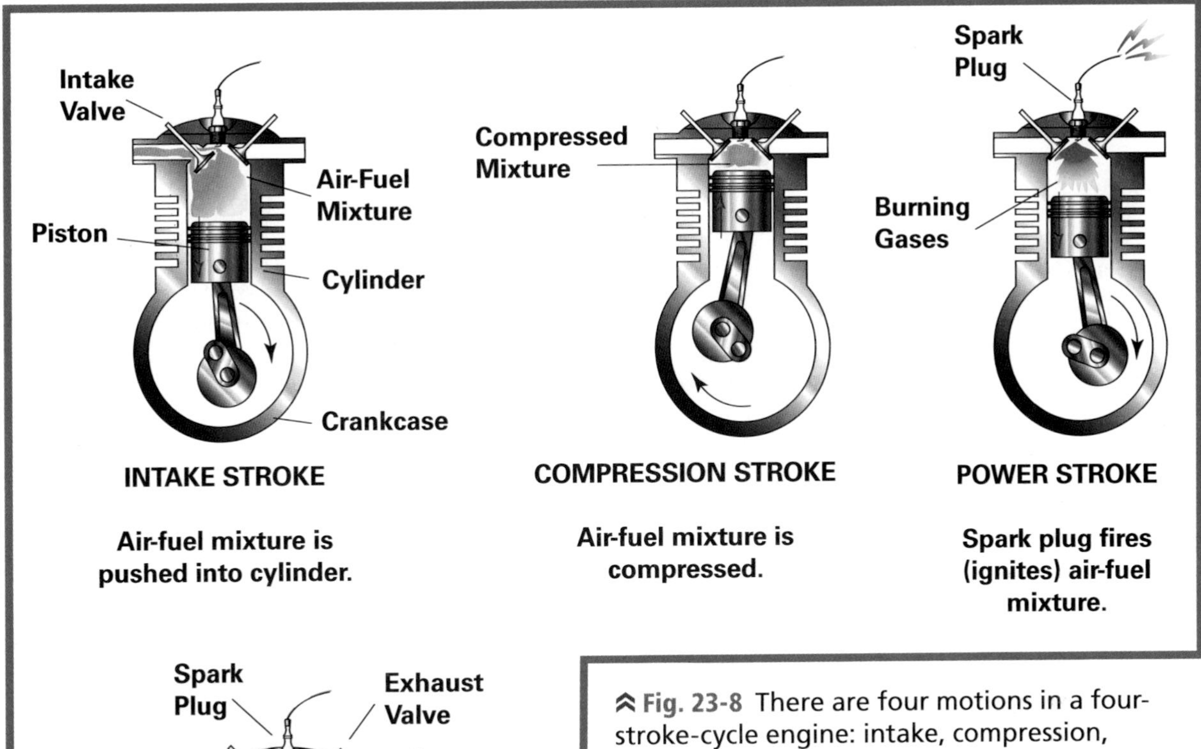

Fig. 23-8 There are four motions in a four-stroke-cycle engine: intake, compression, power, and exhaust.

- **Exhaust stroke.** The piston comes back up, pushing the spent gases out of the cylinder.

Together these four strokes make two revolutions. While the engine is operating, the cycles repeat over and over again.

Two-Stroke-Cycle Engines complete the four operations in only two piston strokes, or cycles. See **Fig. 23-9** on page 456. Small boats and snowmobiles use two-stroke-cycle engines. During the first stroke, the piston goes up. It allows the air-fuel mixture to enter the cylinder and compresses it at the same time. The spark plug fires when the piston is at the top. This pushes the piston down for the power stroke. When the piston is at the bottom of the stroke, the exhaust is let out of the cylinder.

This design means that two-stroke engines have fewer moving parts. The shorter cycle and fewer parts allow two-stroke engines to run at higher RPMs than four-stroke engines. A dis-

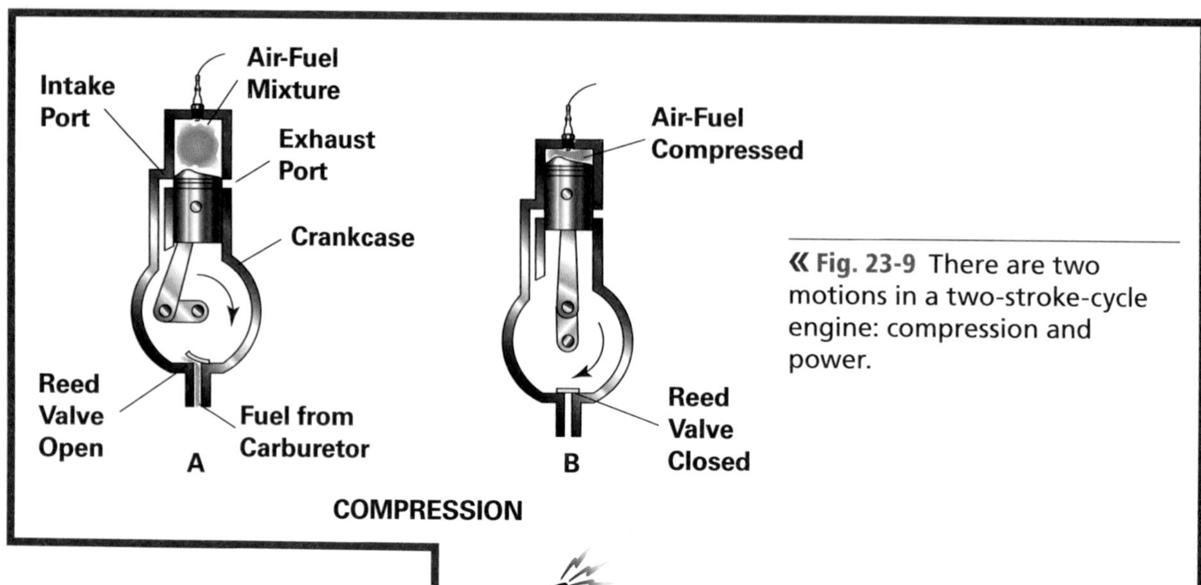

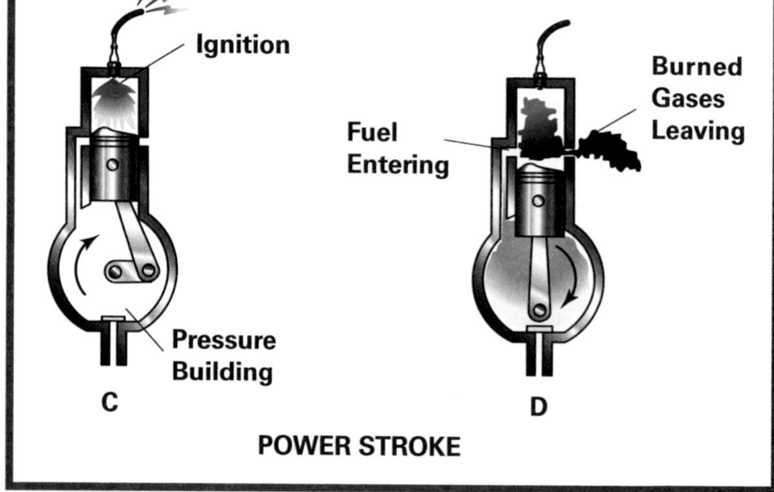

« **Fig. 23-9** There are two motions in a two-stroke-cycle engine: compression and power.

advantage is that a two-stroke engine usually produces more pollution than a four-stroke engine. The lubricating oil added to the fuel is burned, and the smoke is released with the fuel exhaust.

Diesel Engines A special type of two- or four-stroke-cycle engine, called a **diesel engine**, does not need spark plugs. Special low-grade diesel fuel is ignited by heated, compressed air. See **Fig. 23-10**. Because of the high compression necessary, diesel, or heat, engines are usually heavier than gasoline engines.

Today, most highway trucks, about half of all local delivery trucks, and about 30 percent of all personal light trucks are powered by diesel engines. Diesel engines also propel most commercial ships, while burning many different types and grades of fuel. As improvements in manufacturing, design, and electronics are found, a smaller diesel engine may eventually power more family automobiles.

Why are diesel engines in demand? There are several reasons. Diesel fuel requires less refining and costs less. Because it does not explode easily, it needs no ignition system. It burns completely, which means more miles to the gallon, and advances in diesel engineering have decreased diesel exhaust emissions.

Higher manufacturing costs and a need for larger lubricating and filtering systems does make a diesel engine cost more to produce and install. However, a well-maintained diesel engine will last for over 300,000 miles before needing an overhaul. With regular overhauls, many diesel truck engines have logged over 3 million miles.

Math Application

Calculating RPM Revolutions per minute are used to measure many things. For example, a clock's second hand travels at 1 RPM. The number of times a CD turns to go through a track can be measured in RPM.

When people speak of engine RPM, they are referring to how fast the engine is spinning or "turning over." Each revolution is equal to a piston traveling through one full stroke up and down the cylinder. Since most engines are of the four-stroke design, that means that 2 RPM would represent one complete combustion cycle (intake, compression, power, exhaust).

When you are driving, pressing the accelerator causes an increase in RPM. The greater the increase in acceleration, the more cycles your engine will complete. Many engines redline at about 7,000 RPM. Redline refers to the maximum recommended amount of revolutions per minute.

Engines are not the only example of revolutions per minute in transportation. A plane's propeller is also measured in RPM. The number of times that any particular point of the propeller completes a rotation in 60 seconds is a measure of how fast it is going.

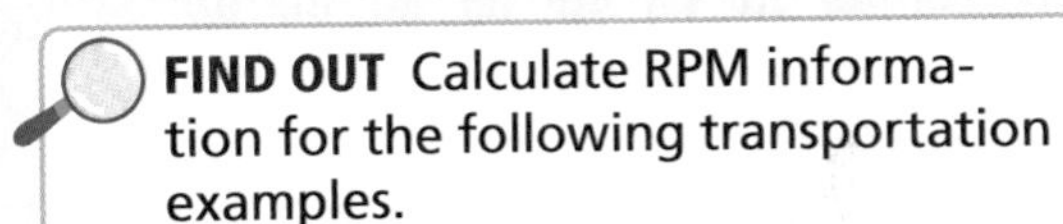

FIND OUT Calculate RPM information for the following transportation examples.

1. At 7,000 RPM, how many complete engine combustion cycles will take place each minute?
2. If a propeller is running for 2 hours and is spinning at 2,000 RPM, how many revolutions did it spin all together?

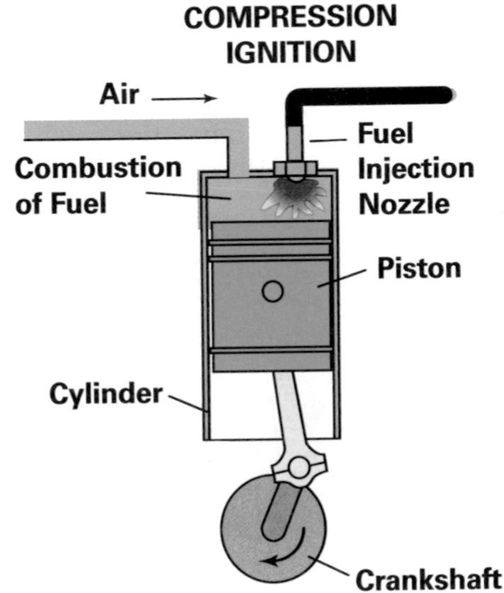

« Fig. 23-10 In a diesel engine, the compression stroke forces the air-fuel mixture together. Heat caused by compression ignites the mixture.

External Combustion Engines

In **external combustion engines**, the fuel is burned outside the engine. They provide reciprocating and rotary motion, but not linear motion.

Steam-Driven Engines

Almost all of the external combustion engines currently used in transportation are steam-driven. Water is heated inside a closed container called a boiler. The water becomes steam under high pressure. This steam is used to drive pistons or turn turbine blades. See **Fig. 23-11**. The size of the steam boiler must be appropriate for the size of the vehicle. Using a heavy boiler in a compact automobile, for example, would not be efficient.

Steam engines used in vehicles operate as a closed system. After steam has provided power, it is allowed to cool. The steam condenses into water. The water is reheated and again becomes steam. This process keeps repeating. There is little noise or water contamination.

The heat needed to boil the water can come from many sources. Early vehicles burned coal and wood. Most modern vehicles burn petroleum products. Some modern military vessels use nuclear fuel. Because very little nuclear fuel is needed, these vessels do not need to refuel often.

External combustion engines that are steam-operated have several advantages. Almost any type of fuel can be used to produce the steam. The boiler and the engine can be placed in different physical arrangements, and many different engine designs are available.

There are also several disadvantages. Boilers require a great deal of space, and their use is limited to large vehicles. Even when handled carefully, steam is dangerous. The steam and water cause metal to break down. Continuous care and preventive maintenance are necessary.

Fig. 23-11 Steam from cylinder A pushes against the piston, causing it to move to the right. Then a valve cuts off the steam to cylinder A. This forces the steam into cylinder B, moving the piston to the left. This back-and-forth motion turns the driven wheel.

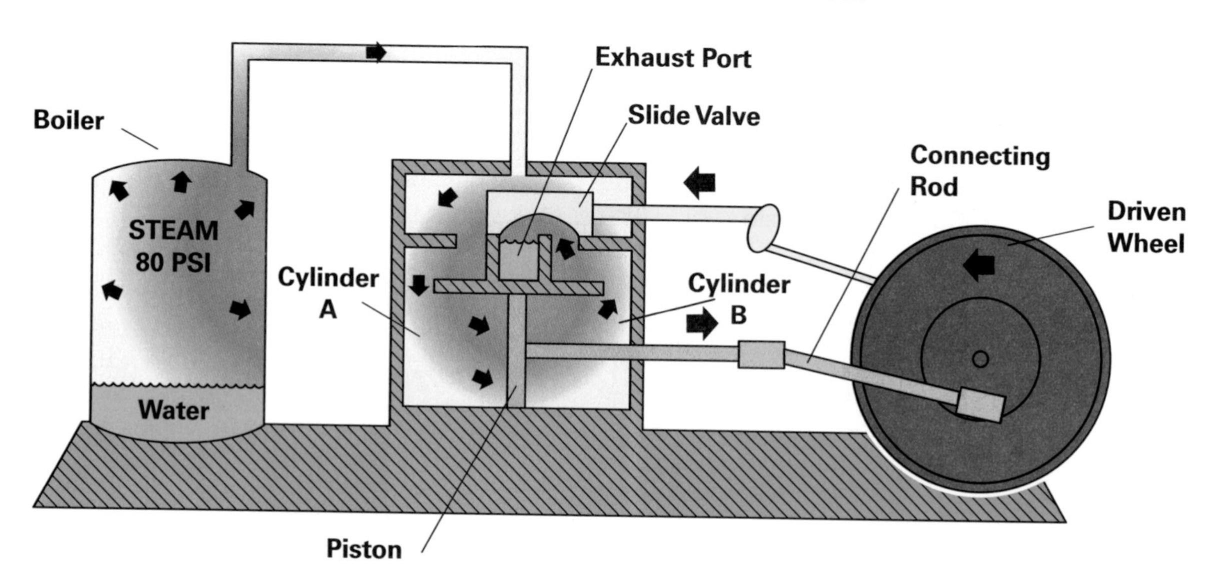

Modern Uses for Steam Power

Along with naval vessels, some modern locomotives use steam to power a rotary engine. The steam turns a turbine rotor. The rotor is connected to an electrical generator. The generator produces the electricity needed to power electric motors. These electric motors then turn the axles and wheels of the locomotive.

Reciprocal and rotary engines are both used to power ships. Ships have room for large boilers. Some still use coal as fuel, but most use oil. The steam produced turns a turbine, which turns the propeller.

Motors

Electric- and fluid-powered motors can be used to move vehicles. Most on-site transportation vehicles, heavy construction vehicles, and subway trains are powered by motors.

Electric Motors

For many years, electric motors have powered urban mass transportation commuter vehicles. An electric streetcar line runs daily on tracks in New Orleans. See Fig. 23-12. Subway trains use a "hot" third rail on the ground.

Fig. 23-12 The oldest operating streetcar is in New Orleans, Louisiana. Power travels through the overhead cable.

Electrically powered passenger and freight trains are relatively common in the northeastern section of the United States and all across Europe. In addition, several large cities, such as Washington, D.C., have electrically powered mass transportation systems.

At the present time, electric cars depend on batteries for power. Batteries currently available are heavy, need frequent recharging, and lose power in cold conditions. New types of batteries are being developed to overcome these drawbacks. Small cars that operate efficiently on batteries for short distances are available now.

There are many advantages to electrically powered vehicles. Although pollution is caused by electrical generating plants, the vehicles do not pollute the environment. Operating costs are comparatively low. The electric motors require less maintenance and repair work than other power systems because they have fewer moving parts. They are not noisy. Electric motors will generally run in either direction. No transmission gears are needed.

Vehicles that operate solely on electricity have one big disadvantage. Supplying electricity to each vehicle is often difficult and inconvenient. This problem is especially serious for vehicles that are operated independently, such as cars and trucks. A combination of batteries with onboard battery chargers is available on a few small cars.

Fluid-Powered Motors

Fluid-powered motors can be hydraulic or pneumatic. They can generate considerable horsepower but not much speed. They are used in heavy equipment and other slow-moving vehicles that require a lot of power.

Fluid-powered motors are controlled by increasing or decreasing the pressure of the fluid by using a pump. Although almost any rotating power source can be used, most fluid-powered vehicles use internal combustion engines to power the pumps. Most fluid-powered motors will also run in reverse.

Pneumatic pressure is used to help move gases, liquids, or slurry through pipelines. The pressure pushes the pipeline contents along. Engines and motors produce the pressure needed.

Motor and Diesel Engine Combinations

Most modern locomotives use diesel-electric power. These locomotives work just like rotary-engine-powered models except that a diesel engine is used to turn an electric generator. The generator produces the electricity needed to power the electric motors that turn the wheels. Diesel-electric engines are very efficient and long lasting. They require much less space than steam engines, so the locomotives can be smaller. Diesel-electric engines also accelerate faster than steam engines. See **Fig. 23-13**.

Diesel-hydraulic locomotives use diesel engines to drive a torque converter instead of a generator. The torque converter, which includes a pump and a turbine, uses fluids

Fig. 23-13 Diesel-electric engines power most locomotives today.

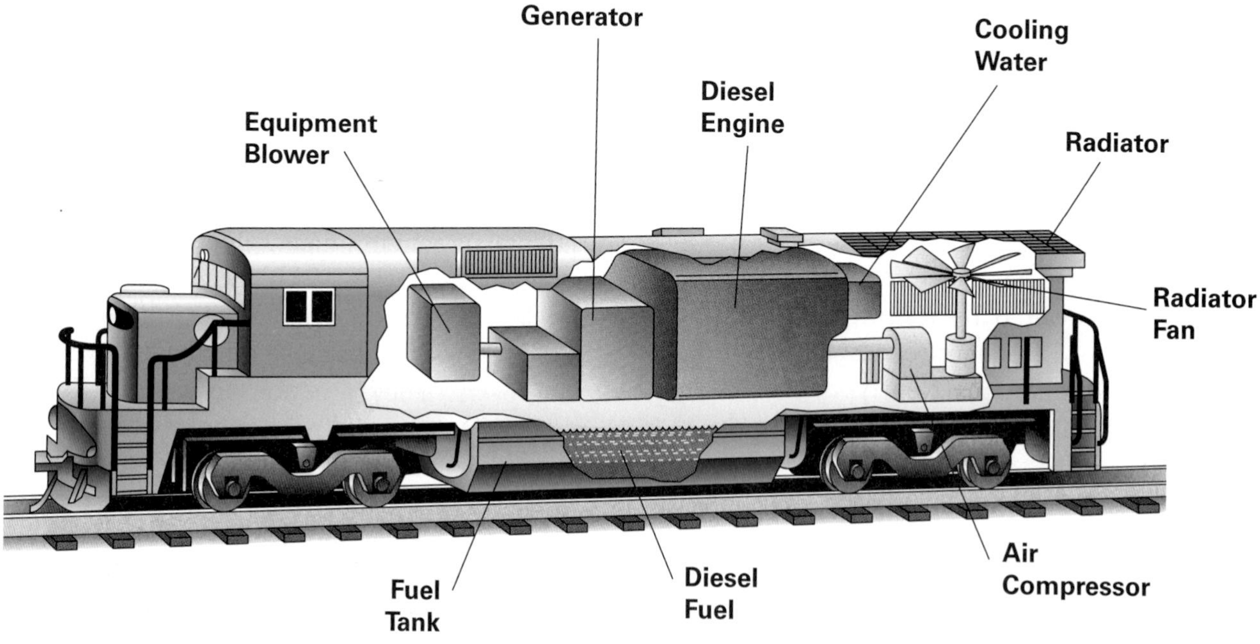

Modes of Transportation	Types of Power Commonly Used
Trucks, buses	Gasoline engine Diesel engine
Cars	Gasoline engine Diesel engine Electric motor
Propeller-driven planes	Reciprocating, internal combustion engine Turbine engine
Jet planes	Linear-motion internal combustion engine Turbine engine
Spacecraft, missiles	Linear-motion internal combustion engine
Locomotives	Steam engine Electric motor Turbine engine Diesel-electric combination engine Diesel-hydraulic combination engine
Ships	Steam engine Turbine engine
Pipelines	Electric motor

Fig. 23-14 More than one type of power may be used for certain types of vehicles.

under hydraulic pressure to transmit and regulate power received from the diesel engine. The pump forces oil against the blades of the turbine. This action causes the turbine to rotate and to drive a system of gears and shafts that moves the wheels. See **Fig. 23-14** for a table showing different modes of transport and the types of power commonly used.

Alternative Power

One of the biggest problems facing transportation is finding inexpensive power sources that are in plentiful supply and that do not harm the environment.

Wind Power

Wind-powered sailing ships traveled the seas long before engines were developed. Today, modern technology is finding new ways to use wind to power oceangoing ships. Turbosails, like those shown in **Fig. 23-15** on page 462, produce almost four times as much thrust as the best traditional sails.

Wind energy is also being turned into electrical energy. Some small sailing yachts have windmills and electrical generators installed on their masts. Even when winds are light, the electricity produced by these windmill units is enough to charge the batteries of the boat. The same designs, on a larger scale, are used to provide additional electricity on larger ships. Power is limited only by battery capacity.

Fig. 23-15 The *Alcyone* vessel is powered by computer-controlled turbosails.

Solar Power

Photovoltaic cells change solar energy into electricity. These have already been used to power experimental vehicles. Some vehicles use the power from the cells to directly power an engine. More often, the electricity is used to charge high-capacity batteries. This allows the vehicle to be used even when the sun is not shining. The charged batteries also produce a steady flow of electricity to the motor. This makes it much easier to maintain constant speeds and gives the vehicle a longer range. See **Fig. 23-16**.

Solar-produced electricity is limited by battery capacities. Most batteries that are powerful enough to store enough electricity to power vehicles are either very heavy or very expensive. Several types of new battery designs that may soon solve this problem are now under development.

Alternative Fuels

The Energy Policy Act of 1992 defined alternative fuel as being a fuel that is substantially not petroleum and that yields substantial environmental benefits and energy security. Some of the current alternative fuels are natural gas, ethanol, methanol, biodiesel, and hydrogen.

Natural Gas The main ingredient in natural gas is methane. Methane is given off when plant or animal waste decays, so methane gas can have many sources, including garbage dumps. Experiments have shown that methane may be used to power many vehicles. Some large garbage dumps already collect methane for fuel purposes.

Like other fuels, methane can be mixed with other gases, decreasing their use and extending the supply of natural fossil-based gas.

Ethanol and Methanol are alcohols that can be blended with gasoline. When the mixture is more than 70 percent ethanol or methanol, it qualifies as an alternative fuel. Some flex-fuel vehicles use E85 (85 percent ethanol), M85 (85 percent methanol), or fuel with even higher alcohol percentages.

Biodiesel Fuel is produced by mixing oil derived from plants (biomass) with fossil diesel fuel. If properly mixed, the advantages seem to far outweigh the disadvantages.

Several metropolitan bus companies report maintenance savings of 50 to 70 percent on just a 10 percent bio-oil mixture. Diesel engines run cleaner and produce fewer exhaust emissions. Bio-oil, generally from soybeans, also adds slickness to the moving parts of the engine, which decreases friction.

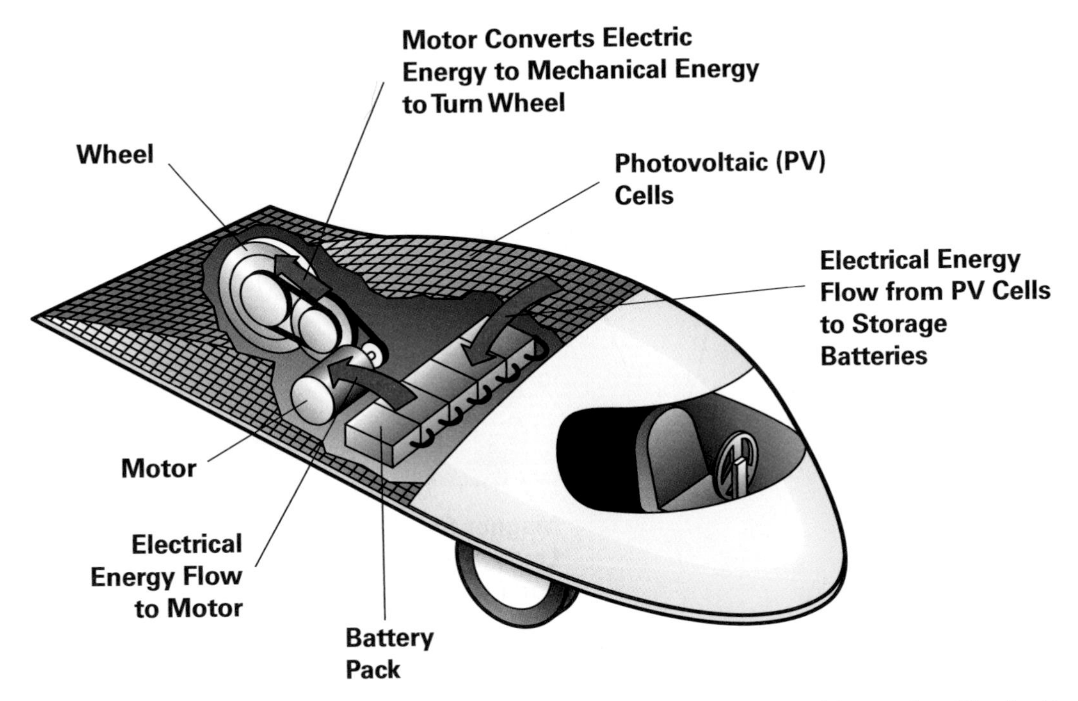

Fig. 23-16 This solar car uses photovoltaic cells to charge special batteries. The batteries then provide electrical power to the motor.

The reported disadvantages include a lack of availability in some regions, an increase in cost per gallon of fuel, and fuel gelling. Gelling (thickening) occurs when the air temperature drops to around freezing. Gelling is reduced when less biodiesel fuel is used.

Hydrogen can be used either in fuel cells (discussed later in this chapter) or in a hydrogen combustion engine. The hydrogen combustion engine works much the same as an internal combustion engine, except that liquid hydrogen is used instead of gasoline or diesel fuel. However, the cost to make such an engine is high.

Recycled Fuels

Some fuel substances can be recycled. Steam-powered engines on boats and trains run well on waste motor oil. Diesel-quality fuel can be refined from used cooking oil. Other fuels may be obtained from old tires or plastic bottles.

Maglev Propulsion

Maglev systems are rail systems that operate on the scientific principle that like poles of a magnet repel each other. Maglev is short for magnetically levitated. (Levitate means to rise or float in the air.)

Magnets in the maglev guideway (rail) repel magnets of like polarity on the bottom of the maglev vehicle. See **Fig. 23-17** on page 464. This causes the train to levitate above the guideway, creating a nearly frictionless riding surface. Magnets are also involved in the vehicle's propulsion. Changing the polarity of the magnets on the train and the guideway at the proper moments speeds up or slows down the train.

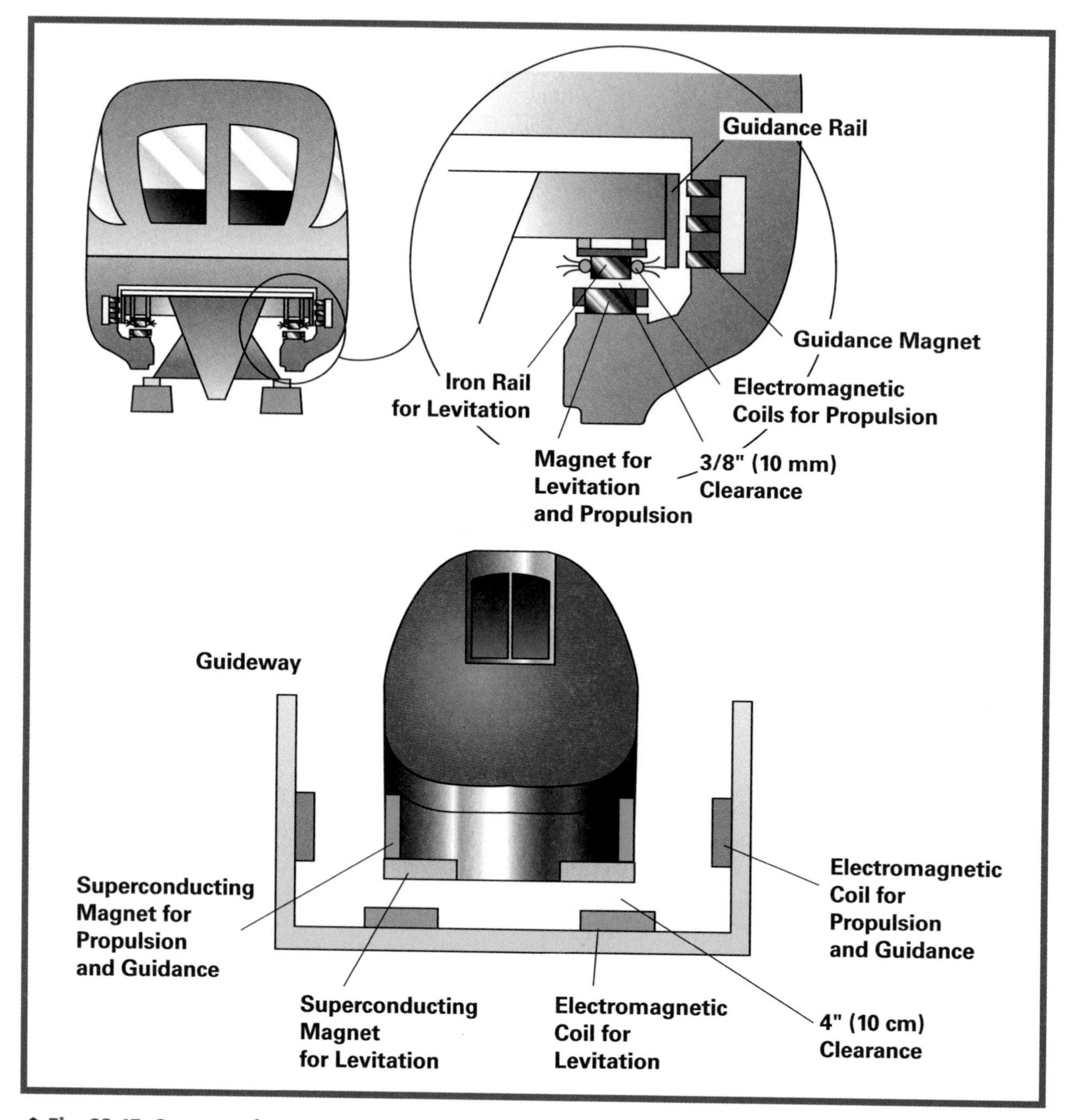

Fig. 23-17 One set of magnets levitates the train. Another set propels it along the track.

Because there is so little friction and the vehicles are designed to reduce air resistance, the trains can easily reach speeds of 300 mph. They glide quietly, smoothly, and swiftly along the guideway, using relatively little energy.

The first commercial maglev is already in use in Shanghai, China. In the United States, maglev trains are being considered for passenger routes between Baltimore and Washington, D.C. However, the expense of building the guideway has made development slow.

Science Application

How Hybrid Vehicles Work The popularity of hybrid power for passenger cars is growing in the United States as gasoline prices stay high and environmental concerns increase. The most common hybrid car is one that operates using gasoline and electricity. How does it work?

In gasoline-electric hybrids, an internal combustion engine uses gasoline fuel for power, and a set of batteries provides electricity to an electric motor. The vehicle is essentially an electric car. An electric motor provides all the power to the wheels, and batteries supply the motor with electricity.

A completely separate gasoline engine powers a generator.

However, all these components add extra weight to the vehicle. The car must carry the weight of the electric motor, the generator, the batteries, and the engine. A full-size electric motor plus a generator can weigh several hundred pounds. Car manufacturers have been working on this problem. Some newer hybrid cars have the gasoline engine provide most of the power, with the electric motor adding power as needed for acceleration. The electric motor can double as a generator when not in use. Without having an actual generator, the load of the vehicle is lightened.

FIND OUT Many other vehicles are examples of hybrid power. One such example is a moped (motorized pedal bike), which uses both a gasoline engine and the rider's leg muscles. Research to find other examples of hybrid power and report your findings to the class.

Fuel Cells

Fuel cells are batteries that generate power by means of the interaction between two chemicals, such as hydrogen and oxygen. Fuel cells are nonpolluting. However, fuel storage causes problems.

Fuel cells have been used by NASA in space, and some city buses are operated with them. Daimler-Chrysler has produced the Dodge Fuel Cell Sprinter, which is being tested for commercial use with the United Parcel Service (UPS). General Motors, Toyota, Honda, Nissan, and other automakers are producing their own models.

Hybrid Power

A hybrid vehicle is any vehicle that combines two or more sources of power. For example, some hybrid vehicles are powered by both gasoline and electricity. At low speeds, the car relies on the electric motor, which gets its power from batteries. At higher speeds, the gasoline engine takes over, recharging the batteries at the same time. The result is increased fuel efficiency. Most automakers are now manufacturing hybrid vehicles.

Power from Vehicle Design

Design can make vehicles more fuel efficient, more economical, and safer. Today, most design considerations involve aerodynamics, size and weight, materials, and electronics.

Aerodynamics

Aerodynamics is a science that deals with the interaction of air and moving objects. Air molecules tend to resist or slow movement. Vehicles meet the resistance when moving through the air. The faster they travel, the greater the resistance. A strong, power-robbing force is created called aerodynamic drag. Design engineers look for ways to reduce this drag to ease the workload of the engine and improve fuel efficiency.

Shape is the critical factor. Engineers describe an aerodynamically efficient shape as "slippery." They try to design vehicles so that air will glide smoothly over, under, and around them. Even a mirror or door handle that sticks out can increase drag. See **Fig. 23-18**. The underside of the car should be smooth, reducing turbulence. Body surfaces can be grooved or dimpled to speed airflow. Outside mirrors can be replaced by TV camera "eyes."

Aircraft must also be aerodynamic. The basic wing (cross section) shape is not likely to be changed. However, length, positioning, angle, and other factors may be changed.

Size and Materials

American cars weigh about 1.5 tons. This extra weight requires a large engine to push it along. Lighter, smaller vehicles are generally more fuel-efficient than large, heavy vehicles.

Design engineers are also looking for ways to reduce weight without reducing size. Lightweight materials such as aluminum or plastic are used instead of heavier materials such as steel.

Plastics Many new, smaller aircraft are made primarily of special fiber-reinforced plastics. They are light but strong. Parts of larger planes are also made of these materials. See **Fig. 23-19**.

Plastics have long been used in car and truck interiors. It is expected that more and more plastics will also be used to form exterior parts. Nearly all new vehicles have some plastic body panels. Advanced car designs are using ultralight technology, which includes components made from carbon fibers, composites, and advanced engineering plastics.

Fig. 23-18 This prototype car has an aerodynamic design. ***Besides a traditional car, what other vehicles may have influenced the designer when creating this prototype?***

Ceramics are made from nonmetallic minerals that have been fired at high temperatures. Ceramics are used in spacecraft because they can withstand extreme temperatures and because they insulate well. The tiles on the space shuttle are ceramic. Some car engines include ceramic parts.

Because the diesel engines used in trucks produce great amounts of heat, the front of a truck must be built large and flat like a wall to allow for cooling airflow. However, this shape increases aerodynamic drag and reduces fuel efficiency. In order to give trucks a more aerodynamic shape, the engine must be redesigned. If ceramics could replace metal parts or protect them from the heat, air-cooling would not be necessary. The design of trucks could be changed.

Before ceramics can be widely used, two major problems must be overcome. Ceramic materials are so hard that special diamond-coated tools must be used to cut them. They are also brittle. A part with even a small blemish may break under stress.

Composites A material that is made by combining two or more materials is a composite. Each component retains its own properties, but the new material has more desirable qualities.

Composites are being used more and more today to make parts for different types of transportation. For example, fiberglass reinforced with plastic is now being used for boat hulls and automobile bodies. Kevlar™, which is a composite that is very difficult to cut, is being used to make reinforcing belts for tires. Carbon/graphite composites are now being used to make lightweight bicycle frames and to produce several new types of aircraft.

Electronics Computers, microprocessors, and other electronic devices are revolutionizing vehicle operations. Electronic controls are used to precisely regulate the air-fuel mixtures in the engine. This reduces the amount of pollutants in vehicle exhaust. It also reduces fuel waste and improves engine performance. Other innovative processes and techniques are used to design improved brakes, power steering, transmission, and heating and air conditioning.

Fig. 23-19 This ultralight prototype plane is made from fiberglass and plastics. It is economical and easily maneuverable.

AMTRAK has introduced a new high-speed train that can take curves at 112 mph, reducing travel times. Microprocessors in each car feed information from sensors to a hydraulic system that tilts the cars. The tilting keeps passengers from falling out of their seats when the train rounds a curve.

Fiber Optics The optical fibers used in vehicles carry signals used to control such items as power windows and door locks. Signals from several controls can be combined, and a pair of optical fibers can replace many wires. This saves space and weight. It also greatly simplifies the wiring system.

Eventually, fiber optics might be used to link electronic operations in engines. However, the fibers currently available cannot take the heat produced by today's engines. As engines are redesigned, their use may increase.

CHAPTER

23 REVIEW

Main Ideas

- Prime movers supply the power in transportation. A prime mover is a basic engine or motor.
- Three kinds of motion are produced by engines and motors.
- Alternative power refers to power sources that are available to use but do not harm the environment.
- Design can make vehicles more efficient, more economical, and safer.

Understanding Concepts

1. Name the three types of motion and give examples of each.
2. What is the main difference between internal and external combustion engines?
3. Fluid-powered motors are used by what type of vehicles?
4. What device can change the energy from sunlight into electricity?
5. How is aerodynamic design used to improve cars?

Thinking Critically

1. How do you think governmental regulations have influenced transportation?
2. If you were going to design a perfect transportation vehicle, what type of engine would you put in it? Why?
3. What sports involve aerodynamics?
4. Should big, electric-powered trucks be designed? Why or why not?
5. Which alternative power source do you think is the most promising? Explain your answer.

Applying Concepts & Solving Problems

1. **Design** Redesign a bike to make it more practical for people traveling to and from work. Diagram all forms of energy conversion taking place.
2. **Transportation Activity** Write a song or jingle to help you remember the stages of a four-stroke-cycle engine (input, compression, power, exhaust).
3. **Use of Technology** The Segway human transport was designed as an alternative to cars in downtown areas. Do you think its design was successful? How might the Segway be more functional in a different way?

Activity: Design Transport for the Future

Design Brief

Increased air and land traffic congestion, poor air quality, road and bridge maintenance needs, and a shortage of space for parking are a few of the transportation-related problems facing communities today. In the future, public and private transportation will have to be different. Solutions such as new vehicle pathways, smart highways, alternative-fuel vehicles, and special rail systems are all being studied. Low-cost, non-polluting alternatives for commuting short distances must be found.

Research and develop an alternative transportation system, including its power needs, for use within your community.

Criteria

- Create a model of your system, showing topography, landmarks, and new system features. Streets and other features must be labeled.
- Your model may be built by hand using conventional materials or with 3D CAD or animation software.
- Write a report describing your system, the power sources used, and the transportation problems it will help solve. This report can be part of your system, such as by means of signs, animated pop-ups, or audio bits.
- Your system must include vehicles, vehicle pathways, guidance systems, and control systems. Guidance systems can include the Global Positioning System. Controls can include such things as signs, signals, and remote sensing.

Constraints

- At least three alternative forms of transportation must be included, such as high-speed trains. All forms must be nonpolluting and use energy efficiently.
- The system must allow a user to travel anywhere within the community using at least one of the alternative forms.
- A conventional model must not exceed 2 square feet in size.

Engineering Design Process

1. **Define the problem.** Write a statement that describes the problem you are going to solve. For example, it might reduce traffic and pollution problems in the downtown area.
2. **Brainstorm, research, and generate ideas.** With your team, discuss possible solutions. Hint: You may want to study similar systems that have been developed in other communities. If possible, interview a member of the local planning board or a civil engineer. Visit your library for maps of your community.
3. **Identify criteria and specify constraints.** These are listed on page 470.
4. **Develop and propose designs and choose among alternative solutions.** Choose the best design that will solve your problem. For example, if a river runs through your community, you might make use of water transport.
5. **Implement the proposed solution.** Decide on the process you will use to make your model. Gather any needed tools or materials. Then make a preliminary design.
6. **Make a model or prototype.** Create a rough version of your preliminary design.
7. **Evaluate the solution and its consequences.** Compare the rough model to your original plan. Does it meet all the criteria and constraints? Will your solution solve the problem? What impacts will result from this solution?
8. **Refine the design.** Based on your evaluation, change the design if needed.
9. **Create the final design.** Once you have an approved design, create your final model.
10. **Communicate the processes and results.** Write your report. Present your finished design to the class. Be prepared to answer questions about possible impacts. Turn in your assignment to your teacher. Be sure to include the name of the activity, your definition of the problem, a description of how you solved the problem, and your report.

SAFETY FIRST

Before You Begin. Make sure you understand how to use the tools and materials safely. Have your teacher demonstrate their proper use. Follow all safety rules.

CHAPTER 24 TRANSPORTATION SAFETY & SECURITY

≈ Being aware of your surroundings is essential to being a safe driver. In this chapter's activity, "Test Your Powers of Observation," you will have the chance to observe an area where transportation occurs. You will then report on your findings.

Objectives

- Identify the five main areas of influence in transportation safety and security.
- Explain how government regulation affects the use of different modes of transportation.
- Describe individual responsibility in using various modes of transportation.

Vocabulary

- electronic stability control (ESC)
- maritime
- metal fatigue
- delamination

Reading Focus

1. As you read Chapter 24, create an outline in your notebook using the colored headings.
2. Write a question under each heading that you can use to guide your reading.
3. Answer the question under each heading as you read the chapter. Record your answers.
4. Ask your teacher to help with answers you could not find in this chapter.

Influences on Safety and Security

Many factors help determine how safe and secure we are when we use transportation. These factors can be grouped into five main areas of influence:

- **Laws and regulations.** As long as people obey the rules that are designed to control how a transportation system works, that system is always much safer.
- **The design and functioning of the vehicle and any path on which it travels.** Vehicle design depends upon the designers and engineers who helped build the cars, trucks, trains, boats, and airplanes we use every day. For example, did you ride to school in a car? What is that car's safety record? Designers and engineers determined how well it holds up in a crash. Other workers design, build, and maintain the roadways on which cars ride.
- **The behavior of the person operating the vehicle.** Is the engineer of a train alert to signals along the tracks? Does the driver of a car obey the rules of the road, keep the tires inflated properly, and use a seatbelt? All these behaviors influence the risk of accidents. See **Fig. 24-1**.
- **The behavior of people with whom the vehicle may come in contact during its travels.** These influences are less predictable. A drunk driver may run a stop sign

« **Fig. 24-1** This small yacht is being lifted out of the water by a crane. The vessel had sunk after a crash with a freight ship.

and hit your car, or an airplane you're riding in may be hijacked, for example. On those occasions, passengers must usually rely on the proper authorities to take steps to minimize everyone's risk.

- **Environmental conditions.** Storms, earthquakes, and other natural events can create unsafe conditions for any mode of travel.

Government Regulation

In the United States, transportation is highly regulated. Some rules and regulations are mandated by local and state governments. However, the greatest number are developed and enforced by the federal government. For some people, regulations cause frustration, but the benefits outweigh the inconveniences. Since September 11, 2001, more actions have been taken to provide safe and secure transportation than during any previous period in U.S. history.

The National Transportation Safety Board (NTSB) is an independent governmental organization responsible for investigating any accidents that occurred in any of the different modes of transportation. The Transportation Safety Administration (TSA) was created in response to the terrorist attacks in New York and Washington, D.C. Now under the Department of Homeland Security (DHS), the TSA is the largest organization for frontline safety and security screening in the world.

In order for laws and regulations to work, however, people must respect them and travelers must cooperate with government efforts. This means that every person is at least partly responsible for his or her own safety, as well as the safety of others.

Highway Safety and Security

Highway safety and security usually involve the movement of cars and trucks. Both people and cargo are affected.

Driver Responsibility

Many people don't take highway safety seriously. They seem to believe that an accident will never happen to them. However, the number of highway accidents is increasing. During some years, this rate increase has approached four percent. This increase is partially due to drinking and driving. More than 40 percent of all highway deaths are linked to alcohol consumption.

Many other accidents are the result of driver carelessness. For example, some drivers pay more attention to other passengers or to a cell phone conversation than to the cars around them.

Car and Truck Design

Each year, the cost of new cars and trucks increases. Part of the increased cost is a result of from redesigning and re-engineering vehicles to protect drivers and passengers. See **Fig. 24-2**. Unsafe designs and materials are "engineered out."

Examples of safety improvements include the following:

- More effective seatbelts
- Automatic airbag systems
- Automatic balanced brake systems
- Stronger fasteners used to attach seats to the frame
- Stronger vehicle frames and bodies
- Improved safety glass
- Automatic headlights

Since 1998, all new cars have been required to have airbags installed. According to studies, airbags reduce the chance of dying in a colli-

Fig. 24-2 Are you a NASCAR fan? Many of the safety features engineered into a NASCAR race car are based on those found in ordinary cars.

sion by 30 percent. Some researchers predict that before long cars will have seven or eight airbags instead of only two. Airbags are effective because, as you ride down a highway, your body moves forward along with the vehicle. If your vehicle hits something, the vehicle stops but the energy of your own momentum throws you forward. Cars and trucks are designed to be crushable so that they absorb the energy of impact. Your airbag then has time to slow you down and keep you from flying through the windshield.

The airbag itself is made of nylon and folded into a small compartment. When there is a collision, a sensor reads the impact and triggers a hot blast of nitrogen gas that inflates the bag, holding your body in place. The bag bursts out of its compartment at speeds up to 200 mph. A second later, tiny holes in the bag allow the gas to escape and the bag deflates so you can move again.

Electronic stability control (ESC) is a new technology designed to prevent cars from rolling over when a driver must swerve at high speeds to avoid an object. Computers within the vehicle will detect slipping of the wheels and automatically compensate. While some vehicles have ESC now, government regulations require all cars to have this feature in the future.

Other design details also protect you. Your seat, for example, curves around your rib cage, which provides support. The seatbelt stretches slightly, which helps absorb some of the forward momentum. The windshield is made from safety glass, which is laminated with a sheet of plastic. Safety glass does not shatter into sharp pieces but cracks and forms fragments instead.

Highway Design and Maintenance

The interstate highway system has made many recent changes. The distance from the pavement to the parallel tree or fence line has been increased. Barrier cables have been added in the median strip between opposing lanes of traffic. The number of rest areas has been increased to allow drivers to take more frequent breaks. See **Fig. 24-3**.

Some states are hoping to experiment with "smart" highways. Cars would be equipped with sensors and wireless communication devices so that they could travel under computer control in small groups like train cars. Vehicles would be in constant touch via computer networks so that information about speed, acceleration, braking, and obstacles in the road would be instantly available. Radar or proximity sensors would signal the computer to speed up or slow down the car when other vehicles were near. Traffic flow and speed would be steady. Rush hour stop-and-go traffic would be eliminated, as would collisions.

Highway surfaces are routinely maintained by road crews who repair holes or do resurfacing. In states that receive large quantities of snow, maintenance crews use plows and other equipment to keep the road surface free of snow and ice.

Highway Security

The movement of cargo is more of a transportation security issue than one of safety. The International Truck and Bus Safety and Security Symposium reviews laws, regulations, and concerns of transportation providers every year. Member organizations share information and experiences.

« **Fig. 24-3** This rest area allows motorists to take a break during a long drive. Getting out and stretching helps drivers stay alert. ***In what other ways might taking a break from driving improve your health and safety?***

A program called Highway Watch is one result. It uses the skills, experiences, and "road smarts" of transportation workers to help protect goods and people. Each volunteer is trained to help spot problems such as homeland security concerns, stranded or stolen vehicles, impaired drivers, unsafe road conditions, and even children identified by Amber Alerts. They then report problems to the authorities. Mothers Against Drunk Driving and other private organizations also contribute to highway safety.

Highway security does not now receive as much attention from governmental agencies as the security of other travel modes. However, the possibility of terrorist activities on roads and highways does exist. Cars and trucks can be driven almost anywhere and used to carry dangerous cargo, such as explosives.

Railroad Safety and Security

The Federal Railroad Administration's Office of Safety promotes and regulates safety and security throughout the nation's railroad system. However, each year hundreds of fatalities occur at railroad crossings.

Individual Responsibility

Many railroad accidents are the result of motorists deliberately going around gates and ignoring other control devices. Also, people walking or playing on or near railroad tracks are hit by trains every day. Together, they account for 96 percent of all railroad-related deaths in the United States. In a recent year, 324 people were killed in motor vehicle grade crossing collisions, and 500 people died after being hit by a train.

Math Application

Highway Design and Speed Limits The speed limit for a highway is the maximum speed allowed by law for motorized vehicles. Speed limits are established based on accident/death rates, laws, customs, and other relevant factors.

One important measure for setting speed limits is the design of the highway. A highway design, for example, may take into consideration mountainous terrain. Highways with many curves warrant a speed limit that is lower than those over flat terrain.

When considering a highway's speed limit, traffic engineers often rely on the 85th percentile rule. It states that the speed limit be set so that 85 percent of motorists drive within legal limits. Studies have shown that most vehicle crashes occur when vehicles are traveling at speeds over what 85 percent of motorists are traveling.

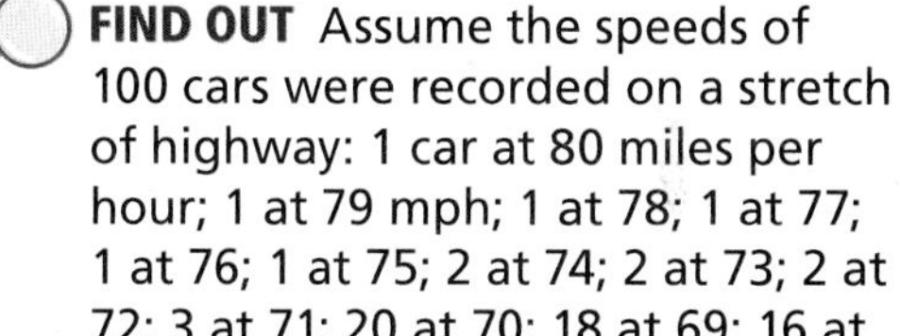

FIND OUT Assume the speeds of 100 cars were recorded on a stretch of highway: 1 car at 80 miles per hour; 1 at 79 mph; 1 at 78; 1 at 77; 1 at 76; 1 at 75; 2 at 74; 2 at 73; 2 at 72; 3 at 71; 20 at 70; 18 at 69; 16 at 68; 15 at 67; 10 at 66, and 6 at 65. What would you set as the speed limit based on the 85th percentile rule?

The average freight train of 100 cars weighs between 12 million and 20 million pounds. Even when using its emergency brakes, such a train takes over a mile to stop. Trains have no steering wheel and must follow the tracks. The engineer cannot swerve out of the way. Even a slow-moving train cannot stop in time to avoid a collision.

What happens when a train hits a vehicle? You've probably seen a soda can that's been run over by a car. The can weighs 12 ounces and the average car weighs 3,000 pounds. The ratio of car to can is 4,000 to 1. That's the same as the ratio of a car weighing 3,000 pounds to a train weighing 12 million pounds. After impact, not much is left of either the can or the car. See **Fig. 24-4**.

Why do people take chances with a train? Because locomotives are so large, they can appear to be traveling much slower than they really are. The parallel lines of the tracks create an optical illusion that makes the train seem farther away than it is. Also, some people do not realize that trains may run on any track in either direction at any time.

People who walk on railroad tracks or fish from railroad bridges are trespassing. Railroad tracks are not public property; they belong to the railroad company. No one has the right to be on them without permission.

Trains, Tracks, and Signals

Although train wrecks are rare, occasionally they do occur. Trains contain many safety devices to prevent accidents. Every wheel has its own brakes, and locomotives are routinely inspected to ensure that the braking systems are working properly. The horn, sidelights, headlight, and bell must all comply with federal regulations. People who live near railroad tracks may find the train horn or whistle annoying. However, the horn is a valuable safety device, especially at night when drivers may not be able to accurately judge the distance of a train at a crossing.

Communication devices are also important, especially when multiple trains use the same tracks and one train must wait on a siding for another to pass. New computer-based electronic systems are being used in some states that automatically control train speed and switching and help protect against human error.

At night or in poor weather, motorists may not realize that a train is moving through a crossing. If the crossing has no gate, they may drive into the side of a moving train. In order to help people be more aware of passing trains, reflective material is placed on the side of freight cars.

Fig. 24-4 While this car has been totaled, the train that hit it received little damage.

During a collision, tanker railcars carrying hazardous materials can be punctured, releasing their cargo into the environment. Such cars must be attached toward the back of a train to reduce damage during impact. See **Fig. 24-5**.

Railroad tracks are routinely inspected to be sure they are aligned properly and that their surfaces remain smooth. Switches and other control devices must also be checked. In the past, devices embedded in the tracks sent signals to crossing gates that opened or closed them as the locomotive passed. Now the signals are generated automatically from the locomotive itself, which is a more reliable method.

Railroad Security

Presently, AMTRAK and local rail transit systems move about 14 million people each day. Railroad passenger security has received more attention since 2001. One reason for concern is that trains carry a large number of walk-on passengers. Most of the time there is no record of who the traveler is.

Railroad workers now receive security training. Public awareness campaigns and increased surveillance are designed to reduce security risks. However, some railroad yards are relatively unsecured and may be fairly easy to enter undetected. This may make any cars filled with hazardous materials easy targets for terrorists.

Fig. 24-5 Tank cars that carry hazardous materials must be checked for leaks or other damage.

Maritime Safety and Security

Maritime usually refers to the sea but is also used when talking about navigation and inland waterways and ports. In the United States, responsibility for safety and security for the maritime industry generally falls to the U.S. Coast Guard, the U.S. Army Corps of Engineers, and several private service industries. Many U.S. ports also have port authorities and emergency agencies.

Individual Responsibility

As with highway transportation, those who travel on the water must follow certain laws and regulations. Did you know, for example, that sailboats have the right-of-way when crossing paths with a motorized boat? That's because it's harder to precisely control a sailboat, which is at the mercy of the wind. Those who plan to use a boat to travel any distance should be sure to learn the rules of the "road."

Even good swimmers should wear a life vest when boating. In case of an accident, the boat or some of its contents can hit them, knocking them unconscious or disabling them so they can't swim. On a large boat or ship, passengers must learn where life jackets, life preservers, and any lifeboats are located and how to use them.

Vessels and Ports

Because ships can sink and passengers can drown, special care must be taken with maritime safety. Reliability of the vessel is important.

The hull of a boat or ship must be continually monitored for leaks and other damage so that it remains seaworthy. Computers, coupled with gauges, are used to measure stresses on the hull of large cargo or passenger ships. An alarm is triggered when stresses become dangerous to hull integrity. Special software may be used to keep a running log of hull readings.

At the harbor, electronic sensors monitor the speed of a ship's approach, its angle of approach, mooring loads, weather conditions, and ocean current data. See **Fig. 24-6**. All of this information is processed and displayed on computers at the harbor office and on board the ship.

Fig. 24-6 Computers are used by port authorities to monitor ships using the facilities.

When the ship is a tanker carrying hazardous cargo, such as oil or gas, special precautions must be taken. Long-range identification and tracking (LRIT) of ships and cargo may soon become common. Computers and special tags are used to identify and track containers anywhere in the world.

Maritime Security

The best way of improving security at sea is by developing international regulations that are followed by all shipping nations. The International Marine Organization, an agency of the United Nations, is responsible for improving the safety and security of international shipping.

You may have read or seen movies about pirates who attacked ships centuries ago. Did you know that piracy still occurs on international waters where no law holds sway? Armed pirates in gunboats surround ships, board them, and overwhelm the crews and passengers. They then steal any valuables or cargo. In a recent year, over 400 ships were attacked in the seas around Indonesia and 72 people killed.

It is illegal for shipping companies to employ armed guards, but crews are sometimes creative in defending themselves. Recently, the crew of a Japanese freighter prevented pirates from boarding by spraying them with water from fire hoses. Eventually, the pirates gave up and turned away. A cruise ship, attacked by pirates off the coast of Somalia, used a sonic weapon to knock them off their feet with air vibrations. The weapon was developed by the U.S. military to defend ships from small boats after the terrorist attack on the *U.S.S. Cole* in 2000.

Security for passenger ships has been increased since 2001. Some people fear that cruise ships present an inviting target for terrorist attacks. However, security experts argue that such an attack would not disrupt the country's economy or have the other far-reaching effects that terrorists strive for. Also, passengers and baggage are thoroughly screened during boarding.

Fig. 24-7 Large ships and busy ports are potential terrorist targets. Thorough inspections are made to decrease the likelihood of an attack.

Ports The dozen major international seaports in the United States are gateways for world trade and potential terrorist targets. See **Fig. 24-7**. Eighty percent of all cargo entering the U.S. undergoes rigorous inspections. Two-thirds of shipping containers are screened for nuclear materials. Not every container is searched; instead, computers are used to help identify contents.

U.S. Coastal Waters When ships enter the coastal territory of the United States, a distance not to exceed 12 nautical miles, the U.S. Coast Guard has authority to maintain safety and security. In cases of emergency, this limit is extended to 200 miles.

Once within this territory or while making a call on a U.S. port, ships may be inspected. In a typical recent year, about 7,300 individual vessels registered in 81 different countries made over 72,000 port calls to the U.S. During this time, the Coast Guard conducted over 11,000 inspections. Less than 2.5 percent of inspections resulted in a ship being detained for safety or security reasons.

Inland Waterways Vessels that travel the inland waterways are under the rules, regulations, and enforcement of the Coast Guard. These vessels are also inspected for safety and security reasons.

Safety and security of land-based structures linked to inland waterways are the responsibility of the U.S. Army Corps of Engineers.

Air Transportation Safety and Security

Airplane crashes always make the news, and airline safety and security receive more attention than those for other travel modes. Yet fewer people are injured or killed as the result of air crashes than highway accidents. As with other modes of transportation, prevention and preparedness are the most effective ways to ensure safety and security.

The Federal Aviation Administration (FAA), under the Department of Transportation, operates the world's largest and safest aviation system. It certifies both aircraft and pilots, operates the air traffic control system, and oversees airport security. FAA's Web site offers safety and security tips as well as rankings of safety records. For example, users can research the oldest and safest airline or the most reliable aircraft.

The U.S. Marshal Service has also increased the number of marshals working on flights. With the help of special weapons, these unidentified "passengers" are trained to respond to most in-flight terrorist emergencies.

Passenger Responsibility

When you purchase a ticket, an airline wants to know who you are. The process of validating your address and other information is instantly begun. On the day of your flight, video cameras, voice and photo identification systems, and travel screeners are used to selectively check passengers. See **Fig. 24-8**.

A faster system of passenger identification, called CAPPS II (Computer-Assisted Passenger Prescreening System) is now evolving. Voice patterns, fingerprints, and retina scan information may soon be gathered and kept on file. These, together with standard information, will form a complete picture of a traveler.

Before a flight begins, airplane personnel inform passengers about emergency exits and oxygen masks located above each seat. Passengers are wise to listen, if only to refresh their memories. Some passengers can create safety and security concerns by their personal conduct. They may refuse to obey the rules and cause problems for airline personnel.

Fig. 24-8 Screening passengers takes time but helps ensure that travelers reach their destinations safely.

Aircraft Design and Maintenance

One reason air travel is so safe is that aircraft are carefully engineered and maintained. Any potential problems are identified and usually remedied. Others require constant vigilance.

Metal Fatigue In 1989, United Airlines flight 232 crashed when one of its engines flew apart, destroying three hydraulic systems. The cause of the failed engine appeared to be **metal fatigue**, structural damage to a metal from repeated stresses. To prevent problems from metal fatigue, airlines routinely carry out rigorous inspections and tests to locate the tiny cracks that are signs of metal failure.

Delamination The separation of a composite material into layers that gradually flake off is **delamination**. Like metal fatigue, it is caused by repeated stresses over time. Because so many modern planes contain composites, delamination often occurs. Since nothing shows on the surface, it must be detected with instruments, such as ultrasound devices.

Navigation Aids The global positioning system and other satellite-nagivation aids make it possible for aircrews to know their position anywhere in the world. GPS can now pinpoint a plane's altitude as well as its longitude and latitude. It is being used increasingly for approaches to airports as well as en-route navigation.

Science Application

GPS and Trilateration The United States and Russia operate global positioning systems (GPS), or space-based radio-navigational systems. As of January 2006, the European Union launched its first satellite as part of its GPS system.

You've heard a lot about GPS, but how does it work? A GPS receiver is like a cell phone. However, instead of using ground-based towers, it communicates with multiple satellites orbiting above Earth. With the help of an onboard atomic clock, each satellite transmits data at the same instant and at the speed of light. This data indicates its location and the current time.

Suppose you are holding a GPS receiver. The receiver determines its location based on the location of the three nearest satellites using trilateration. The receiver draws an imaginary sphere around each of the three satellites. These spheres intersect in two points: one in space and one on the ground—the point where you are standing with the receiver.

FIND OUT Learn more about the method of trilateration. See if you can develop a way to locate a landmark on your school grounds by knowing how far away it is from three other objects. There are four important parts of GPS: the satellites (and their atomic clocks), satellites' orbits, monitor and control stations on Earth, and GPS receivers. Learn more about how GPS works with respect to these components.

Stalling When a pilot flies at too steep an angle, the wings of the plane cannot produce enough lift. This can cause the engines to stall, a potentially serious problem. Most planes are equipped with warning devices that signal the pilot when a stall is likely.

Fires have been the cause of several air disasters in which people died when smoke filled the passenger cabins. These accidents led to the installation of new emergency doors and lights along the floor so people could find their way out of the plane through heavy smoke.

Snow and Ice on runways can cause planes to slide. Collected on a plane's wing, they can reduce the wing's ability to create lift, causing serious problems during takeoff. Modern jet planes are designed to prevent ice or snow from accumulating. One method is to direct heated air from engines over the wings and tail.

Bird Strikes Birds sucked into the engine of a plane can cause it to fail, although modern jet engines can survive bird strikes. The highest risk is during takeoff or landing, and many airports have adopted preventive measures to keep birds from nesting in the area.

Air Traffic Control

The safe and orderly flow of aircraft is managed by an air traffic control system. Controllers at each airport keep aircraft separate from one another and guide planes safely in and out of the airport's airspace. See **Fig. 24-9**. Most controllers watch planes under their control through large windows. However, radar is often used as a backup system. Various controllers monitor the different stages of a plane's approach, landing, and takeoff.

Airport Security

Fences, walls, barriers, and security patrols are the first line of defense at airports. They help prevent hijackers and thieves from getting close to a plane, to baggage handling facilities, or to a fuel depot.

All checked baggage is passed through an X-ray device, and passengers walk through a metal detector. When a metal object passes through the metal detector, a magnetic field is created in the object, and sensors detect the field. If the bag's contents seem suspicious, the luggage may be opened and the contents evaluated before being loaded on an aircraft.

Carry-on baggage also receives a thorough screening and may be opened. Screeners are trained to look for anything suspicious, not only obvious items such as guns but also items that could be used in an explosive device. A special "chemical sniffer" is used to check electronic devices, such as laptop computers, for bombs. A cloth is passed over the electronic item and then placed on the sniffer. The sniffer detects any explosive chemicals and alerts the operator. Onboard personnel are trained to recognize dangerous items and identify suspicious activities or movements.

Pipeline Safety and Security

Pipelines sometimes have safety and security problems. However, when compared to other modes of transportation, pipelines are much safer. Seldom do accidents cause injury and death, but they can cause disruptions in oil and gas supplies.

The Office of Pipeline Safety, housed in the Department of Transportation, has oversight and enforcement responsibilities for the more than 2.2 million miles of U.S. pipelines. It has set minimum standards for management and operation. Like other transportation standards, these have been tightened since 2001. Several states impose safety requirements for intrastate pipelines that are stricter than federal regulations.

« **Fig. 24-9** Air traffic controllers manage the movement of aircraft as they approach, circle, and leave an airport.

Young Innovators

Technology Student Association

The Technology Student Association (TSA) is an organization for middle and high school students with a strong interest in technology. It provides a national program of technology-related activities and competitive events. Members are supported by teachers, parents, and business leaders who promote a technologically educated society.

TSA helps prepare students for the challenges of a dynamic world by promoting technological literacy, leadership, and problem solving. Through TSA, students learn about the potential of technology and gain an understanding of careers in technology.

TSA hosts an annual national conference, where it features middle and high school level competitions. These competitive events, as well as other activities of TSA's national conference, are intended to extend student understanding of the development, impact, and potential of technology and careers in technology.

In addition to its roster of middle and high school level competitive events, TSA has partnered with Autodesk, Denford, and Pitsco to bring the F1 in Schools Challenge to interested students. This unique technology and engineering competition offers student teams the opportunity to design, analyze, make, test, and ultimately race a 1⁄20th scale Formula 1 car.

Teams use 3D CAD/CAM software to design a car and implement its manufacture. Once a team's car has been completed and tested, it is ready to be raced against the competition to determine the fastest car. Races take place on an 80-foot track, where cars can reach speeds of up to 50 miles per hour!

Students involved in the F1 Challenge advance from the state level to the national competition, which is held each year at TSA's annual conference. National winners are invited to participate in the F1 World Championship, where teams from up to 20 countries compete. Recent F1 World Championship events have been held in England and Australia.

CHAPTER 24 REVIEW

Main Ideas

- Most transportation regulations are enforced by the federal government.
- Different government offices help monitor the safety and security of each mode of transportation.
- While the government helps regulate safety and security, every individual must also be aware and responsible.
- The design of vehicles is often influenced by the need for safety and security.

Understanding Concepts

1. List the five areas of influence on safety and security.
2. What is the TSA and why was it created?
3. What is the source of more than 40 percent of all highway fatalities?
4. Why is it illegal to walk on railroad tracks without permission?
5. Describe delamination.

Thinking Critically

1. Describe what "smart" highways of the future might be like.
2. What effects of drinking alcohol might help cause car crashes?
3. Hypothesize why many people think that airplane travel is more dangerous than travel by automobile.
4. How might airbags actually do more harm than good?
5. What might be the effects of a pipeline accident?

Applying Concepts & Solving Problems

1. **Math** Which would be harder to stop—a motorcycle (weighing 600 pounds) traveling at 120 miles per hour or a school bus (weighing 12,000 pounds) traveling at 30 mph?
2. **Science** Demonstrate metal fatigue by bending a paper clip back and forth several times. Describe what happens.
3. **Assess** Safety and security are important in other areas besides transportation. Evaluate your school for safety and security. Write a report on your findings.

Activity: Test Your Powers of Observation

Design Brief

How observant are you? Suppose you are walking down the street and suddenly someone runs out of a store right in front of you, jumps in a waiting car, and speeds away. A store clerk follows, calling, "Stop thief!" You are a witness, and the police ask you to describe what happened. How accurate would you be about what you saw?

Two organizations ask ordinary citizens to keep watch on the nation's highways and waterways for such things as crime and terrorist activities. They are the Highway Information Sharing and Analysis Center (ISAC), which sponsors Highway Watch, and the Coast Guard, which sponsors America's Waterway Watch. Their Web sites offer helpful information about the kinds of things to look for.

Select a street intersection, dock, railroad crossing, or other center of activity for transportation. Observe and take notes and/or make drawings of what occurs at that spot. Make a report based on your findings.

Refer to the

HANDBOOK

Criteria

- You must observe your selected spot for a minimum of 30 minutes.
- Draw and submit a map showing where you were stationed during your observation period.
- During the 30 minutes you must make special note of at least one person and one vehicle, which you will describe in detail.
- Submit a written report of the activities that took place at your location during the observation period.

Constraints

- If possible, select a location that is moderately busy, rather than bustling, so you can keep better track of what's happening.
- Keep in mind that you are not looking for—and probably will not see—suspicious behavior. This is just an exercise in observation. If someone questions you, you should explain your task. If you do happen to see something suspicious, do not get involved. Instead, report the information to the proper authorities.

Engineering Design Process

1. **Define the problem.** For the location you have selected, write a statement that describes the problem you are going to solve. For example, what kinds of vehicles use this location?
2. **Brainstorm, research, and generate ideas.** Visit the Web sites of the organizations listed on page 487 to read examples of observations. Think about the kinds of information the police, for example, might want to know.
3. **Identify criteria and specify constraints.** These are listed on page 488.
4. **Develop and propose designs and choose among alternative solutions.** Choose the best observation method that will solve your problem. Study the lists of things to look for. Review the location. Are you in a good spot for viewing?
5. **Implement the proposed solution.** Decide on the process you will use for making your map and taking notes. Gather any needed tools or materials. In real life, things happen quickly. You might want to create a chart that you can quickly fill in on the scene.
6. **Make a model or prototype.** Do a trial run by observing activity at your home or school for several minutes and taking notes.
7. **Evaluate the solution and its consequences.** Will your methods solve the problem? Will you be able to collect enough information quickly?
8. **Refine the design.** Based on your evaluation, change your methods if needed.
9. **Create the final design.** Note the time you begin. Make your observations. Note the time you finished. Draw your map. Write your report.
10. **Communicate the processes and results.** Turn in your assignment to your teacher. Be sure to include the name of the activity, your definition of the problem, a description of how you solved the problem, and your map and report.

SAFETY FIRST

Before You Begin. Find an observation post that is in a safe spot out of the way of any moving vehicles.

UNIT 7

BIO-RELATED ENGINEERING & DESIGN FRONTIERS

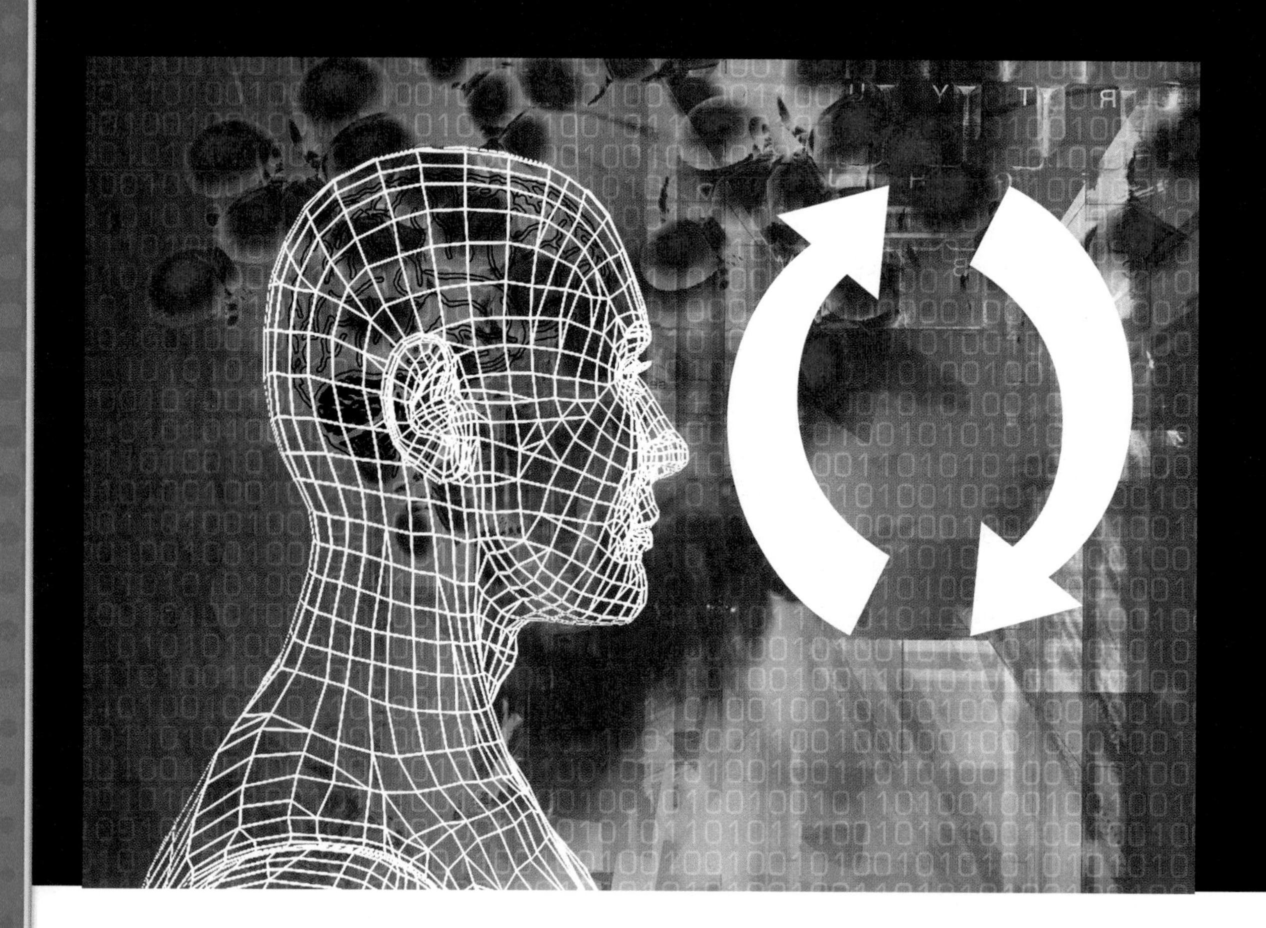

About 100 years ago, doctors began using X-ray technology to "see" inside the human body. Previously, there had been only two ways to view the interior of the human body: perform an autopsy on a dead body or open up a live body during a surgical procedure. X-rays turned out to be a revolutionary medical technology. Today we are on the brink of a new revolution.

The Promise of Molecular Imaging

Researchers are developing sophisticated integrated technologies that allow professionals to monitor living systems at molecular and cellular levels. The new technology, which combines high-powered computing with existing imaging technologies such as CT (or CAT) scan, PET scan, X-ray photography, MRI, and ultrasound, is known as molecular imaging. Some molecular imaging tools are already on the market, and others will soon be available.

Molecular imaging tools will change the way doctors diagnose and treat disease. Today's automotive technicians, using diagnostic computer scans, find potential problems in cars long before travelers risk becoming stranded or having accidents. Similarly, doctors equipped with molecular imaging tools will be able to diagnose disease before signs or symptoms even appear.

Consider cancer. A molecular imaging device pinpoints a "rogue" protein attached to a cell deep inside a patient's lung. The protein's presence indicates that the cell is becoming a cancer. Doctors then devise a treatment to kill the "bad" cell before it divides thousands of times and produces a life-threatening tumor. With traditional imaging tools, such as X-rays or MRI, doctors could find a many-celled tumor, but they would be unable to detect an individual cell likely to start a cancer.

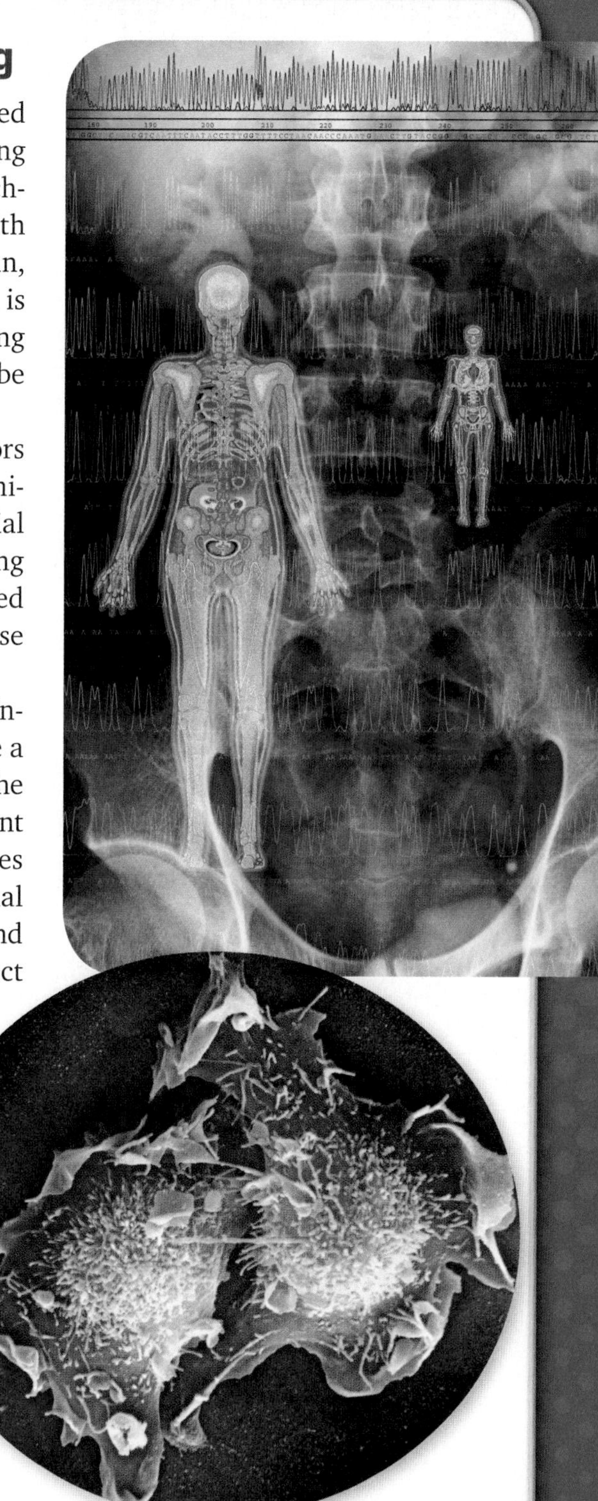

Molecular imaging holds great promise for many other diseases. For example, doctors will be able to detect tiny changes in blood flow to and from a patient's heart, allowing for effective treatment long before a potential heart attack threatens survival.

It is possible, perhaps likely, that you will benefit at some time in your life from molecular imaging technology. Thanks to this breakthrough, your life will probably be better—and longer.

CHAPTER 25 BIO-RELATED FUNDAMENTALS

This wheat is growing under LEDs. Astroculture is a kind of bio-related technology used to produce a closed controlled environment for plant growth in space. In this chapter's activity, "Design an Artificial Ecosystem," you will have the chance to research, design, and build your own controlled environment.

Objectives

- Define *bio-related technology*.
- List seven common bio-related processes.
- Discuss bioethics.
- Identify impacts and effects of bio-related technologies.

Vocabulary

- bio-related technology
- gene
- genome
- propagation
- cloning
- harvesting
- adaptation
- conversion
- bioethics

Reading Focus

1. **Read the title of this chapter and describe in writing what you expect to learn from it.**
2. **Write each term in your notebook, leaving space for definitions.**
3. **As you read Chapter 25, write the definition beside each term in your notebook.**
4. **After reading the chapter, write a paragraph describing what you learned.**

What Is Bio-Related Technology?

As you learned in your science classes, biology is the study of living things. **Bio-related technology**, then, includes all technologies with a strong relationship to living organisms, such as agriculture and health care. When the chef at the pizza parlor uses yeast organisms to make a pizza crust, the chef is using a bio-related technology. See **Fig. 25-1**.

The bio-related technology attracting the most attention today is health care. As researchers learn more about how living organisms work, they hope to find newer, more effective cures for diseases.

How Does Bio-Related Technology Affect Your Life?

Although it is easy to think of bio-related technology as something that goes on in laboratories, it is an important part of our everyday lives.

When you get up in the morning, one of the first things you probably do is get a drink of water. That water has been treated to make it safe for you to drink. If you pour some of the

Fig. 25-1 Yeast organisms cause the pizza dough to rise. Vegetables, as well as animal products such as sausage and cheese, are all results of bio-related technologies.

« **Fig. 25-2** You must keep your body hydrated when exercising. Bio-related research has been used to help develop new drinks to better meet this need.

water down the drain, it might go to a waste management center where it is recycled, another bio-related technology. See **Fig. 25-2**.

The clothes you put on may also be bio-products. Do you wear cotton? It comes from a plant. Do you wear wool? It comes from sheep. Both cotton and wool fibers are processed and turned into yarns that are woven into clothing.

The food you eat for breakfast is the result of other bio-related technologies. See **Fig. 25-3**. If you live in a house made from wood, your house is a bio-product. If the car or bus that took you to school uses biofuels, the fuel came from a plant such as corn, which is a product of agriculture, a bio-related technology. The list goes on and on.

The Development of Bio-Related Technologies

Although many recent breakthroughs have brought bio-related technology to the public's attention, it is not new. Bio-related technologies have been around for as long as humans have been selectively breeding animals and plants, using bacteria to make cheese, and preparing medicines to treat disease. Agriculture itself has been practiced since about 8000 B.C.E. As early as 1557, a book titled *Points of Husbandry* was printed about the breeding of animals. However, we know much more now than we did then.

Fig. 25-3 Breakfast is a very important meal. ***How can you monitor the amount of nutrition you're getting from a bowl of cereal?***

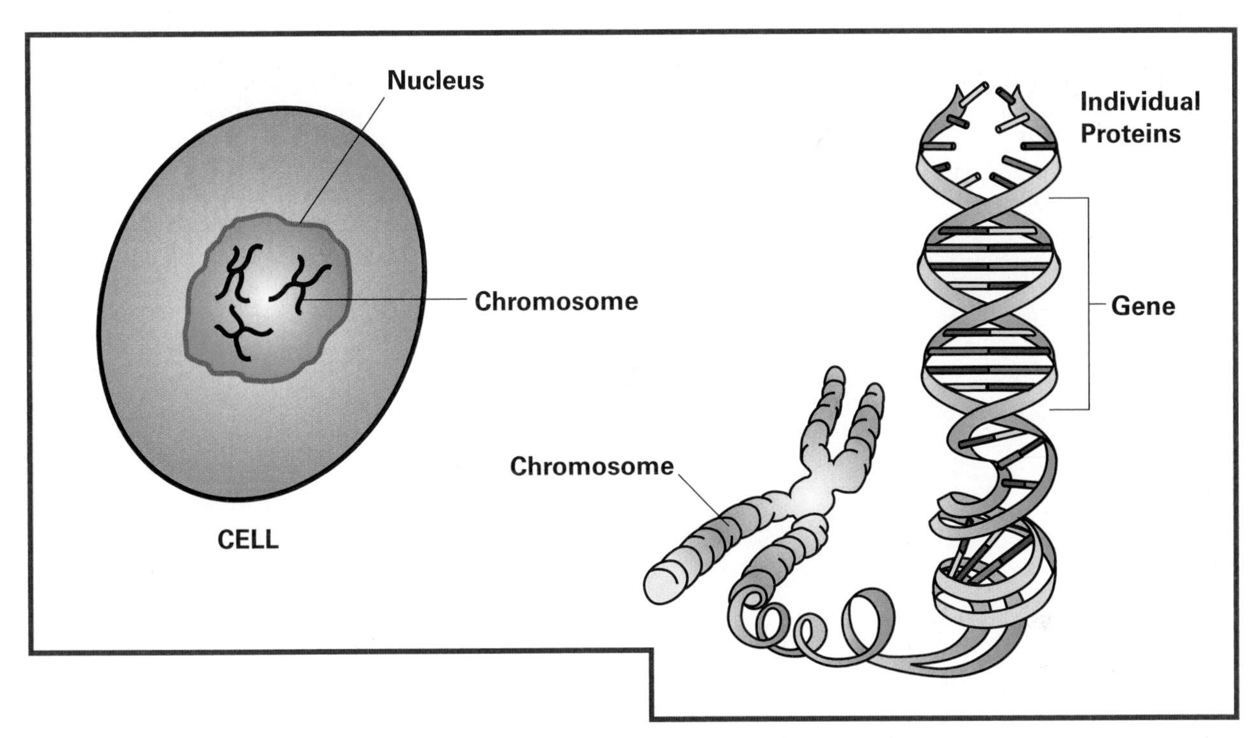

Fig. 25-4 The small segments of the DNA are genes, which are found within the chromosome of every cell.

This increased knowledge has led to a revolution in biology and bio-related technology. In the past, researchers sought to identify organisms and discover how they worked. Today, we have learned to alter those organisms in ways never before imagined. Many scientists have contributed to our growing knowledge. See "Evolution of Bio-Related Technology" on pages 496-497.

Much of the research going on today is related to genetics, the study of how traits (physical characteristics) are passed from an organism to its offspring. All living organisms are made of cells. Within the nucleus of each cell are chromosomes containing long, ladder-like strands of DNA molecules. **Genes** are short segments of DNA that carry specific genetic information. The arrangement of the molecules along the DNA ladder is what determines the traits of the organism. See Fig. 25-4. All the DNA in an organism is called that organism's **genome**.

Genetics is an explosive field today. Researchers in many countries are attempting to decipher the information contained in the genes of a wide variety of living things, from microbes to human beings. In the U.S., the attempt to discover and understand all the human genes is called the Human Genome Project. In June 2000, a working draft of the entire human genome sequence was announced. An analysis was published in 2001. Another phase of the project, called Genomes to Life, will attempt to use DNA sequences to answer questions about the essential processes of living systems.

As genetic information is obtained, scientists are learning how to alter the genetic code of an organism through genetic engineering. While early scientists were busy *identifying* organisms and genetic processes, today we have the ability to *change* those organisms.

EVOLUTION OF Bio-Related Technology

The quest to understand how living things work has led to scientific discoveries and new technologies.

5000 BCE

« By this time, people had learned to use selective breeding to grow new types of wheat, corn, rice, and other plants.

500 BCE

» Even in ancient times, people saw a connection between cleanliness and health. For example, the Greek physician Hippocrates (460–377 BCE) recommended boiling rainwater and filtering it through a cloth before drinking it.

1798

Edward Jenner, a British physician, developed a vaccination method for smallpox.

1857

Alexander Fleming discovered penicillin, the first modern antibiotic produced from living organisms.

Louis Pasteur, a French chemist, discovered that heat-treating milk destroyed bacteria, and the milk stayed fresher longer. The process was named after him, and today milk is still pasteurized before it is sold.

Dolly the sheep was the first mammal to be cloned from an adult cell.

Surgeons began using robotic systems to assist with surgeries.

Francis Crick, James Watson, and Maurice Wilkins determined the molecular structure of DNA (deoxyribonucleic acid).

Bio-Related Technology Systems

We take many of the products and processes of bio-related technology for granted, but they all have an effect on us and our environment. As the ability to tinker with nature increases, it is important to remember that changing one thing can change something else unexpectedly. It is easier to understand these consequences if you think of bio-related technology as a system. Like the other systems you have learned about in this course, it has inputs, processes, outputs, and feedback. See **Fig. 25-5**.

Bio-related technology systems need the same inputs required by any other technology systems—people, information, materials, tools and machines, energy, capital, and time. For example, suppose a food manufacturer that specializes in baked goods wants to develop and market a new flavor of muffins. Food researchers are needed to develop the flavor in the lab. Flour, eggs, and other materials are needed

Fig. 25-5 The parts of a bio-related system are inputs, processes, outputs, and feedback. When all the components work together properly, the result is a finished system.

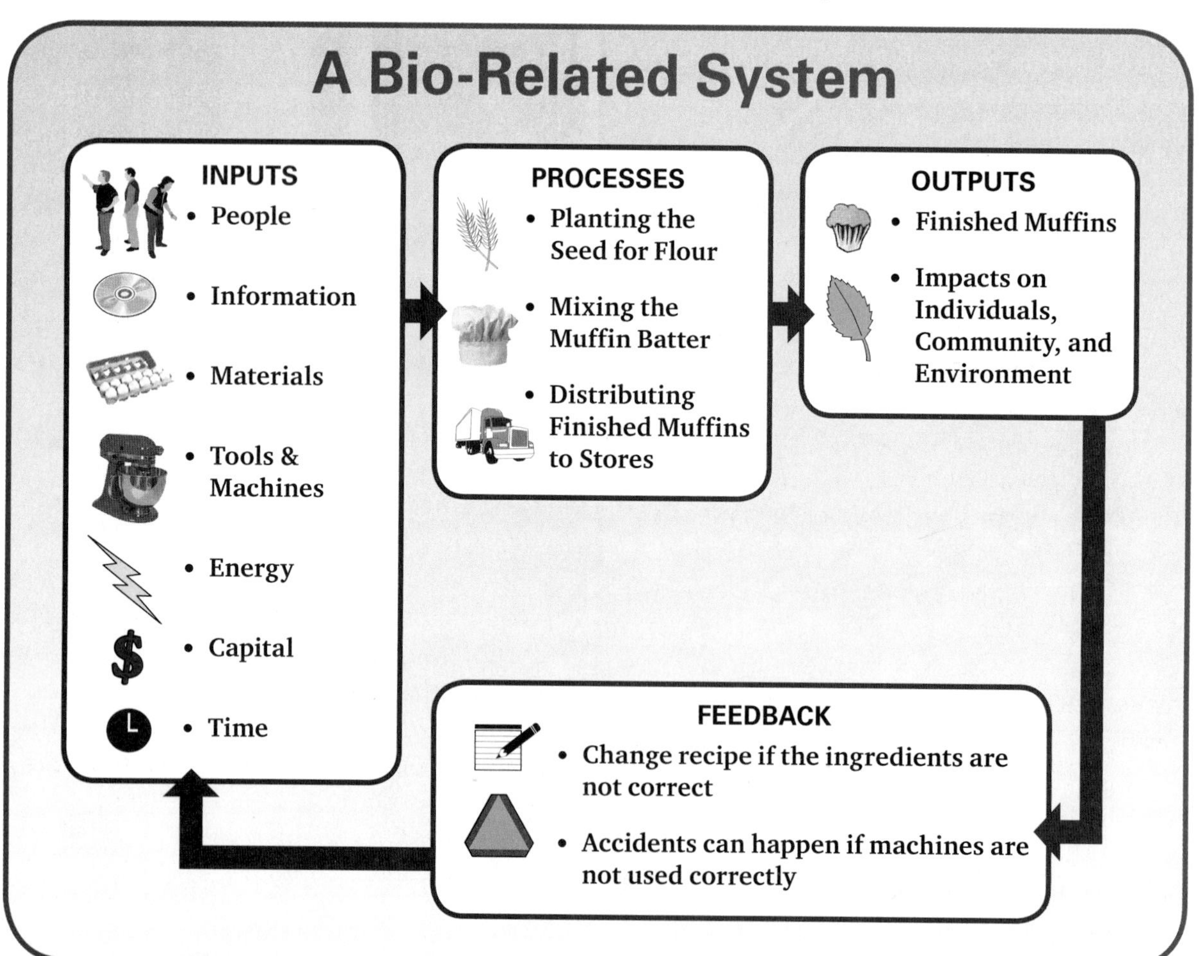

» Fig. 25-6 A greenhouse is a controlled ecosystem designed to maintain plant growth.

to make the mixture. Energy is needed to heat the ovens. Capital is needed to pay the workers.

Processes would include all the tasks going back to planting the seeds to grow the wheat that is ground into the flour. Some processes, such as mixing the muffin batter, would take place inside the manufacturing plant. Still other processes would include distributing the finished muffins to stores and selling them.

As you know, outputs are the results of the system. In our example, the finished muffins would be one desired output. However, not all outputs are positive, and some may be unexpected. Suppose the muffins contained a lot of *trans* fat. For years, food manufacturers have used these fats to increase shelf life and maintain flavor. However, it has been found that *trans* fat has a negative effect on people's cholesterol levels.

Feedback occurs when information about the outputs is put into the system. In our muffin example, the information about *trans* fat would be feedback. This feedback could result in the company's changing its muffin recipe and using only unsaturated fats.

Common Bio-Related Processes

Seven basic processes are common to bio-related technology. They include the techniques necessary to propagate, grow, maintain, harvest, adapt, treat, or convert living organisms. See Fig. 25-6.

Propagation

Propagation is the technique of reproducing a living organism. It includes breeding animals and planting crops.

For example, hybrid plants are those with special characteristics developed by breeding two different varieties. This breeding technique is a type of propagation.

Cloning is also a propagation method. A cell from a plant or animal is used to create a duplicate.

Growth

For a living organism, growth is the period after conception and before maturity. People can alter the growth period of a plant or animal by speeding it up or slowing it down. For example, nutrients are added to water or soil in order to promote plant growth.

Maintenance

Maintenance has to do with providing food and water for an organism, as well as making sure conditions in its environment, or ecosystem, support its growth. An ecosystem is a system of living and nonliving organisms, their interacting relationships, and their surrounding physical environment. A greenhouse is an example of an artificial ecosystem created for plants to support their growth in a cold climate. (See again **Fig. 25-6**.)

Some animals are raised in special housing where workers control what they eat and how they live.

Maintaining environments for humans is also a bio-related technology. Heating or air-conditioning a home is an example.

Science Application

Emerging Viruses Some of the most dangerous microorganisms are viruses. While microorganisms such as bacteria can live independently, a virus depends on a host organism. Outside its host, the virus appears lifeless and cannot reproduce. The host provides the life processes the virus lacks. Once inside its host, it does what viruses do best: duplicates to create additional viruses. Sometimes this process damages or kills the host.

Emerging viruses are being discovered all the time. These viruses have an incredible ability to mutate. They can also move from species to species and help other microorganisms mutate. A virus living in a harmless bacterium that is penicillin-resistant can "jump" to another bacterium that is deadly. By exchanging genetic material, the virus can create a new bacterium that is both deadly and resistant to penicillin. Viruses are also capable of altering the human immune system, turning it against itself.

Emerging viruses pose one of the greatest health dangers in today's world. This danger exists because few viral infections can be treated successfully with drugs. In addition, vaccines, if available, are often not very effective.

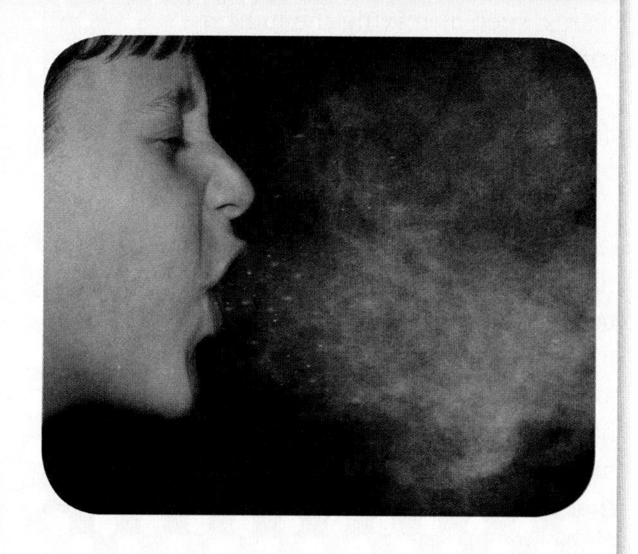

FIND OUT Read one of these books and report on what you learn:

Plagues and People by William McNeil

Emerging Viruses: AIDS and Ebola—Nature, Accident or Intentional? by Leonard Horowitz

Emerging Viruses edited by Stephen Morse

Harvesting

Harvesting is the gathering of organisms and preparing them for use. For example, farmers harvest bean crops by going into the fields with giant machines that pick the beans from the plants and funnel them into a giant hopper. Later the beans are separated from any stalks and other waste products, dried, and sent to a processing plant.

Animal products may be harvested as well. Eggs are gathered regularly from chickens. Each spring, sheep are sheared of their wool.

Adaptation

Adaptation is an organism's ability to change. The change is usually a reaction to something in its environment. Most organisms adapt naturally. A bear, for example, grows a heavier coat of fur for the winter.

Organisms can also be changed artificially. Plants, for example, can be genetically altered to resist certain pests or diseases.

Treatment

Organisms can be treated to improve their characteristics, remedy problems with their growth, or cure them of disease. Have you ever taken prescription medicine for an illness? If so, you were being treated by your doctor.

Animals are also given medicines for illnesses. Plants, too, can be treated in this way. For example, they may be sprayed with pesticides (poisons that destroy insects), or soil conditions may be improved to stop the growth of molds.

Conversion

Conversion is change. In bio-related technology, it means that the form of the organism is altered in some way to prepare it for use. For example, kernels of corn may be washed, flattened, combined with other ingredients, and baked to turn them into breakfast cereal.

Fig. 25-7 Many people are concerned about possible negative impacts and effects of bio-related technologies.

Impacts and Effects of Bio-Related Technology

Most of the impacts and effects of bio-related technologies are positive. Agricultural technologies may increase food output and make some foods more nutritious. Medical technologies can save lives. However, some of the concern people feel about bio-related technologies centers on unexpected negative outputs. See **Fig. 25-7**. They want to be sure that these technologies are regulated and that any potential for harm is discovered early.

Young Innovators

Make a Difference at JETS NEDC

The JETS National Engineering Design Competition allows high school students an opportunity to make a real-world difference by developing solutions to help people with disabilities enter or advance in the workplace. Students are challenged to go beyond paper and pencil to create something that fulfills a real need.

The competition is a three-part, cross curricular activity that involves problem solving, math, science, research, writing, presentation, and drafting/design skills. In the first part of the competition, students conduct an extensive Internet scavenger hunt to gain greater awareness of disabilities and the challenges that they present. The goal is for students to fully understand how disabilities can affect the daily lives of people, particularly those issues surrounding employment. They learn about technology and how it affects the lives of those with severe disabilities and about the potential for new technologies to empower them. After researching a particular disability and workplace scenario, students then use the engineering design process to formulate, design, and create a potential solution.

In round two of the competition, students build a working prototype based their chosen scenario. They also submit a report detailing the steps they took in the engineering design process to create the prototype.

Teams that successfully complete round two of the competition may advance to the third round, the National Finals. In this last phase, all of the research and engineering design that went into the first and second phases of the competition is now pulled together to refine the prototype and final report. Advancing teams are invited to Washington, D.C., to showcase and defend their creations. The team will present its working model that has the potential to empower and improve workplace opportunities for people with disabilities.

As you know, ethics are a set of rules or standards that guide the conduct of a person or group of people. **Bioethics** are those rules or standards that apply to people and companies who engage in bio-related science and technology.

Medical Data

Because scientists are able to map human genomes, they can tell if a certain person is at risk for a particular disease, such as breast cancer. A gene that increases this risk has already been identified. On the surface, this seems like worthwhile information to have. However, the situation is not that simple.

Being at risk for a disease does not mean you will automatically get it. It only means that you have a better chance of getting it than many other people. Should a woman who possesses this gene get a mastectomy (breast removal surgery) rather than take the chance she will get cancer? What if she possesses other, as yet unidentified, genes that will cancel out the risk? What if it turns out that genes play only a small part in getting cancer?

These issues become even more complex when other factors, such as medical insurance, are considered. Insurance companies argue that they need genetic information in order to refuse insurance to people who will cost them a lot of money. Do they have a right to this information? Do they have a right to deny coverage to a person who possesses the breast cancer gene? What other impacts might result from the availability of medical data?

Commercial Use of Living Organisms

A gene for making plastic has been slipped into a type of mustard plant. The gene turned the plant into a natural plastics factory. At least one plastics manufacturer is developing the idea commercially. The same can now be done

Math Application

Scientific Notation Scientific notation is a short way of writing very large or very small numbers. In scientific notation, a number is written as a decimal number multiplied by a power of 10.

For example, forests cover about 2,700,000,000 acres of the earth's surface. To write this number using scientific notation, first move the decimal point (which would ordinarily follow the last zero on the right) nine places to the left so that it follows the first nonzero digit, which is 2. Because you moved the point 9 places to the left, you would write "$\times 10^9$" after the number. Thus,

$$2{,}700{,}000{,}000 = 2.7 \times 10^9$$

What if you want to change 2.7×10^9 back to an ordinary number? The factor 10^9 means you must move the decimal point nine places to the right.

Suppose you have a very small number, such as 0.0000035 meter. To write this length using scientific notation, place the decimal point to the right of the 3, since 3 is the first nonzero digit. Because the decimal point moves six places to the right instead of the left, the exponent on the 10 is negative:

$$0.0000035 = 3.5 \times 10^{-6}$$

FIND OUT Write the following measurements using scientific notation.

1. The length of a soybean field that measures 382 meters.
2. The length of a microbe that measures 0.000275 centimeter.

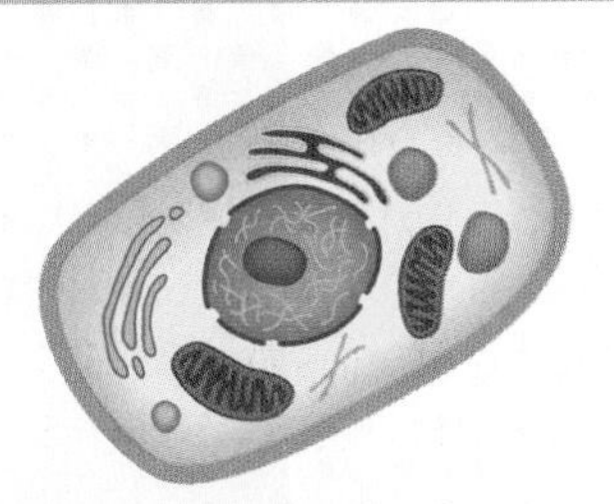

The human body contains about 1×10^{14} cells.

with animals. Cows, for example, can be given the genes for producing certain drugs in their milk, turning the cows into drug factories.

For many years, mature plants have been cloned with the use of their genetic material. Now, several mammal species have been cloned. Researchers hope that cloning will enable farmers to do such things as make copies of a prize dairy cow to create a herd of champion milkers.

We already use plants and animals for food, so many people feel that altering their biology can be no worse. Are they right? What if other, unexpected changes take place later on as a result?

Environmental Interactions

Of special interest to researchers is the use of genetic alterations to remove pollutants from the environment, improve crops, or limit the spread of disease. For example, certain microbes have been designed to eat different types of toxic waste. See Fig. 25-8. Certain species of corn are engineered to resist pests. A genetically altered mosquito that does not transmit malaria well is being studied to help limit the spread of that disease.

Although no one would argue with the need to clean up pollutants, make better crops, or curb disease, the introduction of altered organisms into the environment makes some people uneasy. Researchers claim that such technologies are thoroughly tested, yet it is impossible to test for every possible interaction in the laboratory. Will new or altered organisms behave differently in a natural environment or escape our control? Will organisms that appear beneficial today mutate tomorrow and create worse problems than those they were designed to remedy? These and other questions must be carefully considered as we enter the bio-related technology era.

Fig. 25-8 This bio-related material has been engineered to absorb oil from the water after an oil spill.

CAREERS IN

BIO-RELATED TECHNOLOGY

Bio-related technology is one of the fastest-growing career fields. It includes hands-on jobs (farming, for example), high-tech jobs (engineers), jobs helping people who are ill (medical workers), and jobs on the cutting edge of research (scientists).

Medical Lab Technician • Surgeon • Medical Research Scientist
Nurse • Emergency Medical Technician • Biomedical Engineer
Optometrist • Water Treatment Plant Operator • Ergonomics Engineer
Sanitation Engineer • Pharmacist • Pharmaceutical Salesperson
Veterinarian • Farm Worker • Milk Truck Driver • Combine Operator
Forester • Florist • Butcher • Toxic Waste Technician • Biochemist
Waste Management Plant Supervisor • Environmental Engineer
Food Plant Machine Operator • Food Safety Inspector • Nutritionist

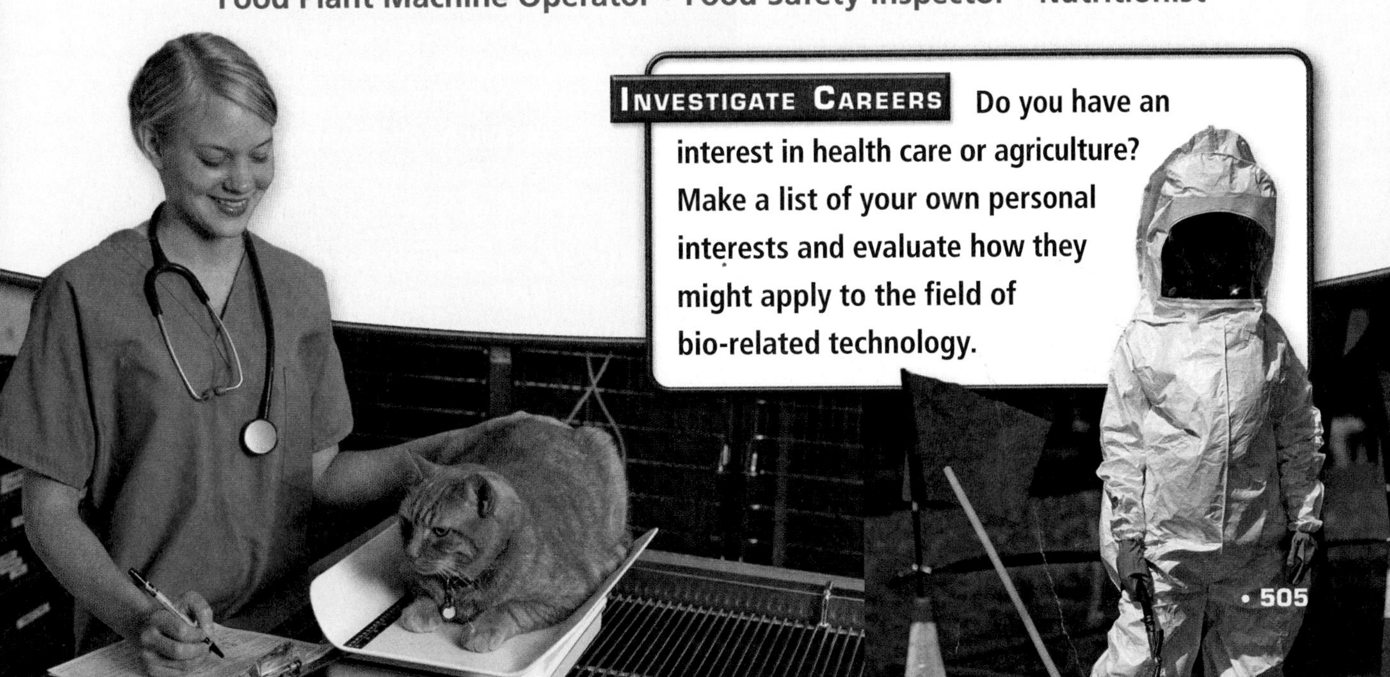

INVESTIGATE CAREERS Do you have an interest in health care or agriculture? Make a list of your own personal interests and evaluate how they might apply to the field of bio-related technology.

CHAPTER 25 REVIEW

Main Ideas

- While there is increased interest in bio-related technology, it has actually been around for a long time.
- Much of the bio-related research going on today involves genetics.
- All bio-related technology systems have an impact on us and our environment.
- Seven basic processes are common to bio-related technology.

Understanding Concepts

1. Define *bio-related technology.*
2. How has the study of organisms changed over the years?
3. What are the seven common bio-related processes?
4. What are bioethics?
5. Name three bio-related technologies that have issues involving bioethics.

Thinking Critically

1. Describe the bio-related technologies you have interacted with today.
2. Imagine you are going to have minor surgery. What inputs, processes, outputs, and feedback might be involved in a minor surgery?
3. How might a greenhouse be less risky than traditional planting?
4. Maintaining your environment is important. How many environments do you interact with each day?
5. Technology is changing at a much faster rate today than it has in the past. Why do you think this is so?

Applying Concepts & Solving Problems

1. **Government** Organize a panel of your classmates to debate whether or not the federal government should regulate the actions of geneticists.
2. **Writing** Imagine you are contacting the federal government about a bioethical concern. Write a letter defending your position on the bio-related technology in question. You should provide ample support in your defense.
3. **Math** The complexity of the Human Genome Project almost defies comprehension. Approximately 800 bases (the smallest of the chemical "building blocks") of a typical DNA sequence are equivalent to 1/3,800,000 of the complete human genome. How many bases would there be all together?

Activity: Design an Artificial Ecosystem

Design Brief

In 2004, President George W. Bush proposed the goal of building a colony on the moon by 2024. Such a colony would be the ultimate challenge for bio-related technologists. It would have to provide almost everything required to sustain human life in a hostile environment. No matter where such a colony might be built, it would require special technologies for food, water, sanitation, waste management, health care, and air supply.

In the 1990s, two artificial ecosystems called Biosphere I and Biosphere II were built to test our ability to construct and live in such environments. Both habitats encountered unexpected difficulties. Research continues.

Research and design an artificial ecosystem that can support human, plant, and animal life in a hostile environment. Make drawings and/or a model of your ecosystem and include a written or audio "tour" of the facility that explains how it works.

Refer to the

HANDBOOK

Criteria

- You must include facilities for growing food for a balanced diet; obtaining and treating water; getting exercise, sunlight, other necessities for health; and managing waste.
- Compensate for conditions in your environment. For example, on the moon, intense ultraviolet light from the sun would pose health problems.
- You must provide factual data about the hostile environment (for example, gravity on the moon is ⅙ that on Earth) that shows evidence of research.
- Create a sketch or model of the exterior of the structure, as well as a general floor plan or model of the interior.
- Provide a written or audio-taped "tour" of your ecosystem that describes the facilities and tells how each will compensate for the hostile conditions and sustain life.

Constraints

- Most of the materials you use to sustain the ecosystem must be derived from what is found in your hostile environment. For example, seeds and starter plants can be shipped in, but not the food eaten on a daily basis.
- If you choose to make a model, it must be no larger than two square feet. If you choose to make an audiotape, it should be no longer than eight minutes.

Engineering Design Process

SAFETY FIRST

Before You Begin. Make sure you understand how to use the tools and materials safely. Have your teacher demonstrate their proper use. Follow all safety rules.

1. **Define the problem.** Write a statement that describes the task you are going to accomplish, as well as any obstacles you will need to overcome (i.e., providing food for organisms).
2. **Brainstorm, research, and generate ideas.** With your team, discuss possible solutions. Hint: You may want to research various artificial environments designed to sustain life, such as the International Space Station, generational space ships, ocean colonization, moon colonies, and Biosphere I and II.
3. **Identify criteria and specify constraints.** These are listed on page 508.
4. **Develop and propose designs and choose among alternative solutions.** Choose the design that will most likely support life in a hostile environment.
5. **Implement the proposed solution.** Decide on the process you will use for making the model or drawing. Gather any needed tools or materials.
6. **Make a model or prototype.** Create your model or drawing. Follow all safety rules.
7. **Evaluate the solution and its consequences.** Does your model work as intended? Will it sustain life?
8. **Refine the design.** Based on your evaluation, change the design if needed.
9. **Create the final design.** After any changes or improvements take place, create your final model or drawing. Create a tour of your environment.
10. **Communicate the processes and results.** Present your finished model or drawing to the class. Use your written or audio tour to guide your classmates through the environment. Hand in all materials to your teacher.

CHAPTER 26 MEDICAL TECHNOLOGIES

Objectives

- Discuss technologies used for the prevention, diagnosis, and treatment of disease.
- Explain the purpose of human factors engineering.
- Give examples of how environmental engineering is used to prevent disease and protect the environment.

Vocabulary

- immunization
- antibody
- vaccine
- pasteurization
- gene therapy
- stem cell
- prosthetics
- telemedicine
- forensic medicine
- environmental engineering

Newer prostheses allow people to be as active as this football player who has an artificial leg. In this chapter's activity, "Create a New Replacement Joint," you will have a chance to design and build a model of a human replacement joint.

Reading Focus

1. **Write down the colored headings from Chapter 26 in your notebook.**
2. **As you read the text under each heading, visualize what you are reading.**
3. **Reflect on what you read by writing a few sentences under each heading to describe it.**
4. **Continue this process until you have finished the chapter. Reread your notes.**

Fig. 26-1 When recovering from knee surgery, or almost any surgery, it is important to get back on your feet and establish a normal lifestyle as soon as possible.

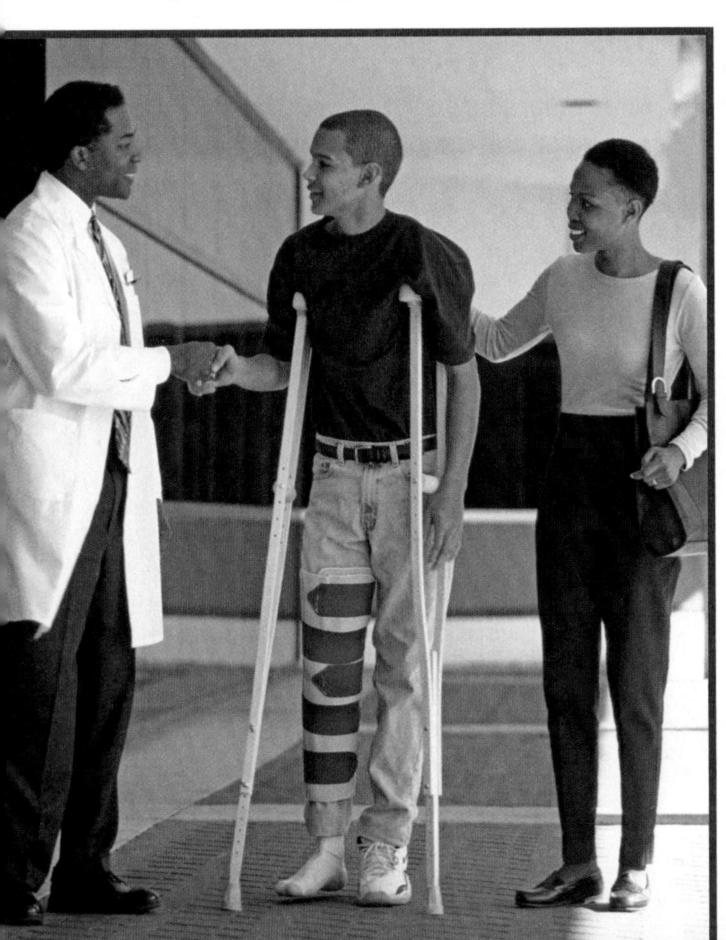

Importance of Medical Technologies

In 1900, the average life span for someone born in the U.S. was 45 years. Today, the average life span is 77 years. One cause of this change is bio-related technology.

Have you ever been treated for a serious illness or injury? You or someone you know may have overcome a problem that was once life-threatening. See **Fig. 26-1**.

Do you live in an area with high pollution? Bio-related technology is also used to monitor and improve our environment. Our surroundings directly affect how we live.

In this chapter, you will learn how improvements in disease prevention, diagnosis, and treatment; human factors engineering; and environmental engineering all contribute to our health and well-being.

Disease Prevention

Have you noticed the words "Vitamin A & D" on milk cartons? The vitamins have been added to help make sure people will get enough. A shortage of these vitamins can cause eye, skin, and bone disorders. Adding vitamins to food is one example of disease prevention.

Immunization

Immunization is the process of making the body able to resist disease by causing the production of antibodies. **Antibodies** are proteins that attack foreign substances in the body, such as pathogens (disease-producing organisms). Many viruses, bacteria, fungi, and protozoa are pathogens.

Immunization can happen naturally. People who have had measles are now immune to the disease. Their blood carries antibodies that attack and destroy measles viruses. It's possible, though, to become immune to measles without first getting the disease. That's because there is a vaccine that can make people immune.

A **vaccine** is made from killed or weakened pathogens or from purified toxins they produced. The vaccine does not cause disease, but it does stimulate the body's immune response. Antibodies are produced. If the person is later exposed to the disease, the antibodies recognize and attack the pathogens. See **Fig. 26-2**.

Scientists are now working on DNA vaccines. These vaccines use a gene from a virus or bacterium rather than the whole organism. It is therefore safer and more effective. When the vaccine is injected into a human being, the immune system will produce a response that protects the person from infection. See **Fig. 26-3**.

Pasteurization and Sterilization

Sometimes disease can be prevented by killing pathogens in food and the environment before people become sick. In **pasteurization**, a food, such as milk, is heated to a temperature high enough to kill bacteria.

Sterilization also prevents disease with the use of heat and/or chemicals to destroy all microbes. (A microbe is any single-celled life form too small to be seen without a microscope.) For example, instruments used in hospitals and doctors' offices are sterilized before they are used. Doctors and nurses scrub to be as free of pathogens as possible before surgery.

Diagnosis of Disease

To diagnose means to identify by signs and symptoms. For many years, doctors had to rely on their own senses to diagnose disease. In 1819, the French physician R.T.H. Laënnec used a perforated wooden cylinder to listen to sounds from a patient's chest. That cylinder was the first stethoscope, and it made diagnosis easier. Today, there is a wide variety of tools for diagnosing disease. This section will discuss a few of them.

Laboratory Tests

Most laboratory tests examine body fluids. For example, a complete blood count (CBC) measures more than a dozen characteristics of blood. Among other things, it calculates the number of white or

« **Fig. 26-2** Researchers are concerned that a virus causing a type of flu common in birds may mutate and become lethal to humans. The ideal solution would be a vaccine to immunize people against the virus.

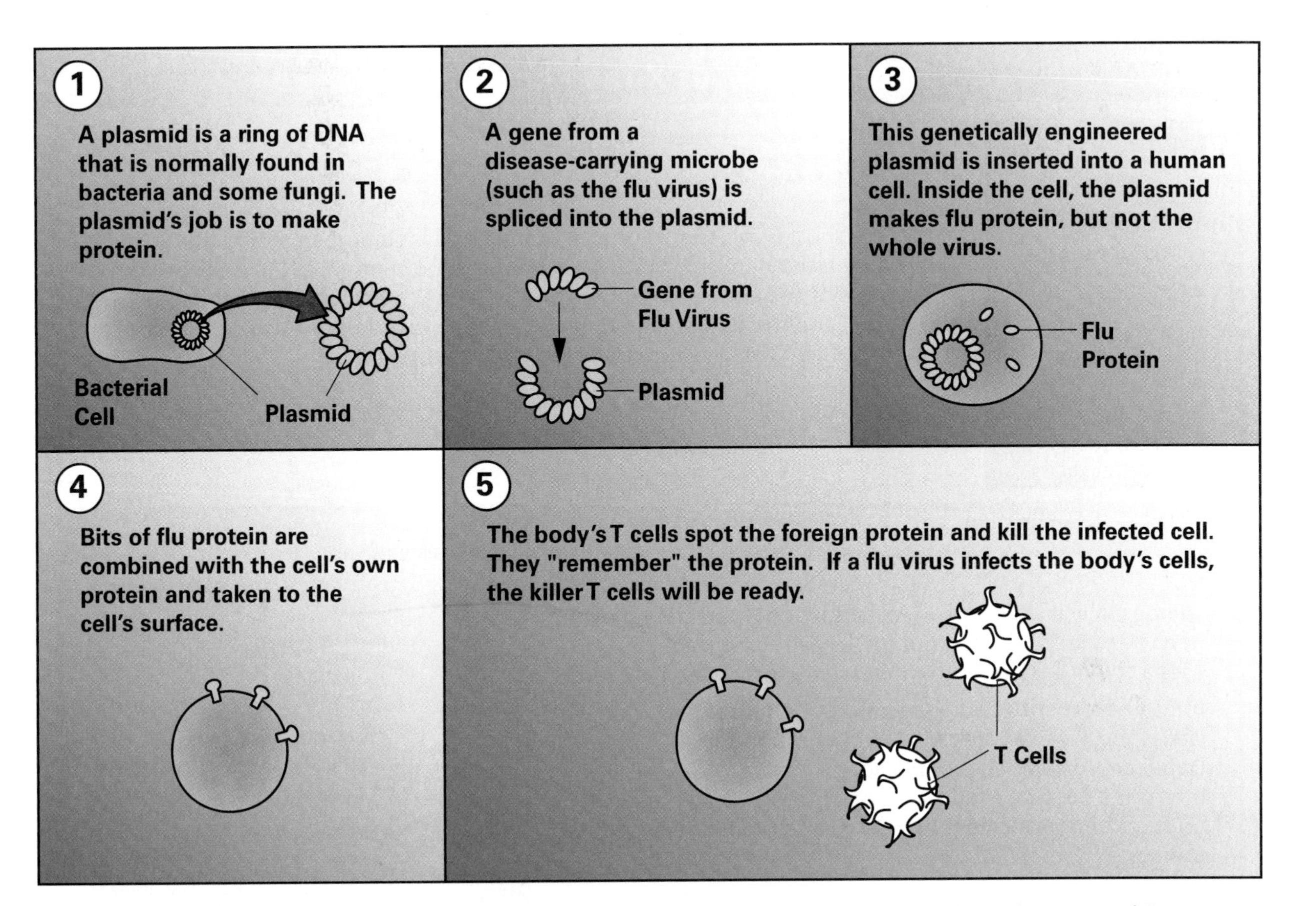

Fig. 26-3 This diagram shows how a DNA vaccine would work. Scientists have been working to develop DNA vaccines against the flu, AIDS, malaria, and even some cancers.

red blood cells in a cubic milliliter, the different types of white cells, and the volume of red cells.

Laboratory tests are valuable, but they are not perfect. All laboratory tests will sometimes yield a false-positive or false-negative result. A false-positive result means the test indicated the presence of disease, but the patient does not really have that disease. A false-negative result means the person has the disease, but the test did not show it. Additional tests, plus the knowledge and experience of the physician, are needed to make the correct diagnosis.

Instrumental Screening

Many instruments can be used to examine the body for disease. One example is the gastroscope. This is a flexible lighted shaft used to examine the inside of the stomach. The patient is given a local anesthetic, and the tube is passed through the mouth and down into the stomach. The gastroscope can be used to check the stomach for ulcers or other disease.

Some instruments measure electrical activity. An electrocardiograph measures electrical activity of the heart. It is used to detect abnormal heart actions. The instrument draws a graph to show the electrical activity. The graph is called an electrocardiogram (ECG or EKG).

Radiography, or X-ray Photography

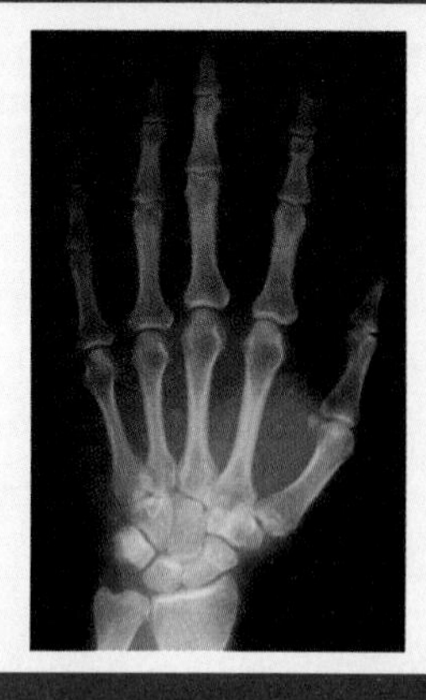

X-rays are short wavelengths of electromagnetic radiation. In radiography, the patient is placed between the X-rays and a sheet of film. The X-rays pass through some parts of the body and expose the film. Other areas, such as bones, block the X-rays, and the film behind them is not exposed. The resulting picture looks like a negative, with areas that block X-rays showing up light.

Computerized Tomography, or CT Scan

Tomography is a technique for obtaining X-ray images of deep internal structures. Computerized tomography (CT) uses a computer rather than film to generate the images.

In CT, the patient's body is scanned by an X-ray tube that circles the patient. The rays "slice" through the patient. Detectors collect data about those X-rays. A computer interprets the data, and the image is graphically displayed on the computer screen. CT is also called computerized axial tomography, or CAT.

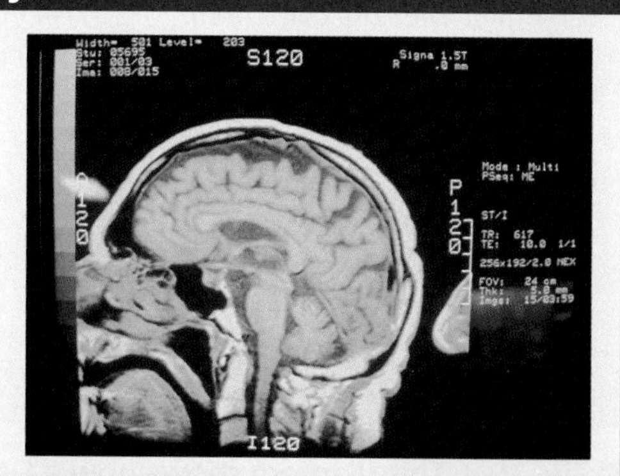

Positron Emission Tomography, or PET Scan

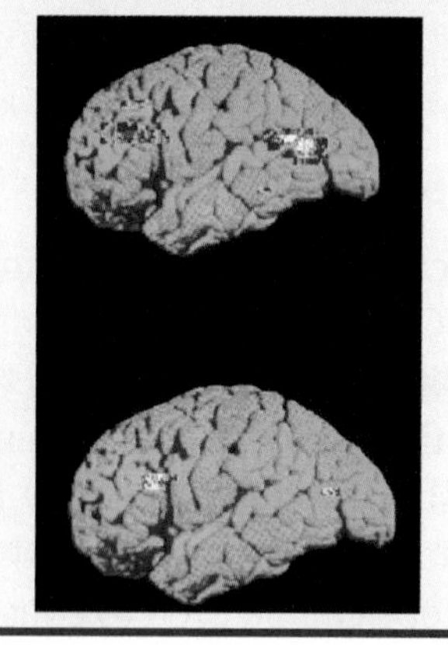

For a PET scan, a chemical containing a short-lived radioactive substance is injected into the body. As the substance decays, it emits gamma rays. The rays are sensed by detectors on opposite sides of the patient. The data from the detectors is analyzed by a computer, and an image is produced. A common use of PET scans is to diagnose disorders of the brain.

Fig. 26-4 Various imaging methods are used in health care.

Medical Imaging

Medical imaging allows us to "see" inside the body using various methods. Which one to use depends on what the health care professional is looking for. What all the imaging methods have in common is that they send energy waves into the patient's body. The waves' reactions to what they encounter produce the images. Several imaging methods are shown in **Fig. 26-4**.

Sonography, or Ultrasound

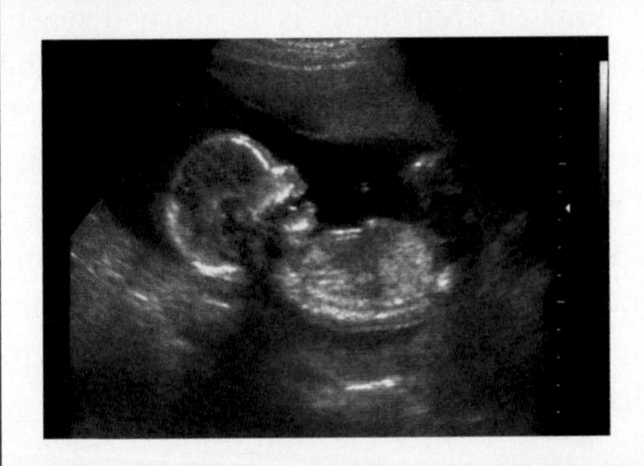

Sonography sends out sound waves and interprets the echoes that return. The sound waves are of a very high frequency, too high for humans to hear.

Different parts of the body reflect the sound waves in different ways. Fluids, for example, echo differently than solids. Echoes from things that are farther away take longer to return. A computer translates data about the strength and position of the echoes into an image called a sonogram.

Sonography is useful for cases in which X-rays or chemicals would be damaging. For example, sonography may be used to determine whether a fetus is developing normally.

Magnetic Resonance Imaging, or MRI

MRI is useful for showing soft tissue. (*Tissue* is a group of cells that form a structural part; for example, skin, bone, kidney.)

For an MRI exam, the patient is placed inside a tube-shaped magnet. The magnetic field generated by the tube causes the hydrogen atoms in the patient to line up in parallel. Radio signals then cause some of the atoms to tip over. When the signal stops, the atoms go back to their earlier position. As they do so, they emit a signal. The MRI machine reads the signals and translates them into images.

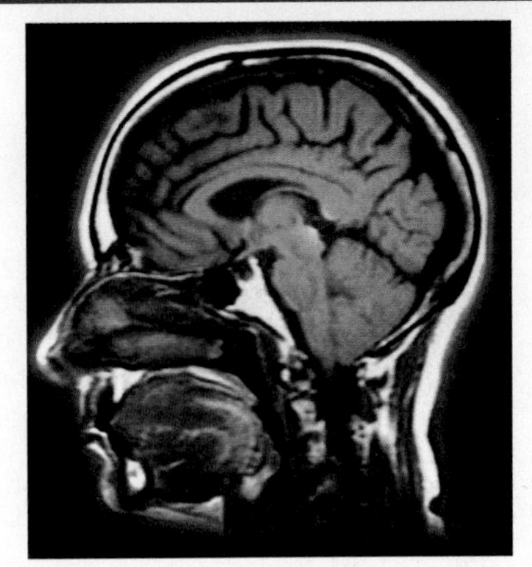

Fig. 26-4 Continued.

Surgical Examination

Some diagnostic techniques require surgery. One type of surgical examination is the biopsy. In a biopsy, a small amount of tissue is removed from the body and examined under a microscope.

If a simple biopsy cannot provide enough information for a diagnosis, exploratory surgery may be required. The surgery is done to examine and/or remove abnormal tissue.

Genetic Testing

Genetic testing is usually done for one of two reasons. The people being tested may want to know whether they have a condition that threatens their health, or they may want to know whether the health of future children is in doubt. For example, a woman with a family history of breast cancer may want to know whether she carries a gene that puts her at risk for developing this cancer. A couple with a family history of cystic fibrosis may want to know whether they are carriers who might produce children with this disorder.

Various procedures are used to do genetic testing. One procedure is chromosomal analysis. Human cells are grown in the laboratory. The cells are stained and sorted, and the chromosomes are then counted and displayed. Abnormalities in the chromosomes can predict the risk of inherited disorders such as Down syndrome.

If people know they carry certain genes for disease, they can make lifestyle adjustments and lower their risk. However, the information may also be turned against them unless laws and ethical standards are established for its use.

Treatment of Disease

One goal of treatment is to cure disease. When the disease is unknown or incurable, the goal of treatment is to relieve symptoms such as pain. Many kinds of treatment are available.

One recent invention, called the Hemopurifier, is a pen-sized filtering device that cleans the blood of pathogens. Following are other current and emerging technologies used to treat disease.

Math Application

Metric Measurements If you look at the record of your last visit to the doctor, you may find your height and weight recorded in metric form. For instance, your height may be written as 168 cm and your weight as 69 kg. Can you figure out from this information how tall you are in feet and inches or how much you weigh in pounds?

By multiplying by a conversion factor, you can easily figure out how different measurements compare to one another. The table below lists conversion factors for three different types of measurements: length (or height), weight, and speed.

For example, if your height is 168 centimeters and you would like to know how tall you are measured in inches, you multiply:

$$168 \text{ cm} \times 0.394 = 66.2 \text{ in}$$

Dividing by 12, you find that that your height is about 5'-6". Your weight of 69 kilograms can be converted to:

$$69 \text{ kg} \times 2.2 = 152 \text{ lbs}$$

Another conversion changes speed from one form to the other. If you run a 5 kilometer race in 30 minutes, your average speed is 10 kilometers per hour. What is your speed in miles per hour?

$$10 \text{ km/h} \times 0.622 = 6.22 \text{ mph}$$

FIND OUT Use the conversion table to calculate each of these conversions.

1. 150 pounds to kilograms.
2. 6 feet, 2 inches to centimeters.
3. 4 miles per hour to kilometers per hour.

To Convert		
From	**To**	**Multiply by**
inches	centimeters	2.54
centimeters	inches	0.394
pounds	kilograms	0.455
kilograms	pounds	2.2
miles per hour	kilometers per hour	1.61
kilometers per hour	miles per hour	0.622

Medications

Many drugs are available to doctors to treat illness. Some, such as blood pressure medicines, promote health and slow down the onset of serious problems. Others, such as painkillers, relieve the symptoms of a disease. Still others kill the pathogens that cause some diseases.

Antibiotics, such as penicillin, are used to kill certain bacteria. However, overuse of antibiotics has helped reduce their effectiveness. Some strains of bacteria have grown resistant to them.

A new science called pharmacogenomics tries to match a person's genes with medications that will be most effective. Doctors may soon be able to customize drugs for a particular group of patients. One genetically targeted medication, Herceptin, is already being used for certain patients with breast cancer.

Antibiotics are generally not effective against viruses or fungi. For example, antibiotics can't cure the common cold, which is caused by viruses. There are, however, other drugs that are antiviral or antifungal.

Recombinant DNA

In health care, recombinant DNA is used to make drugs. One such drug is alpha interferon, which is used in the treatment of hepatitis B and C and some cancers. Alpha interferon is a protein that improves the body's ability to fight tumors and viruses.

Interferon is produced naturally in the body, but in very small amounts. Those amounts may not be enough to fight off serious diseases such as hepatitis. Recombinant DNA makes it possible to produce large amounts of interferon quickly. The human gene for interferon is spliced into the *E. coli* bacterium. The bacterium then produces interferon. Large numbers of the bacteria are grown, and the interferon is harvested from them. See **Fig. 26-5**.

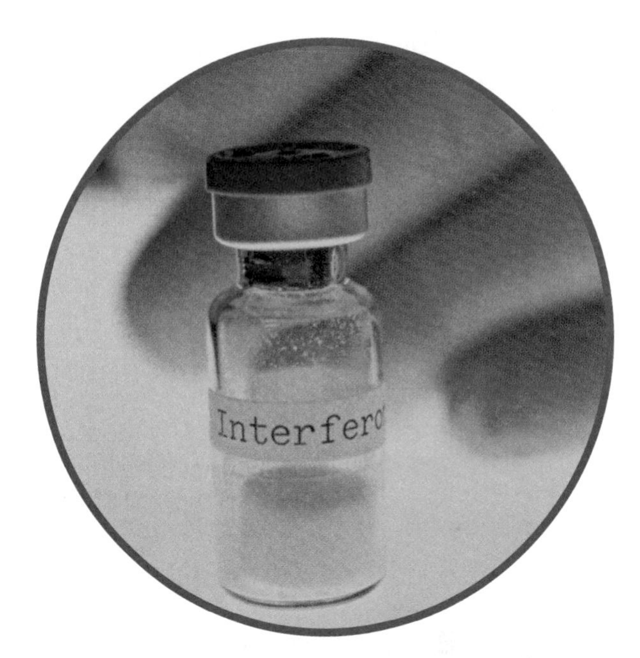

Fig. 26-5 Colonies of bacteria are used to produce interferon in large pharmaceutical plants.

Monoclonal Antibodies

As stated earlier, antibodies attack foreign substances by binding to the substance. Monoclonal antibodies bind to one specific type of foreign substance. (*Mono* means "one.") The antibodies are clones of each other; they are all exactly alike.

In the 1970s, scientists developed a way to produce large amounts of monoclonal antibodies. The fact that these antibodies are very specific makes them useful tools in the diagnosis and treatment of disease. For example, certain monoclonal antibodies bind to cancer cells. Radioactive atoms can be added to these antibodies and given in tiny amounts to a patient. Because the antibodies bind to cancer cells, the radiation reveals the location of the cancer. Monoclonal antibodies are also being used experimentally to deliver drugs that kill cancer cells.

Gene Therapy

Gene therapy involves the transfer of a normal gene into an individual who was born with a defective or absent gene. Cystic fibrosis is a hereditary disease caused by a defective or absent gene. It mainly affects the lungs, making breathing difficult. Experiments have been done in which genetically altered viruses are used to deliver healthy genes to the lungs of people with cystic fibrosis. Thus far, the treatment has been effective only for a few months. One day, there may be a permanent cure.

Regenerative Therapies

Scientists working on regenerative therapies are hoping to reprogram the body to heal itself. You have probably heard of the controversy over stem cell research. **Stem cells** are cells that can turn themselves into several other types of cells that are specialized. For example, a stem cell might turn into a brain cell or a liver cell. Embryonic stem cells, found in the tissues of unborn babies up to eight weeks after conception, can change into any other type of cell. Adult stem cells are more specialized and are limited to becoming only certain other types of cells. So far, the most promising research has occurred using adult stem cells.

Much stem cell research has focused on repairing the brain and the heart. In recent tests, adult stem cells from bone marrow have been injected into damaged areas of patients' hearts. See Fig. 26-6. Results suggest that the cells are causing growth of new heart muscle.

Another new regenerative therapy is tissue engineering. Tissue engineering is the science

Fig. 26-6 Scientists have injected adult stem cells into diseased hearts, helping to restore function.

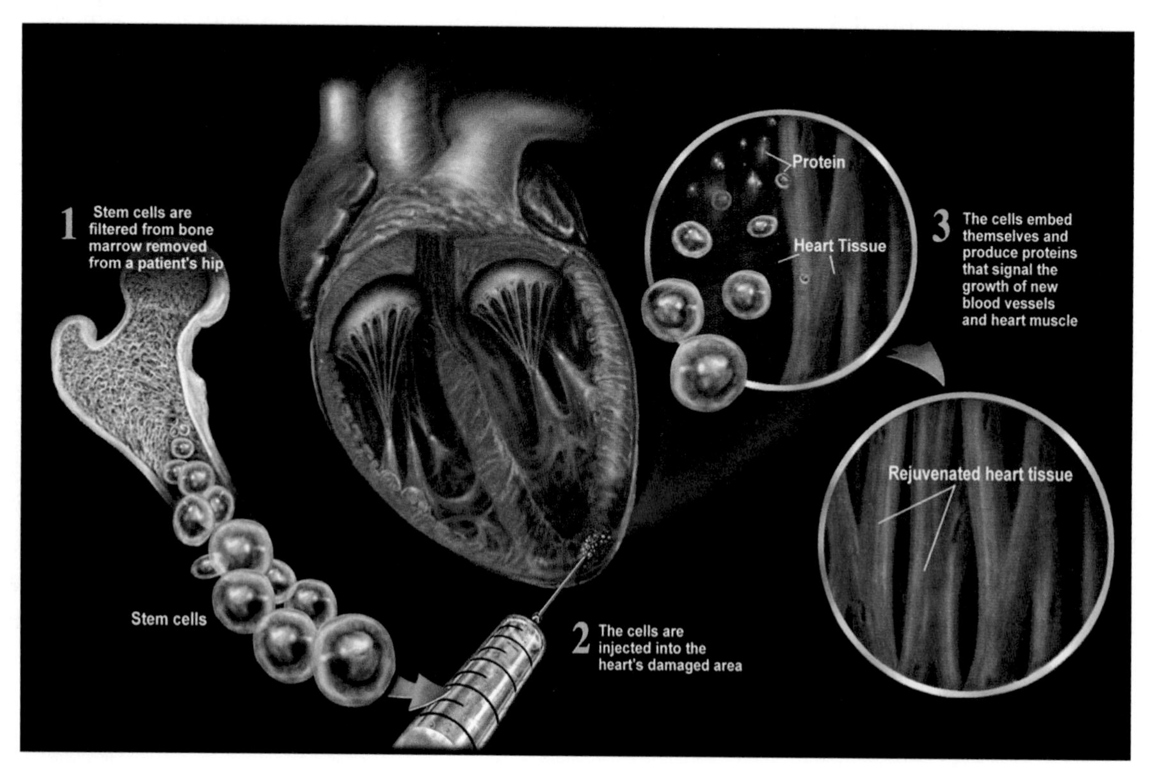

» Fig. 26-7 Surgeons can guide robotic arms to perform precise tasks during an operation.

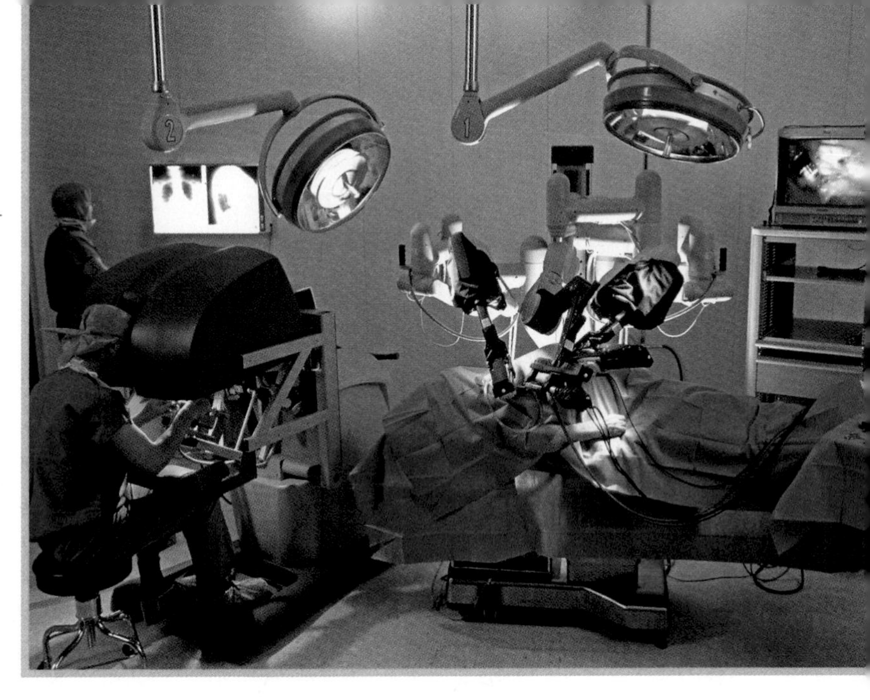

of growing living tissue to replace or repair damaged human tissue, such as skin.

In the laboratory, thin sheets of polymer (plastic) or other material are "seeded" with skin cells. The cells grow and form skin tissue, which is then transplanted into a patient. Skin patches have been grown this way and used on patients with severe burns. The skin patches cover the open wounds while the patient's own skin regrows.

Polymers can also be used as "scaffolding," providing a place for cells to attach themselves and grow. Cartilage and bone have been grown this way in the laboratory. After being implanted in the body, the polymer eventually dissolves, leaving only the cartilage or bone.

Nanotechnology Manipulating materials on a very small scale is done with nanotechnology. Using it, technologists try to create artificial structures as tiny as 1/50,000 the width of a human hair. They hope to use these structures to promote biological repair. They have already developed molecules that self-assemble to form new muscle and blood vessels in living animals. They first develop a plastic scaffold and then seed it with the appropriate cells. Guided by the scaffold pattern, the cells organize themselves into muscle strands complete with arteries and veins. When implanted in the animal, the muscle works normally.

Surgical Treatment

New surgical tools and techniques have been developed. They provide more effective treatment, with less pain and faster recovery, than traditional surgery. A few of these new methods are described here.

- **Laser surgery.** Because lasers can create intense heat, they can be used as a surgical tool. For example, surgeons use lasers to "weld" a detached retina in place or to stop the bleeding of an ulcer.
- **Cryosurgery.** In cryosurgery, extreme cold is used to destroy tissue. For example, liquid nitrogen, whose temperature is about -195 degrees Celsius, is applied to warts and to some skin cancers.
- **Ultrasound.** You've read how ultrasound is used in medical imaging. Sound waves of even higher frequency can be focused on tissue. The sound waves cause the tissue to vibrate and heat, eventually destroying it. Kidney stones can be destroyed with bursts of focused ultrasound.
- **Robotic surgery.** The use of robots for surgery is one of the developments in medical technology that seems like science fiction. However, it is happening right now. See Fig. 26-7. Robotic surgeons are controlled

by human surgeons using a computer. The cutting tool is often a laser and is so accurate that it makes the surgery much safer. For example, during open-heart surgery, the patient's heart is usually stopped. However, a robotic arm used for cardiac cases can operate on a heart while it continues to beat. The patient suffers fewer side effects.

Implants and Prostheses

Technology solves problems and extends human capabilities. This is very evident in the science of **prosthetics**, the artificial replacement of missing, diseased, or injured body parts.

Artificial feet and legs were made as early as 500 B.C.E. Today, there are more than 200 artificial parts, or prostheses, for the human body. Some of them use electric signals to communicate with natural muscles or nerves. This gives patients more and better movement abilities.

For example, a patient who loses a hand may be fitted with a myoelectric hand. This prosthesis looks very similar to a human hand. It is made of plastic and has a flesh-tone glove fitted over it. The myoelectric hand contains electronic sensors that enable it to operate much like a real hand. Newer prostheses include sensors that relay to the wearer a sense of touch and the ability to distinguish hot from cold. See **Fig. 26-8**.

Fig. 26-8 In a myoelectric hand, electronic sensors detect signals from the nerve endings of the person's arm and relay the information to activate the wrist, hand, and fingers.

Prostheses can also help restore the function of a disabled body part. Cochlear implants, for example, improve hearing by translating sounds into electrical signals. The signals stimulate nerve fibers inside the ear. The nerves transfer the signals to the brain.

Other types of prosthetic implants are being developed. One kind delivers electric impulses to the brain to treat the shaking caused by Parkinson's disease. Another helps paralyzed people use their hands.

Biomaterials are materials that are used for prostheses and that come in direct contact with living tissue. Most biomaterials are synthetic polymers, although metals and ceramics may also be used. They are found in artificial heart valves, grafts for blood vessels, and joint replacements.

One major concern with biomaterials is whether they are compatible with the body. Biomaterials must also be strong enough to withstand constant use and last many years.

Telemedicine

Telemedicine is medicine done remotely, at a distance, usually with the aid of computers. It allows doctors in one city or country to communicate or work with other doctors or patients in another city or country. Patient records and X-rays can be sent as data files, enabling the consulting doctor to view them and make recommendations.

Telemedicine also refers to accessing research files and other information in faraway locations. A doctor in India, for example, can refer to research done in California simply by logging on to the Internet or a medical network. Robotic surgery is also a form of telemedicine because the human surgeon does not do the work but controls the robot remotely.

Forensic Medicine

Forensic medicine is the application of medical science and technology to the law and the solution of crimes. Some of the same techniques used to diagnose disease in a living person can be used to determine whether or not someone died of natural causes. Blood chemistry, DNA analysis, and toxicology (the investigation of poisons and drugs) are a few of the tests used.

DNA testing is proving to be one of the most important tools in forensic medicine. Because each person's DNA is unique, DNA evidence taken at a crime scene may be able to identify victims as well as those who committed a crime.

Science Application

Designing Biomaterials

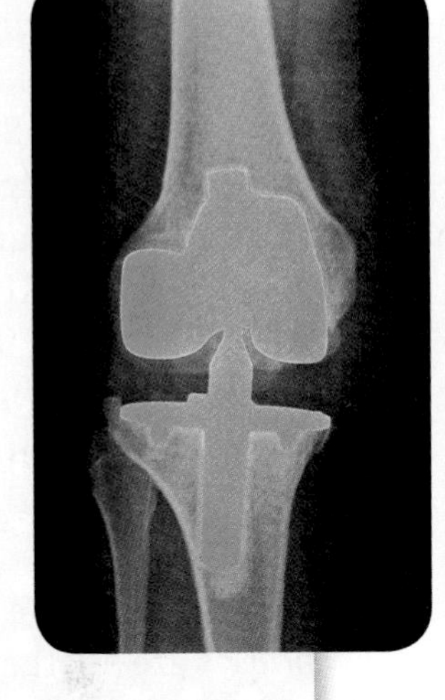

Prosthetic implants have to function as close as possible to the part they are replacing. They must not interfere with tissues or fluids, and they have to last a long time. You really don't want to have surgery every year to replace a hip joint!

One of the most common parts of the human body to be replaced is a joint, such as a hip, shoulder, or knee. Metals, ceramics, and synthetic polymers are often used. The metal parts are usually made of titanium or an alloy of cobalt and chromium. These metals are relatively lightweight and unreactive.

While the metal materials last for a long time, the plastic pieces tend to wear out after about 10 years. Constant friction from moving joints causes gradual deterioration. Biomedical engineers are working to design longer-lasting components for artificial joints.

FIND OUT Make a list of the features that would be important in a material to be used to replace each of the following: teeth, tendons, and arteries.

Human Factors Engineering

Human factors engineering is the design of equipment and environments to promote human health, safety, and well-being. It is sometimes called ergonomics.

Do you use a wrist rest with your computer mouse? That's an example of human factors engineering. The wrist rest helps prevent stress and strain on your wrist. It makes the interaction between human (you) and machine (the computer) easier and more comfortable.

What other measures might make working at the computer more comfortable? How about taking frequent breaks to relieve muscle and eye strain? What about making sure the room is not too hot or too cold? These measures all relate to human factors engineering. This bio-related technology is about the design of machines, work methods, and environments.

Machine Design

If a machine or tool is hard to use, the work takes longer. It may even be dangerous. Human factors engineering seeks to design machines and machine systems that fit the way humans move and think. See **Fig. 26-9**. Consider the way controls are placed on the dashboard of a car. Drivers can see and reach them easily. If they couldn't, it would be much harder to drive the car safely.

People come in many different sizes and shapes. They don't all have the same abilities. Some products therefore are made in different designs for different needs. Left-handed scissors are one example. Easy-open medicine bottles are another.

Work Methods

Sometimes it's not the machines or equipment that causes problems but the way we use them. For example, people who work with desktop

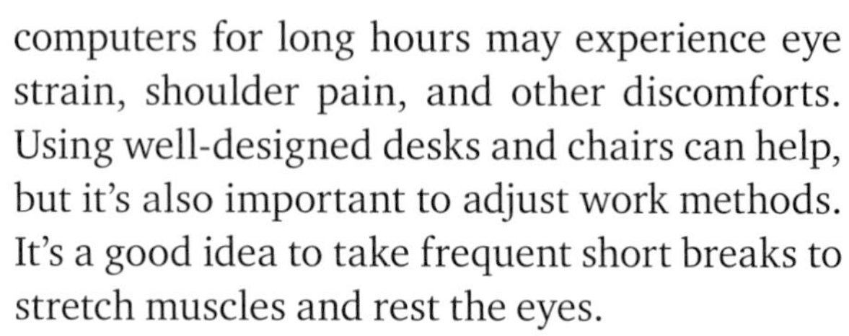

computers for long hours may experience eye strain, shoulder pain, and other discomforts. Using well-designed desks and chairs can help, but it's also important to adjust work methods. It's a good idea to take frequent short breaks to stretch muscles and rest the eyes.

The way in which repetitive tasks are done is also a factor. The placement of items to be put together on an assembly line can affect how tired a worker gets.

Fig. 26-9 These grass clippers have been designed to fit comfortably in the user's hand. ***Can you think of other tools that have been designed for more comfortable use?***

The Built Environment

Human factors engineers design environments that improve comfort and safety. Suppose someone must use a wheelchair to get around. Homes can be designed with wider doorways so that wheelchairs can pass through easily. Kitchens can be designed with lower countertops so that a person seated in a wheelchair can reach them easily.

Today, computers help keep our environment safe and comfortable. In smart buildings, sensors and computerized controls operate the lighting, heating, and security systems. These control systems improve efficiency and reduce the costs of operating the building.

Universal Design

Human factors engineers face the challenge of satisfying many different types of people. Engineers achieve this through universal design. This is the design of all products and structures to be as usable as possible by as many people as possible regardless of age, ability, or situation. For example, the company OXO makes utensils and tools that are comfortable for all people to use. See **Fig. 26-10**.

Environmental Engineering

As you now know, ergonomics involves how human-made surroundings affect our health. Our natural surroundings also affect our health. **Environmental engineering** is the use of science and engineering principles to improve the environment. Environmental engineering can be used to improve the water, land, and air. It includes purifying water and air as well as managing solid, liquid, and hazardous waste.

Fig. 26-10 These kitchen utensils have been designed for comfort. Well-designed utensils are very important to those who spend a lot of time cooking and preparing food.

Water and Air Purification

Communities must provide clean water to ensure the health of the people who live there. Your local water department is responsible for checking that the water you drink is clean and free of pathogens. The water is usually filtered, and chemicals may be added to it.

The air we breathe also needs to be purified. As more air emissions are produced from large urban areas, new methods of air purification are needed.

Car manufacturers are trying to produce cars that emit fewer pollutants. In the transportation unit of this textbook, you learned about vehicles that are designed to save energy and produce fewer air emissions.

Factories also need to control air emissions from smokestacks. Many factories use scrubbers, pollution control devices attached to exhaust systems to help purify the air. See Fig. 26-11.

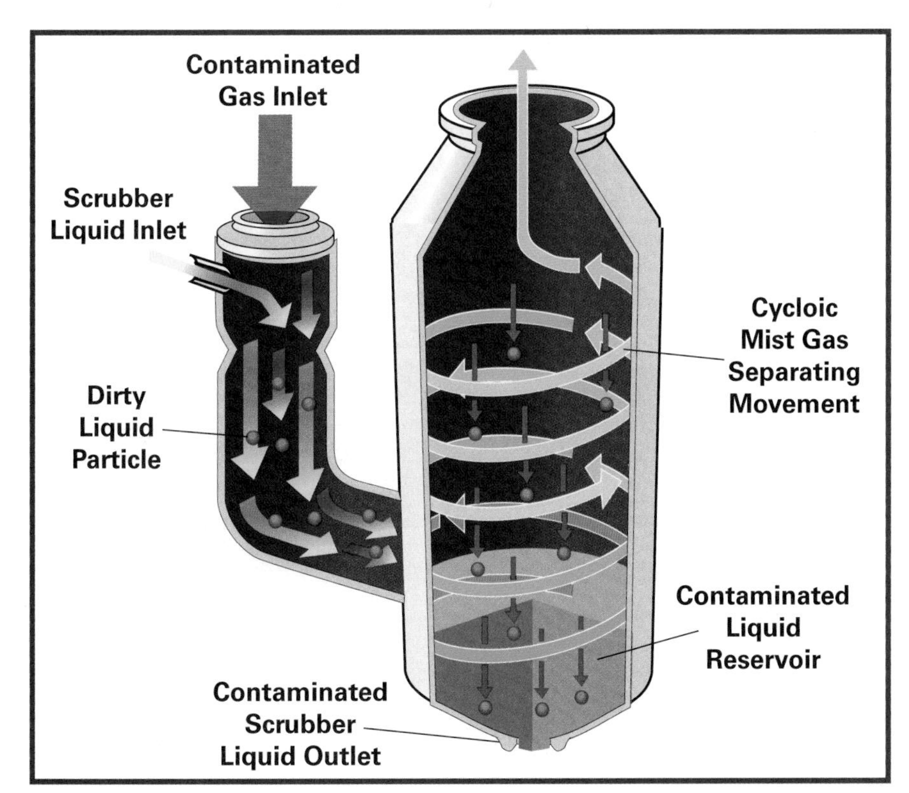

Fig. 26-11 In this air scrubber, a liquid surrounds the dirty particles. As the gas cycles upward, the liquid-covered particles drop down into the reservoir.

Solid Waste Management

Waste management refers to all the operations involved in the collection, storage, and treatment of waste. How waste is managed has an impact on the health of the community and on the environment. Solid waste includes household garbage, construction debris, and other nonsoluble materials. In the United States, we throw out more than 1 billion pounds of solid waste each day. Only about 10 percent is food waste.

Most communities provide regular pickup of solid waste from homes and businesses. In many communities, people are required to sort the waste before pickup. For example, they may have to keep yard waste in a different container than household garbage.

Treatment and storage of solid waste varies.

- Incineration is the burning of waste. One advantage of this treatment is that the heat produced can be used to generate steam and electricity. However, the gases and ash produced by burning can be hazardous. Keeping the gases and ash from polluting the environment can be expensive.
- Composting may be done for yard waste and/or food waste. The waste is stored either outdoors or in large containers. Aerobic bacteria (bacteria that use oxygen) "digest" the waste, turning it into a crumbly product that can be used as a soil conditioner or mulch.
- Many communities bury solid waste in sanitary landfills. The landfill must have a liner at the bottom to keep the waste from polluting surface water or groundwater. Each day, a thin layer of waste is added to the landfill. Heavy equipment compacts it, and the layer is then covered with soil. When no more waste will fit in the landfill, it must be capped with a waterproof cover. See **Fig. 26-12**.
- Recycling of solid waste can greatly reduce the amount that must be burned or sent to a landfill. Recyclable waste (paper, glass, metals, and plastics) can be sorted at a material recycling facility, or MRF (a "murf"). Once sorted, it can be sent to facilities that use it to make new products. Paper, for example, can be recycled to make new paper. What had been waste becomes a resource.

Sewage Management

Sewage includes the liquid and solid wastes that are carried by sewers. It includes waste from homes and businesses, factories, and storms. Sewage is mostly water, and the water can be returned to the environment. First,

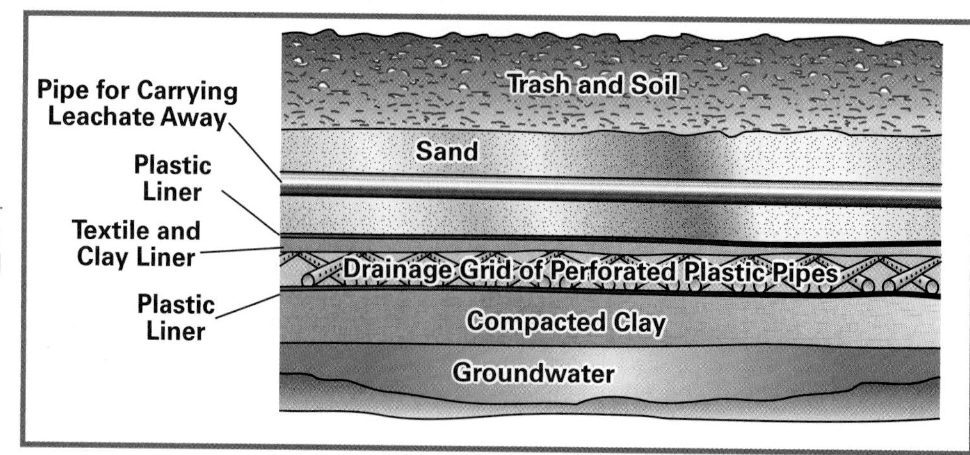

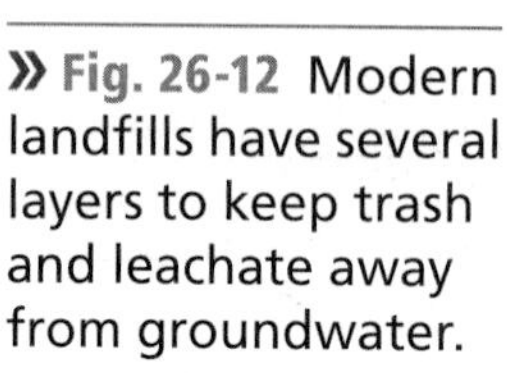

» **Fig. 26-12** Modern landfills have several layers to keep trash and leachate away from groundwater.

though, it must be treated to remove solids and toxic chemicals and to destroy harmful microbes.

The sewage usually goes through several stages of treatment. The purified wastewater is then released into a lake or river. The sludge left behind is treated to make it safe to handle and to remove odors. It may be sent to a landfill or incinerated.

Hazardous Waste Management

Hazardous waste is any waste that can harm human health or the environment. It may be chemical, biological, or nuclear waste.

Hazardous chemical waste is usually a byproduct of industrial processes. Hazardous biological waste can include such things as used bandages or needles that carry harmful germs. Some animal and human waste can also contain drug residues, and sewage treatment plants are not designed to remove them. Trace amounts have been found in some drinking water. Nuclear waste is produced mainly by nuclear power plants. The radiation it emits can cause illness or genetic mutations.

Special measures must be taken when storing, treating, or disposing of hazardous waste. The waste, which might be liquid, solid, or gaseous, must be stored in such a way that it is not released into the environment. For example, some liquid wastes are temporarily stored in ponds built for that purpose. The bottom of the pond has a liner to keep the liquid from seeping into the groundwater.

Some hazardous wastes can be treated to make them harmless. One treatment method, called landfarming, is used for petroleum wastes. The waste is brought to a plot of land, where it is mixed into the surface soil. Microbes that can "eat" the waste are added, along with nutrients. Using living organisms or parts of organisms to change materials from one form to another is called bioprocessing. Sometimes the microbes used are genetically engineered bacteria. When this treatment is used on land that has been previously contaminated, the process is called bioremediation. See **Fig. 26-13**.

Wastes that cannot be treated are either incinerated or buried. In both cases, special precautions must be taken. Remember that a clean environment is as important for health as taking the right vitamins or getting treatment when ill.

Fig. 26-13 Petroleum waste is mixed with soil. Then microbes eat the waste.

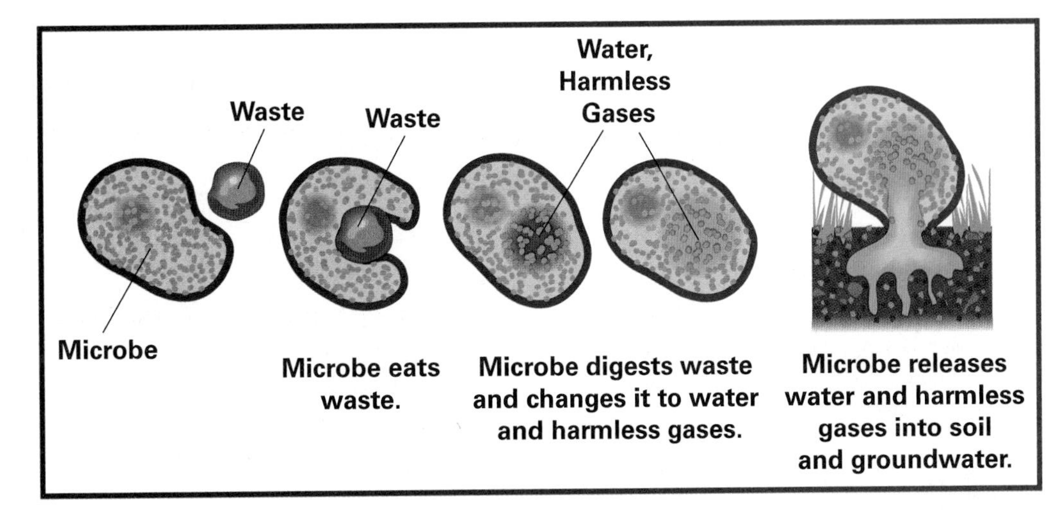

Main Ideas

- Bio-related technologies in health care involve the prevention, diagnosis, and treatment of disease; human factors engineering; and environmental engineering.
- Technologies used for diagnosis include laboratory tests, screening, medical imaging, surgical examination, and genetic testing. Once a diagnosis has been determined, treatment can begin.
- Human factors engineering and environmental engineering are both related to our well-being.

Understanding Concepts

1. Describe how immunization works.
2. What do all medical imaging technologies have in common?
3. What are four methods of surgical treatment?
4. What is the purpose of human factors engineering?
5. Explain how environmental engineering is related to our health.

Thinking Critically

1. What might be a negative impact of telemedicine?
2. How is organization critical to performing laboratory tests?
3. Why must a doctor also have communication and organization skills?
4. How might someone who works in a hospital setting apply math and science?
5. Would you want to be genetically tested for a serious disease that has no cure? Explain your answer.

Applying Concepts & Solving Problems

1. **Math** Imagine you work in a hospital and have learned an accident victim has lost 1.5 pints of blood. Determine how many liters would be needed to replace the lost blood by finding the conversion factor.
2. **Science** What are some recent achievements in "intelligent prostheses"?
3. **Communication** With another student, simulate a conversation between a doctor and a patient in which the doctor must convey serious news to the patient. Take turns being both the doctor and the patient. How can the doctor communicate the necessary information and still be sensitive to the patient's needs?

Activity: Create a New Replacement Joint

Design Brief

In recent years, human joint replacement has become very common. You may know of someone who has received an artificial knee or hip implant. These implants relieve the pain and suffering caused by injured or worn joints. After surgery, patients can lead more active lives.

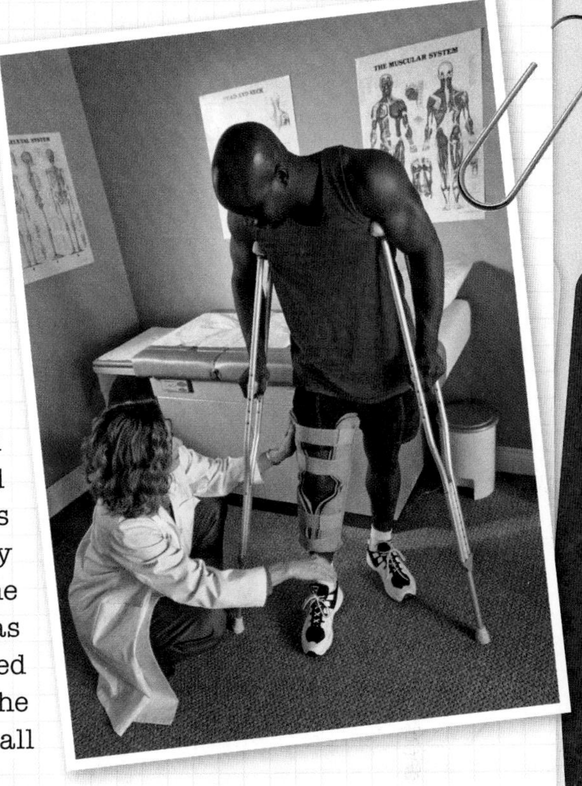

Implants are made of special metal alloys and plastics. The metal is secured to the bone tissue, and the plastic serves as a spacer that fills out the joint the way normal cartilage does. In order for the artificial joint to work almost as well as the natural joint, parts must be machined within critical tolerances. That means the patient must be accurately measured so all the parts fit correctly.

CHALLENGE

Analyze a human joint, design a replacement for it, and create a model.

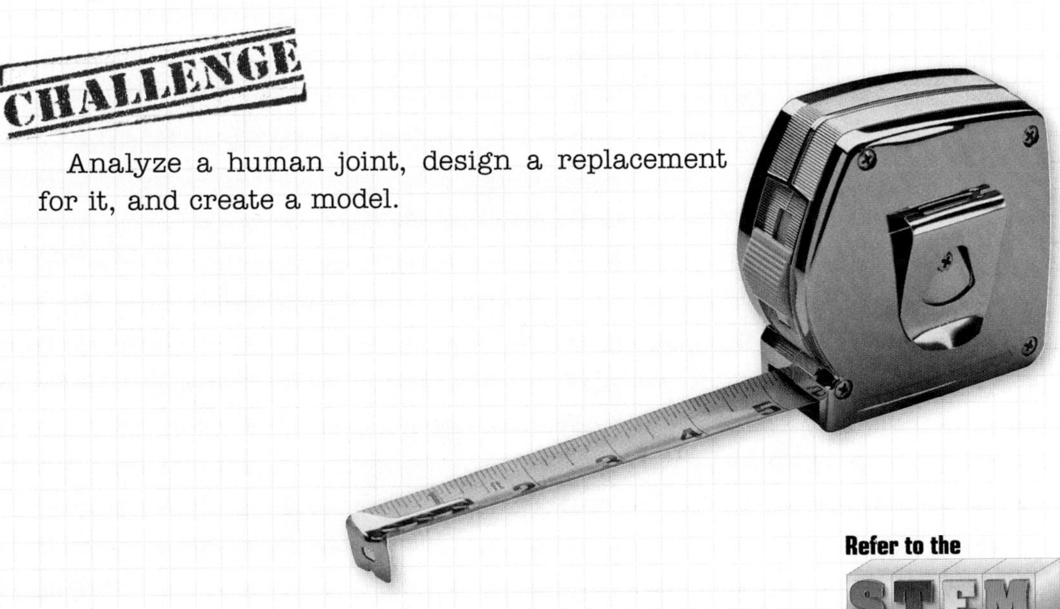

Refer to the STEM HANDBOOK

Criteria

- You must research the anatomy of the joint you have selected.
- Make a sketch of your chosen joint with bones, tendons, and cartilage labeled.
- Measure your own body using a tape measure and calipers to obtain the approximate sizes for your model.
- Write a paragraph explaining what type of forces act on your chosen joint. Indicate how this would influence your choice of materials if you were making a real joint.

Constraints

- Your model must be full scale, and the parts must be labeled.
- All labeling must correctly identify parts, and terms must be spelled correctly.
- You may create a traditional model using such materials as papier-mâché, clay, and plastic or a virtual model using a computer and CAD software.

Engineering Design Process

1. **Define the problem.** Write a statement that describes the task you are to accomplish, as well as any obstacles you will need to overcome (i.e., making sure the joint will fit in your own body).
2. **Brainstorm, research, and generate ideas.** With your team, discuss possible solutions. Hint: If the biology lab or classroom in your school has a model of the human skeleton, you should study it.
3. **Identify criteria and specify constraints.** These are listed on page 528.
4. **Develop and propose designs and choose among alternative solutions.** Choose the design that will work as the best replacement for a missing joint.
5. **Implement the proposed solution.** Decide on the process you will use for making the model. Gather any needed tools or materials.
6. **Make a model or prototype.** Create your model or drawing. Follow all safety rules.
7. **Evaluate the solution and its consequences.** Does your model work as intended? Will it function as a replacement?
8. **Refine the design.** Based on your evaluation, change the design if needed.
9. **Create the final design.** After any changes or improvements take place, create your final model to scale. Be sure to include all necessary labels.
10. **Communicate the processes and results.** Present your finished model to the class. Use your labels to point out how the replacement is connected and how it will function.

SAFETY FIRST

Before You Begin. Make sure you understand how to use the tools and materials safely. Have your teacher demonstrate their proper use. Follow all safety rules.

CHAPTER 27 AGRICULTURAL TECHNOLOGIES

≈ Planning and growing a garden can be a very interesting and fulfilling hobby or vocation. In this chapter's activity, "Experiment with Agriculture," you will have the chance to grow and test your own garden.

Objectives

- Describe technologies related to food production.
- Explain ways in which agricultural technologies play a part in medical treatments, energy production, and biowarfare.
- Describe how agricultural technologies contribute to the health of the environment.

Vocabulary

- hybrid
- genetic engineering
- recombinant DNA
- clone
- controlled environment agriculture (CEA)
- pharming
- biowarfare
- integrated pest management (IPM)
- bioremediation

Reading Focus

1. As you read Chapter 27, create an outline in your notebook using the colored headings.
2. Write a question under each heading that you can use to guide your reading.
3. Answer the question under each heading as you read the chapter. Record your answers.
4. Ask your teacher to help with answers you could not find in this chapter.

Food Production

Agriculture is the practice of producing crops and raising livestock. Although food is its primary product, agriculture produces many other things, from medicines to fuels. Those will be covered later in this chapter.

Plant and Animal Breeding

People have long worked to develop improved varieties of plants and animals. Today's corn, for example, is very different from the wild corn that first grew in America. By selecting individual plants with desirable traits and planting the seeds, Native Americans developed varieties of corn that were larger and produced more kernels. During the 20th century, scientists in the United States developed corn hybrids. A **hybrid** is the offspring of different varieties, breeds, or species. See Fig. 27-1. The hybrid has characteristics of both its "parents."

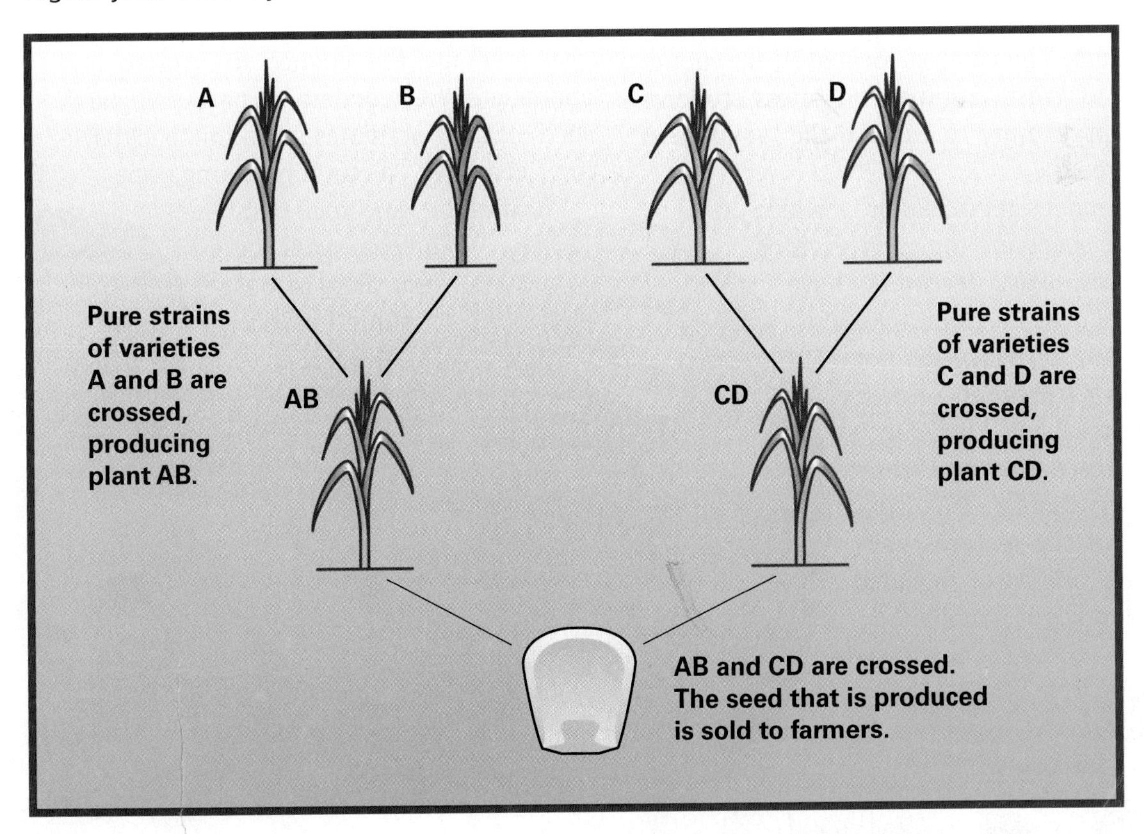

Fig. 27-1 Hybrid seed corn is produced by double crossing. The hybrid seed produces a higher yield than any of the earlier seeds.

Young Innovators

NYSC: The Grand Adventure

The National Youth Science Camp (NYSC) is a three-week summer program open to high school students the summer after they graduate. The NYSC is an all-expenses-paid program. The camp is located in the mountains of West Virginia and offers students the opportunity to explore the wonders of the area.

Attendees get to explore a wide variety of academic topics in exciting lectures. Past lecturers have included Al Bartlett, a widely recognized speaker on population growth and energy resources, and Alan Daly, a field researcher and water conservationist. The subjects of the lectures vary widely and have included such topics as nanobiotechnology, exploring Mars, hydrogen fuel cells, and forensic anthropology.

Students also have the opportunity to participate in hands-on activities and research projects. Directed-study programs allow them to explore topics through hands-on research. Past directed-study programs have included dissection, exploring the technology of renewable energy, investigating astronomical data from radio telescopes, and gene isolation and cloning.

Finally, attendees can listen to and participate in seminars. These seminars are often informal and give delegates the opportunity to learn about subjects outside of science as well as present their own research or share their talents with fellow attendees.

The setting and location of the camp also give students the chance to explore and appreciate the wonders of nature through such exciting activities as hiking, backpacking, camping, mountain biking, spelunking, and rock climbing!

Whatever your interest, there is something for you at the NYSC.

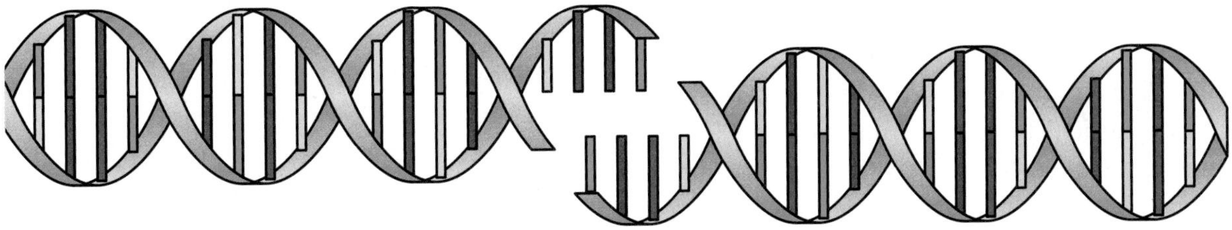

DNA from one organism is cut . . .

. . . and combined with DNA from another organism.

Fig. 27-2 In gene splicing, DNA molecules are cut at certain points. The ends are combined with the cut ends of other DNA molecules to make recombinant DNA. ***What do the different colors represent?***

Hybrids are one example of selective breeding, or artificial selection. In nature, plants and animals reproduce randomly. The offspring may or may not have traits that make them useful to humans. With selective breeding, human beings control the process. Selective breeding has resulted in plants and animals that produce higher yields, are more nutritious, and are resistant to disease and insect pests.

A number of other technologies help humans to control plant and animal production. Among these are genetic engineering and cloning.

Genetic Engineering While selective breeding works with an organism's natural traits, **genetic engineering** gives organisms traits they never had naturally. One way to do this is by altering the organism's normal genes.

Genes can be moved from one organism's DNA into another's. The process is called gene splicing. In gene splicing, enzymes are used to remove the gene from one strand of DNA and glue it into another strand. See Fig. 27-2.

For example, iron is an essential nutrient, and iron deficiency causes many health problems. About 30 percent of the world's population suffers from it. Researchers have developed a genetically engineered variety of rice that provides more iron. Genes from two other plants are spliced into the rice's DNA. One gene provides the iron. The other gene helps the iron be absorbed more readily into the body.

A gene from the California bay tree has been spliced into canola, an herb of the mustard family. The transgenic canola produces laurate, an ingredient used to make soaps, detergents, and shampoos.

The technique of gene splicing makes recombinant DNA possible. **Recombinant DNA** is DNA formed by joining (recombining) two pieces of DNA from two different species. The result is a transgenic plant or animal.

Genetically engineered plants and animals hold great promise. Plants that resist insects and disease will reduce the need for toxic pesticides. New foods will help feed a hungry world. Altered tobacco plants could be used to produce medicine instead of cigarettes. Mosquitoes could be bred with new genes that make them unable to spread disease.

However, there are also risks. There is concern that genetically engineered organisms can become a threat to our environment or our health. A food plant that can tolerate herbicides may be a good thing, but not if the food plant passes along its genes to a weed cousin. A food plant that is made too resistant to insects and disease might become a weed itself, spreading over large areas. As more and more genetically altered plants and animals enter our environment, new pests or new diseases might be accidentally created.

Cloning A **clone** is an individual that is genetically identical to another individual. Cloning (making clones) is not new. It's been common in agriculture for centuries. All the McIntosh apple trees in the world are clones derived from a single plant. In recent decades, however, the technology of cloning has advanced. It is now possible to clone many kinds of organisms, including mammals.

One method of cloning plants is to take a single cell and place it in a medium that contains the right nutrients. This technique is called tissue culture. See Fig. 27-3. The plant that grows from the cell is genetically identical to the plant from which that cell came. One plant can provide enough cells for many clones.

The technology for cloning embryos of animals such as pigs, cows, and sheep has been known since the early 1980s. In the late 1990s, scientists for the first time produced clones from adult mammals. However, animal cloning is not yet widely practiced. It's too costly and risky and raises serious ethical and economic questions.

Science Application

Building Herbicide Resistance If you pass a field of young corn or soybean plants, you usually see the crops growing in orderly rows. There may be some dead weeds among the rows, but it is likely that there are very few living weeds. An herbicide has been applied to the field to kill the weeds that compete with the crops. But why are the corn and soybeans not attacked by the herbicide at the same time?

In many cases, the answer is genetic engineering. When a new herbicide is developed, there are usually some species or varieties of plants that resist its effects. Unfortunately, the herbicide resistance does not necessarily occur in the crop plants you want to grow in the field.

Scientists study the resistant plants to find the gene that protects the plant from the herbicide. This gene is then isolated and implanted into the DNA of a desirable plant, such as corn or soybeans. Once the resistance is genetically engineered into the plant, its seeds will produce plants that are not killed by the herbicide.

This resistance is useful because the farmer can apply herbicide even when the crop is already growing. It is applied only when there is an actual weed problem, saving time and money.

FIND OUT Use the Internet to find out about herbicide-resistant crops. Make a table of the advantages and disadvantages.

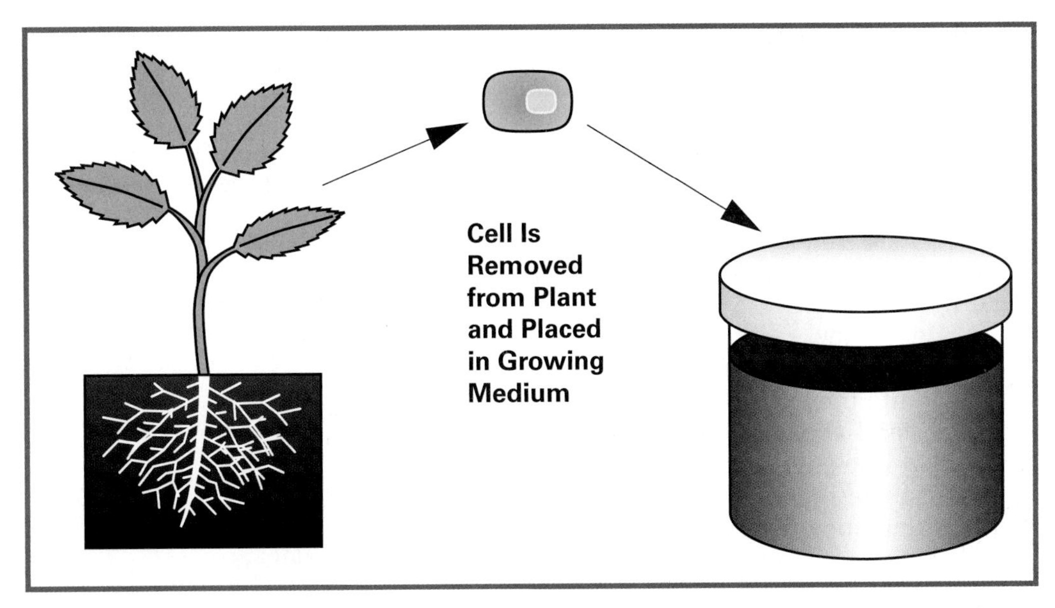

Fig. 27-3 In plant tissue culture, one piece of a plant is used to grow a new plant. If the piece is a single cell, then the new plant will be a clone.

Plant and Animal Maintenance

Plants and animals need care if they are to thrive. They need food, water, the proper conditions for growth, and freedom from disease. Technology often plays a part in maintaining plants and animals. New maintenance technologies include the use of special fertilizers, disease prevention, and organic farming.

Use of Fertilizers As plants grow, they absorb nutrients from the soil. Different plants may absorb different nutrients. If the same plants are grown in the same plot year after year, the soil loses so many of those particular nutrients that crops no longer thrive. Fertilizers are chemical compounds that replace or add nutrients to soil. Today, special fertilizer mixtures can be developed for use with individual crops. Use of nanotechnology to create the fertilizer at the molecular level is being researched for making fertilizers that better target a crop's specific needs. Nanotechnology is also being used to help create soil particles that bind together better and prevent erosion (wearing away of the soil).

Although fertilizers increase crop yields, they have had some negative impacts. Weeds also like the nutrients they provide and grow abundantly. Runoff from fields has carried the fertilizers into lakes and streams and affected wildlife and the growth of algae. Fertilizers have also encouraged monoculture farming—growing only one crop or one species over large areas. When diseases or pests attack that particular crop, an entire region can be affected.

Math Application

Dimensions in Land Surveys Land surveys were once based on temporary features such as a large tree, a stream channel, or a pile of rocks. To define boundaries more clearly, Congress passed the Northwest Ordinance of 1787.

Under this law, lands outside the states that already existed could not be sold or transferred until they had been surveyed. For the survey, land was divided into survey townships. Each township was a square measuring 6 miles on each side. A township was divided into 36 numbered sections, each measuring one mile on each side. A surveyed area could then be described by its section number. A section was divided further into quarter sections and each quarter was divided into quarters again.

Today, land areas are commonly described by the number of acres. One acre measures 43,560 square feet, and there are 640 acres in one square mile, or one section. That means that a quarter is 160 acres. A quarter-quarter section is 40 acres, the smallest unit of agricultural land that is normally surveyed. Terms such as "lower 40" and "back 40" refer to specific quarter–quarter sections on a farm.

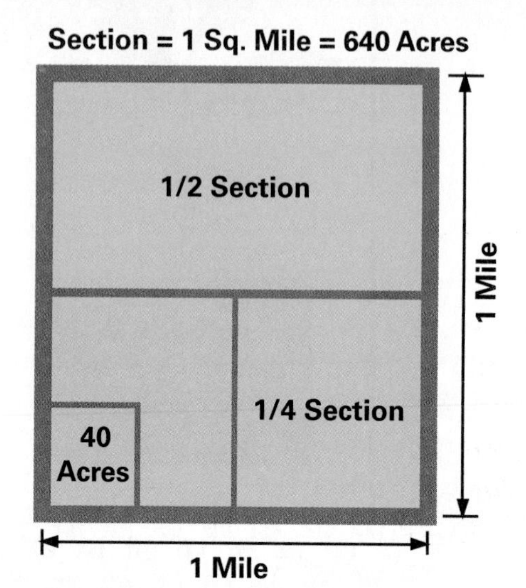

FIND OUT If a piece of land is square, its side length can be determined by taking the square root of its area.

1. Use a calculator to find the dimensions in feet of a square lot with an area of one acre.
2. Without using a calculator, determine the side length of a square parcel of land with an area of 9 sections.

Disease Prevention Farm animals often live under very crowded conditions, and diseases spread quickly. To prevent disease and promote growth, many farmers have been giving their animals antibiotics.

Unfortunately, the meat and other foods produced by these animals contain the antibiotics, which are then passed along to humans. They have also been found in animal wastes, in the soil, and in waterways.

In recent years it has been learned that overuse of antibiotics in this way has helped produce pathogens that are resistant to them. The antibiotics no longer work against those pathogens and the diseases they cause. Both humans and animals are now less safe.

Some research is being done on bacteria that create toxins that kill other bacteria with which they are in competition. Researchers hope that new antibiotics may be derived from those toxins.

Researchers hope that nanotechnology will be able to help detect certain diseases in animals. Tiny particles are being designed that attach themselves to disease-causing proteins like those found in mad cow disease and anthrax and identify them. If the disease is found soon enough, sick animals can be isolated or destroyed, preventing further outbreaks.

Organic Farming is the term used for agricultural practices that avoid the use of synthetic chemicals and bioengineering in favor of naturally occurring fertilizers, pesticides, and other growing aids. Nor can antibiotics and hormones be used in livestock production. The consumer demand for organically raised products is increasing, and organic farms have become very profitable. However, switching from conventional to organic methods is not always easy.

Fields must be made free of chemical fertilizers and pesticides, a process that takes about two years. Pests are controlled by natural methods, such as introducing beneficial insects. Special tilling and harvesting methods are used that protect and preserve the soil. See **Fig. 27-4**. Plant and animal wastes, for example, are used for soil enrichment. Some of the land is returned to woodland to control erosion. Crop yields make organic farming worthwhile. Some farmers report 20 percent increases over yields produced by conventional methods.

The U.S. Department of Agriculture has established standards for organic products. Before a product can be labeled "organic," a government-approved certifier inspects the farm where the food is grown to make sure the farmer is following all the rules necessary to meet USDA organic standards. Companies that handle or process organic food before it gets to supermarkets or restaurants must be certified, too.

Fig. 27-4 This farmer is turning the stubble into the ground. This is done after the harvest in order to return nutrients to the soil.

Mechanization

Using machines to clear and turn the soil, plant seeds, cultivate and harvest crops, and manage livestock has had a great impact on farm efficiency and production. In the U.S., farmers make up less than 2 percent of the population. Because of mechanization, they manage to feed the other 98 percent and still have enough food left over to sell to other countries.

The most recent technology to improve farm mechanization is the global positioning system (GPS). This navigation system, combined with automated steering devices, guides tractors and other large equipment with greater accuracy. GPS monitors on satellites can also tell farmers which fields show signs of a lack of water or fertilizer.

Food Processing and Food Safety

After they leave the farm, most foods require further processing. Milk is pasteurized. See **Fig. 27-5**. Wheat grains are ground into flour. Meat animals are killed and butchered. Some foods are treated with enzymes or microbes. In all food processing, safety is essential.

Enzymes are proteins that catalyze (speed up) a chemical reaction. Enzymes from useful microbes are added to some foods to speed up such processes as fermentation that would take too long otherwise. (Fermentation is a biochemical process in which an organism breaks down a substance into simpler ones.) Enzymes are used to break down proteins in flour so yeast organisms can act on the flour and make bread rise. They are added to fruit juices to make them clear instead of cloudy. Enzymes also help turn milk and cream into yogurt and cheese. See **Fig. 27-6**.

Nuclear Magnetic Resonance (NMR) You read about magnetic resonance imaging in Chapter 26. The same technology is being used to improve the quality and safety of processed foods. NMR can probe foods at the molecular level to learn when the food is likely to spoil. Changes in processing and packaging can then be made to prolong its safe use. NMR can also be used to analyze flavors to determine which ingredients create the most people-pleasing tastes.

Irradiation and High Pressure Not all bacteria found in food are harmful. However, some, such as *E. coli* and *salmonella*, can cause disease and even death. Irradiating foods with X-rays or placing them under high pressure and heat destroys disease-causing organisms. Some people object to irradiation because they fear it makes the food radioactive or destroys its nutritional value. However, years of testing by the Food and Drug Administration show no evidence that the food is harmed. Only the living cells, such as those of pathogens, are destroyed.

« **Fig. 27-5** Before being bottled, milk is heated to a certain temperature. This destroys many harmful organisms and helps maintain freshness in the store and in your refrigerator.

» Fig. 27-6 Enzymes from useful microbes help turn this vat of cream into cheese.

Controlled Environment Agriculture

In **controlled environment agriculture (CEA)**, a plant's or animal's surroundings are carefully monitored and adjusted. The plant or animal receives the right amount of humidity, heat, light, and nutrients for the best growth. Three kinds of CEA are hydroponics, aquaculture, and agroforestry.

Hydroponics Have you ever visited a greenhouse where flowers or vegetables were being grown in pots of soil? That's one type of CEA. Hydroponics is another. Hydroponics is the cultivation of plants in nutrient solutions without soil. The term comes from the Greek words for water (*hydro*) and labor (*ponos*).

Although hydroponics has been practiced for centuries, large-scale use began about 70 years ago. During the 1930s, Dr. William F. Gericke of the University of California experimented with hydroponics. He placed a wire frame over shallow tanks full of liquid nutrients. The roots of the plants on the frame descended through the mesh to feed on the solution below. With this technique, Dr. Gericke grew tomato plants that were more than 25 feet high.

There are several hydroponic systems. The system Dr. Gericke used was water culture. The plants' roots were submerged in water all the time. Another system is aeroponics. In this system, the plant hangs in the air, and its roots are sprayed with nutrient solution. In aggregate culture, the plant grows in a container filled with small pieces of mineral rock or with sand and gravel. Nutrient solution is added to keep the roots moist.

A hydroponic system can be set up much like a factory. The plants can be grown indoors, where temperature, humidity, and light can be controlled. The nutrient solution can be dispensed automatically. Computers can help manage the system, reducing the amount of human labor required. Crop yields are about the same as for soil-grown crops. Today, most of the tomatoes sold in grocery stores are grown hydroponically. See **Fig. 27-7**.

As with other factories, the start-up costs are high. Also, energy is needed to run the system. In areas where the soil and climate are good, there are not many large-scale hydroponic systems.

Fig. 27-7 Nutrients for these tomatoes are contained in a special solution. No actual soil is needed.

Aquaculture is the raising of fish, shellfish, or plants in water under controlled conditions. The aquaculture farm may be a pond or concrete pool. It may be an area along the ocean's coast, with barriers set up to keep the fish in and the predators out. See **Fig. 27-8**. Some shellfish are raised in cages set in the water.

You've probably seen farm-raised catfish in the grocery store. Other fish commonly raised by aquaculture are carp, salmon, and trout. Many aquaculture farms specialize in shellfish such as oysters, shrimp, and crayfish. Various types of red algae are grown commercially. The algae are used to make carrageenan, a white powder used as a binder and thickener in many products, from hot dogs to toothpaste.

Producing food for people is the goal of most aquaculture. However, many government agencies raise fish in order to stock lakes and rivers for sport fishing. There are also aquaculture farms that raise aquarium fish or fish that are used as bait.

Some studies have called into question the nutritional value of farm-raised fish and shellfish. If the food fed to the fish lacks nutrients they could obtain in the wild, the fish them-

Fig. 27-8 Newer fish farms will float on ocean currents and include crew quarters. Fish farms can produce up to 300,000 pounds of fish a year.

selves are not as nutritious for humans. Also, many fish farms are located close to shore where waters are more polluted. The pollutants collect in the flesh of the animals and are passed along to humans.

Agroforestry is a type of controlled-environment agriculture dedicated to the replacement of trees. (This is a type of bio-restoration.) Most agroforestry is done to produce trees used to make paper and lumber products. Genetically engineered seedlings grow fast and quickly replace stock cut down, which makes tree farming more profitable. The farms also mean that fewer wild trees are cut down for commercial purposes.

Fig. 27-9 Sea sponges are being studied as a possible source for new medicines.

Medical Treatments

Living organisms are not only used for food. They are also a source of medical treatments.

Pharmaceuticals

Pharmaceuticals is another term for medicines. Medicines have always been made from plant and animal sources. Bark from a certain tree, for example, was the original source for aspirin. Then synthetic molecules were created to mimic or improve on these natural remedies.

Today, drug manufacturers are working to develop living organisms that will produce pharmaceutical chemicals. When the plants, animals, or microbes have been genetically altered to produce pharmaceuticals, the process is called **pharming**. (The term is a combination of the words *pharmaceutical* and *farming*.) Genetically engineered cows, for example, produce compounds in their milk that can be used to counter diseases such as diabetes and arthritis.

Other researchers are exploring the depths of the oceans for sources of new medicines. Sea sponges, for example, produce toxic chemicals that may one day be used to fight cancer and AIDS. See Fig. 27-9.

Transplant Organs

Many people with life-threatening diseases require organ transplants. However, there are not enough human donor organs available. Researchers are therefore studying the use of organs from animals. An organ or tissue transplanted from one species into another is called a xenotransplant. Using animal organ transplants is not a new science. It was tried unsuccessfully as early as the 17th century.

Much of today's research on xenotransplants is being done with pigs. Heart valves from pigs are being used to replace defective valves in people. Someday the entire heart might be used in people requiring heart transplants.

Researchers are now raising genetically altered animals that have tissues more compatible with the human body so fewer complications will result after surgery. However, xenotransplants are controversial. Many people have ethical or religious objections to the use of animal parts for transplants. There also are concerns that viruses in a transplanted organ could adapt to a human host and become infectious to the general population.

Energy Production

Fuels can be made from agricultural products. You may have seen a label reading "gasohol" or "ethanol" at your local gas station. This fuel is a combination of gasoline and fuel alcohol made from materials such as corn and crop wastes. (These materials are called biomass.) The percentage of fuel alcohol in this mixture is usually around 10 percent. However, flex-fuel vehicles that can run on either plain gasoline or a mixture containing 85 percent ethanol are now available. Questions have been raised, however, about the efficiency of using ethanol fuels. The amount of energy required to produce the fuel may exceed the energy it produces.

Methane gas, which is similar to natural gas and propane, can be produced from plant and animal wastes combined with certain bacteria. Methane gas can be used to power vehicles and other machinery.

Microbes are also being engineered to convert the sun's energy into usable fuel. The green algae that grow on the surface of ponds might one day produce hydrogen fuel that could be used to power vehicles. See Fig. 27-10. Other organisms, such as viruses and yeasts, are being studied for use in batteries and other electronic components.

An interesting new form of energy production called thermal conversion, or thermal depolymerization, can turn plant and animal wastes, as well as old tires and plastics, into high-quality oil. Using heat and pressure, the process also destroys any contaminants. Thermal conversion could eventually solve many energy and waste disposal problems.

« Fig. 27-10 The green algae covering this pond are plantlike organisms that may one day be used to produce hydrogen fuel.

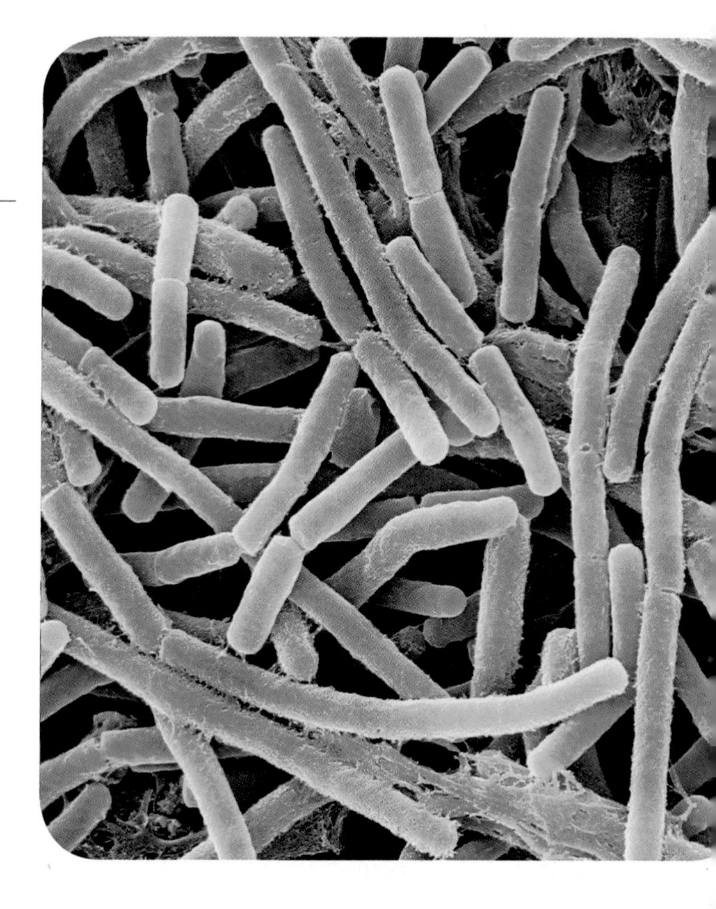

» Fig. 27-11 In 2001, the dangerous anthrax bacterium was used as a bioweapon.

Biowarfare

In 2001, a number of people in the United States received letters contaminated with the anthrax virus. Anthrax is a highly infectious disease that can be deadly. The use of viruses and other biological organisms as weapons is known as **biowarfare**. See Fig. 27-11.

Biowarfare is not new. Centuries ago, corpses of people who had died of plague were thrown over the walls of towns under siege in an effort to infect the enemy townspeople. In more recent times, governments have developed bio-weapons that could be used to infect large numbers of people. Anthrax, smallpox, and genetically engineered "superbugs" have all been part of these weapons programs. In spite of this, however, people in general have been very cautious about employing such weapons. They are hard to control and can infect the very people who choose to use them. In 1972, President Richard M. Nixon signed a treaty with the Soviet Union to ban such weapons except for defense, although the Soviets continued their research secretly.

The anthrax letters and the war in Iraq have focused attention back on bioweapons. Researchers seem to agree that they are more likely to be used against agricultural targets than human ones. The goal would be to create food shortages, raise food prices, and cause unemployment, thus weakening the country economically and politically. Countries that depend on one primary crop or that practice monoculture farming are especially vulnerable. Some researchers are attempting to develop nano particles that could be used to detect bioweapons.

The Environment

Because agriculture has to do with living things, it is linked to the environment. Pest management, conservation, and bioremediation are all environmental technologies related to agriculture.

Integrated Pest Management

Pests such as insects, microorganisms, and weeds destroy billions of dollars' worth of crops every year. See **Fig. 27-12**. Chemical pesticides can be effective, but the pests can eventually build up a resistance. In addition, the chemicals may be harmful to other animals and to humans. One way to reduce use of pesticides is with integrated pest management.

Integrated pest management (IPM) combines various techniques to control pests. The overall goal is not to eliminate all pests but to minimize the damage.

- **Mechanical controls.** For example, reflective aluminum strips can be placed like mulch in vegetable fields to reduce aphid attacks. This technique has been used to protect cucumbers, squash, and watermelons.
- **Cultural controls.** These have to do with the way a crop is grown. Crop rotation is one example. Clearing away fallen fruit from an orchard is another. Such techniques reduce the pests' supply of food or shelter.

Fig. 27-12 Insects, diseases, and weeds are pests that invade crops. Researchers hope to devise natural ways of controlling these pests, such as with beneficial insects.

» **Fig. 27-13** This praying mantis is eating a small aphid, which is a soft-bodied insect that sucks plant juices. Aphids are pests that can be controlled using natural predators.

- **Biological controls.** Bringing in natural predators that attack the pests is an example. See **Fig. 27-13**. Earlier in this chapter, you read about organic farming and genetic engineering to make plants that are resistant to pests. Another technique is to release sterile insects into the wild insect population. The sterile insects mate, but they can't produce offspring. "Friendly" bacteria, called probiotics, are also being used. For example, spraying the bacteria *Lactobacillus acidophilus* on cattle feed reduces the number of dangerous *E. coli* microbes in beef. (The acidophilus bacteria is the type commonly added to yogurt and cheese.)
- **Chemical controls.** When necessary, chemical pesticides are used. The strategy is to use the least amount necessary to do the job.

IPM requires constant monitoring of crops to catch problems early. Monitoring can include such things as inspecting crops for signs of pests, setting insect traps, and checking for weather conditions that are favorable to pests. When pests are found, treatment must be prompt and specific.

Conservation

Conservation is the preservation, management, and care of natural resources and the environment. Biotechnology can aid in conservation efforts.

Recently, for example, it was discovered that a medicine used to treat cattle in India was endangering vultures there. When the cattle died, the vultures ate the carcasses. Drug residues in the meat caused the birds to develop kidney failure. A biotech research team has now developed a new medicine that will help cattle without killing vultures.

Another way in which biotechnology aids conservation is a process called bioremediation. In **bioremediation**, bacteria and other microbes are used to clean up contaminated land and water. Genetically engineered microbes have been used to purify land or ocean water following an oil spill or other deposits of toxic waste. This process is also used in landfills to break down garbage.

CHAPTER 27 REVIEW

Main Ideas

- Agriculture is the practice of producing crops and raising livestock.
- Living organisms are used for food and as a source of medical treatment.
- Fuels can be made from agricultural products.
- Biowarfare is the use of viruses and other organisms as weapons.
- Because agriculture has to do with living things, it is linked to the environment.

Understanding Concepts

1. What is an agricultural hybrid?
2. Define *enzyme*.
3. How do fertilizers help crops to thrive?
4. Why does disease spread so quickly among farm animals?
5. Name three examples of CEA.

Thinking Critically

1. List both positive and negative effects of the U.S. using ethanol as its primary fuel source for vehicles.
2. Humans have often considered themselves at the "top" of the food chain. Is this true? Explain your answer.
3. What are the advantages of CEA?
4. Why do you suppose consumer demand has increased so much for organically grown products?
5. Why does integrated pest management choose *not* to eliminate all pests?

Applying Concepts & Solving Problems

1. **History** Research the origin of hydroponics. Where do historians think it may have begun?
2. **Research** Visit a local supermarket and ask the produce manager what percent of his/her products are organically grown.
3. **Science** A mule is a hybrid animal—a cross between a horse and a donkey. Do some research to learn what characteristics are obtained and how a mule differs from its parents.

Activity: Experiment with Agriculture

Design Brief

As you know, agriculture includes a wide range of technologies. In all cases, these technologies involve living things. Some are concerned with food production. Others involve the environment.

Select one of the following projects, design it, and carry it out:

- Using seedlings and other plants, carry out the restoration of a small area of land.
- Grow food plants hydroponically.
- Conduct a taste test and survey to compare organically grown foods with those grown conventionally.
- Produce yogurt using fermentation processes.

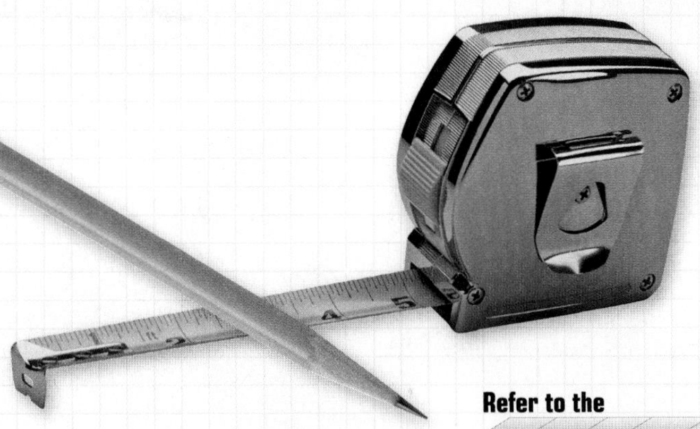

Refer to the

Criteria

- You must research your chosen project to find out what materials you'll need and what steps you should follow.
- If you choose to restore land or build a hydroponic system, you must submit drawings of your design.
- If you choose to compare organic with conventionally grown foods, you must conduct your test and survey with a minimum of 10 people.
- If you choose the yogurt-making project, you must keep a record of times, temperatures, and other variables that affect the results.

Constraints

- You must submit your choice of project, a plan to carry it out, a list of materials needed, and the source you used to your instructor for approval before you begin work.
- Plants grown hydroponically must be an edible variety.
- You must obtain approval from the owner before restoring a plot of land.
- A plot of land should not exceed 10 square feet.
- For safety, you should use a commercial yogurt product as a starter.

Engineering Design Process

1. **Define the problem.** Write a statement that describes the task you are to accomplish, as well as any obstacles you will need to overcome (i.e., trying to make organically grown foods that taste as good as traditionally grown foods).
2. **Brainstorm, research, and generate ideas.** With your team, discuss possible solutions. Hint: Your library is an excellent resource when researching your chosen project. How much can you find out about agricultural technologies?
3. **Identify criteria and specify constraints.** These are listed on page 548.
4. **Develop and propose designs and choose among alternative solutions.** Choose the design that will work the best.
5. **Implement the proposed solution.** Decide on the process you will use. Gather any needed tools or materials.
6. **Make a model or prototype.** Create your experiment. Follow all safety rules.
7. **Evaluate the solution and its consequences.** Does your experiment work as intended?
8. **Refine the design.** Based on your evaluation, change the design if needed.
9. **Create the final design.** After any changes or improvements take place, create your final agricultural experiment.
10. **Communicate the processes and results.** Present your finished experiment to the class. Try to involve your classmates in your presentation. When you involve your audience, they will usually be more interested in what you have to say.

SAFETY FIRST

Before You Begin. Before starting this activity, be sure you understand how to use any tools and materials safely. If you are working with foods, follow instructions for their safe handling.

Safety First

General Safety Rules

In a technology course, you have many opportunities to design and build products. You can apply your creativity and problem-solving skills. It's very important that you also apply common sense and practice safe work habits.

Develop a Safe Attitude

- Read and follow all posted safety rules.
- Take the time to do the job right.
- Consider each person's safety to be your responsibility. Avoid putting others in danger.
- Stay alert. Be aware of your surroundings.
- Work quietly and give your full attention to the task at hand. Never indulge in horseplay or other foolish actions.
- Stay out of danger zones as much as possible. These are usually marked with black and yellow striped tape.
- Put up warning signs on things that are hot and could cause burns.
- If you bend down to pick up an object, use your legs, not your back, to lift up. Keep your back straight. To keep better control, get help to lift or move long or heavy items.
- Handle materials with sharp edges and pointed objects carefully.
- Report accidents to your teacher at once.

» Notice the safety jackets worn by the students working on this solar-powered car.

Have Respect for Tools and Equipment

- Never use any tool or machine until the teacher has shown you how to use it and has checked the setup.
- Before using any tool or machine, make sure you know the safety rules and make sure you get your teacher's permission.
- Use equipment only when the teacher is in the lab.
- Do not let others distract you while working.
- Do not use electrical tools or equipment if the cord or plug is damaged.
- Always use the right tool for the job. The wrong tool could injure you or damage the part you are working on.
- To avoid injury, use the right machine guard for the job. Check with your teacher for the appropriate guard.
- Keep hands and fingers away from all moving parts.
- Before you leave a machine, turn it off and wait until it stops. If you are finished, clean the machine and the area around it.
- When you have finished working, return all tools and unused supplies to their proper places.

Prevent and Control Fires

- Store oily rags in a closed metal container to prevent fire.
- Know where the nearest fire extinguisher is and how to use it, *if that is your school's policy.*

Wear Appropriate Clothing and Protective Equipment

- Always wear eye protection. Special eye protection may be needed for some activities, such as using a laser, welding, or using chemicals.

- Wear hard shoes or boots with rubber soles.
- Use ear protection near loud equipment.
- Do not wear loose clothing, jewelry, or other items that could get caught in machinery. Tie back long hair.
- Do not wear gloves while operating power tools.

Have Respect for Hazardous Materials and Waste

- Products with major health risks should have a Material Safety Data Sheet (MSDS) available. Ask your teacher about the MSDS before you use materials that may be hazardous. Know how to read the MSDS. You can find more information about Material Safety Data Sheets on page 559.

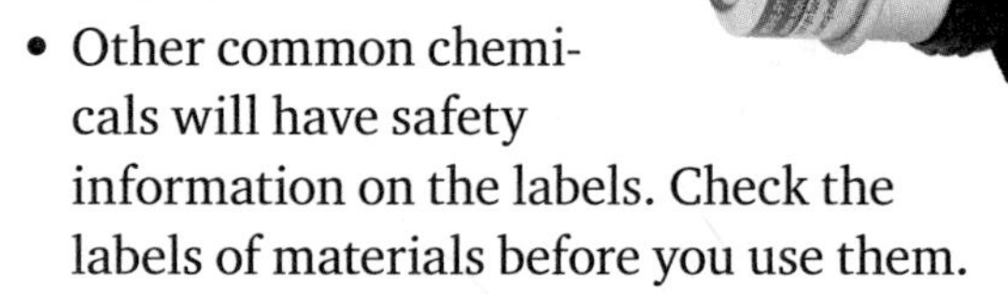

- Other common chemicals will have safety information on the labels. Check the labels of materials before you use them.
- Wear appropriate personal protective equipment (PPE) when working with hazardous materials.
- Work in a well-ventilated area.
- Follow your teacher's instructions for disposal of hazardous materials and waste.

Maintain the Lab

- Keep the work area clean. Keep the floor and aisles clean at all times.
- If a liquid is spilled, clean it up immediately as instructed by the teacher.
- Always use a brush, not your hands, to clean dry materials from a table or piece of equipment.
- Store all materials properly.

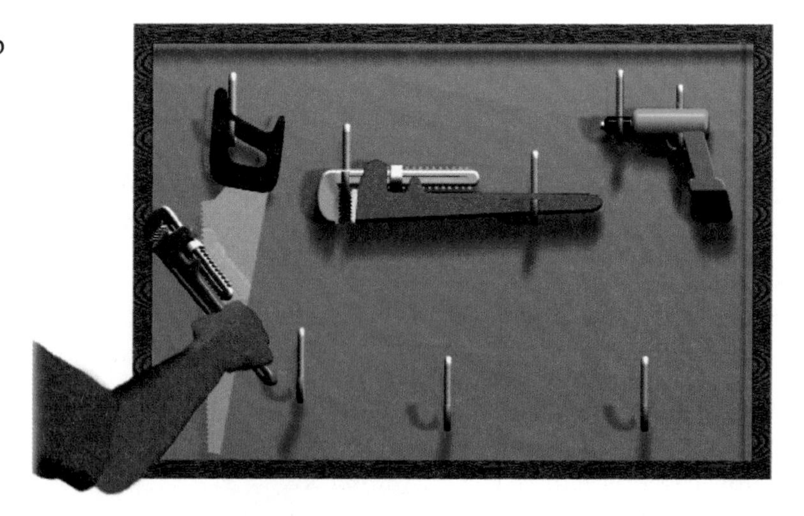

⮝ This carpenter is nailing roof trusses. ***Why should eye protection be worn when using a hammer?***

Safe Use of Hand Tools

Misuse and improper maintenance pose the greatest hazards in using hand tools. Observe the following rules when using hand tools.

- Keep all hand tools clean and in good condition.
- Wipe tools clean before and after use.
- Use only tools that are in good condition. If a tool is damaged, tell your teacher about it.
- Use a hand tool only for the purpose for which it was designed.
- Hold and use the tool in the proper way, following manufacturer's instructions.
- Always wear safety glasses or goggles.
- Cutting tools should be sharp.
- Do not use a screwdriver on a part that is being handheld. The screwdriver can slip and hit your hand.
- Carry sharp or pointed tools with the point down and away from your body.

Safe Use of Electric Power Tools

Use care and common sense when working with electric power tools. Observe the following safety practices.

Prevent Electric Shock

- Do **not** stand in water while working on equipment.
- Make sure that all electrical cords are free of frays and breaks in the insulation.
- Pull the plug, not the cord, when you unplug a tool or machine. Damaging a cord may cause an electric shock.
- Keep electrical cords away from sharp edges.
- Use only power tools that have been properly grounded or double insulated.
- Make sure that all extension cords are the three-wire grounded type.
- Make sure the three-pronged plug is used in a grounded receptacle.

Work Safely

- Do **not** wear loose clothing, ties, or jewelry that can become caught in moving parts of machinery. Be sure to tie back long hair.
- Do **not** wear gloves while operating power tools.
- Wear the appropriate personal protective equipment (PPE), such as safety glasses or a face shield.
- Never set a handheld power tool down while it is running or coasting.
- Avoid accidental startups by keeping fingers off the START switch when carrying a tool.
- When you approach a machine, be sure it is off and that it is not coasting.
- Secure the work piece with clamps or a vise. This will free both hands to operate the tool.
- Disconnect the power source before changing accessories such as bits, belts, and blades.
- Keep tools as sharp and clean as possible for best performance.
- Tell the teacher immediately if the machine doesn't sound right or if you can see that something is wrong.

Safe Use of Cutting Tools

Observe the following safety practices when using electric power tools to cut materials.

Check stock before cutting it. Nails, cracks, or loose knots could cause a power saw to kick back or send pieces flying out at high speed. Before starting to cut, put on safety goggles.

- Allow machines to reach full speed before starting to cut.
- Before working on stock (wood or other workpieces), check it for cracks, loose knots, and nails.
- The shortest piece of lumber that can safely be run through most equipment is 12 inches long.
- Keep your balance; don't overreach.
- Support ends of long stock before cutting.
- Wait until the blade stops completely before removing any scraps. Use a brush, not your fingers.

Safe Use of Pneumatic Tools

Pneumatic tools are powered by compressed air. The air is fed to the tool through a high-pressure hose connected to an air compressor.

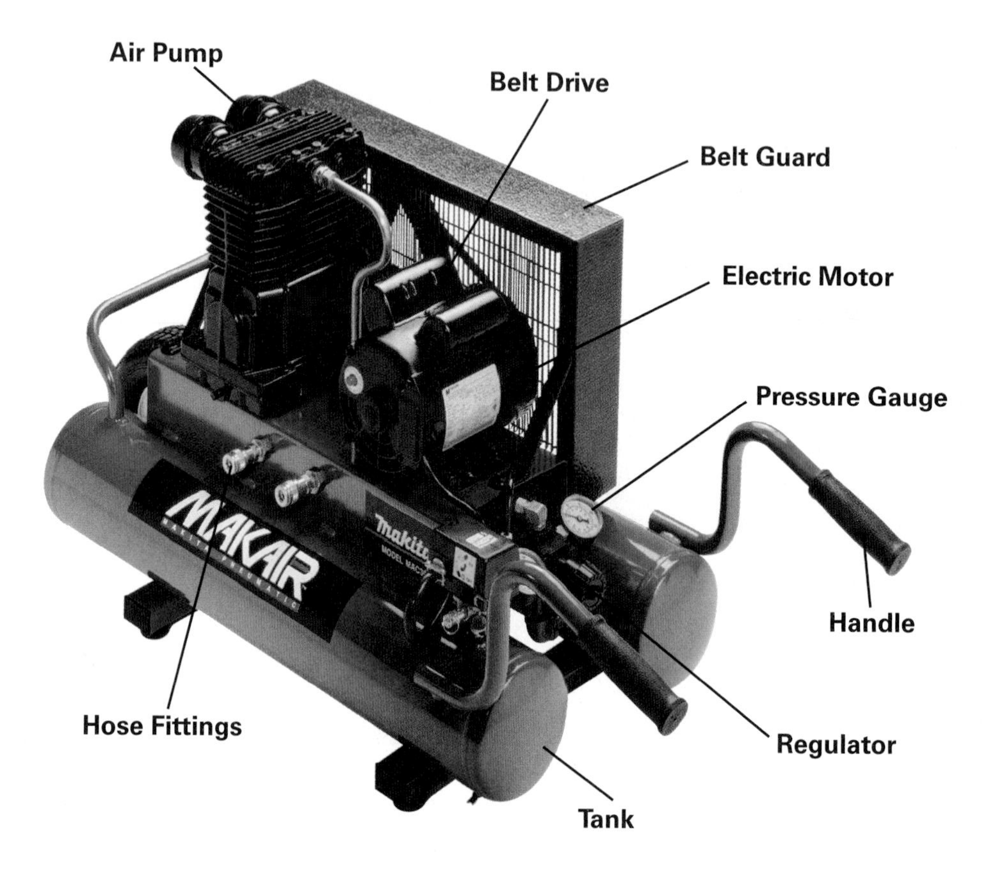

The main parts of an air compressor are shown here.

- When using a compressed-air gun, always wear the correct personal protective equipment (PPE). For example, wear safety glasses or goggles and a face shield. Always direct the airflow away from you.
- Always carry a pneumatic tool by its frame or handle, not by the air hose.
- Make sure all pneumatic tools are securely attached to the compressed-air line.
- Never use a pneumatic tool, such as a compressed-air gun, to remove debris from your clothing or body.
- Securely position a pneumatic tool before operating it. These tools operate at high speeds or under high pressure.

Tool and Machine Maintenance

Tools and machines that are well maintained do a better, faster job and are safer to use.

Hand Tool Maintenance

- Keep sharp tools properly sharpened.
- Check edges, sharp points, and other working surfaces for cracks and other defects.
- Keep working surfaces and handles clean and free of dirt and rust.
- Check that any moving parts or mechanisms are working properly.
- Check for loose handles and other loose parts.
- Check wood handles for splinters and other damage.
- Clean the tool following use each day.
- Inspect the tool for damage at least once each day before use.
- Lubricate any tool parts that require it.
- Preserve wood handles regularly with special oils made for that purpose.
- Store tools in their proper places in an organized way.
- Store sharp tools in a safe place with the cutting edge protected.

Clean brushes after each use to keep the bristles soft and in good condition.

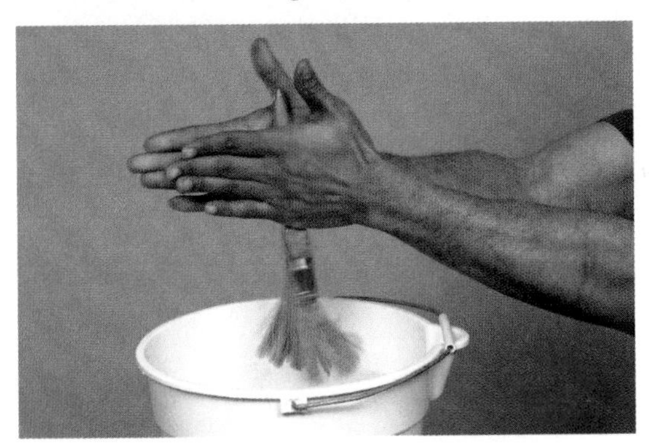

Blades of band saws and other cutting tools should be cleaned regularly to remove resin, pitch, and other debris. A clean blade will cut straighter and smoother.

Power Tool and Machine Maintenance

- Keep sharp tools properly sharpened.
- Check edges, sharp points, hoses, and other working parts for cracks and other defects.
- Keep working surfaces, handles, and hoses clean and free of dirt and rust.
- Check that any moving parts or mechanisms are working properly.
- Check for loose handles and other loose parts.
- Check for broken plugs or lugs removed from grounding plugs.
- Check for split insulation or damaged electrical cords.
- Disconnect and clean the tool following use each day.
- Disconnect and inspect the tool for damage at least once each day before use.
- Lubricate any tool parts that require it.
- Store tools in their proper places in an organized way.
- Store sharp tools in a safe place with the cutting edge protected.

Safe Use of Chemicals

Follow these rules when working with paints, stains, varnishes, paint thinners, adhesives, or other chemicals.

- Read and follow all label precautions.
- Always wear protective clothing and approved eye protection. Use appropriate gloves or tongs when needed.
- Many chemicals produce harmful fumes. Work only in well-ventilated areas. Use a respirator whenever it is required.
- Know where the eyewash station is and how to use it.
- Know where the Material Safety Data Sheets (MSDS) are located.
- Know where the poison-control phone number is located.
- Mix chemicals only as directed. If you need to mix acid and water, get the water first. Then carefully add the acid to it.
- Pay attention to others around you when working with chemicals and report any unusual reactions.
- Clean up spills immediately.
- Clean all tools and equipment properly after using.
- Avoid skin contact with chemicals. Wash thoroughly before leaving the area.
- Many chemicals need to be stored away from heat or away from moisture. Follow label directions.
- Never store chemicals in an unlabeled or incorrectly labeled container.
- Store chemical-soaked rags in an approved container.
- Dispose of chemicals properly.

» Use gloves to protect your hands from contact with chemicals.

Material Safety Data Sheet

OSHA requires that workers be informed about any hazardous chemicals to which they may be exposed. A Material Safety Data Sheet (MSDS) is a form used to communicate information about hazards. The table at the bottom of this page is part of an MSDS for acetylene. There are different types of Material Safety Data Sheets, but they all must include the following kinds of information.

- The **identity** of the hazardous chemical, including both its chemical and common name(s). If the chemical is a mixture, the ingredients are listed.
- **Physical and chemical characteristics**, such as flash point (the lowest temperature at which a vapor may ignite).
- **Physical hazards**, including the potential for fire, explosion, and reactivity.
- **Health hazards**, including signs and symptoms of exposure, and any medical conditions that are generally recognized as being aggravated by exposure to the chemical.
- The **primary route(s) of entry**. For example, a chemical may produce fumes that could be inhaled.
- The **limit of safe exposure**.
- Whether the chemical is **carcinogenic** (cancer-producing).
- Generally applicable **precautions for safe handling and use.**
- Generally applicable **control measures**, such as personal protective equipment (PPE).
- **Emergency and first-aid** procedures.
- The **date** the MSDS was prepared or last updated.
- The contact information of the chemical manufacturer, importer, employer, or other responsible party who can serve as a **provider of additional information** on the hazardous chemical and appropriate emergency procedures, if necessary.

2. Hazard Ingredients and Identity Information

COMPONENT	% VOLUME	OSHA-PEL	ACGIH-TLV	CAS NUMBER
Acetylene	95.0 to 99.6	Not Available	Simple Asphyxiant	000074-86-2
Acetone	Unavailable	1000 ppm TWA	750 ppm TWA 1000 ppm STEL	000067-64-1

3. Physical and Chemical Characteristics

Boiling Point:	-118.8°F	**Specific Gravity:**	0.906	**pH:**	N/A
Melting Point:	-116°F	**Evaporation Rate:**	N/A	**Physical State:**	Gas
Vapor Pressure:	635 psig	**Solubility ($H^{2}0$):**	Soluble		

Appearance and Odor:
Pure acetylene is a colorless gas with an ethereal odor. Commercial (carbide) acetylene has a distinctive garlic-like odor.

How to Detect This Substance: N/A

Other Physical and Chemical Data:
Liquid density at boiling point, 38.8 lb/ft^3 (622 kg/m^3)
Gas density at 70°F 1 atm, 0.0691 lb/ft^3 (1.107 kg/m^3)

Safety Color Codes

Safety signs and labels are color-coded to signify hazards or to identify the location of safety-related equipment.

Color	Meaning	Examples
Red	Danger, stop, or emergency	• Fire-protection equipment • Flammable-liquid container • Emergency stop bars and switches
Orange	Be on guard	• Hazardous parts of equipment or machines that might injure • Safety starter buttons on equipment or machines
Yellow	Caution	• Physical hazards, such as steps and low beams • Waste containers for combustible materials
White	Storage	• Housekeeping equipment
Green	First aid	• Location of safety equipment, such as first-aid kit
Blue	Information or caution	• Out-of-order signs on equipment • Cautions against using out-of-order equipment

Warning Labels

The warning labels on containers of hazardous material may use colors and numbers to indicate the hazard level. There are several labeling systems. One was developed by the National Fire Protection Association (NFPA). Their label consists of squares arranged into a "diamond" shape. Each square has a different color to represent the type of hazard. Red represents flammability, and yellow represents reactivity. Blue is used for health hazards, and white is for special information. A number on a square indicates the severity of the hazard. Numbers range from 0 to 4, with 4 representing the greatest hazard. However, any category rated 2 or higher should be considered potentially dangerous.

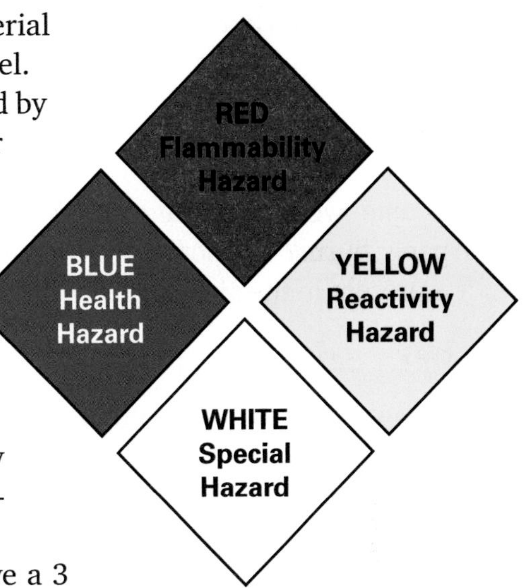

For example, an NFPA label for toluene would have a 3 in the red square, a 2 in the blue square, and a 0 in the yellow square. Toluene is a chemical used in many paints, paint thinners, lacquers, and adhesives. It is flammable, can be harmful if inhaled or swallowed, and is not reactive when mixed with water.

Fire Safety

In order for a fire to burn, three basic things must be present.

- A source of heat
- Oxygen
- Fuel

For example, what if a piece of paper comes in contact with a hot wire and the paper burns? The hot wire provides heat, the paper is the fuel, and oxygen is present in the air.

Most fires can be extinguished by

Reducing the heat The most common way to reduce the heat of a fire is to throw water on it. This has cooling action and also produces steam. The steam helps to exclude oxygen. However, water should not be used to extinguish some fires, such as grease or electrical fires.

Preventing oxygen from reaching the fire A fire can be deprived of oxygen by spraying it with an inert gas, such as carbon dioxide. Carbon dioxide is contained in many fire extinguishers. Keeping oxygen away is also the method used when a person's clothing catches fire. The person is wrapped in a blanket, which smothers the fire.

Removing the source of fuel If a gas or liquid, such as gasoline, is feeding a fire, it can often be turned off in some way. That removes the fuel. In some cases, the fuel may be allowed to burn until it is used up and the fire goes out.

Class of Fire	Type of Flammable Material	Type of Fire Extinguisher to Use
Class A	Wood, paper, cloth, plastic	Class A Class A:B
Class B	Grease, oil, chemicals	Class A:B Class A:B:C
Class C	Electrical cords, switches, wiring	Class A:C Class B:C
Class D	Combustible switches, wiring, metals, iron	Class D
Class K	Fires in cooking appliances involving combustible vegetable or animal oils and fats	Class K

Developing a Fire Emergency Plan

Your lab should have a plan for use in a fire emergency. If your teacher has not explained it to you, ask about it. Be sure you know where all the exits are, where the fire extinguishers are kept, and what your responsibilities are in case a fire occurs. Then follow these steps to record the fire emergency plan.

1. Draw a floor plan of the lab.
2. Locate all the fire exits and label them on the floor plan. If the lab has no direct exit to the outside, draw a map of your section of the school. Show the location of the lab and draw arrows from the doors in the lab to the nearest fire exits.
3. Locate and label any windows that could also be used for escape. (Check to be sure the opening will be large enough for an adult to pass through.)
4. Locate all the fire extinguishers and label them on the floor plan. Indicate on the plan the class of fire for which each extinguisher can be used.
5. Locate places where flammable or explosive materials should be stored and label them. Be sure they are in a location far away from any source of heat and that their cabinet is fireproof.
6. Determine where you and your classmates should meet after you have left the building. This is important so that someone can check to be sure everyone has escaped safely.

7. Ask your teacher for procedures to follow if a fire occurs. If time allows, this may include such things as closing windows, turning off equipment, and grabbing the first-aid kit. Someone should be responsible for reporting the fire. However, human safety is the most important consideration. Fires can spread quickly, and smoke can be just as deadly as flames. Your most important responsibility is to get out alive.
8. Be sure everyone has a chance to study the emergency plan. Then, with your teacher's approval, post it in a prominent place where it can be seen easily.
9. Be sure that you and your classmates are informed as to housekeeping duties that help prevent fires, proper handling of any materials that could catch fire, how to treat burns, and where to report a fire.

Troubleshooting

To "troubleshoot" means to follow a systematic method to find the cause of a problem and fix it. Owner's manuals often contain troubleshooting charts to help users find and correct problems.

Troubleshooting begins by looking for the most likely causes. Often, the problem is a simple one that can easily be fixed. If that turns out not to be the case, then the troubleshooter looks for more unusual causes. It is also a good idea to look at a system's parts and test each part one by one. For example, when your computer is unable to read a new CD, you might follow a procedure like the one outlined in this table.

Symptom	Possible Cause	How to Check
Computer cannot read CD.	1. CD is not formatted for your computer system.	1. Read label of CD for hardware and software requirements.
	2. CD is not properly inserted in drive.	2. Reinsert CD into drive, being careful to place it correctly.
	3. CD is damaged.	3. Insert a different CD into the drive to see whether it will work.
	4. CD drive damaged.	4. Replace or repair CD drive.

Here is a simple troubleshooting table to use in the school lab.

Symptom	Possible Cause	How to Check
Electric tool is not working.	1. Cord is not plugged in.	1. Plug in the cord.
	2. Switch is turned off.	2. Turn on switch.
If the tool still does not work, report the problem to your teacher. The tool may need repairs, or there may be a problem with the electric power supply.		

HANDBOOK

Science, Technology, Engineering, and Math offer broad career pathways into the high-growth industries competing in the global marketplace. The jobs in these areas are offered by businesses on the frontline of technological discovery. With a solid background in STEM, you'll be following a career path that will offer the opportunity to participate in the development of high-demand and emerging technologies.

With advances in science driving innovations in technology, materials and procedures researched in the lab are now the backbone of many new processes. Science is used in the design of cell phone circuits, the construction of surfboards, and the processing of food products. While science begins with careful observation, research, and experimentation, it involves more than lab work. Some opportunities will take you into the field. You might also be involved in quality control or technical writing. The range of science career opportunities is extremely large.

Technology joins science skills with those in engineering and math. It is essential in managing communication and information systems. In manufacturing, technology is used to focus human resources and materials in the production of products ranging from electronic devices to footwear. If you want a career that combines science, math, and engineering, consider a career in technology. In that field, you'll work closely with individuals who have a hands-on knowledge of these subjects. As part of their team, you'll help make the decisions needed to bring a product from design through manufacturing to market.

A career in engineering will involve you in the technological innovations that flow from discoveries in science and math. You'll be incorporating these new discoveries in products that expand our technological reach in many areas, including communications. If you think that engineering is unexciting, name a product or service of interest to you that was introduced within the last year. Then identify the main benefit of that product or service. That benefit probably resulted from an engineering effort that built on developments in science, technology, and math. A career in engineering could place you in a position to participate in the development of such products and services.

Math input is essential to product creation and development—whether the product is real, such as a DVD, or virtual, such as an online video game. If you want to be a valued member of the teams providing the math input for the development of products, you will need solid math skills. You'll need these skills for high-level problem solving and innovation. As you advance along the career pathway, you'll be challenged to apply mathematics to real-world problems. There is a wide variety of math careers from which to choose.

HANDBOOK

SCIENCE

Machines

Although machines make work easier, they don't actually reduce the amount of work that has to be done. How, then, do machines help us? Recall the formula for work:

$$\text{Work} = \text{force} \times \text{distance}$$

In this equation, if you increase the force, you can decrease the distance and still do the same amount of work. If you increase the distance, you can decrease the force. Machines help us by increasing distance, increasing force, or changing the direction of force. Complicated machines are combinations of a few basic machines. These basic machines are shown and described below.

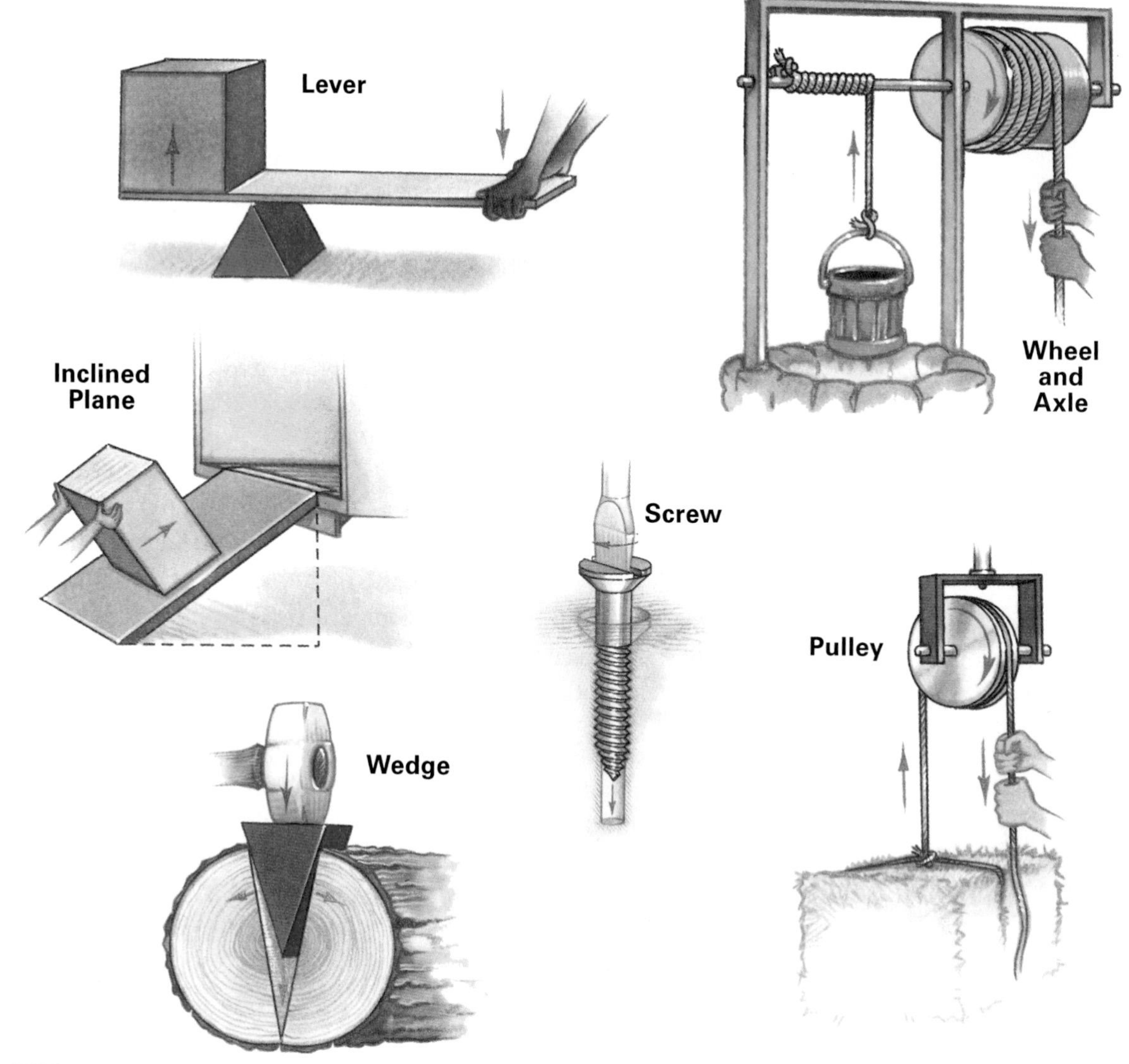

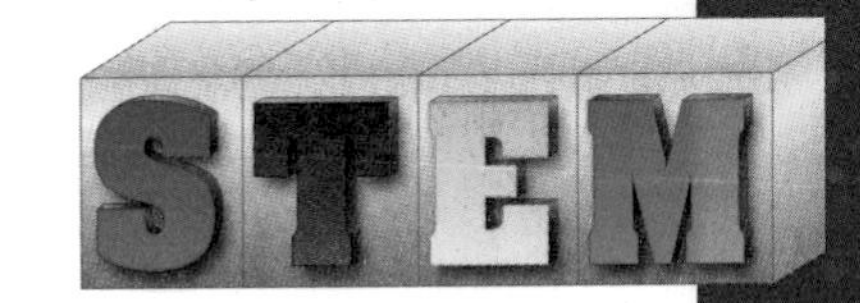

Mechanical Advantage

Some machines are more helpful than others. The measure by which a machine increases force is known as its mechanical advantage (MA). This can be calculated by the following formula:

$$MA = \frac{F_r}{F_e}$$

In this equation, F_r stands for resistance force (load). This is the force applied by the machine. For example, a screwdriver exerts a force on a screw. F_e stands for effort force, the force applied to the machine. An example is the force you apply to the screwdriver. If a small effort force results in a large resistance force, the machine has a high mechanical advantage. The greater the mechanical advantage, the more helpful the machine.

For machines that increase distance, we can calculate velocity ratio. The velocity ratio is the distance the load moves divided by the distance the effort moves.

Efficiency

Efficiency compares the work a machine can do with the effort put into the machine. The energy output of a machine is always less than the energy input. That's because some of the energy input is used to overcome friction between moving parts. Efficiency is usually stated as a percentage. Machines with a higher percentage are more efficient.

$$\text{Efficiency} = \frac{\text{work output}}{\text{work input} \times 100\%}$$

What if you applied all of a machine's energy output to the machine itself? Is it possible to build a perpetual motion machine—a machine that will run forever? Unfortunately, no. All machines need a continuous outside supply of energy. If no energy is brought into the system, the machine will eventually stop. Also, all machines use more energy than they output. No machine is 100 percent efficient.

Power

Power is work done within a certain period of time. To calculate power, determine the amount of work done. Then divide the amount of work by the time it took to do the work.

$$\text{Power} = \frac{\text{work}}{\text{time}}$$

Gears and Cams

Gears are intermeshing toothed wheels and bars that transmit force and motion. They can alter the force's size and the motion's speed and direction. A cam is an offset wheel. Cams are connected to rods. A second rod, or follower, rests on the top of the cam. As the cam turns, the top rod moves up and down.

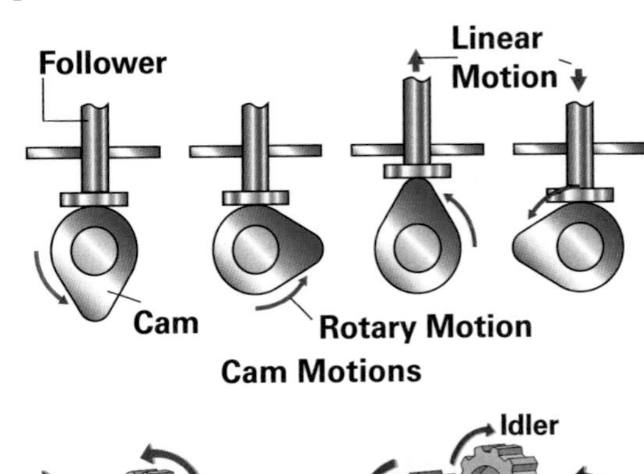

Cam Motions

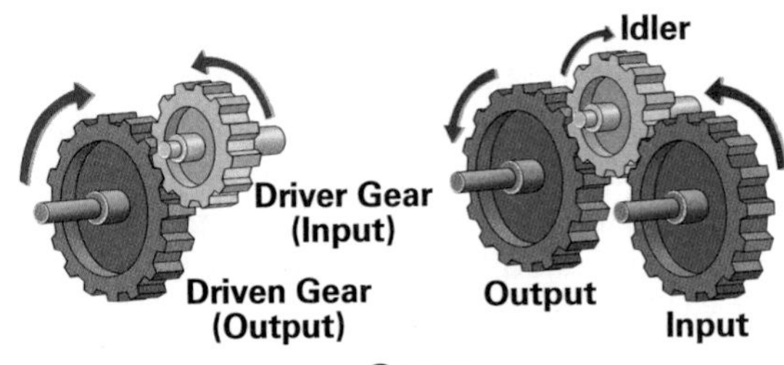

Gears

Forces and Motion

A force is a push or a pull. Forces are everywhere. The force of gravity keeps your feet on the ground. Motion takes place when a force causes an object to move. Think about machines. They make work easier by applying or changing forces.

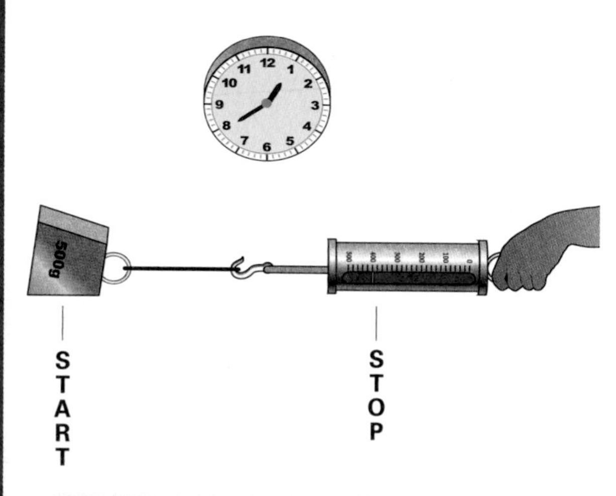

Basic Physical Forces

The four basic physical forces in the universe are gravity, electromagnetism, the strong nuclear force, and the weak nuclear force. You are already familiar with gravity and electromagnetism. The strong nuclear force holds the nuclei of atoms together. The weak nuclear force involves radioactivity and particle decay.

I feel it in my GUT . . .

Scientists have already proven that two of the forces—electromagnetic and weak nuclear—are different aspects of one force, called the electroweak force. Now they're working to prove the theory that gravity and the strong nuclear force are also related to the electroweak force. This theory is called the Grand Unified Theory, or GUT.

Forces can act from a distance, without objects touching one another. Earth's gravity affects the moon, even though the moon is nearly 250,000 miles away. The moon's gravity also affects the earth, causing the tides.

The "weightlessness" of astronauts as they orbit the earth in the space shuttle is not the result of their distance from the earth. The astronauts float because they are falling toward the earth at the same rate as the shuttle. However, the shuttle is also moving forward at a fast speed. This forward motion, combined with the downward pull of gravity, keeps the shuttle in a circular path around the earth.

Gravity

Gravity is a force that pulls objects toward each other. Gravity increases as mass (amount of matter) increases. Gravity decreases as distance increases.

Weight is the amount of force exerted on matter by gravity. While mass remains constant, weight varies, depending on the amount of gravitational attraction. For example, gravity on the moon is only one-sixth of that on Earth.

Terminal Velocity

Gravity accelerates a falling object at the rate of 32.2 feet per second, per second. Does that mean an object will just keep falling faster and faster? No, because the object encounters air resistance. As the object falls faster, air resistance increases. Eventually, the air resistance equals the force of gravity. The object then falls at a steady speed. This speed is the object's terminal velocity.

Pressure

Pressure is the weight (force) acting on a unit of area. When your *Technology: Engineering & Design* textbook is flat on your desk, it exerts pressure on the desktop. The book covers an area of approximately 80 square inches. It weighs about 2.75 pounds. To calculate how much pressure the book exerts on the desktop, use this formula:

$$\text{Pressure} = \frac{\text{force}}{\text{area on which the force acts}}$$

The book exerts 0.034 pounds of pressure per square inch. If you place the book upright, it will touch 8 square inches of the desktop. How much pressure per square inch will it exert?

Can you see that concentrating force on a smaller area increases the pressure per square inch? How might this knowledge help you if you were designing a machine and wanted to get a lot of pressure from a small force? What if you were designing a building? Would you want the force concentrated or spread out?

Vectors

What happens when two or more forces act on an object at the same time? When the forces are balanced, they cancel each other out. There's no change. When forces are unbalanced, there's change in movement.

A vector is a mathematical expression of a force. Forces can be expressed mathematically as vectors. A vector has a direction and size. When two vectors are combined, the result is called the resultant. To find the resultant of two forces acting at an angle to each other, draw two lines to represent the size and direction of the two forces. (You'll need to draw the lines to scale and at the correct angle to each other.) Next, draw two lines parallel to these. You've drawn a parallelogram. The diagonal of this parallelogram represents the size and direction of the resultant.

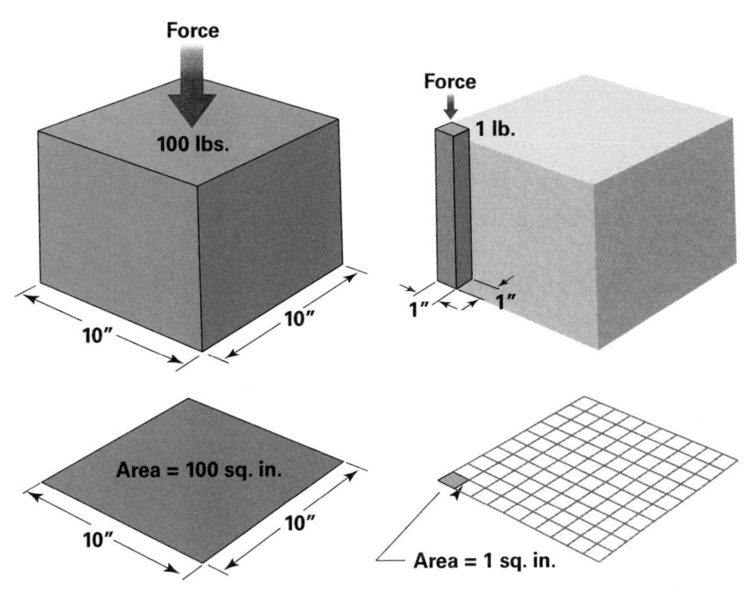

Newton's Laws of Motion

In the seventeenth century, the English scientist Isaac Newton described three laws of motion.

Newton's First Law Sometimes called the "law of inertia," Newton's first law states that an object at rest will stay at rest unless a force acts on it. If an object is in motion, it will stay in motion unless a force acts on it. For example, have you ever noticed that when riding in a car that stops quickly, your body will continue to move forward and tighten up against the seatbelt? This is because your body is trying to stay in motion, but the force of the seatbelt is stopping you. The car, too, is trying to stay in motion, but the force on the brakes is slowing it down.

Newton's Second Law When a force acts on an object, the result is a change in speed or direction. The object will start to move, speed up, slow down, or change direction. The greater the force, the greater the change. For example, if you and a friend push on a stalled car, the car will move. Once the car begins to move, the harder you push, the faster the car will go.

Newton's Third Law For every action, there is an equal and opposite reaction. For example, when you bounce on a trampoline, you exert a downward force on the trampoline. The trampoline then exerts an equal force upward, sending you into the air.

Speed, Velocity, and Acceleration

Speed, velocity, and acceleration are related, but each is distinct.

Speed is a measure of how fast an object is moving.

Velocity is the speed of an object in a certain direction. The velocity of a car on the highway might be 65 miles per hour. If the car moves onto an exit ramp, there is a change in velocity.

A change in velocity is called acceleration. The greater the mass of an object, the greater the force needed to accelerate it. If the change is a decrease in speed, it is negative acceleration, or deceleration.

Forces and Fluids

Hydraulics is a branch of science that deals with the motion of liquids. Aerodynamics deals with the motion of gases. Liquids and gases are fluids—substances that can flow. The properties of fluids make them well suited for certain technological applications.

Force, Mass, and Acceleration

Force (a push or pull) causes a mass (amount of matter) to accelerate (change speed and/or direction). The formula for finding force is

$$\text{Force} = \text{mass} \times \text{acceleration}$$

In the metric system, force is measured in newtons (named for Isaac Newton). Mass is measured in kilograms. Acceleration is measured in meters per second, per second. In other words,

$$\text{Newtons} = \text{kilogram} \times \frac{\text{meter}}{\text{second}^2}$$

In the customary system, force is measured in pounds. Acceleration is measured in feet/second^2. The unit of mass is the slug. One slug equals 14.6 kilograms.

- When a force is applied to a confined fluid, the pressure is transmitted throughout the fluid (Pascal's principle).
- Pressure increases with depth.

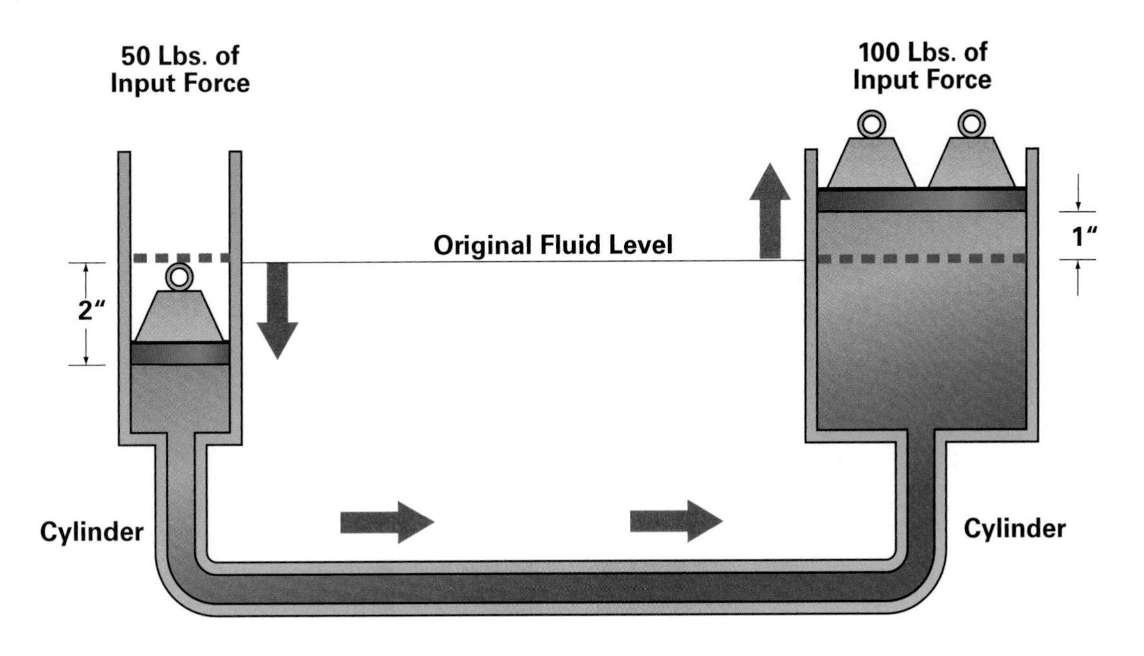

- A fast-moving fluid has a lower pressure than a slow-moving fluid (Bernoulli's principle). That's why airplanes can fly. The air moves faster above the wings than below them. The difference in air pressure enables the airplane to stay aloft.

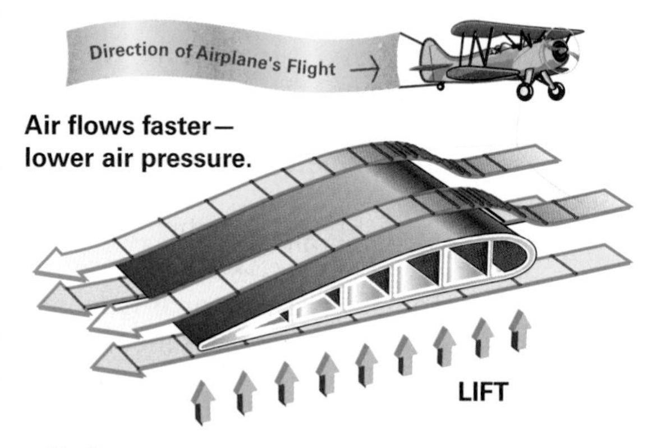

Energy and Work

Potential and Kinetic Energy

Energy is the ability to do work. Although it can take many forms, there really are only two kinds of energy: kinetic and potential. Kinetic energy is energy in motion. Potential energy is energy at rest.

Consider what happens when you use a hammer to drive a nail. As you lift the hammer, you do work. You exert force that moves an object over a distance. Some of your energy is transferred to the hammer. That energy is stored in the hammer as potential energy. When you move the hammer toward the nail, the potential energy becomes kinetic energy. When the hammer head hits the nail, energy is transferred from the hammer to the nail. The hammer stops, but the nail moves.

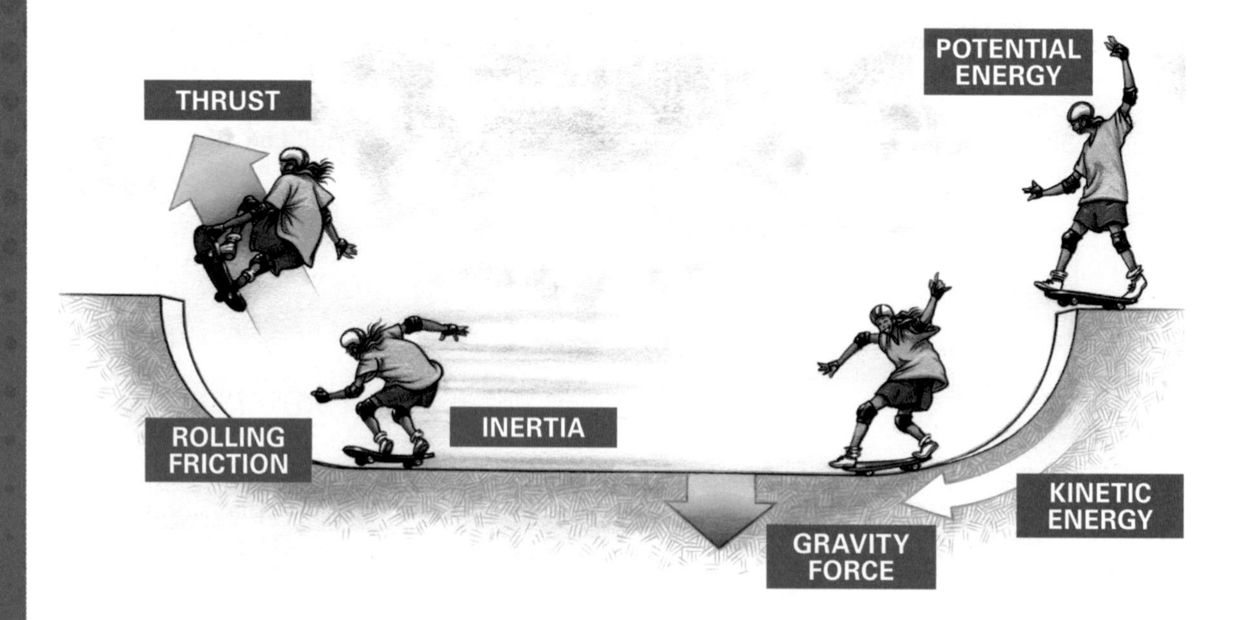

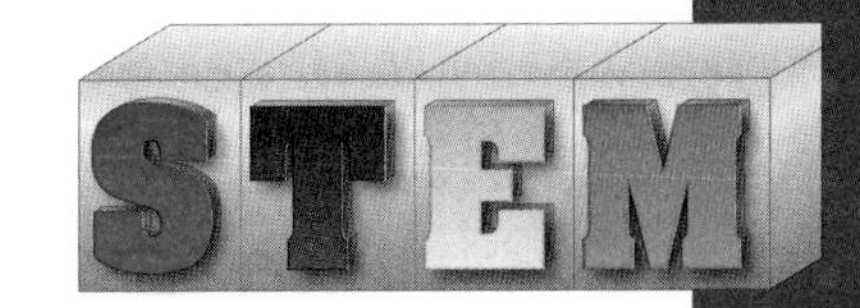

Measuring Kinetic Energy

The kinetic energy (ek) of a moving object is found by using the following formula:

$$E_k = \frac{1}{2}mv^2$$

In this formula, m is the mass of the object, in kilograms, and v is the velocity in meters per second.

From this formula, you can see that kinetic energy increases with the mass and speed of a moving object. An object with twice the mass has twice the kinetic energy. An object with twice the speed has four times the kinetic energy. That's why automobile accidents at high speeds cause more damage. There is more kinetic energy.

Types of Potential Energy

Potential energy is the energy an object has because of its position or condition.

- Gravitational potential energy is stored when work lifts an object against the force of gravity, as in the example of lifting the hammer.
- Elastic potential energy is stored when work twists or stretches an object or changes its shape. A stretched rubber band is an example of stored elastic energy.
- Electrical potential energy is the stored energy of electric charges.
- Magnetic potential energy is the stored energy of a piece of iron near a magnet.

Conservation of Energy

Energy does not disappear. The total amount of energy stays the same. This principle is known as the conservation of energy. It states that energy cannot be destroyed. However, energy can change form. It can move from object to object.

Converting Mass to Energy

Can energy be created? Albert Einstein showed that, in a nuclear reaction, some mass is converted to energy.

$$E = mc^2$$

In this formula, E is energy, m is mass, and c is the speed of light. Since the speed of light is 300,000 kilometers per second, you can see that even a small amount of mass can release a huge amount of energy.

Forms and Uses of Energy

Technology converts energy from one form to another to accomplish work. Here are a few examples.

- Chemical energy stored in a flashlight battery is converted into light and heat energy.
- Mechanical energy (movement of objects) is converted to electrical energy by generators.
- Heat energy from the earth is converted to electrical energy in geothermal power plants.
- Light energy is converted to electrical energy in a solar cell.

HANDBOOK

- Sound (a type of mechanical energy) is converted to electrical energy in a telephone transmitter. At the other end of the line, it is converted back to sound.
- Electrical energy is converted to sound and light in a television set.
- Nuclear energy is released when the nuclei of atoms are split. This splitting, or fission, yields large amounts of heat energy that nuclear power plants convert to electricity. Nuclear energy is the strongest known force. If you could release the nuclear energy in 1 kilogram [2.2 pounds] of coal, you'd have the energy obtained from burning 3 million kilograms [6.6 million pounds] of coal.

Work

When a force acts on an object and moves that object in the same direction as the force, then work has been done. Work involves a transfer of energy. For example, energy from gasoline or diesel fuel is transferred through a car's engine to the wheels. The car moves.

It's important to remember that work involves both force and motion. If applying a force does not result in motion, then no work has been done. Also, the motion must be in the direction of the force.

Measuring Work

The amount of work done is found by multiplying force times distance.

$$W = f \times d$$

In the metric system, force can be expressed in newtons and distance in meters. The unit for work is the newton-meter, or joule. One joule is equal to the work done when a force of one newton moves something a distance of one meter in the direction of the force.

Several kinds of units are used for measuring energy and work. The units used depend on the amount of work being measured. You've already learned about the joule. Other units are the erg, electronvolt, calorie, Btu, and foot-pound.

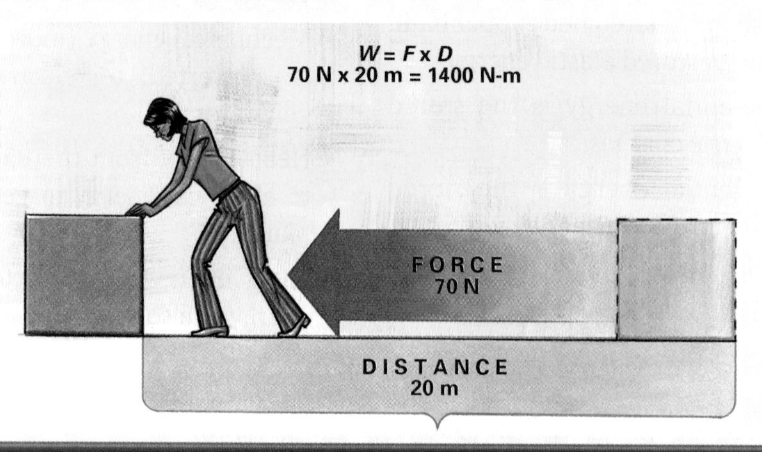

Matter

Our world consists of energy and matter. Energy is the ability to do work. Matter is what energy does work on. Matter is the "stuff" that things are made of. Anything that takes up space, has mass, and is made of atoms—whether it's wallboard or water vapor—is made of matter.

An atom is the smallest unit of matter, so small that trillions of atoms fit on a pencil point. An atom has a positively charged core called the nucleus and one or more negatively charged electrons. These orbit the nucleus (opposites attract, as with the poles of a magnet). However, electrons also respond to nearby atoms to form bonds that define the matter we know and work with.

For example, atoms may donate electrons to neighboring atoms. The steel-blunting hardness of quartz comes from this kind of bond. Plastics, on the other hand, are built on electron-sharing (covalent) bonds. Metals have their own systemwide form of electron sharing, known as the metallic bond. All these bonds, involving attraction of electrons to neighboring atoms, are chemical bonds.

Mass and Density

Mass is the measure of the amount of matter in an object. It is measured in kilograms. Density is a measure of mass per unit of volume. Put simply, if mass is a measure of how much "stuff" there is in an object, density is a measure of how tightly that "stuff" is packed.

Archimedes' Principle

Archimedes was a Greek mathematician. He discovered the law of buoyancy, which is sometimes called Archimedes' principle. This law states that the buoyant force on a completely or partially submerged object always equals the weight of the displaced fluid.

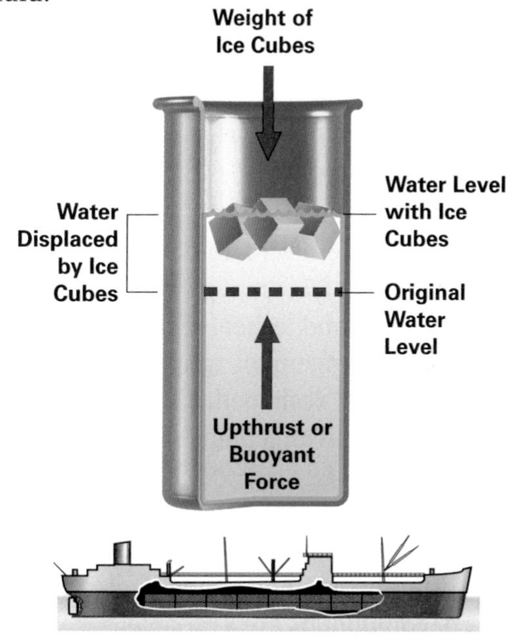

Elements

An element is a substance containing only one kind of atom. One kind of atom differs from another according to its number of protons, particles in the nucleus that give the nucleus its positive charge. An atom of hydrogen (H) has one proton. Helium (He) has two protons.

Compounds

Elements exist in pure form, such as pure oxygen. More often, elements combine with other elements to form another substance that may be quite different from its ingredients. For example, two gases (hydrogen and oxygen) combine to form water, a liquid. A substance that combines two or more elements in specific proportions is called a compound.

The specific proportions determine how two or more elements will bond and how the resulting compound will behave. Two atoms of hydrogen combine with one atom of oxygen to form water (H_2O). Two hydrogen atoms combine with two oxygen atoms to form hydrogen peroxide (H_2O_2). This is a bleaching agent that has been used as rocket fuel.

The proportions affect the arrangement assumed by the atoms in a compound. An interlocking-matrix arrangement gives hardness to glass and other ceramics. A long, chainlike arrangement gives plastics the capacity to be reshaped. Certain solvents have a ring arrangement of carbon and hydrogen.

Elements can combine in many ways—forming different kinds of chemical bonds and different arrangements of atoms. Millions of compounds are available.

Mixtures

A mixture is a combination of elements or compounds that does not involve a chemical bond. In other words, the ingredients hold onto their atomic identity. As a result, the ingredients of a mixture can be separated and recovered by ordinary mechanical means. These include filtering, settling, and distilling.

The ingredients in a mixture generally keep their characteristics. Sugar dissolved in water, for example, is both wet and sweet even though the sugar disappears from sight.

An alloy is a mixture made by mixing two or more pure metals or by mixing metals with nonmetals. The mixture is stronger than pure metal. For example, mixing small amounts of carbon with iron makes steel. Differences between compounds and mixtures include the following:

- Mixtures can be separated mechanically. Separating a compound requires breaking chemical bonds.
- In a mixture, ingredients keep their original properties.
- In a mixture, the proportion of ingredients does not have to be exact. Brass, for example, may have varying amounts of copper and zinc.

Properties of Materials

Designers, engineers, and builders must know the properties of materials in order to select the right ones for the job. Materials have many properties. The following explanations and drawings describe common properties.

- Sensory properties are those we can see, hear, smell, taste, or feel. Colors may be used to please the eye or to grab attention. Materials for blankets and infants' clothes are chosen for their softness. Scents are added to cleaning products to emit fresh, clean odors.
- Optical properties determine how a material reacts to light. Window glass is transparent—light passes through it, and you can see through it clearly. Frosted glass is translucent—light passes through it, but you cannot see through it. Opaque materials do not allow light to pass through. Some opaque materials, such as aluminum foil, also reflect light.
- Thermal properties determine how a material reacts to heat. Most metals are good heat conductors. This is why pots and pans are frequently made of metal. Fiberglass

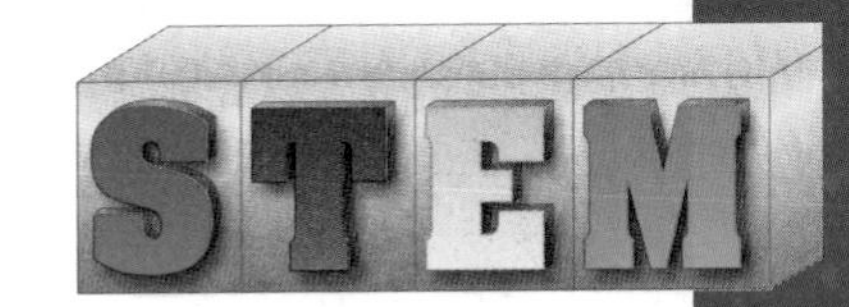

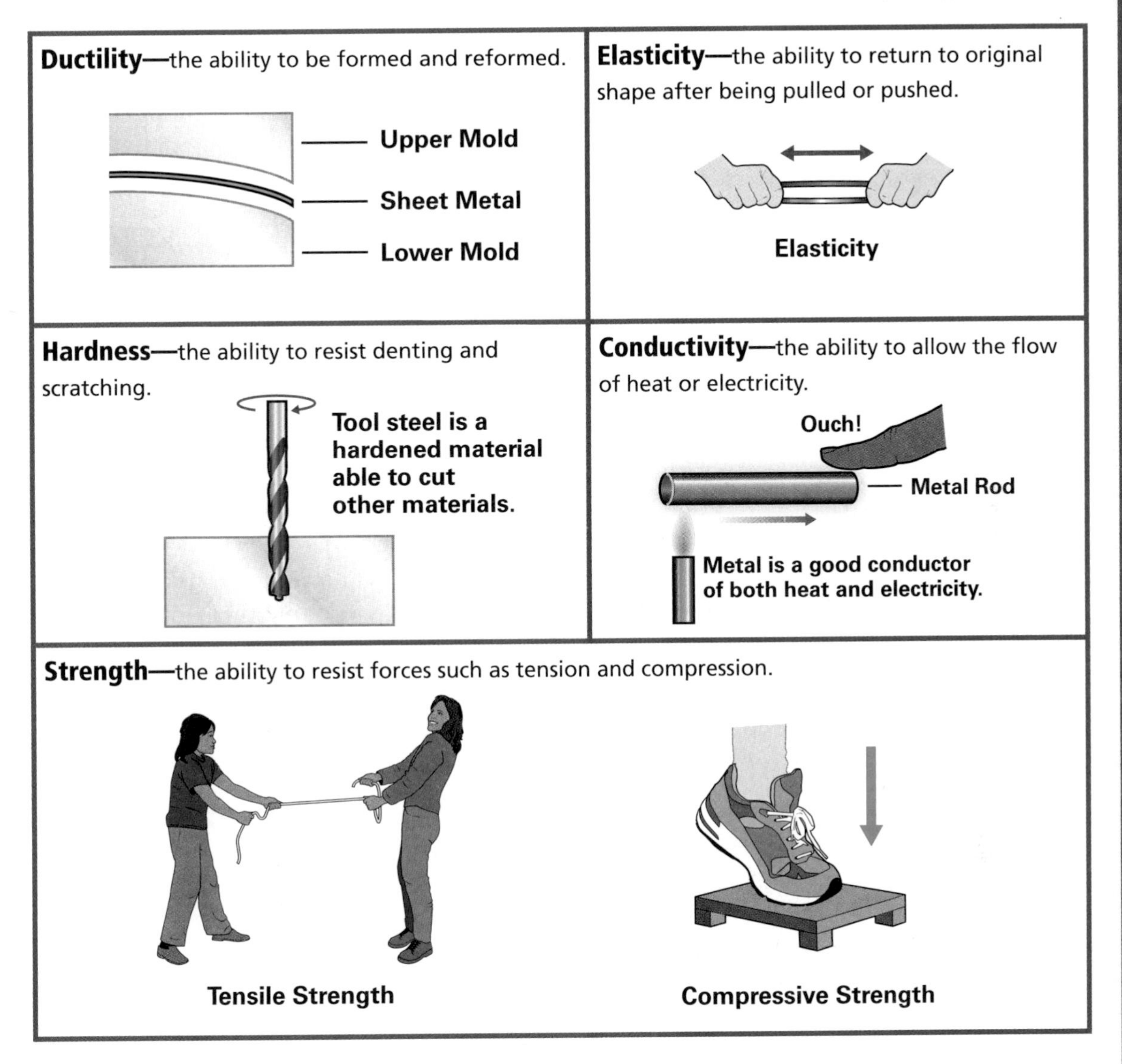

and Styrofoam™ resist the flow of heat. Therefore, these materials are commonly used to insulate structures and products.

- Electrical properties determine whether a material is a conductor, an insulator, or something in between (a semiconductor). Copper is a good electrical conductor, so it is frequently used in electrical wiring. Rubber blocks the flow of electricity. It is frequently used to cover the copper wire in electrical cords for safety.
- Magnetic properties determine how a material reacts to a magnet.
- Chemical properties deal with things such as whether a material will rust or whether it can dissolve other materials.
- Mechanical properties are those that describe how a material reacts to forces.

HANDBOOK

Electricity and Magnetism

Electricity

Electricity exists because most elementary particles of matter (such as electrons and protons) have a negative or positive charge. Like charges (two positives or two negatives) repel each other. Unlike charges attract. This electrical force is responsible for many natural phenomena. Technology has found many uses for it.

Electricity can be classified as static or current. In static electricity, the charges are "at rest." They remain in certain positions on objects. Static electric charges can build up when two different materials rub together, as when you walk on a carpet. If you then touch metal, the built-up charges suddenly flow away. They are no longer static, and you feel a shock. Lightning is another example of the sudden discharge (flowing away) of static electricity.

A flow of electricity is a current. A material through which current flows easily is called a conductor. A material through which almost no electricity can flow is called an insulator. Some materials can behave either as conductors or insulators, depending on what other materials are mixed with them. These are called semiconductors. They can be used to control the flow of electricity. For example, semiconductors are used to make computer chips.

Series and Parallel Circuits

The path electricity takes as it flows is called a circuit.

In a series circuit, individual components are connected end to end to form a single path for current flow. Series circuits have two major disadvantages. First, when connected in series, each circuit has to have its own switch and protective device. Also, if one component is open, the entire circuit is disabled.

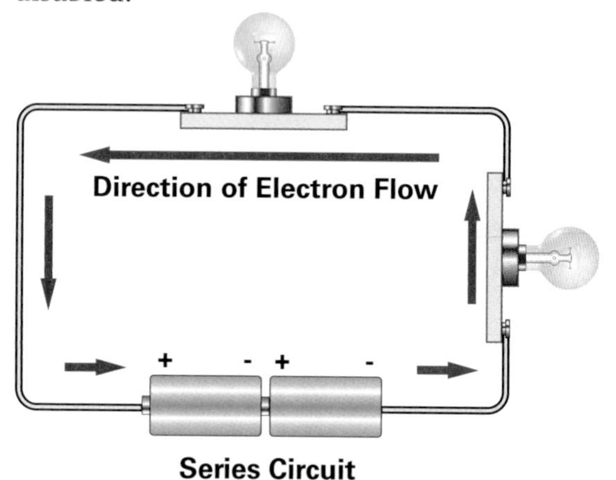

Series Circuit

In a parallel circuit, two or more loads are connected in separate branches. In most cases the parallel circuits are connected in series with a common switch and protective device. Equal voltage is applied to each

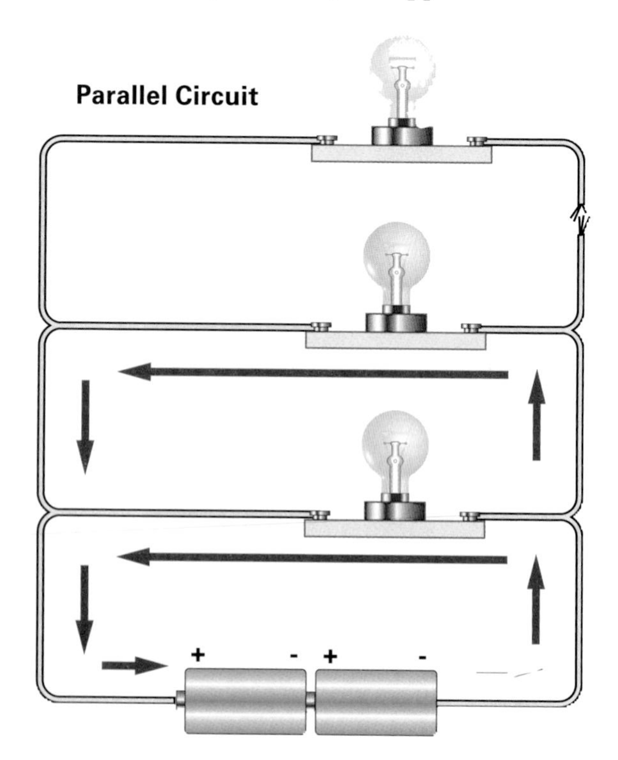

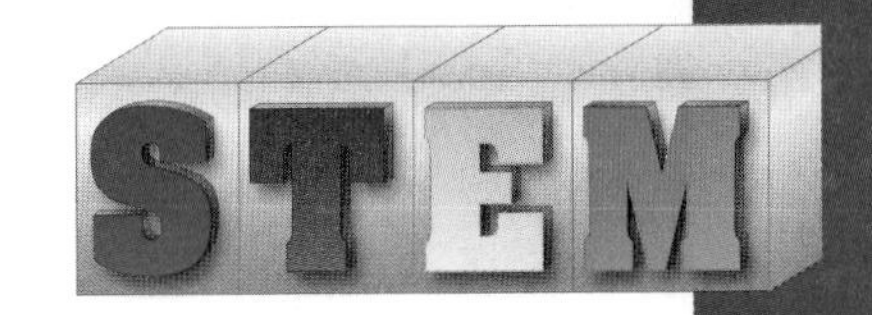

branch of a parallel circuit. Current flow divides as it reaches the parallel path. The amount of current flowing through each branch depends on the resistance in that path only.

A flashlight has a simple circuit. Turning on the switch completes the circuit. This allows electricity to flow from the battery, through the switch, then through the bulb and back to the battery.

Ohm's Law

Current is the amount of charge that flows past a point in a circuit during a given time. Current is measured in amperes (A).

For a current to exist in a conductor, there must be an electromotive force (emf), or potential difference, between the two ends of a conductor. This electromotive force is measured in volts (V).

Even conductors resist the flow of current somewhat. The greater the resistance, the less current that flows. Resistance is measured in ohms (Ω).

The voltage, amperage, or resistance of an electrical circuit can be calculated by using Ohm's law (named for German physicist George S. Ohm). This law states that electric current equals the ratio of voltage and resistance.

$$I = E/R$$

In this formula, I stands for the intensity of the current, measured in amperes; E stands for electromotive force, measured in volts; and R stands for resistance, measured in ohms.

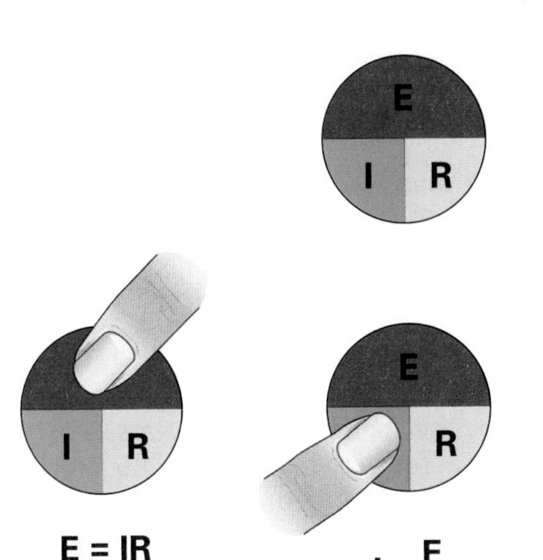

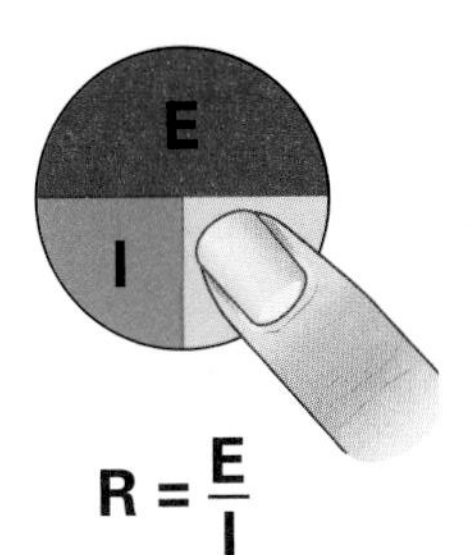

Magnetism

A pole is an area where magnetic density is concentrated. All magnets have two poles. These are designated as the north and south poles. The north pole of one magnet will attract the south pole of another magnet, while it will repel the other magnet's north pole.

Around the magnet is an invisible force field. When certain materials, such as iron, enter this force field, they become magnets, at least temporarily.

The planet Earth is a magnet with a force field extending into space. The force field, called the magnetosphere, protects us from harmful solar radiation. The poles of the earth magnet are not in exactly the same place as the geographic poles.

There is a scientific explanation for the action of magnets. Atoms produce magnetic fields. The magnetic field around a single atom is very weak. If the magnetic fields of a group of atoms align (point in the same direction), the force is much stronger. Such a group of atoms is called a domain. If the domains in a material align, that material becomes a magnet.

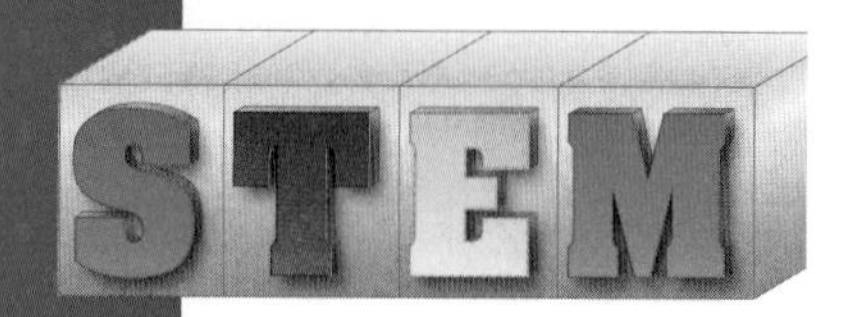

HANDBOOK

Electromagnetism

Magnets are not the only source of magnetic fields. Electric currents also produce magnetic fields. Magnetism produced by electricity is called electromagnetism. A magnet created by electricity is called an electromagnet. Electromagnets are very useful in many devices. Unlike other magnets, they are easily turned on and off. Their strength can be controlled by controlling the strength of the current.

Uses for Electricity and Magnetism

The relationship of electricity to magnetism has led to many developments in technology.

- Electromagnets make possible the magnetic levitation in maglev trains.
- A magnetic resonance imaging (MRI) machine contains a large coil that creates a strong magnetic field. This magnetic field affects the atoms in the human body. The machine detects the patterns of energy absorption.
- The electric current used in most homes and businesses is produced by generators at a power plant. Generators use magnets to convert the energy of motion (from steam, falling water, or wind) into electricity.
- Electric motors use magnets to change electricity into motion. They work by the interaction of a permanent magnet and an electrically induced magnetic coil.
- Computers, VCRs, and tape players make use of magnetic tapes and disks. Recording heads change electrical signals into magnetic patterns on the tape or disk. On playback, the process is reversed.

Light

Light is a form of energy. Sometimes it behaves like waves, sometimes like a stream of particles. The study of light has helped us develop technologies that make use of light as energy.

Properties of Light

Light travels in waves. The behavior of light waves is similar to the behavior of sound or water waves. For example, light waves bend as they pass through a small opening and spread out on the other side of the opening. This behavior is called diffraction. However, unlike water or sound waves, light waves can travel through a vacuum (an absence of air).

The speed of light is 300,000 kilometers [186,000 miles] per second in a vacuum. It travels at slower speeds through air, water, glass, or other transparent materials. When moving through an empty vacuum, light travels in a straight line. When it meets an object, the light may reflect (bounce off), be absorbed, or be transmitted (travel through).

Index of Refraction

As you've read, the speed of light changes when it passes from one medium (such as air) into another (such as glass). The index of refraction expresses the relationship between the two speeds. In the example of the glass, the index of refraction would be the speed of light in air divided by the speed of light in glass. Most glass has a refractive index of 1.5 compared with air. This means light travels 1.5 times faster in air than in glass.

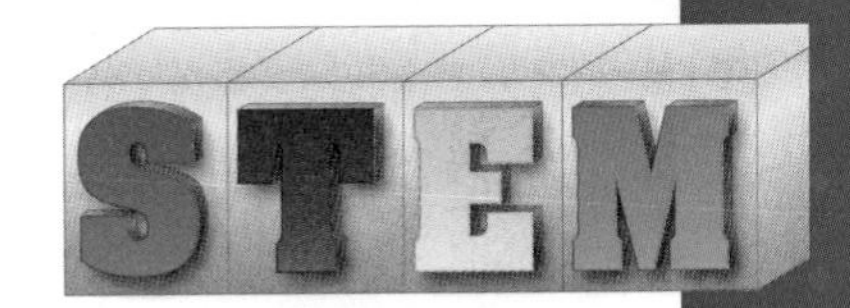

Light and the Electromagnetic Spectrum

A vibrating electric charge produces an electric field. The electric field produces a magnetic field. The magnetic field produces an electric field, which produces a magnetic field, and so on. This combination of constantly changing electric and magnetic fields is called an electromagnetic wave. Wave properties include frequency, speed, amplitude, length, and direction.

Frequency is the number of waves that pass a given point in one second. Amplitude is the strength (or height) of a wave.

Speed is the frequency times the wavelength. The speed of all electromagnetic waves is the same. Waves of higher frequency have shorter lengths. Waves of lower frequency have longer lengths.

Visible light is a type of electromagnetic wave and part of the electromagnetic spectrum. This spectrum is a range of wavelengths that also includes radio and television waves, radar (microwaves), infrared rays, ultraviolet rays, X-rays, and gamma rays.

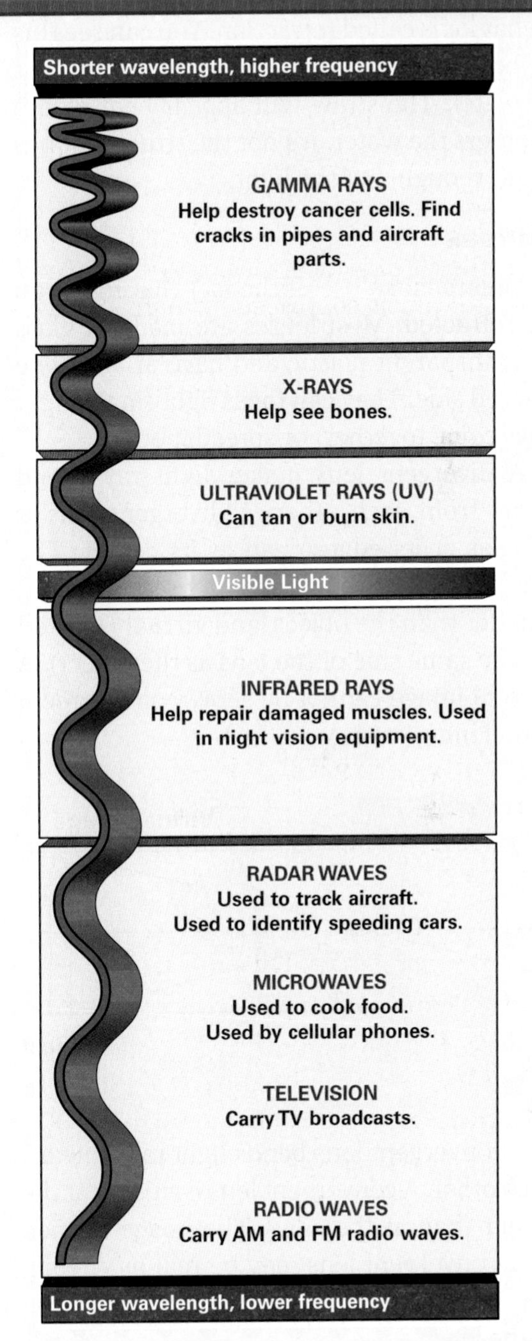

HANDBOOK

When light passes through a material, it slows down and changes direction. This behavior is called refraction. You can see this effect if you put a straight straw into a glass of water. The straw will appear bent where it enters the water. It's not the straw that has bent, though, but the light.

Lenses

Lenses make use of the fact that light can be refracted. Most lenses are made of glass or transparent plastic and have at least one curved side. They can focus light (make light rays come together) or spread it out.

A divergent lens makes light rays bend away from each other. A divergent lens is thicker at its edges than at its center. The image seen through this lens is upright, smaller than the object, and virtual (located on the same side of the lens as the object). A virtual image cannot be shown on a screen. A real image can.

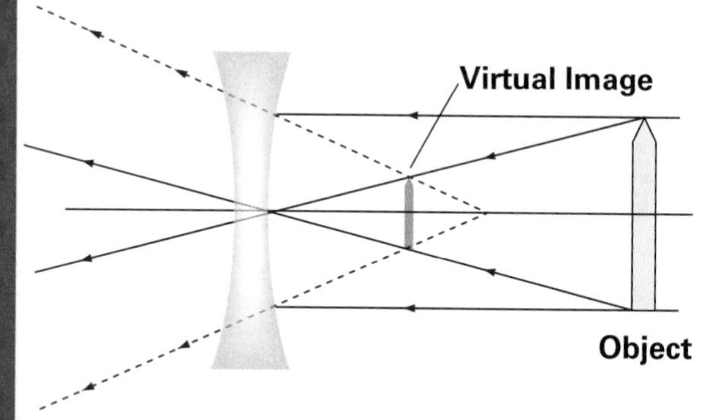

A convergent lens bends light rays toward each other. A convergent lens is thicker at the center than at the edge. The image formed by a convergent lens may be real or virtual, depending on the object's position in relation to the lens.

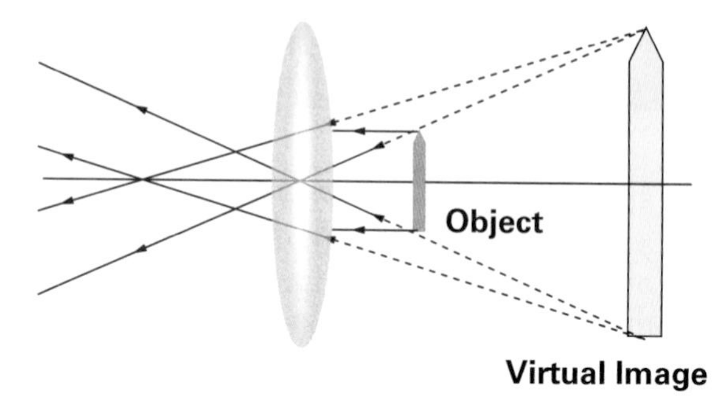

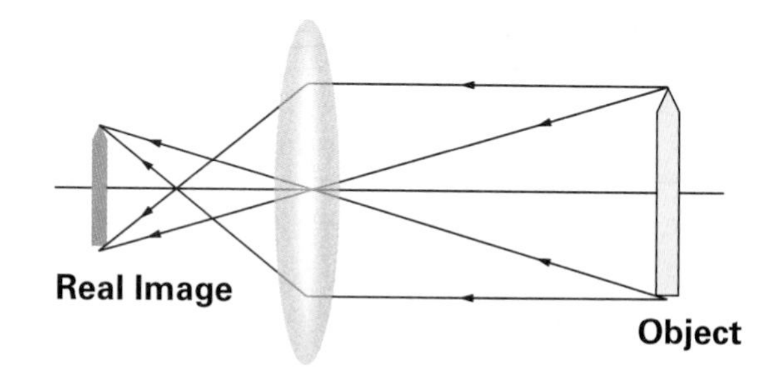

Uses of Light

Light finds various uses in communication and energy production.

- Pulses of light can be transmitted through fiber-optic cables. These pulses can carry telephone conversations, TV signals, or computer data.
- Photography (discussed in Chapter 7) is the control and recording of light. In videophotography, light is recorded as magnetic patterns.
- Photovoltaic cells turn light energy into electrical energy. Some experimental vehicles are solar-powered.
- Lasers produce a very precise beam of light. The light waves are "in step" with each other. Laser uses include surgery and reading compact discs.

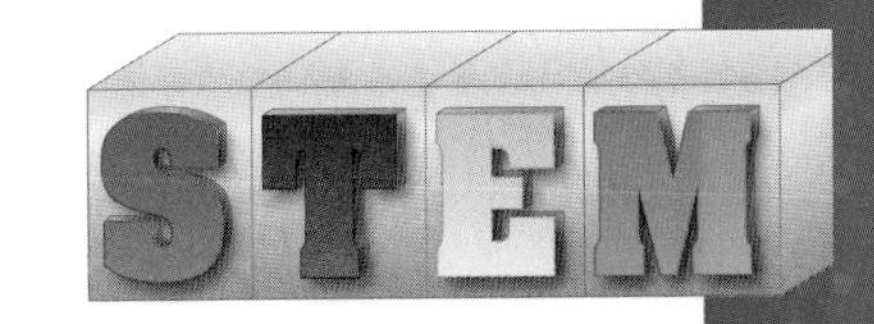

HANDBOOK

Cells

All living things are made of cells. Most cells are microscopic. An egg is an example of a very large cell.

Creatures such as bacteria consist of only one cell. Others are made up of many kinds of cells that have specialized functions. In the human body, for example, hair cells differ from liver cells. Both differ from muscle cells. In a plant, cells that make up the roots differ from the cells in the stalk.

Despite their variety, all cells share certain basic traits. Cells live by taking in food, converting it into energy, and eliminating waste. Cells reproduce by dividing in two. To carry out these functions, cells regulate their chemical reactions.

The chemical reactions that occur inside a cell depend on long, chainlike molecules called proteins. Even the simplest cell contains many different kinds of proteins. Each has a highly specific job to perform. DNA, for example, is a complicated combination of proteins. A technologist can cause a cell to produce something or stop producing something. The tool to make it happen would be a protein.

Plant and animal cells have certain features in common. A double-layered cell membrane made of a lipid (fat) keeps the watery contents of a cell in. It keeps almost everything else out. The membrane is dotted with proteins that act as gatekeepers. If a substance pushes against a gatekeeper protein and offers the right chemical fit, a passage opens. This is how a cell lets food in and discharges waste.

- Cytoplasm is a gel-like substance that fills the cell. It provides a medium for internal transport of materials.
- The nucleus is the core of the cell, containing its chromosomes. Chromosomes are made of DNA. This provides "templates" for assembling the right chemicals in the right way to carry on the life of the cell. Certain single-celled organisms, such as bacteria, do not have a nucleus.
- Mitochondria are structures (organelles) that provide energy to the cell. They break down nutrients with enzymes.
- Vacuoles are pockets for storing nutrients. Some animal cells do not have vacuoles. They may have other structures for storing materials until they are needed.
- Plant cells have two features not found in animal cells: chloroplasts and a thick cell wall.
- Chloroplasts create energy for plant cells. They use sunlight to convert water and carbon dioxide into simple sugars.
- A thick, fibrous cell wall surrounds the membrane of a plant cell. This provides rigidity to the overall structure of a plant.

H A N D B O O K

TECHNOLOGY

Technology consists of processes and knowledge that people use to extend human abilities and to satisfy human needs and wants. The chapters of this textbook discuss the nature of technology, its effect on our lives, and the ways technology is used in the designed world.

Technology can be classified into six broad areas.

Communication

Energy & Power

Manufacturing

Transportation

Construction

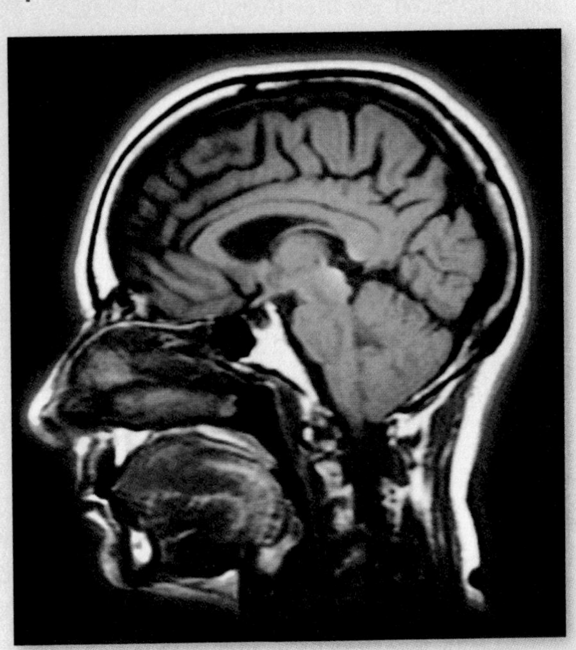

Bio-Related

ENGINEERING

Design and Drawing for Engineering

Suppose you received the following instructions: "Go north on Route 6. Turn right at Hendricks and park by the pool." These instructions sound clear enough, don't they? But we all know that even well-chosen words can prove inadequate. The drawing below, on the other hand, is much clearer.

If you want to be sure someone understands your message, use a picture. As the old saying goes, it is worth a thousand words.

Drawing pictures is a way of "thinking on paper." Pictures aid problem solving. If you're not sure whether the jewelry box you're making would look better with one drawer or two, sketch different designs until you find one you like.

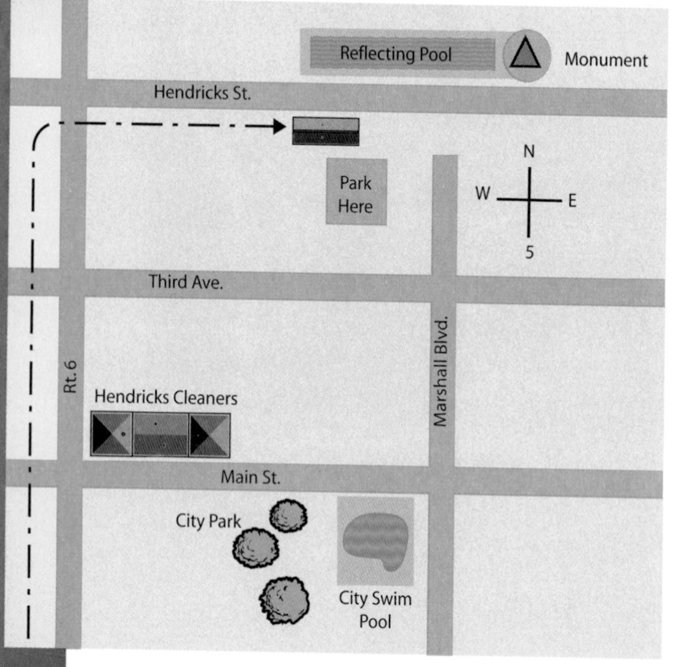

This section of the STEM Handbook will help you use a few graphic methods to communicate and work out solutions to problems. It is not a course in mechanical drawing. However, it will help prepare you for activities in this student text.

If you have access to a CAD system, it is the tool of choice for drafting. Can a computer draw by itself? No. Will it turn a poor student into a good student? No. So what good is it? Perhaps the best thing about CAD is that it allows the user to modify a drawing, just as word processing allows the editing of text. You can cut and paste and save and recall files. You can quickly perform functions that might otherwise take much time and effort.

Sketching

The best way to express yourself graphically when time is short is by sketching. You will rapidly discover what a difficult task interpreting verbal description is. Yet a single sketch may clarify complex thoughts. You can draw "good" sketches by practicing. Few people make attractive sketches at first. Repeated practice over an extended time period, coupled with attention to constructive criticism, is the best approach.

Sketching Tools

The pencils shown here are the different kinds used for drawing. Pencils are classed by the hardness of their lead. Pencils with hard leads are used when accuracy is needed, such as for charts and diagrams. Medium pencils are used for technical drawing and sketching. Soft pencils are used for general artwork and quick sketches.

Is It a Good Design?

Before you begin to design a product and during each design stage, ask yourself these questions:

- Is it appropriate to the need you're trying to fill?
- Do people really need it or think they need it? (If they don't, they won't use it.)
- Can it be made under the present conditions? (For example, if you have only a week to make it, can you make it in that time?)
- How much does it cost to make? (If it's too expensive, you'll have trouble marketing it.)
- Is it durable? Will it last?
- Is it attractive?

Use a soft pencil. Hold it about 1½ inches from the point. Work big. Don't make small, cramped pictures. A good sketch should not look stiff and "perfect" like a mechanical drawing. The lines should be free and loose.

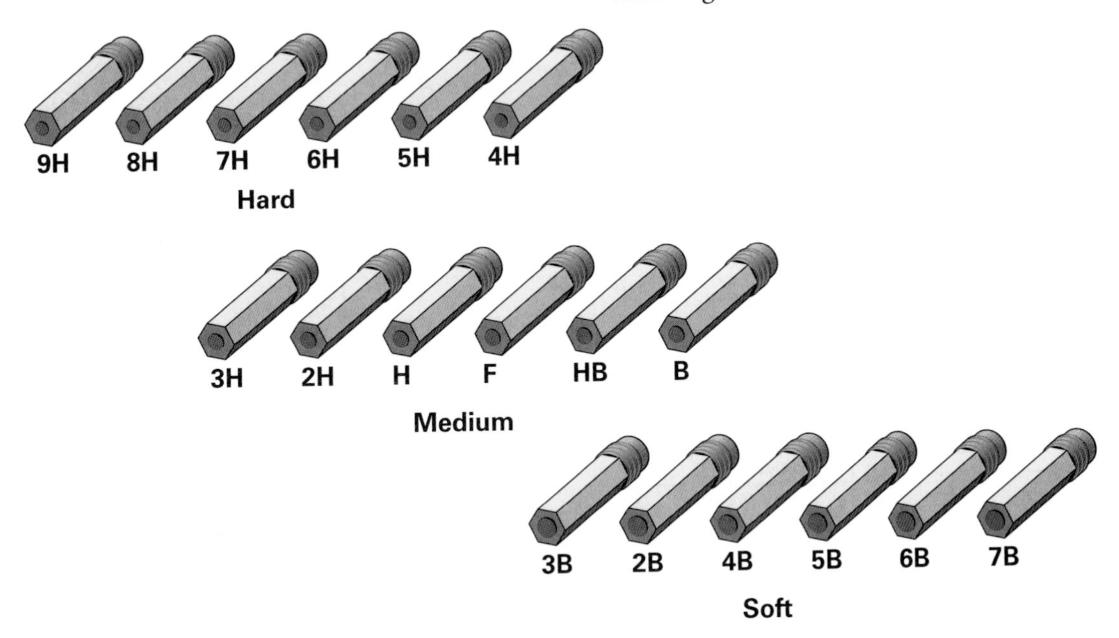

When sketching straight lines, spot your beginning and end points (A). Keep your eye on the point toward which you are drawing.

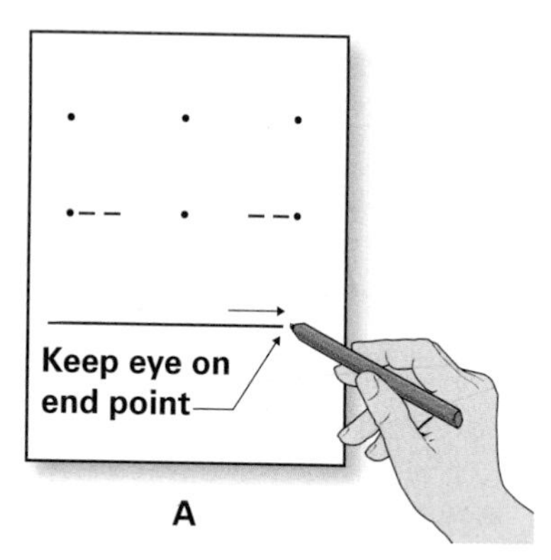

A

Draw vertical lines downward (B).

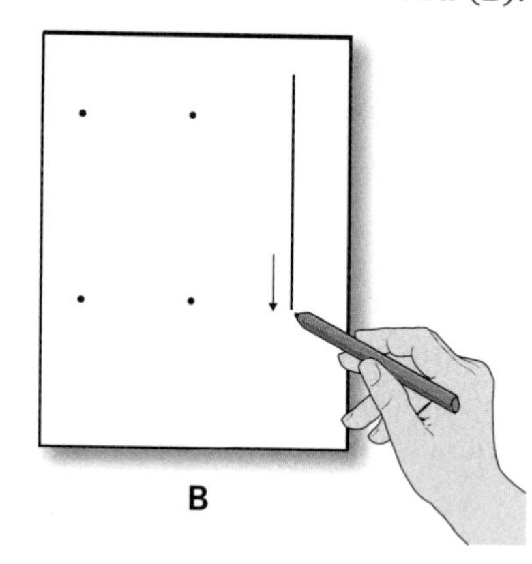

B

If an inclined line is to be nearly vertical, draw it downward (C and D).

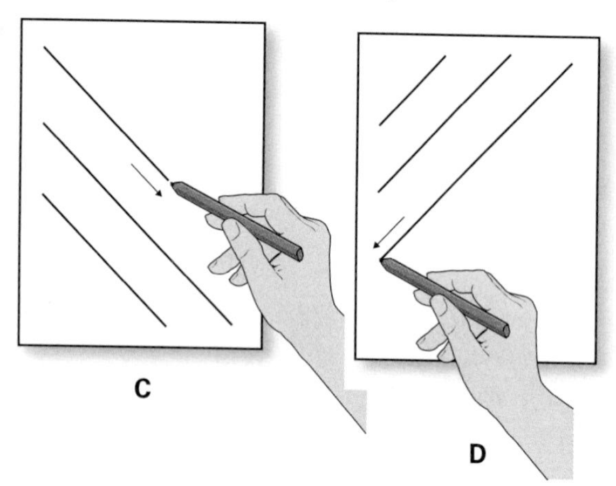

C

D

If an inclined line is to be nearly horizontal, draw it to the right (E).

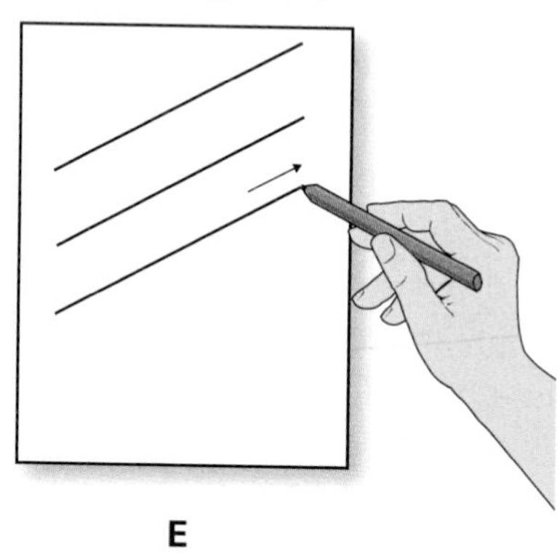

E

Inclined lines can also be drawn as if they were vertical or horizontal by simply turning the paper (F and G).

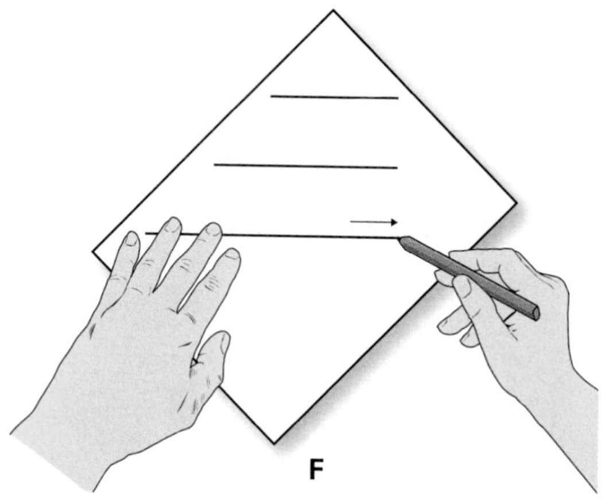

F

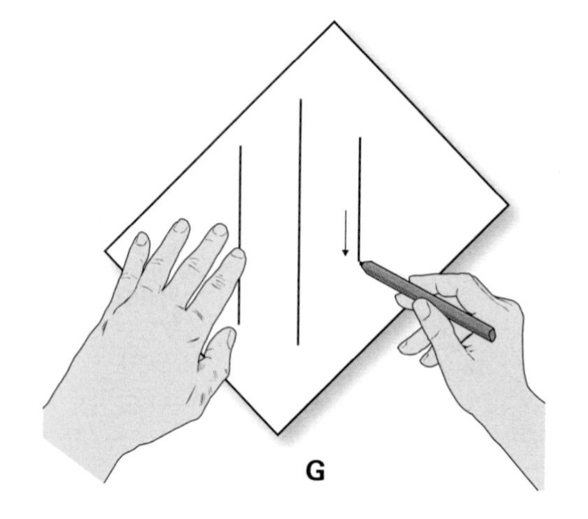

G

Sketching a Circle

Circles are hard for many people to draw. Many times they are lopsided. There are tricks you can use, however, that make drawing circles easier.

First, lightly sketch a square that is the diameter of your circle (1). Then mark the midpoints of the sides. Next, draw diagonal lines (2) and mark where the line of the circle will cross them. Finally, sketch in the circle (3). (Oval shapes can be drawn the same way, using a rectangle instead of a square.)

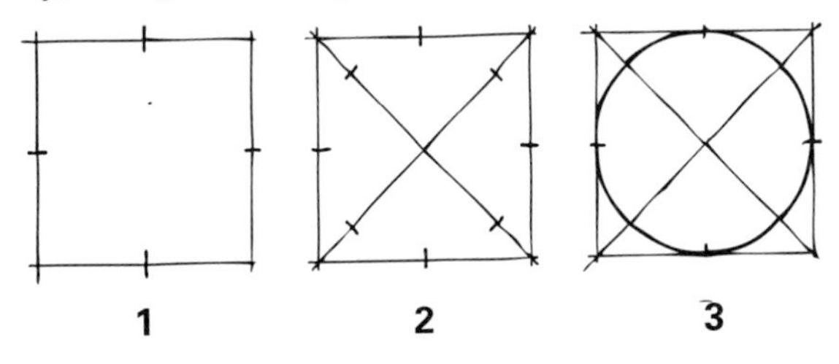

You can also draw a circle with the aid of centerlines. First, sketch two centerlines (1). Then add light radial lines, or "spokes," between them (2). Make marks where the line of the circle will cross. Finally, sketch in the circle (3).

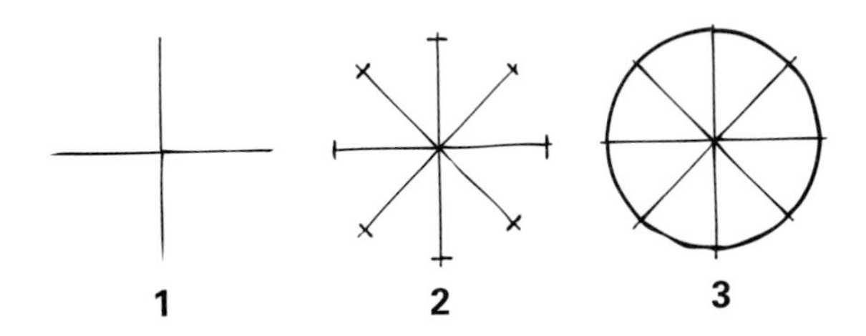

Arcs can be sketched using the same general methods as for circles. First, lightly sketch a square of the same size as the arc. Draw a diagonal line, and mark off the size of the arc. Finally, sketch in the arc itself.

Sketching Practice

These are the steps to follow in making a sketch.

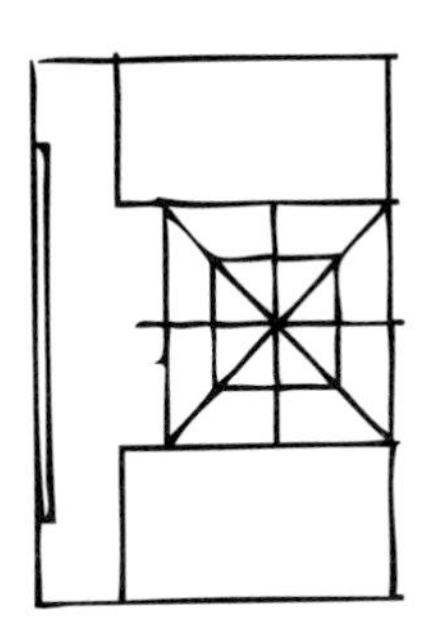

1. Block in main shapes.

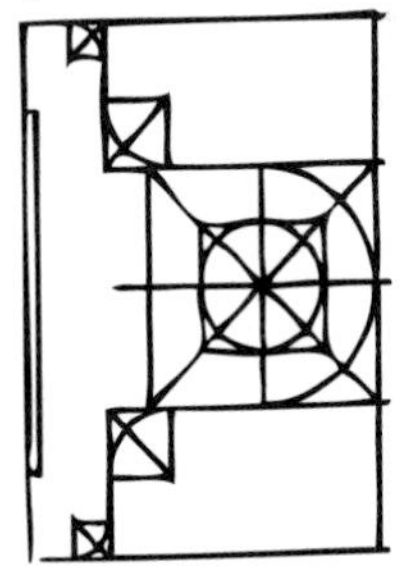

2. Sketch arcs & circles.

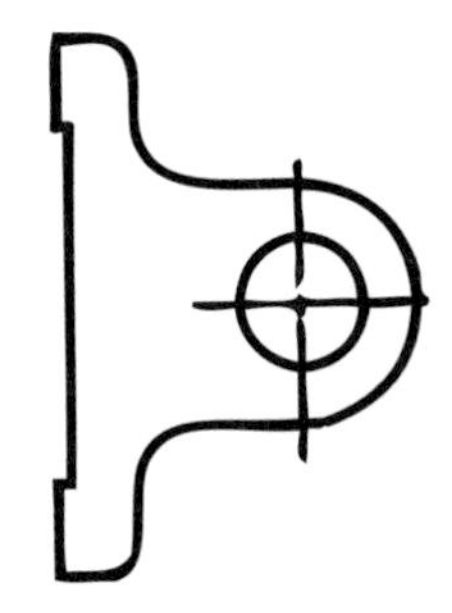

3. Draw heavy final lines.

HANDBOOK

Building from Basic Shapes

The square, the rectangle, the circle, the cylinder, the triangle, and the cube can be used to draw other objects. Identify which of them have been used to help create the objects shown here.

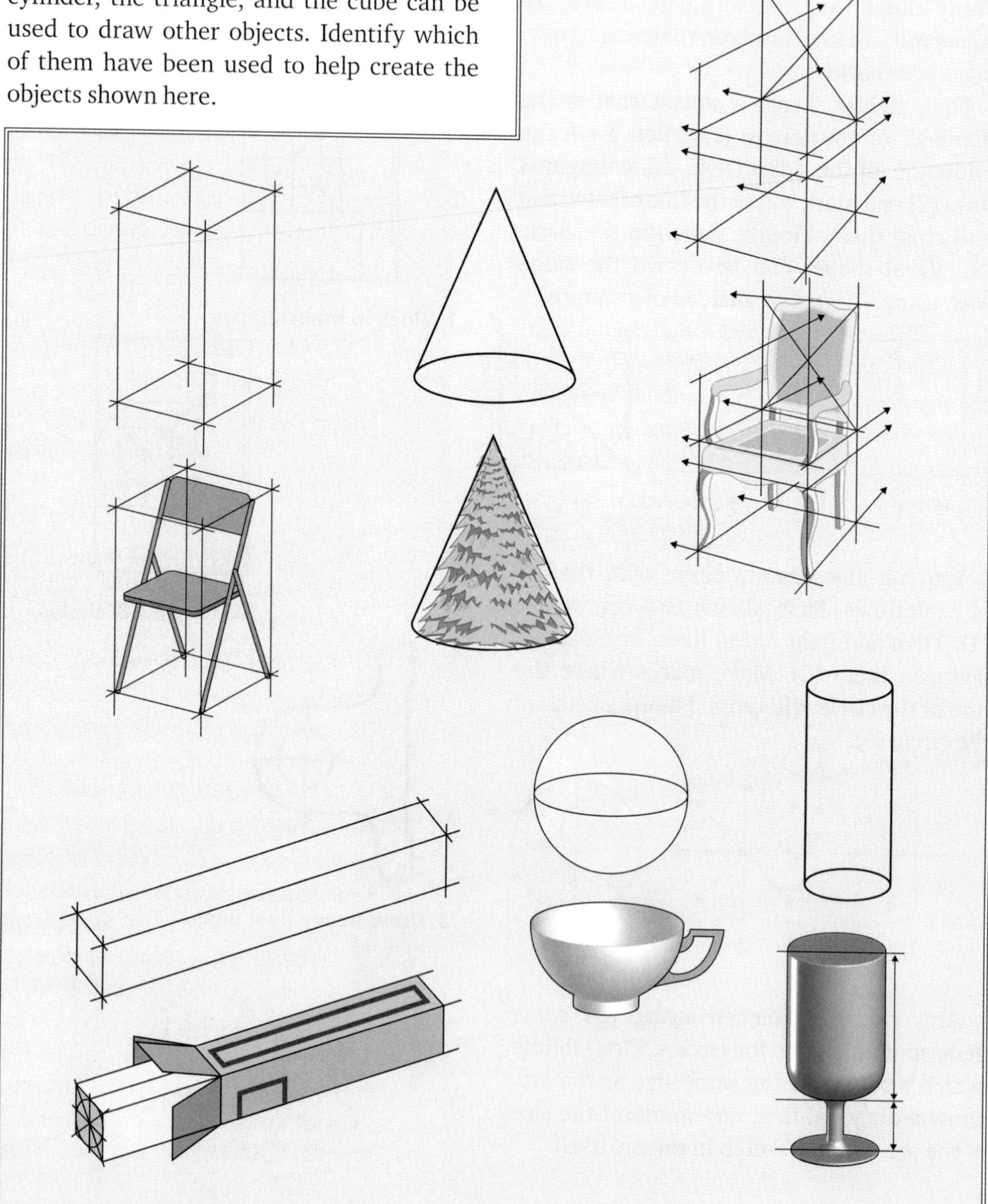

Using a Protractor

The protractor is used for measuring angles.

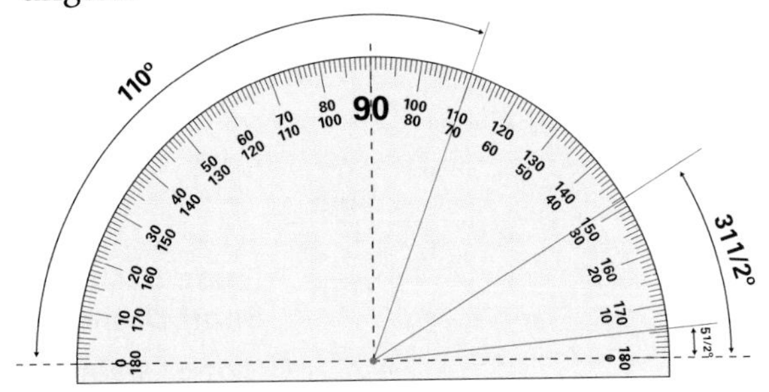

Drawing to Scale

Some objects cannot be drawn full-size. They are too big or too small. For this reason they are drawn smaller or larger than actual size. A building, for instance, may be shown on a map 1/96 size. Drawing objects larger or smaller than actual size but in the correct proportions is called drawing to scale.

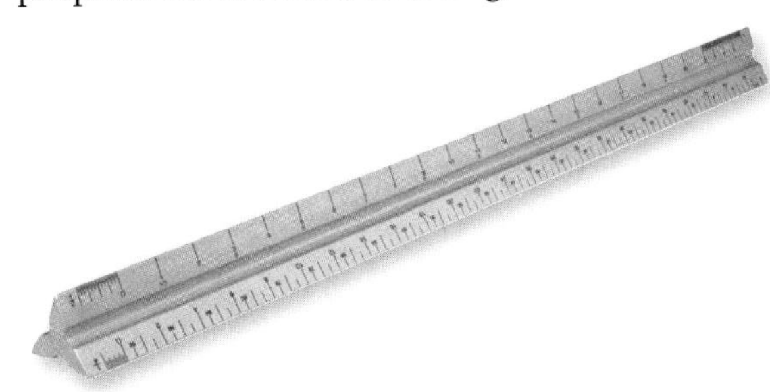

The scale to which a drawing is made should be indicated on the drawing. For example, "Scale: Half size" or "Scale: 1 in. = 1 ft."

Tools called scales are used by architects and engineers to quickly calculate scale reductions or enlargements. The architect's scale includes reduced-size scales in which fractions of an inch represent feet.

Dimensioning

Dimensions on a drawing give sizes. They are important when you or someone else must make an object from a drawing. There are many lines and methods used in dimensioning.

You will need to be able to recognize the various kinds of lines. The most important lines are object lines, hidden lines, dimension lines, extension lines, and leader lines.

- Object lines show the visible lines of the object.
- Hidden lines show where a line would be if you could see through the object.
- Dimension lines usually have the dimension written in the center. They end in arrows.
- Extension lines extend from the object to the dimension line and a little beyond it. They show the boundaries of the area being measured.
- Leader lines are usually used to give information or to dimension interior details, like holes. They lead from the dimension to the part referred to. They end in an arrow that touches the part.

All dimension figures should align with the bottom of the drawing. Never place a dimension directly on the object itself unless there is a good reason for it. The largest dimension goes on the "outside" of any shorter dimensions. Diameters of round objects are indicated with a symbol, which means "diameter." It is written before the dimension number. The symbol for a radius is R. It, too, is written before the number.

STEM

HANDBOOK

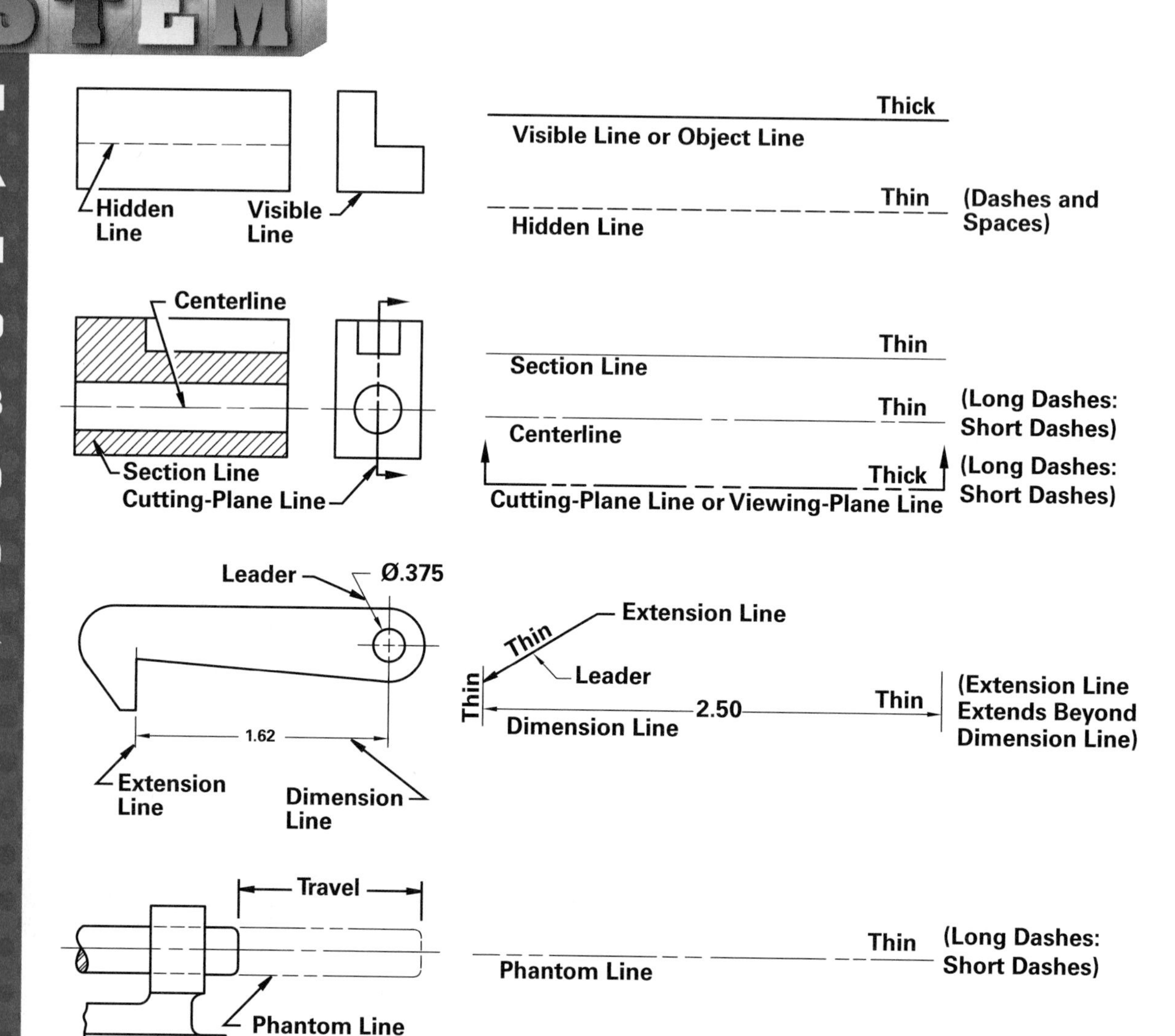

A dimension on a drawing is exact. It is usually not possible to make an object that is exact to within thousandths of an inch. For that reason dimensions are often given tolerances. A tolerance is the amount a given measurement can vary. For example, in the drawing of the five holes, the diameter of each hole must be drilled to a size of .250 inch. With the tolerance given, the holes could be any size between .255 inch and .245 inch. In other words, there is a tolerance of plus or minus .005 (five-thousandths) of an inch.

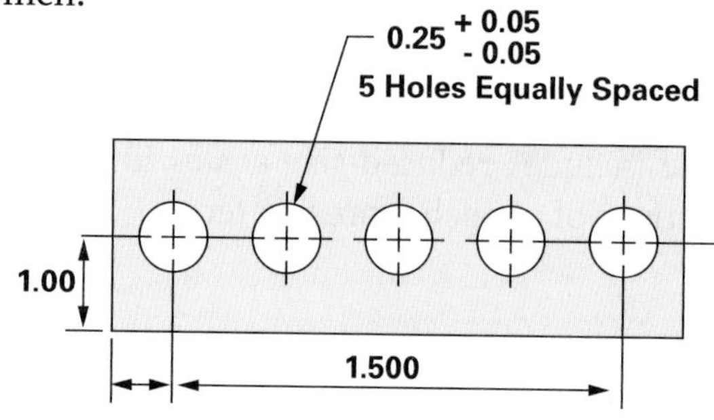

HANDBOOK

Working Drawings

Working drawings, or multiview drawings, show how to make a product. All dimensions are given. The object is shown from as many angles as needed to indicate its width, height, and depth.

Working drawings show "head-on" views of the top, front, back, sides, and bottom of an object. Ordinarily only three views are shown—top, front, and right side. They are produced by means of orthographic projection. "Ortho" means straight or at a right angle. In orthographic projection the pictures (views) of the object are of surfaces at right angles to one another.

Working drawings may require dimensioning. Most are drawn to scale. Drawing A shows the steps to follow in making a three-view working drawing.

A

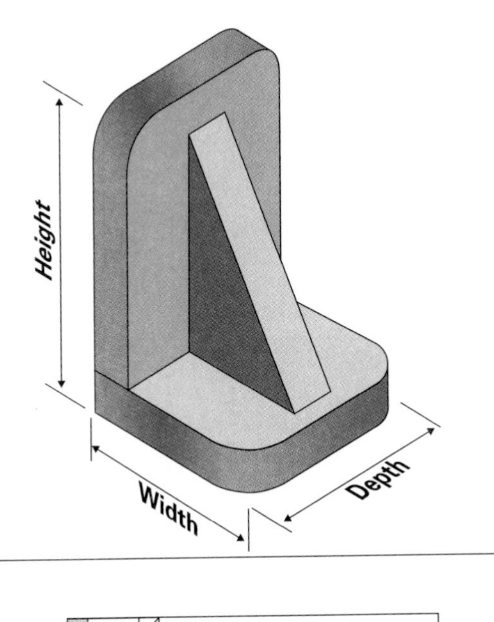

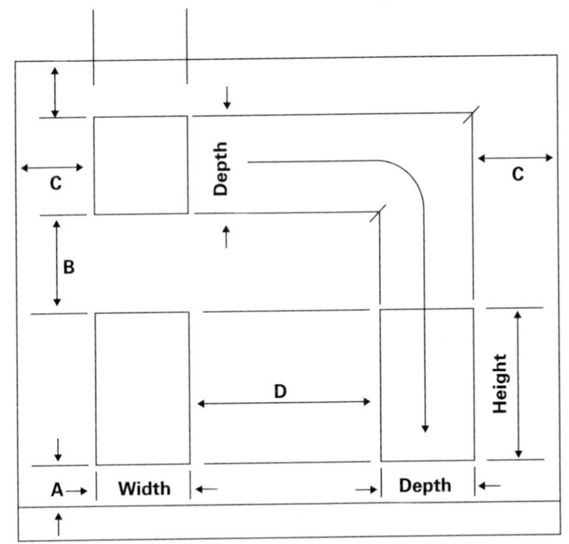

1. Block in boxes for views in proportion.

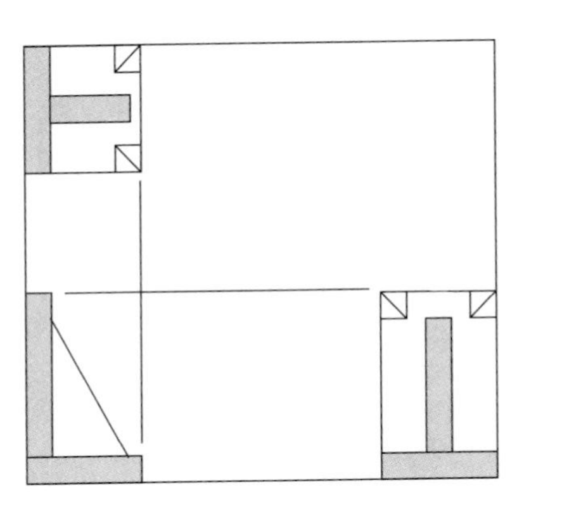

2. Block in construction for arcs, circles, etc.

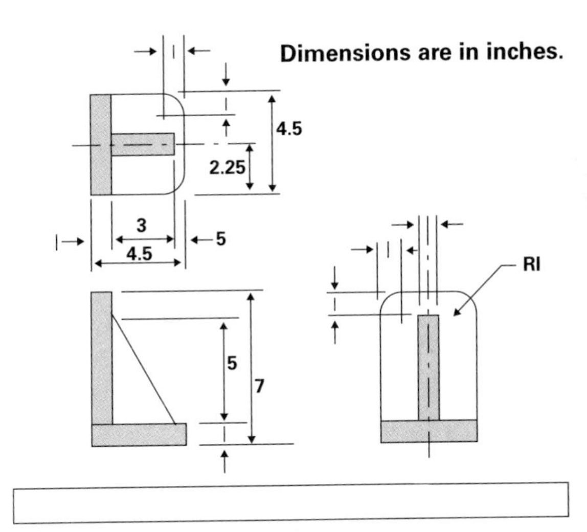

3. Complete details.
4. Add dimensions.

HANDBOOK

Orthographic projection is used to make a working drawing. Orthographic projection is the technique of showing several surfaces of an object "head on." Because many objects have six sides, as many as six views are possible. Usually, however, only three views are shown. These show the front, top, and right side. Note where the observer is standing when viewing the house.

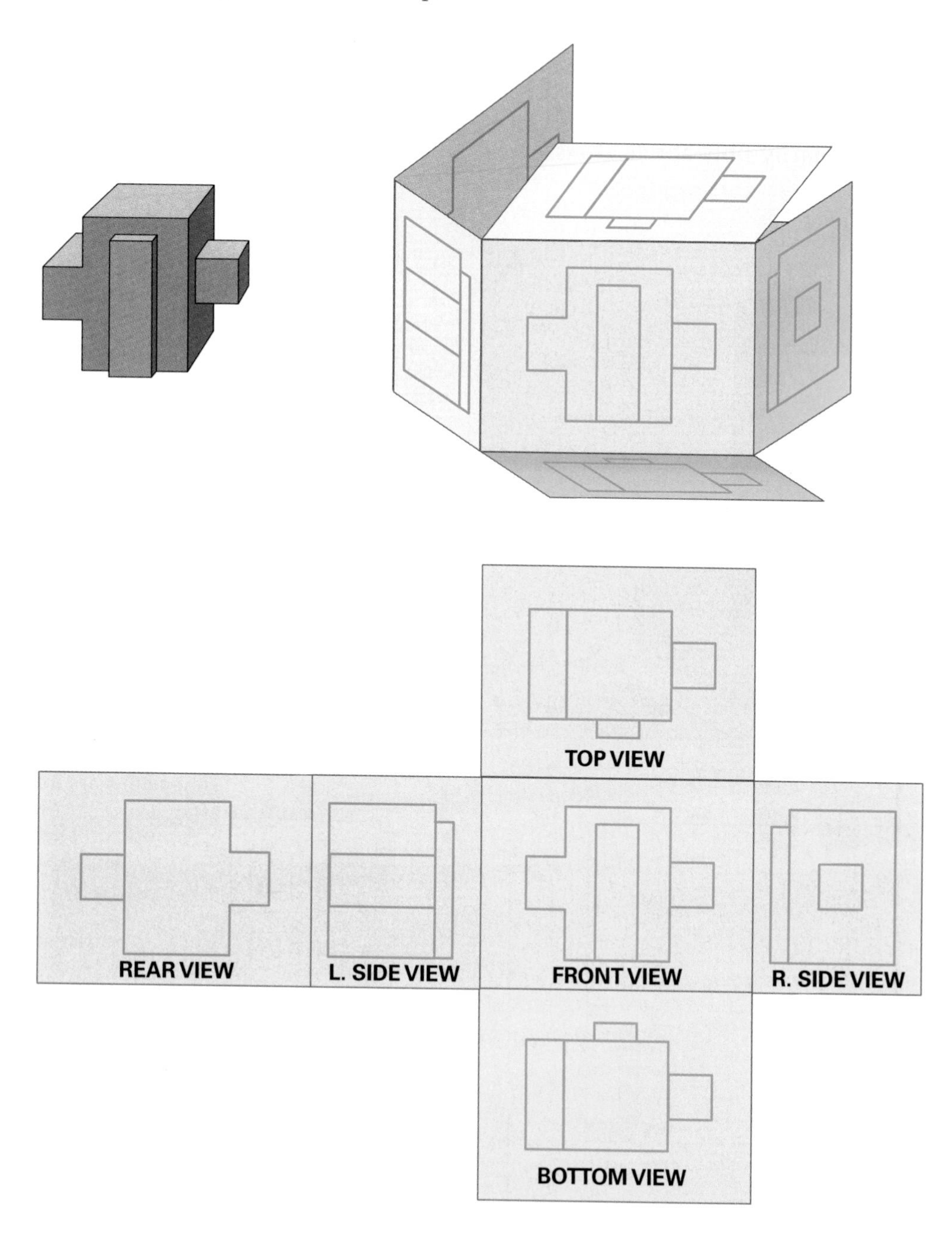

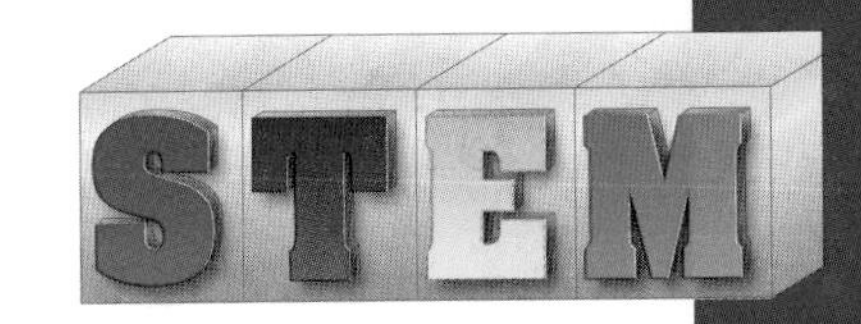

Engineering Design

Engineering Design Process

These are the main steps in the engineering design process.

1. Define the problem.
2. Brainstorm, research, and generate ideas.
3. Identify criteria and specify constraints.
4. Develop and propose designs and choose among alternative solutions.
5. Implement the proposed solution.
6. Make a model or prototype.
7. Evaluate the solution and any of its consequences.
8. Refine the design.
9. Create the final design.
10. Communicate the processes and results.

Defining the Problem

The design problem must be clearly stated before design work begins. It must include all of the important design requirements. You may need to clarify and restate the design problem. The problem will determine the design.

Checking the Design

Defining and refining the design is just the first part. A good design depends on the proper relationship of several elements. This requires careful planning. A good design depends on good planning. Check the design at each stage to ensure that the design will work.

Elements of Design

The elements of design are the main parts of the design. These elements include:

- **Line.** This can be thin, thick, straight, or curved, among other forms.
- **Shape.** Is it a rectangle, a triangle, or a circle?
- **Mass.** What is the size of the object?
- **Color.** This element can be used to highlight function or add interest.
- **Pattern.** What is the appearance of the object's surface? Is it plain or does it carry a design?

Principles of Design

The effectiveness of a design can be judged against certain principles. These principles include:

- **Balance.** Pleasing integration of various elements.
- **Proportion.** Proper relation of parts to one another.
- **Harmony.** Pleasing arrangement.
- **Unity.** Ordering of parts to create an undivided total effect.

The acronym DEAL is a good way to remember the main steps in the basic problem-solving process.

D: define the problem
E: explore all possible solutions
A: act on the best solution
L: look back and evaluate

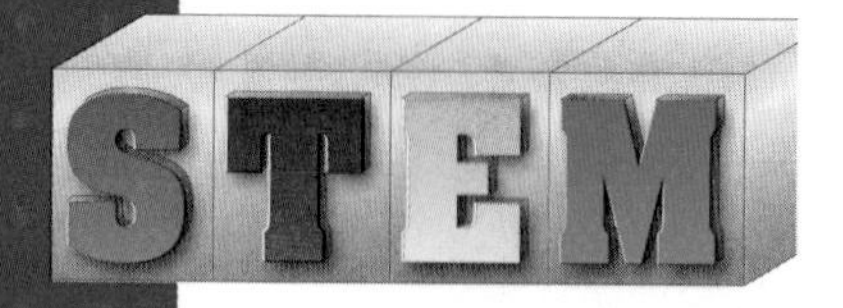

HANDBOOK

MATH

Linear Measurement

A linear measurement is one made along a line. To make a linear measurement, use a rule or a tape measure.

Working with Fractions and Decimals

In the customary system of measurement, you'll often be working with fractions, especially when measuring materials. Most customary rules are divided into inches. A one-foot rule has 12 inches. A yardstick has 36 inches. The inches are divided into smaller parts (fractions). Typically, there are marks to show ½ inch, ¼ inch, ⅛ inch, and sometimes 1/16 inch.

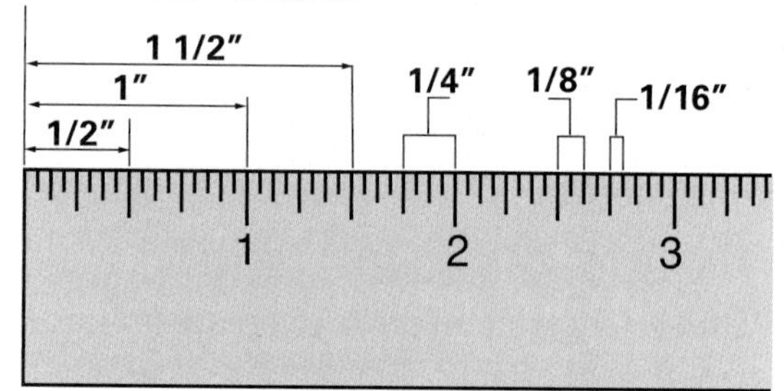

Adding Fractions

If the denominators are the same, add the numerators. Simplify the sum if necessary.

Example:

$$\frac{1}{8} + \frac{1}{8} = \frac{2}{8}$$
$$\frac{2}{8} = \frac{1}{4}$$

If the denominators are different, you must rename the fractions with a common denominator before you can add them.

Example:

$$\frac{1}{4} + \frac{3}{8}$$
$\frac{1}{4}$ is equal to $\frac{2}{8}$
$$\frac{2}{8} + \frac{3}{8} = \frac{5}{8}$$

If the numbers are mixed, add the fractions. (If necessary, rename them with a common denominator first.) Add the whole numbers. Rename and simplify.

Example:

$$1\tfrac{3}{4} + 2\tfrac{1}{2}$$
$2\tfrac{1}{2}$ is equal to $2\tfrac{2}{4}$
$$1\tfrac{3}{4} + 2\tfrac{2}{4} = 3\tfrac{5}{4}$$
$3\tfrac{5}{4}$ is simplified to $4\tfrac{1}{4}$

Subtracting Fractions

If the denominators are the same, subtract the numerators.

Example:

$$\frac{3}{8} - \frac{2}{8} = \frac{1}{8}$$

If the denominators are different, you must rename the fractions with a common denominator before you can subtract them. Simplify if necessary.

Example:

$$\frac{5}{16} - \frac{1}{8}$$
$\frac{1}{8}$ is equal to $\frac{2}{16}$
$$\frac{5}{16} - \frac{2}{16} = \frac{3}{16}$$

If the numbers are mixed, subtract the fractions. (If necessary, rename them with a common denominator first.) Subtract the whole numbers. Rename and simplify.

Example:

$$4\tfrac{1}{8} - 2\tfrac{1}{16}$$
$\frac{1}{8}$ is equal to $\frac{2}{16}$
$$4\tfrac{2}{16} - 2\tfrac{1}{16} = 2\tfrac{1}{16}$$

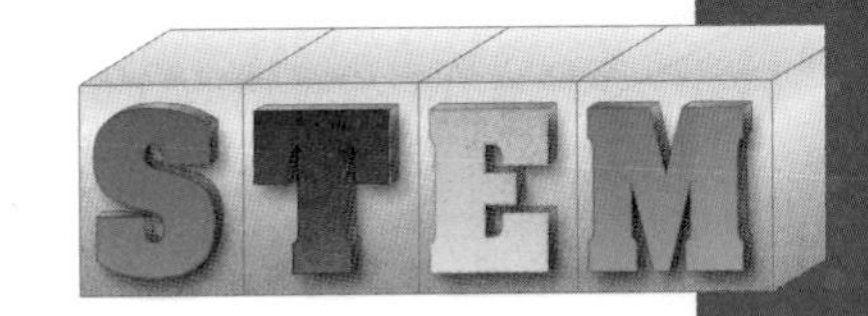

Multiplying Fractions

Multiply the numerators and then multiply the denominators. Simplify if necessary.

Example:

$$\frac{2}{16} \times \frac{1}{2} = \frac{2}{32}$$

$$\frac{2}{32} = \frac{1}{16}$$

If the numbers are mixed, rename them as improper fractions. Then multiply the fractions. Rename and simplify as needed.

Example: Suppose you have 3 ¼ pounds of cement. You need to use ⅔ of it. How many pounds would that be?

$$\frac{2}{3} \times 3\frac{1}{4}$$

$$\frac{2}{3} \times \frac{13}{4} = \frac{26}{12}$$

$$\frac{26}{12} = 2\frac{1}{6}$$

Dividing Fractions

To divide by a fraction, multiply by its multiplicative inverse. In other words, invert the fraction and then multiply. Simplify if necessary.

Example: You have a collection of CDs. Each one, in its case, is ⅜" thick. If you make a CD holder that is 6" wide, how many of your CDs will fit in it?

$$6 \div \frac{3}{8}$$

$$6 \times \frac{8}{3} = \frac{48}{3}$$

$$\frac{48}{3} = 16$$

Converting Fractions to Decimals

To convert fractions to decimals, divide the numerator by the denominator.

Example: To convert the fraction ⅜, divide 3 by 8.

$$3 \div 8 = .375$$

Converting Decimals to Fractions

To convert decimals to fractions, write the decimal as a fraction and simplify.

Example 1:

$$.25 = \frac{25}{100}$$

$$\frac{25}{100} = \frac{1}{4}$$

Example 2:

$$.375 = \frac{375}{1000}$$

$$\frac{375}{1000} = \frac{3}{8}$$

The greatest common factor of the two numbers is 125. Dividing 375 by 125 = 3. Dividing 1000 by 125 = 8.

Angles

An angle is the figure formed when two lines or two surfaces originate at the same point. Angles are measured in degrees. The following drawings show some common angles.

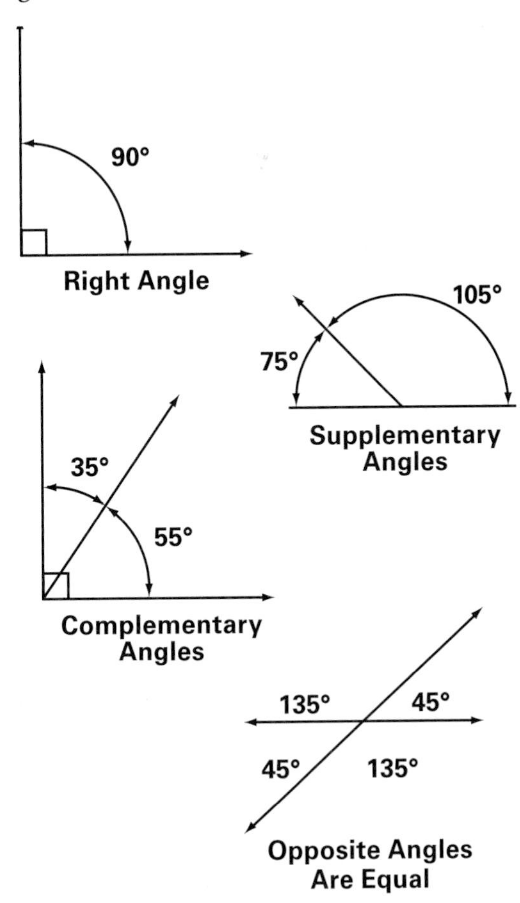

Polygons

A polygon is a closed figure with straight sides. It is classified by the number of sides. In a regular polygon all sides are equal in length and all angles are equal.

3 Sides = Triangle

4 Sides = Rectangle

4 Equal Sides = Square

5 Sides = Pentagon

6 Sides = Hexagon

8 Sides = Octagon

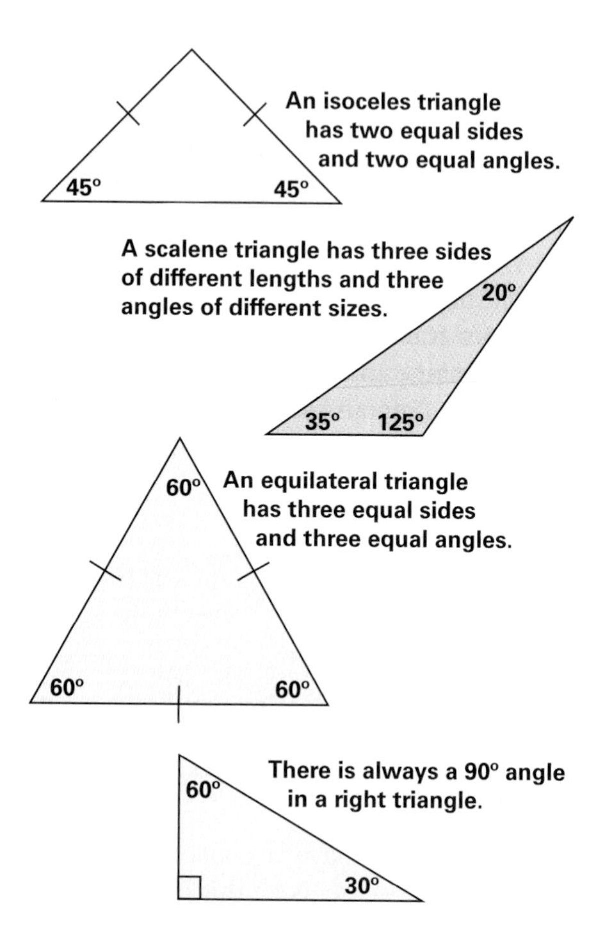

Types of Triangles

Triangles are classified according to the length of their sides.

- **Isosceles.** A triangle having two equal sides.
- **Equilateral.** A triangle having all sides equal.
- **Scalene.** A triangle having three sides of unequal length.

Triangles are also classified according to the size of their largest internal angle. The right triangle has one internal angle of 90°, which is a right angle.

Rectangles and Triangles

A rectangle is a four-sided shape with four right angles (90°). The diagonals of a rectangle are always equal. They divide the rectangle into two right triangles. A right triangle has three sides. The longest side is called the hypotenuse. These facts are important when laying out a perfect rectangle. It can be used when constructing a simple box or even when laying out a large building.

$A^2 + B^2 = C^2$

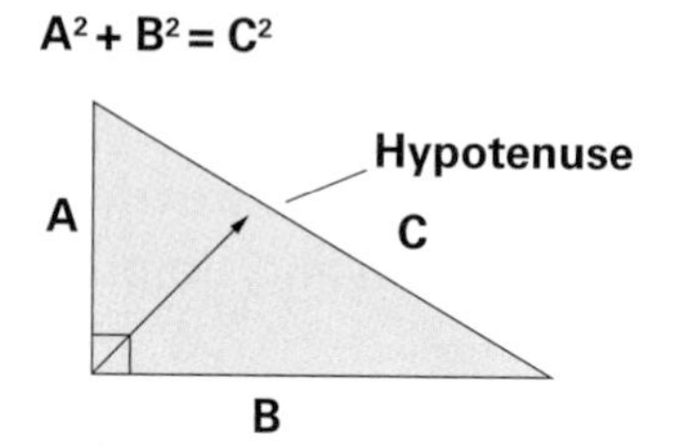

Using Right Triangles

If you know the length of two sides of a right triangle you can figure the length of the third side by using the equation:

$$a^2 + b^2 = c^2$$

c = the hypotenuse, the side opposite the right angle

a and b = the two other sides

This equation is called the Pythagorean theorem. Here's a practical application. Let's say you want to build a storage shed that is 6' × 8'.

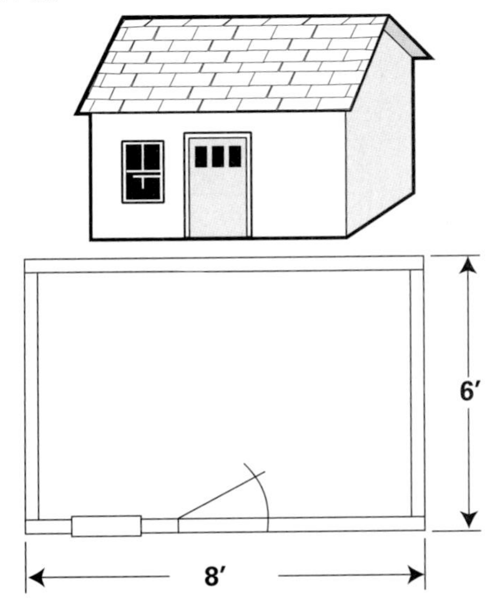

1. First establish the 8' side of the shed. Place a stake in the ground at each end of this side.

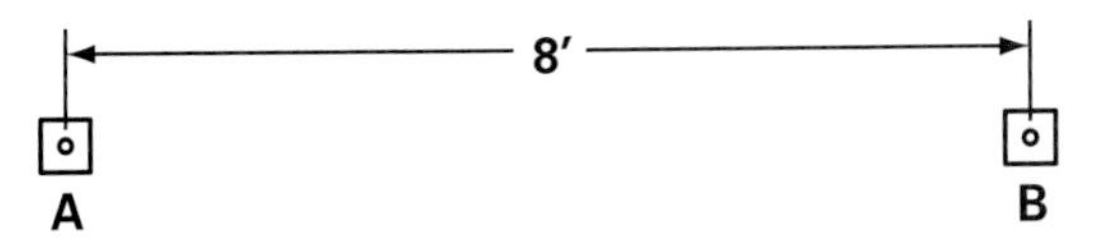

2. Using a calculator, enter 6. Then push the x^2 key. If you do not have that key, multiply 6 × 6. The result should be 36. Write down the answer or store it in the calculator's memory.
3. Repeat Step 2 using the 8' dimension.
4. Add the answers from Steps 2 and 3. It should be 100.
5. With the answer of 100 still showing, press the square root key. The answer should be 10. This is the length, in feet, of the diagonal for your shed.
6. From the stakes you placed in Step 1, and using two tape measures, measure 6' from one stake and 10' from the other stake. Where the two measures intersect is where one of the back corners of the shed is located.

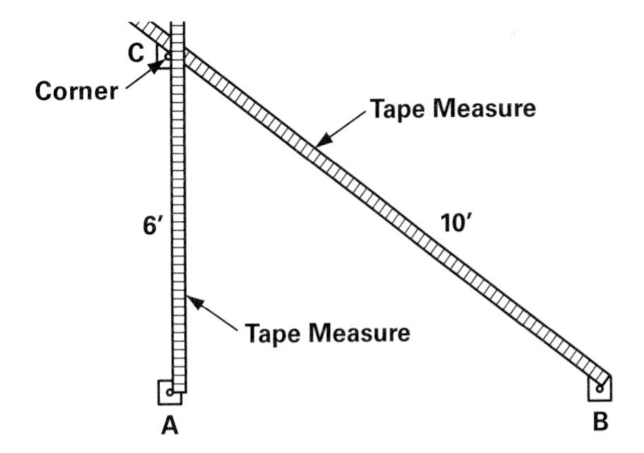

7. Repeat Step 6 from the other stake on the 8' line. Because diagonals of a rectangle are equal, you will have the layout of a perfect rectangle for your shed.

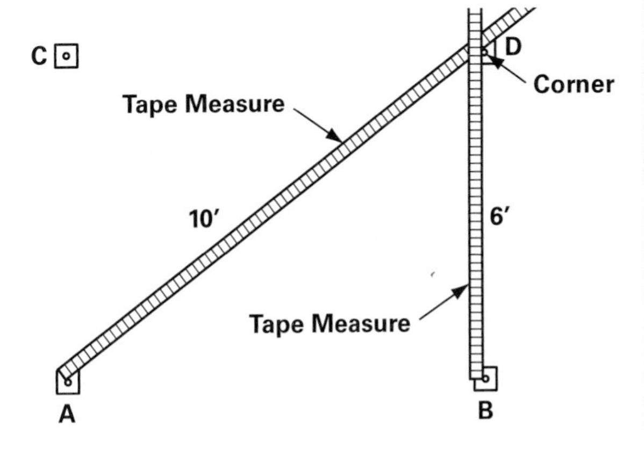

HANDBOOK

Surface Measurements

Surface measurements measure either the distance around a shape or the area that the shape covers.

Perimeter of a Rectangle or Square

The perimeter of a rectangle or square is the sum of all the sides. Since rectangles have two pairs of equal sides, you can find the perimeter (P) by adding twice the length (l) plus twice the width (w).

$$P = 2l + 2w$$

This formula works also for a square, but there's a simpler way to find the perimeter of a square. All four sides of a square are of equal length. Multiply one side by 4.

Area of a Rectangle or Square

To find the area (A) of a rectangle or a square, multiply one side (length, or l) by an adjacent side (width, or w). **Note:** Area is measured in square units.

$$A = lw$$

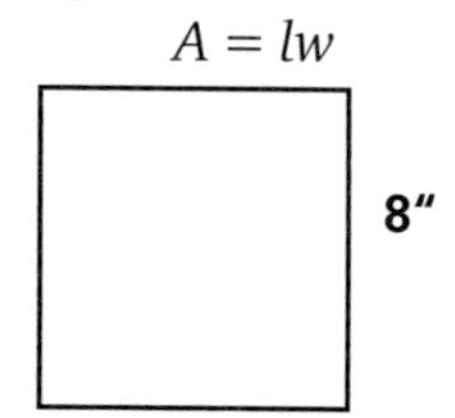

Circumference of a Circle

The circumference is the distance around a circle. To find the circumference (C) of a circle, multiply the diameter (d) by the value of π, which is 3.14.

$$C = \pi d$$

The diameter of a circle is equal to twice its radius. You can also find the circumference if you multiply twice the radius (r) by π.

$$C = 2\pi r$$

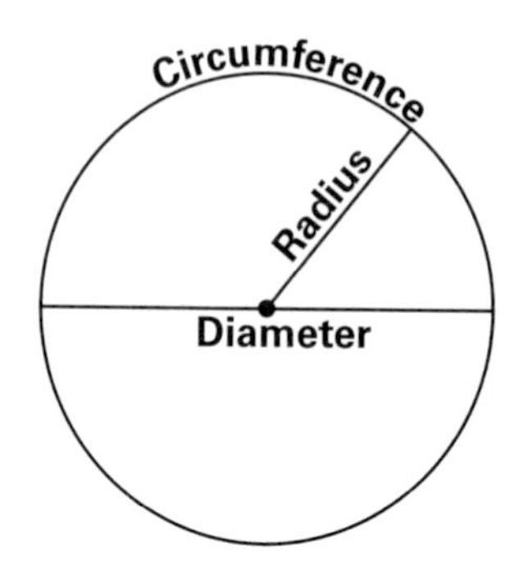

Area of a Circle

To find the area (A) of a circle, multiply π by the radius (r) squared.

$$A = \pi r^2$$

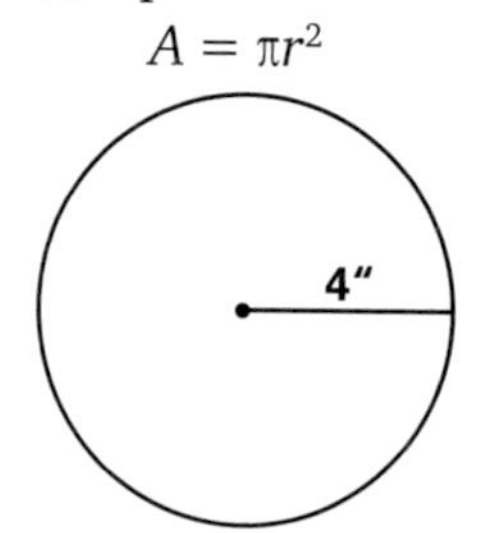

Perimeter of a Triangle

To find the perimeter of a triangle, add the lengths of the three sides.

$$P = a + b + c$$

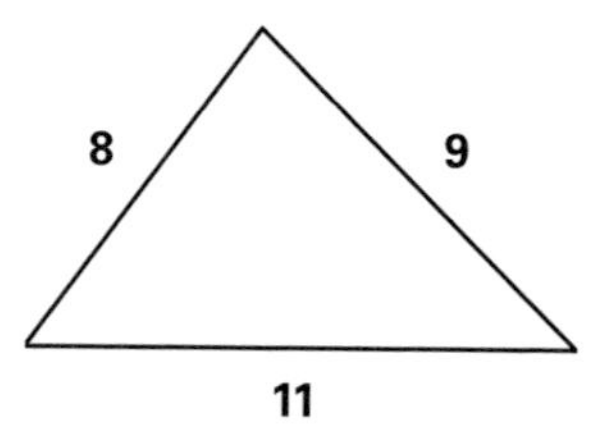

Area of a Triangle

The area (A) of a triangle is one-half the product of the base (b) and the height (h).

$$A = \tfrac{1}{2} bh$$

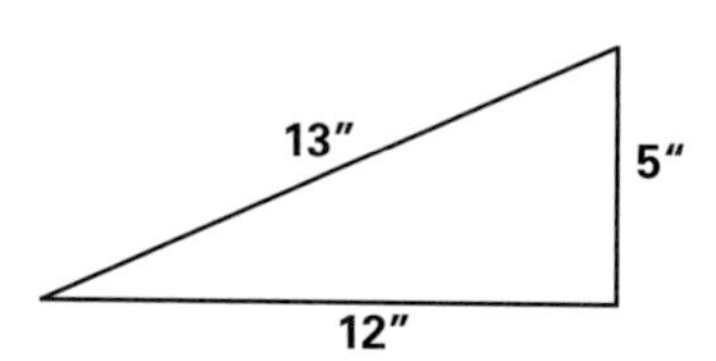

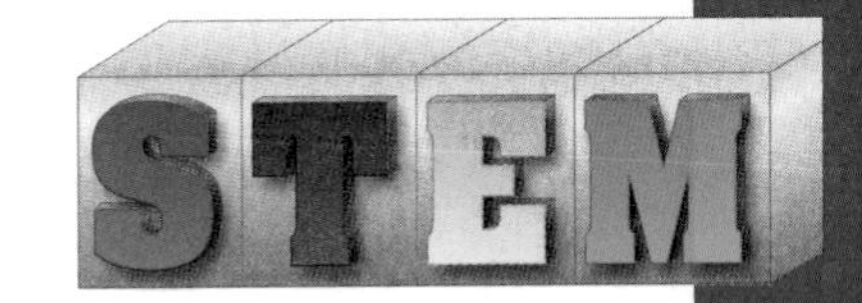

Volume Measurements

Volume measurements measure the space inside three-dimensional shapes. Volume is measured in cubic units.

Volume of a Rectangular Prism

To find the volume (V) of a rectangular prism, multiply the length (l), the width (w), and the height (h).

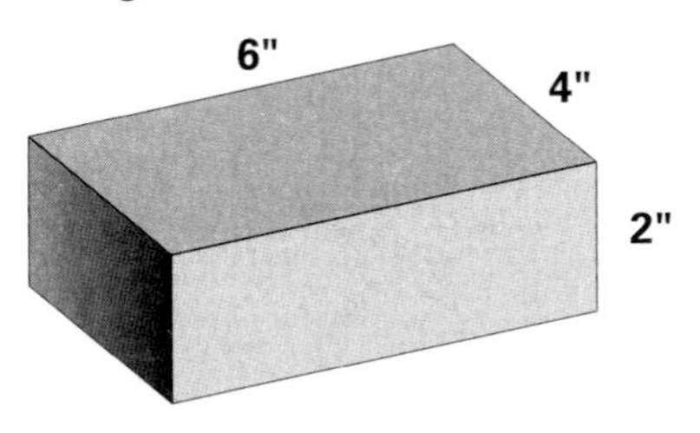

$$V = lwh$$

Volume of a Cylinder

The base of a cylinder is a circle. To find the volume of a cylinder, multiply the area of the circular base by the height. (Remember, to find the area of a circle, multiply π times the radius squared.)

$$V = \pi r^2 h$$

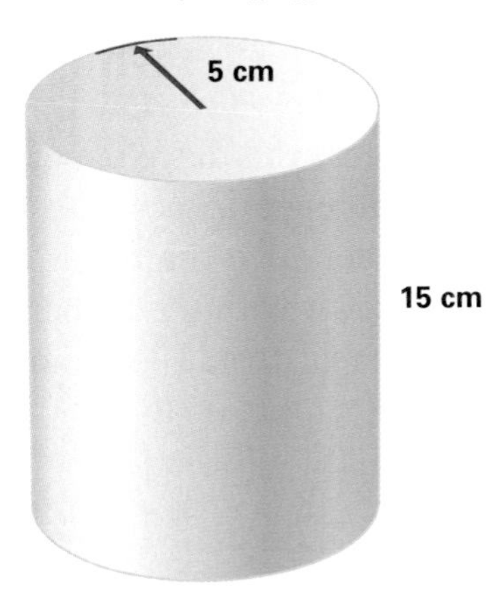

Ratios

A ratio is a way to compare two numbers, using division. For example, directions for mixing concrete may call for 120 pounds of sand and aggregate and 20 pounds of portland cement. The ratio of sand-aggregate to cement can be expressed as 120:20 or 120/20.

The ratio from this mix yields about one cubic yard of concrete. To find the right amounts to make 5 cubic yards of concrete, multiply both terms of the ratio by 5.

$$120 \times 5 = 600 \text{ pounds sand-aggregate}$$
$$20 \times 5 = 100 \text{ pounds cement}$$

To make a smaller batch—say, one-tenth of a cubic yard—divide both terms of the ratio in the same manner.

$$120 \div 10 = 12 \text{ pounds sand-aggregate}$$
$$20 \div 10 = 2 \text{ pounds cement}$$

The three ratios presented here are equivalent ratios. This is easy to see when you work the ratios as division problems.

$$120 \div 20 = 6$$
$$600 \div 100 = 6$$
$$12 \div 2 = 6$$

It's useful to reduce equivalent ratios to their simplest form. This means dividing until there is no number except one that goes into both terms evenly. Both terms of the ratio 12/2 can be divided by 2. To find the simplest form of 12/2, divide:

$$12 \div 2 = 6$$
$$2 \div 2 = 1$$

The simplest form of 12/2 is 6/1. Knowing the simplest form makes for the easiest conversions. With the concrete mix, we can say simply that a batch of any size needs a ratio of six parts sand-aggregate to one part cement.

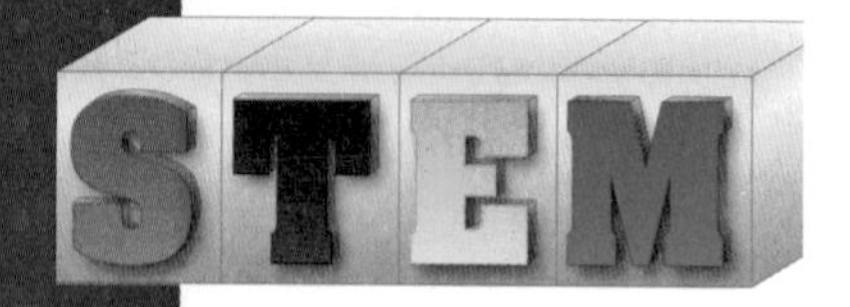

Proportion

Proportion is the relation of one part to another. A proportion is an equality of two ratios. A proportion might be expressed as a is to b as c is to d.

Scale

A scale is the ratio of the number 1 to some larger number. Generally, the larger the second number, the smaller the scale. For example, if 1″ on a drawing is the equivalent of 50″, that would be expressed as 1:50.

Working with Percentages

A percentage is a ratio that compares a number to 100. For example, if 80 percent of household waste is sent to landfills, that means 80 out of every 100 pounds goes to a landfill.

Equations involving percentages include the percentage, the base, and the rate. If you know two of these, you can find the third.

$$\text{Rate} \times \text{base} = \text{percentage}$$
$$\text{Percentage} \div \text{base} = \text{rate}$$
$$\text{Percentage} \div \text{rate} = \text{base}$$

Integers

An integer is any of the whole numbers, as shown here on a number line. A fraction, since it is not a whole number, is not an integer. Integers can be positive or negative. Integers to the left of zero on the number line are negative. Integers to the right of zero are positive.

Negative numbers may represent values such as temperatures below zero or a project's costs figured against expected income. Knowing how to solve equations that have negative numbers helps in figuring out the net effect of opposing forces. Examples would include torque versus rolling distance and thrust versus drag.

To make calculations with negative numbers, it is useful to refer to absolute value. The absolute value of an integer is the integer with its plus or minus sign taken away. Absolute value indicates how many steps an integer is away from zero. It is symbolized by a vertical bar before and after the integer. The absolute value of -3 is $|3|$ and the absolute value of +8 is $|8|$.

To add integers that are both positive or both negative, add their absolute values and give the sum the same sign as the integers. For example, to add +8 and +4:

$$+8 + +4$$
$$|8| + |4| = |12|$$
$$+8 + +4 = +12$$

To add two negative numbers, -5 and -3:

$$-5 + -3$$
$$|5| + |3| = |8|$$
$$-5 + -3 = -8$$

To add integers that have opposite signs, find their difference in absolute value by subtracting. Then give the result the same sign as the integer with the higher absolute value. For example, to add +16 and -2:

$$+16 + -2$$
$$|16| - |2| = |14|$$
$$+16 + -2 = +14$$

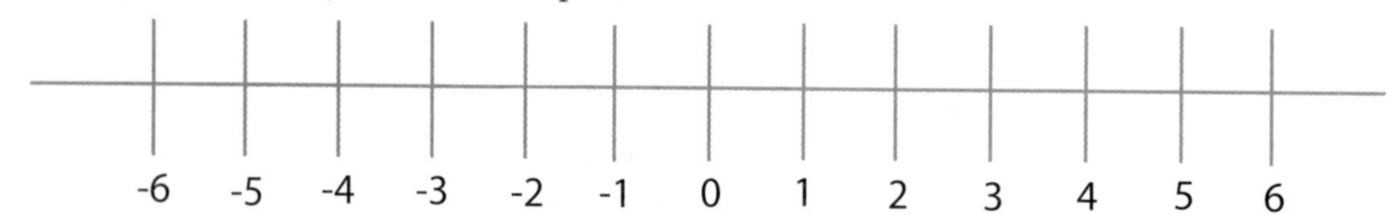

After finding a difference in absolute value of |14|, we next wanted to attach a plus or minus sign to the answer. We attached a plus because the integer with the higher absolute value, 16, was positive.

As another example, let's add -9 and +4:

$$-9 + +4$$
$$|9| - |4| = |5|$$
$$-9 + +4 = -5$$

Given integers with opposite signs, we subtracted and found a difference in absolute value of |5|. To decide whether the answer was plus or minus, we looked to the integer with the higher absolute value, -9.

Cartesian Coordinates

If you overlay two number lines perpendicular to each other so that they intersect at 0, the result is a Cartesian coordinate system. The usefulness of a coordinate system lies in its ability to express a series of related numbers as a graphic. For example, the work life of a machine tool might be represented as a line that slopes gently and then falls off sharply. This would indicate that the tool has passed a critical point of wear.

The intersection marked zero is called the origin. The horizontal and vertical number lines are labeled *x* and *y*, respectively. To plot the points that will form a line or curve, use ordered pairs of coordinates such as (3, 1). The first coordinate, called the abscissa, locates the point horizontally (on the *x*-axis). The second coordinate, called the ordinate, locates the point vertically (on the *y*-axis).

In the example, the *x*-axis might represent the days of operation. The *y*-axis might show how far from nominal (ideal) the tool is in a particular dimension. Let's say design specifications allow the tool to vary from nominal by plus or minus one unit. At first the tool measures consistently as 1 unit above nominal. By day 5, the tool has worn down to the nominal measurement (5, 0). On day 6, the tool is still within 1 unit of specification (6, -1), but by day 7 the tool has worn out (7, -4).

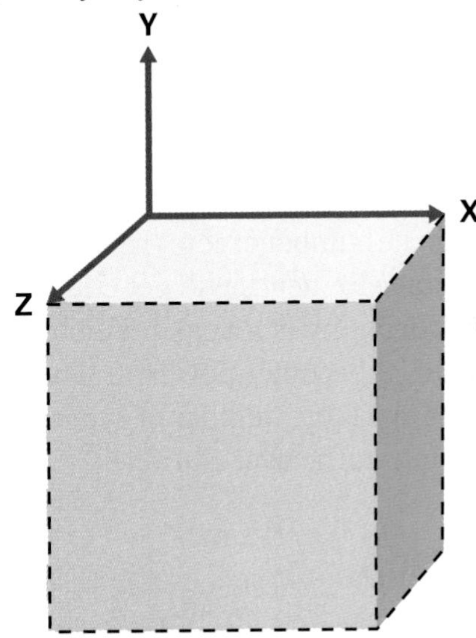

Probability

The probability that a game die will come up 5 on any given roll is 1/6. Rolling a 5 is one outcome among six possible outcomes. The expression 1/6 is equivalent to .17 and may be stated as "a probability of 17 percent." A formula for finding the probability (*P*) of a specific outcome is

$$P = s/t$$

In this formula, *s* is a specific outcome (rolling a 5) and *t* is the total number of possible outcomes (the six sides of a die).

The probability formula relies on the assumption that all possible outcomes are equally likely. If the die is weighted on one side, then a factor other than chance will have a decisive effect. Suppose you are calculating the likelihood of rain tomorrow. The nearness of a massive storm would be a fac-

HANDBOOK

tor other than chance affecting tomorrow's weather. Probability formulas tell you only about the effect of chance on outcomes.

If you flip a true coin five times and it comes up tails every time, people may say that your chance of coming up heads on the next flip has improved. This idea is incorrect. For any one flip of the coin, the probability of coming up heads is always 1/2, or 50 percent.

Looking at a record of many coin flips, you will find the number of heads to be around 50 percent. As the number of coin flips increases, the percentage for heads will get closer to 50 percent. Probability is a weak predictor for a single event. It becomes powerful when you are looking at a large number of events, such as estimating traffic over a bridge.

Statistics

Statistics is the analysis of data, usually in search of patterns that show trends, cause-effect relationships, or other useful factors. Data refers to numerical measurements. Examples include temperatures, copper prices, and real estate values.

Statistics typically relies on sampling. This is the collecting of a small number of measurements to represent the whole. Television ratings, for example, typically rely on a sample of about one thousand households. Statisticians have ways of confirming that a sample is valid. This is important. The alternative—taking all possible measurements—would be extremely expensive.

The use of statistics is particularly effective in two areas: identifying a norm within the data set and identifying change. Global warming and increased productivity in agriculture are examples of change identified by comparing previous data with current data. Such comparisons do not necessarily predict the future. However, they do give a basis for making plans.

Identifying the norm helps us understand what is typical in a data set. Statistics uses three concepts for finding mid-range value that is meaningful in a given application. These three concepts are mean, mode, and median.

Mean

The statistical mean, sometimes called the arithmetic average, is the sum of all the values in a data set divided by the number of items in the data set. Using the table of ABC Company sick days as an example, we find the statistical mean to be:

$$51 \div 10 = 5.1$$

The mean gives a representative view when values are well distributed from one end of the range to the other.

Mode

The mode is the value that appears most often in a data set. In the ABC Company data, the mode is 3. Mode is a choice where in-between values obtained by averaging don't make sense, as in the number of children in a household.

Median

The median is the middle number in a data set when the data is arranged in numerical order. In the ABC Company example, the data set would be arranged as follows: 2 2 3 3 3 5 6 8 9 10. Since the data set has an even number of items, average the two middle numbers (3 and 5) and fix the median at 4. The median is the most useful when a few very high or very low values in the data would tilt the average in a misleading way.

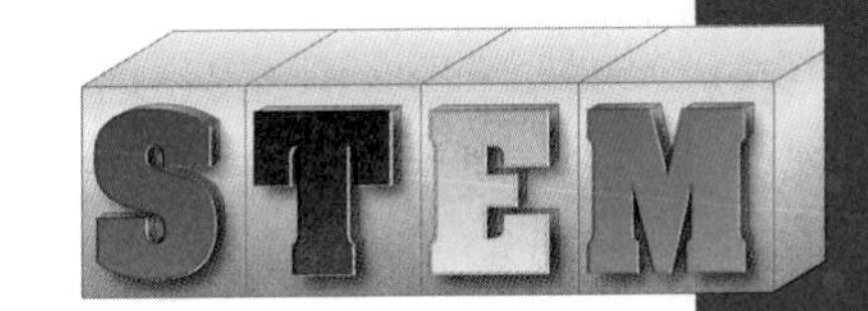

ABC Company Workers Out Sick on Each of the Last Ten Workdays	
Workday	**Number of Workers Sick**
1	2
2	3
3	2
4	5
5	9
6	10
7	6
8	3
9	3
10	8

Using a Calculator

Calculators can help you do math operations faster. Of course, you still need to understand math to know which operations to do. Here are some tips.

- Estimate the answer before you work the problem on the calculator. After you've used the calculator, compare the answer with your estimate. If the two differ, there has been an error. Work the problem again.
- After you have entered the number on the calculator, check the display. If you've made an error, press the "clear entry" (CE) key to remove the number.
- You do not need to enter zero before a decimal point. You do not need to enter final zeroes after a decimal point. Enter the number 0.327000 as .327.
- Remember that the answer displayed on the calculator may not be in its final form. For example, suppose you were solving the problem: $45 is what percent of $90? On a simple calculator, the answer would appear as .5. However, you should write this answer as 50 percent. (Recall that a percent is a ratio that compares a number to 100. The answer .5 is 50/100, or 50 percent.)

Metric System of Measurement

Most countries use the metric system. The United States still uses the customary, or standard, system. However, many industries in the United States use the metric system. It is a good idea to learn to use both systems. The modernized metric system is known as the International System of Units. This is sometimes shortened to SI (for *Système International d'Unités*). It uses seven base units. In everyday life, only four units are in common use:

- The meter, the unit of length, is a little longer than a yard (39.37 inches).
- The kilogram, the unit of weight (mass), is a little more than two pounds (2.2 pounds).

Common Metric Units (SI base units are indicated by *)		
Unit	**Symbol**	**Quantity**
ampere*	A	Electric current
candela*	cd	Luminous intensity
degree Celsius	°C	Temperature
kilogram*	kg	Weight (mass)
kelvin*	K	Thermodynamic temperature (for scientific use)
liter	L	Liquid capacity (volume of fluids)
meter*	m	Length
mole*	mol	Amount of substance
second*	s	Time
volt	V	Electric potential
watt	W	Power

- The liter, the unit of liquid capacity or volume, is a little more than a quart (about 1.06 quarts).
- The degree Celsius is the unit for measuring temperature.

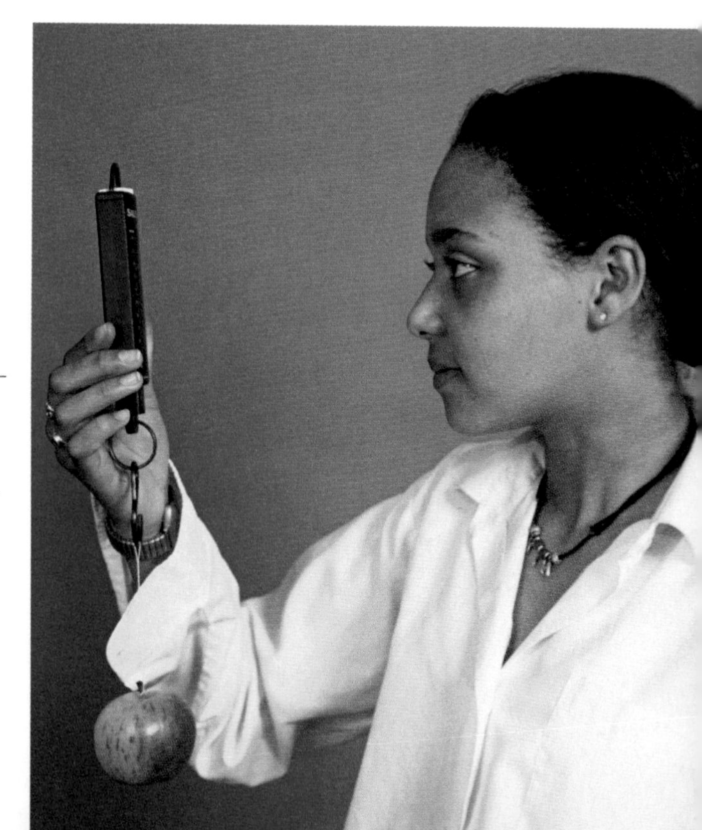

» This student is using a metric spring scale to weigh an apple. The scale measures weight in newtons (N). The apple weighs about 1 N, or 0.1 kg.

The metric system is a decimal system that uses seven base units. All units larger and smaller than the base units are based on multiples of ten, with no fractions. To indicate these larger and smaller units, prefixes are added to the term for the base unit. For example, larger and smaller units of length are indicated by adding such prefixes as *kilo-*, *centi-*, and *milli-* to the word *meter*. A kilometer is 1000 times larger than a meter. A centimeter is 100 times smaller than a meter (one-hundredth of a meter). A millimeter is 1000 times smaller than a meter (one-thousandth of a meter). These three prefixes are the most common. They are used for nearly all units of measurement. Both the prefixes and the names of the units can be shortened by using symbols.

Metric-customary conversions are shown in the table on page 608.

Common Metric Prefixes

Prefix	Symbol	Meaning
nano	n	One-billionth of
micro	μ	One-millionth of
milli	m	One-thousandth of
centi	c	One-hundreth of
kilo	k	One thousand times
mega	M	One million times
giga	G	One billion times

Metric-Customary Conversions

	When you want to convert:	Multiply by:	To find:
Length	inches	2.54	centimeters
	centimeters	0.39	inches
	feet	0.30	meters
	meters	3.28	feet
	yards	0.91	meters
	meters	1.09	yards
	miles	1.61	kilometers
	kilometers	0.62	miles
Mass and Weight	ounces	28.35	grams
	grams	0.04	ounces
	pounds	0.45	kilograms
	kilograms	2.20	pounds
	tons (short)	0.91	tonnes (metric tons)
	pounds	4.45	newtons
	newtons	0.23	pounds
Volume	cubic inches	16.39	cubic centimeters
	cubic centimeters	0.06	cubic inches
	cubic feet	0.03	cubic meters
	cubic meters	35.31	cubic feet
	cubic miles	4.17	cubic kilometers
	cubic kilometers	0.24	cubic miles
	liters	1.06	quarts
	liters	0.26	gallons
	gallons	3.785	liters
Area	square inches	6.45	square centimeters
	square centimeters	0.16	square inches
	square feet	0.09	square meters
	square meters	10.76	square feet
	square miles	2.59	square kilometers
	square kilometers	0.39	square miles
	hectares	2.47	acres
	acres	0.40	hectares
Temperature	Fahrenheit	5/9 (°F - 32)	Celsius
	Celsius	9/5 °C + 32	Fahrenheit

Weight as measured in standard Earth gravity

GLOSSARY

A

acceptance sampling The process of randomly selecting a few typical products from a production run, or lot, and inspecting them to see whether they meet the standards. (Ch. 14)

adaptation The ability of an organism to change by reacting to its environment in such a way that the organism's ability to survive improves. (Ch. 25)

aerodynamics A science that deals with the interaction of air and moving objects. (Ch. 23)

agriculture The process of producing crops and raising livestock. (Ch. 27)

alloy The combination of two or more metals or a metal and nonmetal. (Ch. 11)

alternating current (AC) Current flow in which electrons reverse direction at regular intervals. (Ch. 10)

amperage The rate at which current flows through a conductor; it is measured in amps. (Ch. 8)

amplitude The height of a wave measured from the top of its peak to the bottom of its trough. It is an indicator of the strength of a wave. (Ch. 6)

AMTRAK The AMerican TRavel trAcK (AMTRAK) is a long-distance rail passenger service operated by the U.S. government. The government owns the engines and cars, but the rail lines are privately owned. (Ch. 22)

analog signal A signal with an infinite number of levels or variations; they are continuous. A sine wave is one example. (Ch. 5)

analysis The act of breaking a particular subject into different parts to achieve clarity of understanding. (Ch. 3)

animatronics The construction of robots to look and act like living things. (Ch. 10)

antibodies Proteins that attack foreign substances in the body. (Ch. 26)

aperture The variable diaphragm opening in a camera's lens that allows you to regulate the amount of light reaching the sensor. (Ch. 7)

aquaculture The raising of fish, shellfish, or plants in water under controlled conditions. (Ch. 27)

arch bridge A bridge design in which the load is transferred to the bridge abutments by means of an arch. (Ch. 20)

Archimedes' principle The principle that states that the buoyant force is equal to the weight of the fluid displaced by a submerged body. (Ch. 22)

artificial intelligence The capacity of computers to make decisions and solve problems in a manner similar to that used by human beings. (Ch. 5)

assembling The process of putting parts together. (Ch. 11)

assembly A collection of components put together in a planned way. (Ch. 14)

assembly line A production process in which the part or product moves down a line from one station to the next while parts are added. (Ch. 11)

automated storage and retrieval system (AS/RS) A specialized set of tall racks with a computerized crane that travels between them. (Ch. 15)

automatic factory A factory in which almost all work is performed automatically by machines. (Ch. 15)

automatic guided vehicle system (AGVS) A computer-controlled transportation system that uses special driverless carts (AGVs) that follow a wire path installed in the floor. (Ch. 15)

B

beam bridge A type of bridge design in which the main spans are supported by beams resting on piers. (Ch. 20)

Bernoulli effect The decrease in the relative pressure of a fluid as its velocity increases. (Ch. 20)

bid The price quote that a contractor gives to the customer for the amount that the contractor will charge to perform services. (Ch. 18)

bill of materials A list of the materials or parts needed to make a product. (Ch. 13)

binary code An instruction set for computers in which data is represented as a series of 1s and 0s. (Ch. 5)

bioconversion A method of obtaining energy from waste products. (Ch. 9)

bioethics The set of rules and standards governing people and companies engaging in bio-related science and technology. (Ch. 25)

biomaterial Any material used for prostheses and that comes in direct contact with living tissue. (Ch. 26)

biometrics The measurement and analysis of a person's physical or behavioral information. (Ch. 5)

bio-related technology Technology, such as agriculture and health care, with a strong relationship to living organisms. (Ch. 25)

bioremediation The use of bacteria and other microbes to clean up contaminated land and water. (Ch. 27)

biowarfare The use of viruses and other biological organisms as weapons. (Ch. 27)

Boyle's law The law that states that if the pressure of a gas increases, the volume decreases (provided that the temperature remains constant). (Ch. 10)

brainstorming An activity in which two or more people try to think of as many possible solutions to a problem as they can. Ideas are not evaluated or criticized during this time. (Ch. 3)

break bulk cargo Goods consisting of single units or containers of freight. (Ch. 21)

building code A local or state law that specifies the methods and materials that can or must be used for every aspect of construction. (Ch. 17)

bulk cargo Large quantities of loose cargo.(Ch. 21)

buoyancy The upward force a fluid places on an object. (Ch. 22)

burn in A method of testing a product by putting the product to actual use before placing it into service. (Ch. 14)

C

cable-stayed bridge A bridge design similar to a suspension bridge but having cables connected directly to the roadway. (Ch. 20)

canal An artificial waterway. (Ch. 20)

cantilever bridge A bridge design in which beams called cantilevers extend from each end of the bridge. They are connected by a section called a suspended span. (Ch. 20)

capitalism An economic system in which most or all businesses are privately owned and operated for profit. (Ch. 2)

central processing unit (CPU) A specialized processor that resides directly on the motherboard of a computer. It can be thought of as the brain of a computer. It processes the data; also referred to as a microprocessor or a "computer within the computer." (Ch. 5)

charge-coupled device (CCD) A very small solid state panel that contains light sensitive photosites or photodiodes. The CCD uses the photosites to convert light into electrical signals that are then transformed into an image. (Ch. 6)

circuit The path by which electric current or pulses flow. (Ch. 5)

city planner A person who has studied all aspects of community development and oversees the planning process of city projects. (Ch. 17)

classification yard A rail yard where trains are taken apart and put together; also known as a switch yard. (Ch. 22)

clone An artificially created individual organism that is genetically identical to another individual organism. (Ch. 27)

cloning A propagation method in which a cell from a plant or animal is used to create an exact duplicate of the plant or animal. (Ch. 25)

closed-loop system A system in which outputs are monitored and feedback is used to make sure the system is solving the intended problem. (Ch. 1)

combining The process of putting two or more materials or parts together. (Ch. 11)

commercial construction A type of construction concerned with the design and building of structures intended for use by businesses. (Ch. 16)

communication channel The path by which a message travels from the sender to the receiver. (Ch. 4)

communication technology The knowledge,tools, machines, skills, and other things that people make and do to send and receive information. (Ch. 4)

component An individual part of a product. (Ch. 14)

composite A material made by combining two or more materials; it is usually stronger and more durable than the individual materials. (Ch. 11)

compression A squeezing force exerted on an object. (Ch. 20)

computer An electronic device that can store, retrieve, and process data. (Ch. 5)

computer numerical control (CNC) The use ofa computer to control an individual machine. (Ch. 15)

computer-aided drafting The use of a computer system to produce technical drawings and/or design a product. (Ch. 7)

computer-aided manufacturing (CAM) The process of using linked computers to control machines and plan and control production processes. (Ch. 15)

computer-integrated manufacturing (CIM) The computers, special software programs, and networks used in manufacturing; also known as high-performance manufacturing. (Ch. 15)

computerized tomography (CT) A technique for obtaining X-ray images of deep internal structures. (Ch. 26)

conditioning The process of changing the structural properties of a material. (Ch. 11)

conservation The preservation, management, and care of natural resources and the environment. (Ch. 27)

constraint Restriction or limit placed on a product. (Ch. 2)

construction technology Technology related to the design and building of structures. (Ch. 16)

containerization The process of loading cargo into large boxes or other containers before shipping. (Ch. 22)

continuous production The production system used for mass producing products. It is also called line production or mass production. (Ch. 11)

contractor A person or business that accepts responsibility for providing materials and performing work on a construction project. (Ch. 18)

control unit A computer unit that guides processes and the flow of information. (Ch. 5)

controlled environment agriculture (CEA) A type of agriculture in which a plant's or animal's surroundings are carefully monitored and adjusted. (Ch. 27)

conversion Alteration of an organism's form to convert it for use. (Ch. 25)

creativity Using one's imagination to develop new and original ideas or things; includes the ability to use resources and visualize solutions. (Ch. 3)

criteria Standards that a product must meet to be accepted. (Ch. 2)

critical thinking The process of turning abstractions of thought into concrete ideas. (Ch. 3)

custom production A production system in which products are made one at a time according to customer's specifications. (Ch. 11)

D

dam A structure built across a river for the purpose of controlling or blocking the flow of water. (Ch. 20)

delamination The separation of a composite material into layers that flake off. (Ch. 24)

design Creating things through the process of planning. (Ch. 2)

design for manufacturability Designing a product in such a way that it is easily manufacturable. (Ch. 12)

design for x (DFx) A design system in which a process (x) is created to improve the design, manufacture, or use of a product. (Ch. 15)

diesel engine An engine that burns diesel fuel. The fuel is ignited by heated, compressed air rather than by spark plugs. (Ch. 23)

digital signal A discrete signal with a finite number of levels. (Ch. 5)

digital workflow The use of computers in each stage of the design process. (Ch. 7)

direct current (DC) Current flow in which electrons move in only one direction. (Ch. 10)

direct sales A type of sales in which the manufacturer sells its product directly to the end customer. (Ch. 12)

drafting The process of creating technical drawings that describe the size, shape, and structure of an object. (Ch. 7)

E

earth station A large, earthbound, pie-shaped antenna that transmits and receives signals to and from satellites. (Ch. 6)

efficiency The ability to achieve the desired result with as little wasted energy and effort as possible. (Ch. 8)

electricity Energy created by the flow of free electrons through a conductor. (Ch. 9)

electrocardiograph A device that measures the electrical activity of the heart. (Ch. 26)

electrolysis The passage of electric current through an electrolytic fluid for the purpose of separating molecules. (Ch. 9)

electromagnetic wave A wave created by electric and magnetic fields; it travels through the air like ripples in a pool of water. (Ch. 6)

electronic communication A type of communication involving the use of electricity as its primary method of transmission. (Ch. 6)

electronic stability control (ESC) A technology designed to prevent cars from rolling over when a driver swerves. (Ch. 24)

electrostatic printing A printing process based upon the principle that opposite electrical charges attract, whereas like charges repel. (Ch. 7)

emerging technology A new technology just coming into general use. (Ch. 1)

end effector A device at the end of a robotic arm that the robot uses to manipulate material. (Ch. 15)

energy The capacity to do work or exert effort. (Ch. 8)

engine A machine that produces its own energy from fuel. (Ch. 23)

engineering The combination of knowledge of science, mathematics, technology, and communication to solve technology problems. (Ch. 2)

engineering design process The process that engineers use to fulfill a need or solve a problem. (Ch. 3)

environmental engineering The use of science and engineering principles to improve the environment. (Ch. 26)

entrepreneur Someone who starts a business. (Ch. 2)

ethics The moral principles and values that guide conduct within a group of people. (Ch. 1)

evaluation The judgment of the end result based upon given criteria. (Ch. 3)

exhaustible energy source A source of energythat cannot be replaced. (Ch. 9)

external combustion engine An engine in which the fuel is burned outside the engine. (Ch. 23)

F

failure analysis Examination of the factors that lead to the failure of a product prototype during testing. (Ch. 12)

feedback Any information about the output of a system that is returned to the system for the purpose of determining whether or not the system is functioning as it should. (Ch. 1)

finishing The final step in the production process; it may include steps such as painting the product. (Ch. 11)

finite element analysis (FEA) A type of simulation that predicts how a component or assembly will react to environmental factors such as force, heat, or vibration. (Ch. 15)

floor plan A drawing that shows the locations of rooms, walls, windows, and other features of a structure. (Ch. 17)

force Any push or pull on an object. (Ch. 8)

fluid power The use of pressurized liquids or gasses to control and transmit power. (Ch. 10)

forensic medicine The application of medical science and technology to the law and the solution of crimes. (Ch. 26)

forming The process of changing the shape of a material without the addition or removal of material. (Ch. 11)

fossil fuel Fuel derived from the remains of plants and animals that have been subjected to heat and pressure over millions of years. These remains eventually form coal, oil, and natural gas deposits. (Ch. 9)

foundation The part of a structure that rests upon the earth and supports the superstructure. (Ch. 19)

free enterprise A system in which individuals or businesses may buy, sell, and set prices for goods and services. (Ch. 2)

frequency The number of waves that pass through a given point in one second. It is commonly measured in a unit called a hertz (Hz). One hertz is equal to one complete wave cycle in one second. (Ch. 6)

fuel cell A battery that generates electricitythrough the interaction between two elements, such as hydrogen and oxygen. (Ch. 23)

functional analysis Testing a product's operation by using it in its intended environment and capacity. (Ch. 12)

G

Gauge R&R A test used to check and recalibrate a measuring tool. The test is used both to determine the accuracy of the tool and the operator's interpretation of the data. (Ch. 14)

gene Short segments of DNA that carry specific genetic information. (Ch. 25)

gene therapy The process of transferring a normal gene into an individual who was born with a defective or absent gene. (Ch. 26)

genetic engineering Deliberately altering an organism's genes to give the organism traits that it never had naturally. (Ch. 27)

genetics The study of how traits are passed from an organism to its offspring. (Ch. 25)

genome The sum total of the DNA in an organism. (Ch. 25)

geothermal energy Heat generated within the earth as the result of the decay of radioactive materials beneath the earth's crust. (Ch. 9)

global market Trade of products and services on a global scale. (Ch. 11)

global positioning system (GPS) A tracking system receiving data from 27 satellites orbiting the earth at an altitude of 11,000 miles. An earthbound receiver picks up signals from the satellites. (Ch. 21)

graphic communications The technology field focused on sending messages and other information using visual means. (Ch. 7)

gravure printing The transfer of images from plates that have sunken areas that hold ink. It is also known as intaglio printing. It is the opposite of relief printing. (Ch. 7)

green building A form of construction designed to have the least negative effect possible on the environment. (Ch. 19)

gridlock A situation in which no vehicle can move. (Ch. 21)

group technology A method for identifying and grouping similar parts that the company manufactures. (Ch. 13)

H

hard disk The long-term storage mechanism of a computer. It holds the computer programs used for word processing, playing games, and other tasks. (Ch. 5)

harvesting The gathering of organisms. (Ch. 25)

high-performance manufacturing The combination of highly skilled and empowered workers, advanced technology, and new work methods to achieve high levels of quality, efficiency, and customer satisfaction. (Ch. 15)

holography The use of lasers to record realistic images of three-dimensional objects. (Ch. 7)

horsepower A unit used to measure power; it is equal to the amount of work one horse can do in a given amount of time. (Ch. 8)

human factors engineering The design of equipment and environments to promote human health, safety, and well-being; sometimes referred to as ergonomics. (Ch. 3)

Human Genome Project The attempt to discover and understand all the human genes. (Ch. 25)

hybrid The offspring of two plants (or animals) of different varieties, breeds, or species. (Ch. 27)

hybrid vehicle A vehicle that combines two or more sources of power. (Ch. 23)

hydraulics The use of pressurized liquids as the motive force. (Ch. 10)

hydroponics The cultivation of plants in nutrient solutions without soil. (Ch. 27)

hypothesis A possible explanation for a set of observations. (Ch. 3)

I

identity theft The unlawful acquisition of someone's personal information for use without their consent. (Ch. 4)

immunization The process of making the body able to resist disease by causing production of antibodies. (Ch. 26)

impact The effect of a system's processes and/or output on the system's environment. This effect can be either positive, negative, or in some cases both. (Ch. 1)

industrial construction A type of construction concerned with the design and building of industrial structures. (Ch. 16)

industrial material A material that has been refined from its raw state and is in a form suitable for use in the manufacturing of a product. (Ch. 11)

inexhaustible energy source An energy source that will always be available. (Ch. 9)

information technology All of the methods, techniques, tools, and equipment that enable computers to create, manage, store, send, and receive information. (Ch. 5)

infrastructure The basic framework of a system. (Ch. 17)

inkjet printing A printing process in which ink jets spray precise amounts of ink onto a substrate. The nozzles are controlled by computer data. (Ch. 7)

innovation The alteration of an existing technology or product to solve a new problem. (Ch. 1)

innovative technology A new or improved technology that is still being refined in the laboratory. (Ch. 1)

input A resource that is entered into a system. (Ch. 1)

insulation A material applied to framing surfaces to help keep heat from entering a building during the summer and leaving during the winter. (Ch. 19)

integrated circuit (IC) A tiny silicon wafer that contains thousands of interconnected circuits; also referred to as a microchip or chip. (Ch. 5)

integrated pest management (IPM) A pest control method that uses various techniques to minimize damage. (Ch. 27)

interchangeable parts Parts manufactured to be identical so that any part may be used in any product designed for that part. (Ch. 12)

interdisciplinary engineering The interaction between engineering and other disciplines. (Ch. 2)

intermittent production A production system in which a limited quantity of a part or product is made. After the production run, the machines are retooled or changed over to accommodate the production of a different part or product. (Ch. 11)

intermodal transportation The combination of multiple types of transportation. (Ch. 22)

internal combustion engine An engine in which the fuel is burned within the engine itself. (Ch. 23)

International Organization for Standardization (ISO) An international organization whose purpose is to promote and coordinate worldwide standards for many things. (Ch. 14)

Internet A noncommercial computer network. (Ch. 5)

invention A new technology or product created to solve a new problem. (Ch. 1)

inventory The quantity of items, such as components, on hand. (Ch. 14)

isometric A drawing in which the object is tilted forward 30 degrees and rotated 30 degrees so that its edges form equal angles. (Ch. 7)

J

joist One of a series of parallel, evenly spaced, horizontal boards that support floors and ceilings. (Chs. 17, 19)

just-in-time (JIT) manufacturing A manufacturing process in which materials are delivered in the required quantity at the time when they are immediately needed. (Ch. 15)

K

kinetic energy Energy in motion. (Ch. 8)

L

large-scale construction Construction that involves massive projects such as highways, dams, canals, tunnels, pipelines, bridges, and construction in space. (Ch. 20)

laser printing A type of printing process in which lasers are used to produce the image; the process is similar to electrostatic printing and operates on the same principles, but uses a different light source. (Ch. 7)

law of conservation of energy The law that states that energy can neither be created nor destroyed; it can only be changed from one form to another. (Ch. 8)

lean manufacturing A type of manufacturing whose aim is to reduce the amount of waste created as a by-product of production as well as increase the efficiency of production. (Ch. 15)

lens A precisely ground, glass element that directs light rays so that they are correctly positioned on the sensor. (Ch. 7)

load The amount of resistance the power system must overcome or, alternatively, the amount of force output by the power system; also can refer to weight or pressure. (Chs. 10, 20)

local area network (LAN) A computer network connection within a plant or office. (Ch. 15)

M

maglev system A rail system that operates on the principle that like poles of a magnet repel each other. Maglev is short for magnetically levitated. (Ch. 23)

manufacturing The use of technology to make things that people want or need. (Ch. 11)

manufacturing cell A small group of machines and people working together as a unit to produce a product from start to finish. (Ch. 15)

manufacturing resource planning (MRP II) The organization of resources, usually with the aid of computer software, to ensure the smooth flow of production. (Ch. 13)

maritime Relating to sea navigation, inland waterways, and ports. (Ch. 24)

materials-handling system An automated system for handling, moving, and storing parts and materials. (Ch. 13)

mechanical advantage The use of simple machines to multiply the output force of a mechanical system. (Ch. 10)

metal fatigue Structural damage to a metal from repeated stresses. (Ch. 24)

microbe A single-celled organism too small to be seen with the unaided eye. (Ch. 26)

mock-up A three-dimensional model of a proposed product. (Ch. 12)

modular design A type of design in which a product is composed of modules (basic units that are pre-made and often interchangeable) that are assembled later. (Ch. 12)

modulation The superimposition of sine waves on electromagnetic carrier waves. (Ch. 6)

motherboard In a computer, the primary circuit board to which all internal components are connected either directly or indirectly. (Ch. 5)

motor A machine that uses energy supplied by another source. (Ch. 23)

movable bridge A type of bridge in which a portion of the roadway can be moved to allow large water vessels to pass underneath. (Ch. 20)

multimedia The use of more than one medium to communicate. (Ch. 5)

multiview drawing A drawing that shows two or more different views of an object. (Ch. 7)

N

nanotechnology The field of knowledge relating to materials as they exist on a molecular or an atomic level. (Chs. 11, 26)

National Aeronautics and Space Administration (NASA) The government agency that operates the U.S. space program. (Ch. 22)

North American Free Trade Agreement (NAFTA) An agreement that permits free trade among member nations. The decrease in trade barriers helps manufacturers find larger markets for their products. (Ch. 11)

nuclear fission The splitting of the nucleus of an atom. (Ch. 9)

O

oblique A drawing that shows one perfect, undistorted face of an object while showing the other sides of the object at an angle. (Ch. 7)

Occupational Safety and Health Administration (OSHA) Federal agency established by the federal government to set safety standards for the workplace and ensure compliance with those standards. (Ch. 18)

Ohm's law The law that states that it takes one volt to force one amp of current through a resistance of one ohm. (Ch. 8)

on-site transportation Transportation within a limited area, such as a building or construction assembly yard. (Ch. 21)

open-loop system A type of system in which feedback is not entered into the system and therefore does not affect the cycle of the system. (Ch. 1)

optical fiber A thin, flexible fiber made from pure glass covered in a reflective cladding and an outside protective coating; the fiber is capable of carrying pulses of light. (Ch. 6)

optimization The process of creating the most effective and functional product or process while meeting all the criteria and constraints. (Ch. 2)

organic farming Agricultural practices that avoid the use of synthetic chemicals and bioengineering in favor of naturally occurring fertilizers, pesticides, and other growing aids. (Ch. 27)

output The result(s) of a system's processes applied to a system's input. (Ch. 1)

P

parallel circuit A circuit in which electricity flows along more than one path. (Ch. 10)

part print analysis Examining the working drawings to find the most efficient and effective way of producing the part. (Ch. 13)

pasteurization The heating of a food product to a temperature high enough to kill microorganisms. (Ch. 26)

patent A government document granting the exclusive right to produce or sell an invented object or process for a period of time. (Ch. 1)

pattern The full-size outline of an object. (Ch. 7)

perspective A drawing that shows an object as it would appear in real life. Parts of the object that are further away appear smaller and parallel lines disappear into the distance at the drawing's vanishing point. (Ch. 7)

pharming The deliberate genetic alteration of plants, animals, or microbes to create organisms that produce pharmaceuticals. (Ch. 27)

phosphor A substance that emits light when energized. (Ch. 6)

photographic printing A printing process in which light is projected through a semitransparent plate that contains an image (usually called a negative) on to a light-sensitive material. The image appears after processing. (Ch. 7)

photovoltaic cell A device that converts sunlight directly into electricity. (Ch. 9)

pictorial drawing A type of drawing that shows objects as they appear to the human eye. (Ch. 7)

pilot run A practice run of a production system in which all parts of the system are operated together before production actually begins. (Ch. 13)

pixel A single point of light in an electronic device that creates an image when arranged in a pattern with other points of light. (Ch. 6)

planographic printing Any printing process that involves the transfer of a message from a flat surface. (Ch. 7)

plant layout The arrangement of machinery, equipment, materials, and traffic flow in a plant. (Ch. 13)

pneumatics The use of compressed gas as the motive force in a system. (Ch. 10)

porous printing A printing process in which ink or dye is passed through an image plate or stencil and transferred on to a substrate. (Ch. 7)

potential energy Stored energy or energy at rest. (Ch. 8)

power The measure of work done over a period of time as energy is converted from one form to another or transferred from one place to another. (Ch. 8)

prefabrication The practice of building components of structures or even whole structures within a factory and then transporting the pieces to the job site for final assembly. (Ch. 19)

pressure A measure of the amount of force spread out over a given area. (Ch. 8)

principles of design The formal rules that guide the design process. (Ch. 7)

private sector The part of the economy comprised of ordinary citizens. (Ch. 17)

problem-solving process A six-step process used to solve everyday problems. (Ch. 3)

process A series of operations that includes all of the activities that must be performed in order for the system to yield the expected result. (Ch. 1)

process chart A chart detailing the manufacturing sequence. (Ch. 13)

product configuration An arrangement of the parts of a product; includes color, size, additional components, etc. (Ch. 15)

production The multistep process of making parts and assembling them into products. (Ch. 14)

productivity The comparison of the amount of goods produced (output) to the amount of resources (input) that produced them. (Ch. 11)

profit The money a business makes after all expenses have been paid. (Ch. 11)

programming language A language, used to write computer programs, that can easily be translated into machine language. (Ch. 5)

propagation The technique of reproducing living organisms. Breeding animals and planting crops are forms of propagation. (Ch. 25)

proportion The relationship between the sizes of various parts of a design. (Ch. 7)

prosthetics Artificial replacement of missing, diseased, or injured body parts. (Ch. 26)

prototype A working model of a product used to help clients visualize the finished design. (Ch. 3)

public sector The part of the economy comprised of municipal (city), county, state, and federal governments. (Ch. 17)

public works construction A type of construction concerned with the design and building of structures intended for public use. (Ch. 16)

purchasing agent The person who is responsible for obtaining the proper materials for a construction project at an acceptable price; also known as a buyer. (Ch. 18)

Q

quality assurance The process of making sure the product is produced according to plans and meets all specifications. (Ch. 14)

quality control The process of ensuring that the finished product, service, or system meets the specified criteria and constraints. (Ch. 2)

R

rafter One of a series of sloping roof framing members cut from individual pieces of dimension lumber. (Chs. 17, 19)

rapid prototyping The use of 3D CAD drawings to create real models of a product within a short time. (Ch. 7)

raw material Material that has not been refined to the point where it is suitable for use in manufacturing. (Ch. 14)

receiver A unit that collects waves and decodes the information contained within those waves. (Ch. 6)

recombinant DNA DNA formed by joining (recombining) two pieces of DNA from two different species. The result is a transgenic plant or animal. (Ch. 27)

relief printing A printing process in which the image is transferred to the substrate from a raised surface. (Ch. 7)

rendering A drawing that shows the designer's final ideas about the appearance of the product. (Ch. 12)

renewable energy source A source of energy that can be used indefinitely if properly managed and maintained. (Ch. 9)

research and development (R&D) The process of gathering information and using that information to develop a product. (Ch. 12)

residential construction The building of structures in which people will live. (Ch. 16)

resistance Opposition to the flow of current; it is measured in ohms. (Ch. 8)

RFID A wireless system that can be used to track goods or vehicles. (Ch. 6)

robot A mechanism guided by automatic controls. (Ch. 15)

roof truss A preassembled triangular unit used to frame the roof instead of rafters. (Ch. 19)

route The ways, paths, or roads a vehicle travels to get from one place to another. (Ch. 21)

S

satellite A device that orbits the earth and receives messages from one location and transmits them to another. (Ch. 6)

schedule A plan of action that details what must be done, the order in which things must be done, when things must be done, and who must perform each task. (Ch. 18)

scientific method A method of objective examination that involves making observations, collecting information, forming hypotheses, testing hypotheses, analyzing results, and repeating the process to ensure the consistency of the results. (Ch. 3)

scientific notation A method of writing very large or very small numbers in abbreviated form. (Ch. 25)

sea-lane A shipping route that runs across anocean. (Ch. 22)

sensor A device capable of detecting (sensing) changes in its environment; light and movement are examples. (Ch. 4)

separating The process of changing the shape of a material by removing some of the material. (Ch. 11)

series circuit A circuit in which electricity flows along a single path to more than one electrical device. (Ch. 10)

shear A force exerted on an object in such a way that it pushes the parts of an object in opposite directions. (Ch. 20)

sheathing A layer of material that is placedbetween the framing and the finished exterior. It is often made of plywood or insulating board. (Ch. 19)

shutter In a camera, a device that regulates the amount of time that light is allowed to reach the sensor or film. (Ch. 7)

simple machine A device that creates a mechanical advantage. Complex machines are based on six simple machines in various combinations. (Ch. 10)

simulation The imitation, as closely as possible, of the real-life circumstances in which a solution is intended to be used. (Chs. 3, 13)

site plan A drawing that shows where a structure will be located on a lot; it includes boundaries, roads, and utilities. (Ch. 17)

slurry A rough solution made by mixing liquid with solids that have been ground into small particles. (Ch. 22)

smart growth The creation of more livable, attractive, and economically strong communities. (Ch. 17)

statistical process control (SPC) The use of statistics and statistical methods to ensure that the finished product(s) fall within the specified tolerances. (Ch. 14)

stem cell A cell that is capable of becoming another type of specialized cell. (Ch. 26)

stick construction A manner of building in which small, lightweight pieces of wood are assembled into a framework. This framework is then covered with other building materials to make walls, floors, and a roof. (Ch. 16)

strength A material's ability to resist forces such as tension and compression. (Ch. 12)

structural materials Materials that can be used to support heavy loads or maintain the rigidity of a structure. (Ch. 17)

stud One of a series of parallel, evenly spaced, vertical boards that frame interior and exterior walls. (Chs. 17, 19)

subcontractor A contractor that specializes in a certain type of construction. A subcontractor is hired by the contractor to work on a specific part of a construction project. Electrical systems are usually installed by an electrical subcontractor. (Ch. 18)

subsystem A small, independent system embedded within a larger system. (Ch. 1)

superstructure The part of a structure that rests on the foundation; it usually consists of everything above ground. (Ch. 19)

supply chain Both the path that goods take in moving from the manufacturer to the customer and the sequence of suppliers and processes necessary to deliver finished products to the customer. (Chs. 12, 15)

survey A drawing that shows the exact size and shape of the piece of property, its position in relation to other properties and to roads and streets, the height (elevation) of the property, and any special land features (rivers, streams, hills, gullies, trees, etc.). (Ch. 17)

suspension bridge A type of bridge in which two tall towers support main cables secured by heavy anchorages at each end. The roadway is supported by suspender cables dropped from the main cables. (Ch. 20)

synthesis The act of putting different things together to form a new product or idea. (Ch. 3)

system A group of parts that work together to accomplish a goal. The basic parts of a system are input, process, and output. (Ch. 1)

T

target market The group of customers most likely to want and use the product. (Ch. 12)

team A group of people working together to accomplish a common goal. (Ch. 2)

technology Processes and knowledge that can be used to extend human abilities as well as satisfy human needs and wants. (Ch. 1)

technology assessment Studying the effects of a given technology to determine its impact.(Ch. 1)

telecommunication Communication over a long distance. (Ch. 6)

telemedicine Practicing medicine remotely, usually with the aid of computers. (Ch. 26)

tension A force exerted on an object in such a way that it attempts to stretch or pull the object apart. (Ch. 20)

time and place utility A change in value caused by transportation. (Ch. 21)

ton mile A unit of measurement equivalent to moving one ton a distance of one mile. (Ch. 22)

tooling-up Preparation of tools and equipment in order to begin production. (Ch. 13)

torque A turning or twisting force. (Ch. 8)

torsion A force exerted on an object in such a way that it tries to twist the object. (Ch. 20)

total quality management A management system in which employees are expected to meet a performance standard for their jobs. (Ch. 11)

tractor-semitrailer rig A combination of a tractor and a semitrailer. (Ch. 22)

transistor A miniature electronic switch that is the main component in an integrated circuit. (Ch. 5)

transmitter A device that encodes (changes) signals into sine waves for broadcast on radio waves or microwaves. (Ch. 6)

transportation system An engineered and organized way of moving goods, people, and vehicles. (Ch. 21)

trend A popular movement. (Ch. 1)

truss bridge A type of bridge that uses a series of triangular trusses for support. (Ch. 20)

U

uniform resource locator (URL) An address that directs a computer to a specific Web page.(Ch. 5)

universal design A design process by which the design of all products and structures is made as usable as possible by as many people as possible regardless of age, ability, or situation. (Ch. 26)

universal product code (UPC) The striped code, or barcode, that uniquely identifies the product. (Ch. 5)

utility A service system within a building. (Ch. 19)

V

vaccine A substance composed of dead or weakened pathogens or of the purified toxins they produce that stimulates the body's immune response. (Ch. 26)

value analysis The analysis of a product to determine the best material for the least cost.(Ch. 12)

virtual reality The use of 3D graphics to create a realistic simulation, immersing the user in an artificial environment. (Ch. 5)

voltage The motive force (pressure) that pushes current through a conductor; it is measured in volts. (Ch. 8)

W

waste management All the operations involved in the collection, storage, and treatment of waste. (Ch. 26)

way Specific area set aside for use by transportation systems. (Ch. 21)

wedge Two or more inclined planes meeting at a single edge or point that can be used to create a mechanical advantage. (Ch. 10)

wide area network (WAN) A computer network connection between plant or office locations. (Ch. 15)

work The use of force to act on an object in order to move the object in the same direction as the force. (Ch. 8)

work envelope The limited work area in which a robot can raise, lower, stretch, or turn itself; the robot can use its end effector anywhere within the work envelope. (Ch. 15)

working drawings Detailed drawings that show exact sizes, shapes, and other details critical to the proper manufacture of a product. (Ch. 12)

X

X-ray A short wavelength of electromagnetic radiation. (Ch. 26)

Z

zoning law A law that dictates what kinds of structures may be built in certain areas. (Ch. 17)

CREDITS

Cover, Interior, and STEM Design: Jerome G antner/Mazer Creative Services, **Cover Photos:** Jon Hicks/CORBIS, WireImageStock/Masterfile, Andrew Douglas/Masterfile, NASA, NASA/CNT **5,** Art MacDillos/Gary Skillestad, Kevin May Corporation **6,** Ian Worpole, Glencoe **7,** Arnold & Brown, Bots IQ **8,** Arnold & Brown, College of Engineering of the University of Cincinnati **9,** NASA **10,** Jeff Stoecker **11,** NASA/CNT **20,** NASA/CNT **21,** Kevin May Stock Photos **22,** NASA **23,** BEST Robotics **24,** Rachel Epstein/PhotoEdit **25,** Ian Worpole **25,** NOAO/AURA/NSF **26,** Gustao Tomsich/CORBIS **26,** Roger Viollet/gettyimages **27,** John Howard/Science Photo Library **27,** Ian Worpole **27,** Dorling Kindersley/gettyimages **28,** Kenneth Garrett/gettyimages **28,** Gianni Dagli Ortis/CORBIS **28,** Philip Spruyt/©Stapleton Collection/CORBIS **28,** Adam Woolfitt/CORBIS **29,** Moses C. Tutle/©Minnesota Historical Society/CORBIS **29,** CORBIS/CORBIS **29,** NIDUS CORP. **30,** Steve Karp **31,** Ken Clubb **32,** Ian Worpole **33,** NIDUS CORP. **34,** Shannon Fagan/gettyimages **35,** Kevin May Corporation **35,** Kevin Cooley/gettyimages **36,** Bill Aron/PhotoEdit **37,** Green industry/PhotoDisc **38,** New Home/PhotoDisc **39,** Digital Vision/gettyimages **39,** Research & Technology/DigitalVision **39,** Michael Rosenfeld/gettyimages **39,** Modern Technology/PhotoDisc **39,** Hulton Archive/gettyimages **41,** Jeff Greenberg/alamy **42,** Jose Luis Pelaez, Inc./CORBIS **44,** Technology Student Association **45,** Brownie Harris/CORBIS **46,** Ian Worpole **47,** Art MacDillos/Gary Skillestad **48,** Eric Audras/gettyimages **49,** Keith Brofsky/PunchStock **50,** Artiga Photo/CORBIS **51,** Bob Daemmrich/PhotoEdit **51,** Michael Newman/PhotoEdit **52,** NASA **53,** Bob Gelberg/Masterfile **54,** John Madere/CORBIS **57,** Walter Hodges/CORBIS **57,** Brownie Harris/CORBIS **57,** Photodisc/gettyimages **58,** Lowell Georgia/CORBIS **60,** CORBIS/CORBIS **61,** Kevin May Corporation **62,** Glencoe **62,** Humanscale Corporation **63,** Steve Skjold/alamy **64,** Z-CoiL Footwear **65,** Z-CoiL Footwear **65,** David Young-Wolff/PhotoEdit **66,** Homes & Gardens/PhotoDisc **67,** Rolf Bruderer/CORBIS **68,** Green Industry/PhotoDisc **69,** Frenado Buean/gettyimages **70,** Doug Steley/alamy **70,** Glencoe **70,** NASA/MSFC **71,** Modern Technologies/PhotoDisc **71,** Dynamic Graphics Group/IT Stock Free/alamy **73,** Ian Worpole **74,** Microvision **76,** Vivo Metrics, Inc. **77,** Ambient Devices, Inc. **77,** Martin Riedl/gettyimages **78,** Robin Nelson/PhotoEdit **79,** Araldo de Lucas/CORBIS **80,** Bohemian Nomad Picturemakers/CORBIS **80,** North Wind Picture Archives **80,** Bettmann/CORBIS **81,** Royalty-Free/CORBIS **81,** Earl Scott/Photo Researchers **81,** Nokia **81,** Ian Worpole **82,** Spencer Grant/PhotoEdit **83,** Ian Worpole **84,** Jeff Stoecker **84,** Glencoe **85,** Masterfile/Masterfile **85,** Paul Hardy/CORBIS **85,** Royalty-Free/CORBIS **85,** Hitachi Kokusai Electric America, Ltd. **85,** Jochen Tack/Das Fotoarchiv/Black Star/alamy **85,** Jose Luis Pelaez, Inc./CORBIS **85,** Jeff Stoecker **86,** Kevin May Corporation **87,** Bob Daemmrich/PhotoEdit **87,** Kim Kulish/CORBIS **88,** Royalty-Free/CORBIS **89,** stockbyte/PunchStock **89,** David Young-Wolff/PhotoEdit **89,** Hemera Technologies/jupiterimages **89,** Brand X Pictures/PunchStock **89,**Bill Aron/PhotoEdit **91,** Davis Instruments/Colby Design **92,** Kevin May Corporation **94,** Daivd Young-Wolff/PhotoEdit **95,** Objects of Business/PhotoDisc **96,** Glencoe **96,** Kevin May Corporation **96,** Ian Worpole **96,** Office Supplies/PhotoDisc **97,** Zane Smith/CORBIS **98,** Research & Technology/DigitalVision **98,** Art MacDillos/Gary Skillestad **99,** Medicine & Technology/PhotoDisc **100,** Trotec Laser Systems **101,** Glencoe **101,** Ian Worpole **102,** Art MacDillos/Gary Skillestad **103,** Raster Builders Inc. ©2006. All rights reserved. **104,** Alex Grimm/Reuters/CORBIS **105,** Jeremy Sutton-Hibbert/alamy **105,** James King-Holmes/W Industries/Photo Researchers **106,** Art MacDillos/Gary Skillestad **107,** Kevin May Corporation **108,** Glencoe **111,** George Steinmetz/CORBIS **112,** Image Source/Super Stock **114,** Reuters/CORBIS **115,** Art MacDillos/Gary Skillestad **116,** Art MacDillos/Gary Skillestad **117,** Ian Worpole **118,** Art MacDillos/Gary Skillestad **119,** PE Reed/gettyimages **119,** Alan Schein/zefa/CORBIS **119,** Royalty-Free/CORBIS **119,** Art MacDillos/Gary Skillestad **120,** Art MacDillos/Gary Skillestad **121,** Art MacDillos/Gary Skillestad **122,** NASA/JPL **124,** Ian Worpole **125,** Art MacDillos/Gary Skillestad **127,** James Noble/CORBIS **128,** TK-Ian Worpole **128,** James Leynse/CORBIS **129,** Don Farrall/gettyimages **129,** Bill Aron/PhotoEdit **131,** Ron Chapple/alamy **132,** Temmpler/zefa/CORBIS **134,** Technology Student Association **135,** The NIKE name and Swoosh Design **135,** The McGraw-Hill Companies **135,** Alan Schein Photography/CORBIS **135,** Ken Clubb **136,** Art MacDillos/Gary Skillestad **137,** Ken Clubb **138,** Steve Karp **138,** Glencoe **139,** Art MacDillos/Gary Skillestad **140,** Art MacDillos/Gary Skillestad **141,** Art MacDillos/Gary Skillestad **142,** Johnsonville Sausage **143,** Ian Worpole **143,** Ian Worpole **144,** Art MacDillos/Gary Skillestad **146,** Ian Worpole **147,** Glencoe **148,** Arnold & Brown **149,** Homes & Gardens/PhotoDisc **149,** Kevin May Corporation **150,** Kevin May Stock Photo **150,** Art MacDillos/Gary Skillestad **151,** Ken Clubb **152,** Jill Seiler **153,** Jim Seiler **154,** Jill Seiler **155,** Columbia Corrugated Box Co. **156,** Eye of Science/Science Photo Library **157,** Alfred Pasieka/Photo Researchers **157,** Rosenfeld Images LTD./Science Photo Library **157,** Monopoly® & 2006 Hasbro, Inc. Used with permission. **159,** Paul Seheult:Eye Ubiquitous/CORBIS **160,** C McIntyre/Photolink/gettyimages **162,** Orjan F. Ellingvag/Dagens/CORBIS **162,** Jack Star/Photolink/gettyimages **163,** PhotoEdit **163,** John F. Martin/CORBIS **163,** Roy Ooms/Masterfile **164,** Royalty-Free/CORBIS **165,** Steve Karp **166,** Fujio Nakahashi/gettyimages **166,** North Wind Picture Archives **166,** North Wind Picture Archives **167,** Hulton-Deutsch Collection/CORBIS **167,** Bettmann/CORBIS **167,** Green Industry/PhotoDisc **167,** Cheryl Clegg/Index Stock **168,** Glencoe **169,** Chris Clinton/gettyimages **169,** Business & Industry/PhotoDisc **169,** Elmtree Images/alamy **169,** Business & Industry/PhotoDisc **169,** Ian Worpole **169,** Andrew Lambert Photography/Photo Researchers **170,** Steve Karp **171,** Ian Worpole **172,** Glencoe **172,** Steve Karp **172,** Ian Worpole **172,** Glencoe **172,** Steve Karp **173,** Ted Spiegel/CORBIS **174,** Roger Ressmeyer/

CORBIS **175**, Kevin May Stock Photo **177**, Ian Worpole **178**, Ian Worpole **179**, Ian Worpole **180**, John Rowley/gettyimages **181**, Joeseph Sohm-Visions of America/gettyimages **181**, David R. Frazier Photolibrary, Inc./alamy **181**, Science, Technology & Medicine/PhotoDisc **181**, Theodore Anderson/gettyimages **181**, Digital Vision/gettyimages **183**, Dept. of Energy **184**, Green Industry/PhotoDisc **186**, Roger Ressmeyer/CORBIS **187**, Industry & Environment/PhotoDisc **188**, Industry & Transportation/PhotoDisc **188**, Industry & Environment/PhotoDisc **189**, Glencoe **189**, R. Ian Lloyd/Masterfile **190**, Industry & Transportation/PhotoDisc **191**, Ian Worpole **191**, Intel Corporation **192**, Glencoe **193**, David Nunuk/Photo Researchers **195**, Industry & Transportation/PhotoDisc **196**, U.S. Dept. of Interior **196**, Steve Karp **197**, Modern Technologies/PhotoDisc **197**, Ian Worpole **198**, Ian Worpole **199**, Ian Worpole **200**, Ann Garvin **201**, Goodshot/jupiterimages **203**, Glencoe **204**, Philip Bailey/CORBIS **206**, Ian Worpole **207**, Ian Worpole **208**, Ian Worpole **209**, LWA/gettyimages **209**, David Madison/gettyimages **209**, Industry & Transportation/PhotoDisc **209**, Ian Worpole **210**, Tim Fuller **210**, Arnold & Brown **210**, Gehl Company **210**, Ian Worpole **211**, Ian Worpole **212**, Ian Worpole **213**, Steve Karp **214**, Ian Worpole **215**, Ian Worpole **215**, Sam Ogden/Photo Researchers **216**, Steve Karp **217**, Glencoe **217**, Ian Worpole **218**, Arnold & Brown **219**, Blend Images/PunchStock **221**, Ian Worpole **222**, Freedom of Creation **224**, Freedom of Creation **225**, Z Corporation **225**, AP Images **226**, Greg Pease/gettyimages **227**, William Henry Pyne/gettyimages **228**, Nancy Carter/North Wind Picture Archives **228**, Archive Holdings Inc./gettyimages **229**, Bettmann/CORBIS **229**, Chad Slattery/gettyimages **229**, Volker Steger/Sandia National, Laboratory/Photo Researchers **229**, Steve Karp **230**, Glencoe **230**, Jeff Stoecker **231**, Toyota **232**, Keren Su/gettyimages **233**, James Worrell/gettyimages **234**, Industry & Environment/PhotoDisc **235**, Astrid & Hanns-Frieder Michler/Photo Researchers **236**, Jeff Stoecker **237**, Autobelli Lamborghini Holding Spa PIVA **240**, Andy Sacks/gettyimages **241**, Tanya Constantine/gettyimages **241**, Alan Levenson/gettyimages **241**, Alister Berg/gettyimages **241**, Lester Lefkowitz/gettyimages **241**, David R. Frazier Photolibrary, Inc. **243**, Arnold & Brown **244**, Zigy Kaluzny-Charles Thatcher/gettyimages **246**, Alan Klehr/gettyimages **247**, Yellow Dog Productions/gettyimages **248**, Terra Fugia, Inc. **249**, Arnold & Brown **250**, B. Busco/CORBIS **251**, Steve Karp **251**, Ian Worpole **252**, Glencoe **253**, Kevin May **254**, Ed Kashi//CORBIS **255**, Rosenfeld images, Inc./Photo Researchers **255**, Jeff Stoecker **255**, Jeff Stoecker **256**, TRL Ltd./Photo Researchers **257**, Danny Lehman/CORBIS **258**, Michael Newman/PhotoEdit **259**, Ed Bock/CORBIS **260**, Bob Rowan/CORBIS **261**, Arnold & Brown **263**, Arnold & Brown **264**, Louie Psihoyos/CORBIS **266**, Glencoe **267**, Art MacDillos/Gary Skillestad **269**, Art MacDillos/Gary Skillestad **270**, Waltraud Grubitzsch/CORBIS **271**, Mattias Clamer/gettyimages **271**, AMP, Inc. **272**, John McLean/Photo Researchers **272**, Maximillian Stock Ltd./Photo Researchers **273**, Philippe Psaila/Photo Researchers **274**, Stockbyte Platinum/alamy **274**, Arnold & Brown **275**, Arnold & Brown **276**, Art MacDillos/Gary Skillestad **276**, Dept. of Treasury **276**, Cincinatti Milacron **277**, Lawrence Manning/CORBIS **277**, Kevin May **277**, Steve Karp **279**, /PhotoDisc /Objects of Business **279**, Art MacDillos/Gary Skillestad **280**, Dave Bartruff/CORBIS **282**, Okata Design Inc. **283**, Emerson Electric Co. **284**, Kevin May **284**, Glencoe **285**, Lester Lefkowitz/CORBIS **285**, Art MacDillos/Gary Skillestad **286**, Ed Kashi/CORBIS **286**, Royalty-Free/CORBIS **287**, Dr. James F. Fales **288**, Jeff Stoecker **288**, Glencoe **289**, Jeff Stoecker **290**, Royalty-Free/CORBIS **291**, Art MacDillos/Gary Skillestad **291**, NVision, Inc. **292**, Lester Lefkowitz/ CORBIS **292**, CORBIS **295**, Glencoe **296**, Courtesy of Intel Corporation **298**, Andy Sacks/gettyimages **299**, Glencoe **300**, Monnel Douglas **301**, Ted Kawalerski Photography Inc./gettyimages **302**, Steve Karp **303**, Dell Computers Inc. **303**, Philippe Psaila/Photo Researchers **304**, Caterpillar Inc. **305**, Pascal Goetgheluck/Photo Researchers **306**, Art MacDillos/Gary Skillestad **307**, Sandia National Laboratories **307**, Bots IQ/Greg Munson **308**, Steve Karp **309**, Digital Vision/PunchStock **310**, Steve Karp **311**, Steve Gschmeissner/Photo Researchers **311**, Justin Guariglia/gettyimages **312**, Bruno de Hogues/gettyimages **313**, Steve Karp **314**, George Steinmetz/CORBIS **315**, Tyler Olsen/alamy **317**, Purestock/PunchStock **318**, CORBIS **320**, Systemarchitects **320**, Systemarchitects **321**, Jose Fuste Raga/CORBIS **322**, Panoramic Images/gettyimages **323**, Ken Clubb **324**, Allan Davey/Masterfile **324**, Vanni Archive/CORBIS **324**, Tom Bean/CORBIS **324**, John Muzzarelli **325**, age fotostock/Superstiock **325**, Yanni Behrakis/Reuters/CORBIS **325**, Philip James Corwin/CORBIS **325**, Glencoe **326**, Industry & Transportation/PhotoDisc **327**, Panoramic Images/gettyimages **328**, Mazer Corporation **329**, Ed Quinn/CORBIS **329**, Derek M. Allan; Travel Ink/CORBIS **330**, Jean Pierre Amet/CORBIS **330**, Danny Lehman/CORBIS **331**, NIDUS CORP. **331**, Art MacDillos/Gary Skillestad **332**, Lance Nelson/Stock Photos/zefa/CORBIS **333**, David Young-Wolff /PhotoEdit **333**, Donna Day/gettyimages **333**, IFA Bilderteam/eStock **333**, age fotostock/Super Stock **333**, Jim Reed/gettyimages **335**, Kevin May Stock Photo **336**, Tools of the Trade/PhotoDisc **336**, Michael Robinson/CORBIS **338**, David Sailors/CORBIS **339**, Barry Lewis/CORBIS **340**, Rogelio Solis/AP images **341**, David Frazier **342**, Stefan Sollfors/alamy **343**, Art MacDillos/Gary Skillestad **344**, Bob Child/AP images **345**, Rogelio Solis/AP images **346**, John Norris/CORBIS **347**, George Steinmetz/CORBIS **347**, Schieler & Rassi Quality Buiders, Inc. **348**, David R. Frazier Photolibrary, Inc. **350**, Royalty-Free/CORBIS **351**, David R. Frazier Photolibrary, Inc. **352**, John MAdere/CORBIS **353**, James A. Finley/AP images **355**, Tim Garcha/zefa/CORBIS **356**, Glencoe **357**, Matthias Kulka/CORBIS **358**, Royalty-Free/CORBIS **359**, Glencoe **360**, Art MacDillos/Gary Skillestad **361**, Tom Brakefield/gettyimages **362**, Jose Luis Pelaez, Inc./CORBIS **363**, Chitose Suzuki/AP images **363**, Brent Phelps **364**, Jeff Stoecker **365**, Helen King/CORBIS **367**, Arnold & Brown **368**, Arnold & Brown **372**, Robert Llewellyn/zefa/CORBIS **373**, Bruce Forster/gettyimages **374**, Art MacDillos/Gary Skillestad **375**, Art MacDillos/Gary Skillestad **376**, David R. Frazier Photolibrary, Inc. **377**, Art MacDillos/Gary Skillestad **378**, Art MacDillos/Gary Skillestad **378**, David McNew/gettyimages **379**, Steve Karp **379**, Art MacDillos/Gary Skillestad **380**, Hisham F. Ibrahim/gettyimages **381**, Arnold & Brown **382**, Landmark Homes & Land, Inc. **383**, Carlos Osorio/AP images **384**, Art MacDillos/Gary Skillestad **385**, Larry W. Smith/CORBIS **385**, Kevin May **387**, NASA-MSFC **390**, Steve Karp **391**, Bettmann/CORBIS **391**, Photo Researchers **392**, Remi Benal-iCORBIS **393**, Art MacDillos/Gary Skillestad **394**, Art MacDillos/Gary Skillestad **395**, G.A. Rassati, PhD, College **396**, Ian Worpole **397**, AP images **398**, Andrew K/epa/CORBIS **399**, Raymond Gehman/

CORBIS **399,** Art MacDillos/Gary Skillestad **400,** Mark W. Lipczynski/AP images **401,** Ed Quinn/CORBIS **402,** Questor Corp. **402,** NASA-MSFC **403,** Art MacDillos/Gary Skillestad **405,** RJ Muna/Tesla Motors **408,** G Brad Lewis/gettyimages **408,** AP images **409,** Toyota Motors **409,** Moteur Development International **409,** John A. Rizzo/gettyimages **410,** Ryan McVay/gettyimages **411,** Glencoe **412,** Steve Karp **412,** Andrew Fox/CORBIS **413,** Mike Zens/CORBIS **413,** Michaela Rehle/CORBIS **414,** Ian Worpole **415,** Royalty Free/CORBIS **415,** Royalty Free/CORBIS **416,** Peter Adams/Index Stock **417,** North Wind Picture Archives **418,** Bettmann/CORBIS **419,** CORBIS/CORBIS **419,** NASA-KSC **419,** Toyota **419,** Alexis Rosenfeld/Photo Researchers **420,** Jim Sugar/CORBIS **421,** Jaguar Cars and Wieck Media Services, Inc. **422,** Digital Vision/gettyimages **423,** ColorBlind Images/gettyimages **423,** Frank Wing/gettyimages **423,** David Sailors/CORBIS **423,** Royalty-Free/CORBIS **423,** Gabe Palmer/CORBIS **425,** Art Stein/ZUMA/CORBIS **426,** Creasource/CORBIS **428,** Jim Ross/NASA Dryden Flight Research Center **429,** Mazer Corporation **429,** Kevin Flemming/CORBIS **430,** Laura Rauch/AP images **431,** NASA Kennedy Space Center **432,** NASA Marshall Spaceflight Center **432,** Tim Brakemeier/dpa/CORBIS **433,** Sergio Pitamitz/CORBIS **433,** Dave G. Houser/CORBIS **434,** Free Agents Limited/CORBIS **435,** Douglas Kirkland/CORBIS **436,** Robert Harding World Imagery/CORBIS **436,** Royalty Free/CORBIS **437,** Royalty Free/CORBIS **438,** Steve Karp **438,** Kennan Ward/CORBIS **439,** Glencoe **439,** T.T. Williamson, Inc. **440,** Royalty Free/CORBIS **441,** Owen Franken/CORBIS **442,** Caroline Brown/The MATE Center **443,** Ghislain & Marie David de Lossy/gettyimages **448,** Art Maillo's/Gary Skillestad **449,** Art MacDillos/Gary Skillestad **450,** Art MacDillos/Gary Skillestad **451,** Art MacDillos/Gary Skillestad **452,** Glencoe **453,** American Honda Motor Corp. **453,** Jeff Stoecker **454,** Art MacDillos/Gary Skillestad **455,** Art MacDillos/Gary Skillestad **456,** Art MacDillos/Gary Skillestad **457,** Glencoe **457,** Jeff Stoecker **458,** Bob Krist/CORBIS **459,** Art MacDillos/Gary Skillestad **460,** Glencoe **461,** Alexis Rosenfeld/Photo Researchers **462,** Jeff Stoecker **463,** Art MacDillos/Gary Skillestad **464,** Eugene Hoshiko/AP images **465,** Toru Hanai/Reuters/CORBIS **466,** Sandor Ujvari/AP images **467,** Digital Vision/gettyimages **469,** Panoramic Images/gettyimages **470,** David Frazier **472,** Michael Probst/AP images **473,** Robert E. Klein/AP images **475,** Dennis Maonald/PhotoEdit **476,** Steve Karp **477,** Janie Slaven/AP images **478,** Brownie Harris/CORBIS **479,** Chip East/CORBIS **480,** Frank Schwere/gettyimages **481,** Reuters/CORBIS **482,** CORBIS **483,** Bob Sacha/CORBIS **484,** Technology Student Association **485,** Rudi Von Briel/PhotoEdit **487,** Spencer Grant/PhotoEdit **488,** Royalty-Free/CORBIS **490,** Don Farrall/gettyimages **491,** Science Source/Photo Researchers **491,** NASA-MSFC **492,** Jason Smalley/alamy **493,** David Ewing/alamy **494,** Masterfile **494,** Jeff Stoecker **495,** Business & Industry/PhotoDisc **496,** Steve Karp **496,** North Wind Picture Archives **496,** Phil Degginger/gettyimages **497,** North Wind Picture Archives **497,** Business & Industry/PhotoDisc **497,** Modern Technology/PhotoDisc **497,** Intuitive Surgical, Inc. ©2006 **497,** Ian Worpole **498,** Glencoe **498,** Familiar Objects/PhotoDisc **498,** Tools of the Trade/PhotoDisc **498,** Green Industry/PhotoDisc **499,** William Radcliffe/gettyimages **500,** Joel W. Rogers/CORBIS **501,** JETS NEDC **502,** Glencoe **503,** Joseph Sohm:ChromoSohm Inc./CORBIS **504,** Research & Technology/DigitalVision **505,** Dan Tardif/Corbis **505,** Getty Images/gettyimages **505,** CORBIS/PunchStock **505,** Business & Occupations/PhotoDisc **505,** NASA-JLP **507,** Glencoe **508,** Brad Mangin/CORBIS **510,** Ron Chapple/gettyimages **511,** Alava Local Authorities/HANDOUT/CORBIS **512,** Jeff Stoecker **513,** Science, Tech & Medicine 2/PhotoDisc **514,** Science, Tech & Medicine /PhotoDisc **514,** WDCN/Univ. College London/Photo Researchers **514,** Josh Sher/Photo Researchers **515,** Science, Tech & Medicine 2/PhotoDisc **515,** Arthur Tilley/gettyimages **516,** James King-Holmes/Science Photo Library **517,** Amit N. Patel, M.D., M.S., Director of Cardiac Cell Therapy, The Heart, Lung and Esophageal Surgery Institute-UPMC **518,** Peter Mentel/Science Photo Library **519,** Peter Menzel/Science Photo Library **520,** PHOTOTAKE, Inc./alamy **521,** Glencoe **522,** Ian Worpole **523,** Kevin May Corporation **523,** Art MacDillos/Gary Skillestad **524,** Ian Worpole **525,** LWA-Stephen Welstead/CORBIS **527,** Richard T. Nowitz/CORBIS **528,** Roy Morsch/CORBIS **530,** Jeff Stoecker **531,** National Youth Science Foundation **532,** Jeff Stoecker **533,** Richard Hamilton Smith/CORBIS **534,** Jeff Stoecker **535,** Glencoe **536,** Fremont Stevens/ CORBIS **537,** Lester Lefkowitz/CORBIS **538,** Owen Franken/CORBIS **539,** Lowell Georgia/CORBIS **540,** Natalie Fobes/CORBIS **540,** Lawson Wood/CORBIS **541,** Andre Jenny/alamy **542,** Dennis Kunkel/alamy **543,** Scimat/Photo Researchers **544,** Glencoe **544,** Frank Greenaway/gettyimages **545,** Jeff Greenberg/PhotoEdit **547,** Landmann Patrick/CORBIS **548,** Bill Daemrich/PhotoEdit **550,** Glencoe **551,** Aearo Company **551,** Arnold & Brown **551,** Aearo Company **552,** Glencoe **552,** Bill Frymire/Masterfile **553,** Arnold & Brown **554,** Arnold & Brown **555,** Makita Corporation **556,** Arnold & Brown **557,** Delta Inc **557,** Arnold & Brown **558,** Glencoe **561,** Glencoe **562,** Glencoe **563,** Arnold & Brown **564,** Ken Clubb **566,** Steve Karp **567,** Ian Worpole **567,** Steve Karp **568,** Ian Worpole **569,** TRL Ltd./Photo Researchers **570,** Jutta Klee/Corbis **570,** Grace/Zefa/Corbis **570,** Steve Karp **571,** Ken Clubb **572,** Steve Karp **572,** Ken Clubb **574,** Steve Karp **575,** Steve Karp **577,** Steve Karp **578,** Ian Worpole **579,** Steve Karp **581,** Steve Karp **582,** Ken Kulish/CORBIS **584,** Phillip Bailey/CORBIS **584,** Andy Sacks/gettyimages **584,** David McNew/gettyimages **585,** Art Stein/Zuma/CORBIS **585,** Science, Tech & Medicine 2/PhotoDisc **585,** Steve Karp **586,** Lon C. Diehl/PhotoEdit **586,** Steve Karp **587,** Steve Karp **588,** Steve Karp **588,** Steve Karp **589,** Steve Karp **590,** Steve Karp **591,** Glencoe **591,** Steve Karp **592,** Steve Karp **593,** Ian Worpole **594,** Steve Karp **596,** Ian Worpole **597,** Steve Karp **598,** Steve Karp **599,** Jeff Stoecker **599,** Steve Karp **600,** Steve Karp **601,** Glencoe **602,** Steve Karp **603,** Andrew Lambert Photography/ Photo Researchers, Inc. **606,** Steve Karp **608**

INDEX

A

B

C

D

E

F

G

H

I

N

O

P

Q

R

S

T

U

V

W

X

Y

Z